Lecture Notes in Artificial Intelligence 1317

Subseries of Lecture Notes in Computer Science
Edited by J. G. Carbonell and J. Siekmann

Lecture Notes in Computer Science
Edited by G. Goos, J. Hartmanis and J. van Leeuwen

AF740753

Springer
Berlin
Heidelberg
New York
Barcelona
Budapest
Hong Kong
London
Milan
Paris
Santa Clara
Singapore
Tokyo

Marc Leman (Ed.)

Music, Gestalt, and Computing

Studies in Cognitive and Systematic Musicology

Reading committee:
A. Camurri J. Louhivuori
R. Parncutt A. Schneider

With support from:
European Union (Caleidoscope 96/411063)
Fonds voor Wetenschappelijk Onderzoek - Vlaanderen (FWO-V)
International Society for Systematic and Comparative Musicology

Springer

Series Editors
Jaime G. Carbonell, Carnegie Mellon University, Pittsburgh, PA, USA
Jörg Siekmann, University of Saarland, Saarbrücken, Germany

Volume Editor

Marc Leman
University of Ghent, IPEM
Blandijnberg 2, B-9000 Ghent, Belgium
E-mail: Marc.Leman@rug.ac.be

Cataloging-in-Publication Data applied for

Die Deutsche Bibliothek - CIP-Einheitsaufnahme

Music, gestalt, and computing : studies in cognitive and systematic musicology / Marc Leman (ed.). - Berlin ; Heidelberg ; New York ; Barcelona ; Budapest ; Hong Kong ; London ; Milan ; Paris ; Santa Clara ; Singapore ; Tokyo : Springer, 1997
(Lecture notes in computer science ; 1317 : Lecture notes in artificial intelligence)
ISBN 3-540-63526-2

LIBRARY
12 MAR 2002
LANCASTER UNIVERSITY

CR Subject Classification (1991): I.2, J.5, H.5.1, I.5.4

ISBN 3-540-63526-2 Springer-Verlag Berlin Heidelberg New York

This work is subject to copyright. All rights are reserved, whether the whole or part of the material is concerned, specifically the rights of translation, reprinting, re-use of illustrations, recitation, broadcasting, reproduction on microfilms or in any other way, and storage in data banks. Duplication of this publication or parts thereof is permitted only under the provisions of the German Copyright Law of September 9, 1965, in its current version, and permission for use must always be obtained from Springer -Verlag. Violations are liable for prosecution under the German Copyright Law.

© Springer-Verlag Berlin Heidelberg 1997
Printed in Germany

Typesetting: Camera ready by author
SPIN 10545808 06/3142 – 5 4 3 2 1 0 Printed on acid-free paper

01 06904

Preface

This book contains a selection of papers presented at JIC96-Brugge, the Joint International Conference on Cognitive and Systematic Musicology, held at the College of Europe in Brugge (Belgium) from 8 to 11 September 1996. JIC96-Brugge was a joint meeting within the framework of two conference cycles: (i) the International Symposium on Systematic and Comparative Musicology which, previously, took place at Moravany/Bratislava (Slovakia) in 1993, at Hamburg (Germany) in 1994, and at Zeilern (Austria) in 1995, and (ii) the International Conference on Cognitive Musicology, the first of which was organized in Jyväskylä (Finland) in 1993. The meetings were established independently in the early 1990s to foster systematic musicology in Europe.

The conference was organized by IPEM, the Institute for Psychoacoustics and Electronic Music of the Department of Art, Music and Theater Sciences of the University of Ghent (Belgium), in collaboration with the International Society for Systematic and Comparative Musicology. The colleagues Jukka Louhivuori (Jyväskylä), Antonio Camurri (Genova), and Franz Födermayr (Vienna) involved in administrative preparations. In many practical issues concerning conference organization and preparation of this book, Albrecht Schneider (Hamburg) joined me on behalf of the International Society for Systematic and Comparative Musicology. The selection of papers and the review work was done in collaboration with the members of the Conference's Reading Committee: Antonio Camurri, Jukka Louhivuori, Richard Parncutt (Keele), and Albrecht Schneider.

This book has been set in LaTeX by Bald Herreman, with assistance from Dirk Moelants (both at IPEM). Herman Sabbe, head of our department, supported the project.

Major financial support was provided by the European Union (Caleidoscope 96/411063) and the "Fonds voor Wetenschappelijk Onderzoek - Vlaanderen" (FWO-V).

I wish to thank all collaborators, colleagues, and sponsors for their enthusiasm and help in both the organization of JIC96-Brugge and the post-conference activities concerning this book.

June 1997 Marc Leman

Table of Contents

Introduction

This book aims to give a state-of-the-art in systematic and cognitive musicology. The past decades indeed have witnessed a dynamic growth of research activities in different music-related fields such as sound synthesis and sound analysis, psychoacoustics, cognitive psychology and applications in artificial intelligence and multimedia. The developments have been largely due to technological innovations, in particular: the widespread introduction of digital computers. Computers are currently used for data analysis, sound synthesis, simulation of musical perception and cognition, and even intelligent music/dance behavior. They allow a control over different aspects of music down to the deepest level of the sound. This, obviously, provides a most fruitful context for a discipline which aims to study the relationships between acoustics, human information processing, and culture.

Systematic musicology has a peculiar interest in cognitive issues, mainly because musical abilities are learned (at least to some extend) and above all: context-based. Music is also associated with forms, figuration and global pattern processing. For that reason, the Joint International Conference on Cognitive and Systematic Musicology, held at Brugge in 1996 had a focus on the following two topics: (i) Gestalt concepts revisited – from metaphor to cognitive model, and (ii) Software development, modelling and simulation. The first reflects the idea that Gestalt-based approaches nowadays contribute much to the development of empirically testable models of musical perception, cognition and action, thereby going beyond metaphorical and programmatical thinking. The second reflects the interest in computers as a powerful method and tool for theory construction, theory testing, and the manipulation of musical information or any kind of data manipulation related to music. Both topics relate to the pillars of a cognitive systematic musicology which is both computationally-based and grounded in naturalism. The topics reflect the current interest in methodological innovation and epistemological re-foundation of the old (European) systematic musicology.

The papers contained in this book are organized in different chapters. Some papers, however, do address the topic of more than one chapter. The structure of the book, therefore, is based on personal choices. I prefered to structure the book according to topical issues, rather than the method used. As such, the reader will find the use of sonological methods or neural networks in more than one chapter. The first chapter of the book is devoted to issues concerned with general topics related to theoretical, historical and programmatical issues of the *Music, Gestalt, and Computing*. The other chapters deal with problems or solutions to problems in domains related to pitch perception, harmony, rhythm, organization, expectancy, timbre, texture, expression, and application oriented issues such as hybrid computational architectures, and interactive computer systems. The sound examples (if any) used for illustration are described in an appendix at the end of each paper. At the end of the book, a complete list of all sound examples on the CD is given.

Chapter I: Gestalt Theory Revisited

Chapter I is concerned with theoretical, historical and programmatical issues related to cognitive and systematic musicology, Gestalt perception, and its associated theory. The epistemological and methodological heritage of the Gestalt approach is critically evaluated and reconsidered in the light of new approaches and developments in computer technology.

The paper by M. Leman and A. Schneider, on the origin and nature of cognitive and systematic musicology, may serve as a tutorial to the main topic of *Music, Gestalt, and Computing*. It provides a brief overview of the history of systematic musicology, the development of Gestalt theory and issues related to methodology. The authors argue that computers nowadays form a methodological backbone for a naturalistic research programme. In the last decades, this programme, once started by H. v.Helmholtz and C. Stumpf in the 19th century, has been gradually transformed into a modern cognitive systematic musicology. It results in an approach that can be fully integrated with the most advanced fields of modern science.

J. Louhivuori deals with classificatory relationships between cognitive, systematic, and historical musicology. He argues that cognitive musicology, which developed within the research tradition of cognitive science, may broaden the research field of systematic musicology in a significant way by adopting methods from anthropology or by applying the naturalist methodology to historical issues. He expects that cognitive musicology will soon integrate with the traditional systematic approach and believes that the resulting cognitive systematic approach holds some promising capabilities for the future.

M. Weber, provides a historical background for a better understanding of the tendencies that underlay cognitive systematic musicology. He considers the notions of empiricism, Gestalt qualities and determination of style from a historical point of view by discussing the relationships between some key figures that worked in this field at the end of the 19th and beginning of the 20th Century. The paper has a peculiar focus on G. Adler, who made an important classificatory distinction between systematic and historical musicology. Adler seems to have shared the interest in exploring the application of Gestalt theoretical and empirical foundations of musicology to issues in music history, but a prolific collaboration with colleagues could not be established at the time.

In his paper on Gestalt concepts and music, M. Reybrouck discusses possibilities and limitations of the Gestalt approach. The paper gives an overview of some of the major classical programs and concepts of Gestalt theory. The author argues in favour of a reappraisal of the older insights within the actual context of formal and operational approaches. The possibilities of the cognitive approach within an interdisciplinary context are stressed from a programmatic viewpoint.

The paper by R. Eichert, L. Schmidt, and U. Seifert discusses issues related to logic, Gestalt theory, and neural computation in the light of research on auditory perceptual organization. The authors give an analysis of the concepts of Gestalt quality and functional whole and the related vague and apparently mysterious notion of emergent or holistic attributes of objects. The latter concept is argued to have a merely heuristic value for scientific research, rather than an explanatory value. Consequently, the Gestalt idea of psychoneural isomorphism needs to be revised within a functional systems approach. The authors argue that the problem of perceptual organization should be revised using a Gestalt approach based on modern computational theory including auditory models and artificial neural networks.

In his paper on shapes of musical objects and sound-producing actions, R.I. Godøy, argues that thinking in terms of emergent qualities of musical objects is a privileged mode of human mental representation. His approach is action-oriented in that he considers computer technology as a tool to improve our means of shaping musical Gestalt images of dynamic unfolding, thus helping to explore their features by the production and comparison of variants. The paper shows how the Gestalt-based phenomenological way of thinking nowadays can be integrated within a computational approach. (Sound examples are provided.)

E. Ungeheuer discusses the action-oriented approach within a historical context, focusing on the use of Gestalt concepts in serial music. The author argues that the notion of evolutionary forms in the electro-acoustic avant-garde music of the early 50ies is highly related to the use of statistical concepts in musical thinking. Stockhausen's serial composition *Gesang der Jünglinge* shows how serial composing of mass phenomena was orientated towards the creation of evolving Gestalts on a statistical basis. Such Gestalts generate emergent qualities being referential figures for the composer as well as for the listener.

Chapter II: From Pitch to Harmony

Chapter II has a focus on pitch, context, and harmony. In this paper, the traditional focus on pitch perception is extended with some novel theoretical as well as practical insights that draw upon Gestalt perception and computation. It starts with a reconsideration of Stumpf's theory of tonal fusion in the light of recent neurophysiological findings. Tonal fusion forms the basis of a physiologically inspired computer model of perceptual learning. The properties of context-dependent pitch perception are then further explored in two theoretical approaches to harmony, and one practical approach which explores inharmonic sounds. A novel method for describing the succession of harmonies is proposed and the chapter closes with a neural network model for the study of post-tonal music.

A. Schneider revisits the relationships between the concepts of "Verschmelzung", tonal fusion, and consonance in the work of C. Stumpf. An experiment is set up to test Stumpf's assumption that dissonance and roughness can be separated from each other. The author concludes that Stumpf, in this issue as well as in others, adopted a nativistic point of view though he also stressed the importance of factors of learning and adaptation. This point of view seems to come close to current neurophysiologically based views on pitch perception. According to Stumpf, cognitive functions are necessary to fully apprehend musical structure as well as to understand our own context-dependent processes relevant to listening to, and enjoying, music. (Sound examples are provided.)

The next paper, by M. Leman and F. Carreras, can be seen as an attempt to test the relationship between physiologically based processes and learned processes in context-dependent pitch perception using large-scale computer simulations. The authors develop a physiologically inspired model which gives an account of how perceptual schemata for tonal music emerge by self-organization from listening to Western tonal music, and how these schemata, once they are established, may respond to stimuli or other musical input. The results of the computer simulations are in line with the hypothesis that the brain is capable, within the neurophysiological biased preference for consonant pitch intervals, of extracting an invariant statistics from tonal music and keep it in internal memory, and that the brain responds to context-dependent pitch configuration in an ordered and highly systematic way, reflecting this invariant statistics in terms of circles of fifths. (Sound examples are provided.)

The contribution by N. Cufaro Petroni and M. Tricarico is connected to the previous work in that it provides a consistency test for some of the basic assumptions of the simulation. In particular, the authors test the idea that the regularity of patterns which emerge in the schema after perceptual learning could be the effect of the preprocessing and of circular transpositions of original signal images, rather than of their informative content referred to the tonal relations among the chords. They conclude however that, "beyond any reasonable doubt", the importance of the pitch models in this sort of investigations is central.

The phenomenon of context dependent pitch perception is further explored by R. Parncutt. He presents a cognitive theory that predicts the perceptual root of chords embedded in the context of any chord progression. In addition to previous work based on models of virtual pitch perception, the present model is apt in taking into account the effects of bass note movements (voicing), and the context effects of chords operating in a particular tonality. This work is carried out in the tradition of music theory, but it provides an example of the current concern for empirical testing and modelling adequacy.

D. Werts, in attempting to justify the psychological ratings determined in an experiment on chord progression (by J. Bharucha and C. Krumhansl), adopts a similar theoretical perspective. He provides a musical rationale for the ratings based on six musical properties, puts them in a regression formula, and then tries to approximate the values obtained in the psychological data. A music theoretical weight attributed to individual chords accounts for .541 of the explained variance in the data. Remaining factors are related to chord-order, root progression, and leading note, among others.

The paper by J. Keuler considers the perception of inharmonic spectra within their proper tonal context. The author shows that the perceived similarity between sounds, as well as their consonance, and dissonance, depends on their spectral configuration as well as on their use in a context of other similar sounds. The author argues by means of sound examples that the sounds are perceived as Gestalts in that their pitch and timbre belong to the whole sound, and not to the partials that constitute the sound. (Sound examples are provided.)

R. Eberlein argues in favour of an objective method for describing the succession of harmonies. The author criticizes the conventional description of harmony for being too much involved with speculative interpretation. He therefore develops a system for encoding chords as a succession of harmonic interval combinations and melodic bass steps on an objective basis, restricting himself to a mere description of the harmonic succession. Such an approach can be connected with interpretations that aim to understand harmony in terms of learned patterns or Gestalts of harmonic and melodic intervals.

E. Isaacson uses neural networks to test basic assumptions of the pitch-class set theory. This theory assumes that essential aspects of atonal music can be captured in terms of pitch-class sets or score-based reductions of perceived pitch. The approach is first used to extract abstract constructs from pcsets by learning and hence to test this theory as an abstract perceptual theory. In a second part of the paper, the author shows how to use neural networks to explore segmentation possibilities of scores.

Chapter III: From Rhythm to Expectation

The papers in this chapter are related to the phenomenon of rhythm, with a particular focus on the perception of the beat, meter, issues of grouping and segmentation, organization, and expectation.

The paper by M. Franek and J. Mates deals with experimental research in the field of tempo performance. The authors aim to test the hypothesis, put forward by D. Epstein, that in a piece of music all tempo relations can be expressed by low order whole number ratios. The results provide evidence, however, that certain limits might exist in the operational principles of internal clocks with respect to the tempo proportionality hypothesis but strong limits in favour of proportionality have not been found. The authors conclude that the relations between neighboring tempi could perhaps be better described in terms of categories relating to tempo changes, the average of which is close to whole-number ratios.

D. Moelants gives an analysis of the traditional concept of meter from the point of view of perception and performance. He suggests to broaden the definition of meter in terms of a temporal framework in which basic beat patterns, metric hierarchies, and metric microstructures are contained. He argues that a symbolic description is not rich enough to deal with the metric diversity of performed music and concludes that a subsymbolic approach is needed which would take into account aspects of musical expression as well.

E. Cambouropoulos presents a formal model for the grouping of rhythms in melodies, starting from a score representation. The model detects boundaries in the surface of the melody using rules that are related to the Gestalt principles of proximity and similarity. The author shows to what extend both the accentuation and metrical structure can be inferred from these local grouping structure.

C. Krumhansl reports an experiment which aims to test the implication-realization model of E. Narmour. The author examines the role of perceptual organization in generating melodic expectancies, and attentional processing directed at different levels of hierarchical structure in music. Gestalt principles, such as proximity, similarity, and good continuation are shown to play a role in determining expectations. The results support the model and it is concluded that music draws on the dynamics between several competing principles.

In his paper, S. Larson describes the results of a study in melodic expectation. The author argues that we should regard the melodic expectations of subjects as expectations not so much for continuations, but as expectations for entire melodic completions. The latter represent musical forms shared by experienced musicians and are probably rule-driven. In this cognitive performance, key determination is central to the process of generating melodic expectations.

Chapter IV: From Timbre to Texture

This chapter deals with aspects of timbre and texture. The subject and methods applied in this chapter are illustrative for the great variability in approaching this difficult topic. The first paper is about the use of neural networks and auditory models for classifying timbres, the second has a focus on a sonological analysis of the particular sound color of nasality. The third paper studies the effect of formants in segregation and the fourth deals with aspects of timbre synthesis. The last chapter deals with textural phenomena.

P. Toiviainen studies the effect of using different types of auditory images and distance metrics for the classification of timbres. Computer modelling is based on an auditory model and the use of a self-organizing neural network for classification. The author discusses the role of frequency transition and gradient images but the results indicate that the onset portion of the tones is particularly important. He concludes that the frequency and time resolutions of the images obtained by means of the auditory model are insufficient for extracting quick transitions in the onset portions of the musical tones. (Sound examples are provided.)

M. Rusko explores the phenomenon of nasality and defines this concept using a sonological method. Different types of nasality are distinguished and related to expressive meaning in speech, singing, and musical acoustics. Spectral analysis is used to deduce sonological properties of nasality. The author concludes that the perception of nasality is not merely based on spectral properties of the sound, but that it also depends on transients, similar to onset and dynamic spectral changes in speech.

The paper by C. Reuter tests K.E. Schumann's laws of formant positions and relates it to stream segregation. It is found that in tone sequences with alternating timbres based on different formant positions, two melodic streams are perceived. Segregation also holds when timbres characterized by formants alternate with timbres characterized by fluctuations. When timbres with similar formant positions alternate, then segregation is more difficult to perceive. The same holds with timbres that have the same intensity of fluctuation. The author concludes that Schumann's laws of timbre can be satisfactorily integrated into the investigations of stream integration. (Sound examples are provided.)

T. Rochebois and G. Charbonneau use the principal component analysis to extract valuable spectral and temporal features of a sound. They then explore cross-synthesis for combining timbral parameters from two distinct sounds to obtain a new hybrid sound. The paper shows how an action oriented Gestalt approach at the level of signal manipulation can be implemented. (Sound examples are provided.)

D. Cohen and S. Dubnov propose a new definition of texture that covers principles of organization that need qualitative or statistical tools in order to be described. The authors argue that texture should be conceived of as a complementary contrast to both timbre and learned schemata that represent Gestalt rules. Texture is viewed as a super parameter that is related to various combinations of other musical parameters. As such they may also evoke various emotional associations.

Chapter V: From Musical Expression to Interactive Computer Systems

This chapter is concerned with musical expression both from a performance as well as from a perception point of view. The last three papers deal with hybrid and interactive computer systems.

The paper by A. Tanguiane's is concerned with interpretation and expressive pulses. Interpretation is understood as finding a structure in a musical text based on segmentation and determination of hierarchically related Gestalt segments to which he attributes different intonation patterns. The author develops a system of rules for the segmentation and interpretation which he tests both in a violin performance of a pupil appropriately instructed, as well as with a computer model. (Sound examples are provided.)

The paper by J. Fyk has a focus on a particular aspect of violin performance that is related to anticipation. The author has investigated the change in frequency of tones followed by tones that are either higher or lower. She found that the last state of the tone is an announcement of a frequency change in the next tone and concludes that the performer's anticipation, which is called protention, is carried out to obtain an homogeneous sound unity. Intonational protention is argued to be a true reflection of the Gestalt processes that are at work during a performance.

In the next paper, S. Canazza, G. De Poli, S. Rinaldin and A. Vidolin give a sonological analysis of clarinet expressivity. They have recorded seven performances of different expressivity of a fragment of music by W.A. Mozart and have measured the sonological parameters of the recorded sounds. The aim was to identify which physical parameters were subject to modifications when the expressive intention of the performer was varied. The authors show that the various expressive performances have their own sonological characteristics. (Sound examples are provided.)

In the subsequent paper, S. Canazza, G. De Poli, and A. Vidolin use a complementary approach to the previous paper and perform a perceptual analysis of the musical expressive intention in clarinet performances using analysis techniques of statistical nature. The results show that the performer's intentions and the listeners' impressions, in general, agree. In one of the experiments, the authors use the results of the sonological analysis to synthesize the expression of a piece. The perceptual analysis shows that the subjects' replies agreed reasonably well, which is an argument in favour of the usefulness of the sonological analysis technique for the study of expressivity. (Sound examples are provided.)

The paper by E. Rapoport deals with aspects of expression in singing, based on a sonological analysis of lied and opera singing. The spectrogram analysis, which displays the frequency analysis in the form of a time diagram is considered to reflect the activity of the vocal folds. On this basis, the author relates one cycle of frequency increase followed by frequency decrease, so typical for vibrato, to one cycle of the tension increase, followed by tension release of the vocal folds. The author then argues that the different modes of expressive singing, such as excitement and calmness, relate to brain commands of tightening of the vocal folds: strong and rapid tightening commands corresponds to excitement, whereas weak and slow tightening corresponds to calmness or relaxation. (Sound examples are provided.)

The last papers deal with applications of artificial intelligence techniques for the development of hybrid systems, and interactive computer music systems that rely on Gestalt concepts. The paper by U. Mattusch explores the combination of symbolic and subsymbolic information processing in one single integrated approach. The model is based on a combination of a neural network with methods of inductive learning in artificial intelligence and has a focus on the problem of recognizing musical transposition.

P. Modler deals with interactive computer-music systems and concepts of Gestalt and he discusses the input and output options that are needed in order to achieve a convenient control of technical and aesthetical processes within the context of live performances. Gestalt-theoretical concepts are introduced to characterize the basic properties of interactive computer music systems. The author concludes that the control of sound synthesizers will lead to new Gestalt concepts and revised musical frameworks.

In the last paper of this book, A. Camurri and M. Leman discuss Gestalt-based composition and performance in the context of multimodal environments. These environments provide digital extensions for different human activities, such as movement, thinking, composing, listening, planning, etc. The authors formulate a number of requirements for the application of multimodal environments in music and art and discuss one particular application (called HARP). They conclude that the communication of the causality of movement and sound during a performance is a difficult problem. The paper points to possible practical and industry-oriented applications of cognitive systematic musicology.

M. L.

I

Gestalt Theory Revisited

Origin and Nature of Cognitive and Systematic Musicology: An Introduction

Marc Leman[1] and Albrecht Schneider[2]

[1] IPEM, University of Ghent, Blandijnberg 2, B-9000 Ghent, Belgium
[2] Institute for Musicology, University of Hamburg, Neue Rabenstr. 13, D-20354 Hamburg, Germany

Abstract. This paper gives an introduction to the origin and nature of cognitive and systematic musicology. A brief overview of the history of systematic musicology is presented and the developments are sketched that contributed to the rise of cognitive musicology. It is argued that the breakthrough of computing has offered many new opportunities to revisit the old problems of Gestalt perception and its associated theory. The paper closes with a perspective on future developments.

1 Introduction

Systematic musicology has traditionally been conceived of as an interdisciplinary science, whose aim it is to explore the *foundations of music* from different points of view, such as acoustics, physiology, psychology, anthropology, music theory, sociology, and aesthetics. Since its emergence in the 19th century, research into the foundations of music has furthermore been complemented with a *comparative approach* which aimed at combining *in-depth* investigations with ethnological studies (Graf, 1980; Födermayr, 1983; Elschek, 1992). The systematic approach took into account also the psycho-social and historical context of music and music making but, with just a few dozen scholars representing this broad field, this ambitious project could only partially be realized.

In the first part of this paper, we give an overview of the historical development of the research programme of systematic musicology, with a focus on the origins of the Gestalt approach. In the second part, we discuss the methodological innovations and basic paradigmatic issues that contributed to the present revival of the discipline. The third part provides a perspective on future developments of cognitive systematic musicology.

2 A Short History of Systematic Musicology

The early systematic musicology (before 1900) had, besides its background in philosophy (E. Kant, E. Mach), a main focus on auditory physiology and psychology of music. This empirical orientation was new, compared to the intuitive and purely numerical approaches of the past. In the ancient times (attributed to Pythagoras, ca. 500 B.C.), experiments with the length of strings had led to the discovery that consonant pitch intervals could be represented by simple ratios between the lengths of strings. This was a first step towards a rational and *systematic* understanding of the relationship between man and his musical environment, and a finding which influenced the music theory of the Western culture very profoundly–although at times misleadingly (Chailley, 1967). Great mathematicians such as R. Descartes (1596-1650), C. Huygens (1629-1695) and L. Euler (1707-1783) were fascinated by this idea and tried in turn to found music perception on arithmetics.

The advent of empirism in the 19th Century was therefore a milestone for the emergence of systematic musicology as an independent discipline yet it could hardly have come into existence without the strive for *systematization* as developed in philosophy and science (Schneider, 1993c, 1993b). Systematization of knowledge derived from both lab experiments, fieldwork and other empirical observations, as well as from reasoning and even speculation, was believed necessary to develop more global theories, and to lay down the scientific foundations for music theory and music aesthetics. This becomes very obvious in writings of, for example, Hostinský (1879), and later by Wellek (1963) and Broeckx (1981).

In this respect too, one has to remember that the systematic approach to music theory, in the 19th century, had more normative implications than it perhaps has now. Both H. v.Helmholtz and H. Riemann – though from a quite different point of view – maintained that music theory embodied a normative theory that could (and needed to) be grounded in physiological and psychological evidence pertaining to the listening process. It is well-known that v.Helmholtz tried to explain the sensation of consonance and dissonance referring to the then brand new findings in anatomy and physiology of the human ear, while Riemann, who was related to Neo-Kantian idealism in many ways, first put forward the fundamentals of what he called *musical logic*, around 1870. His treatises on harmony are quite normative as he claimed to present not only a coherent body of music theory, but one that should be able to guide actual compositional practice. By the end of his life, however, Riemann dropped some of his more dogmatic views, and took a *cognitive* turn, offering his "Ideen zu einer Lehre von den Tonvorstellungen" (Riemann, 1914/15, 1916). Taking into account the content of Riemann's essays, its title could be aptly translated into English as *Prolegomena for a theory of tonal semantics*.

At about 1930, music psychology had already gained a solid base of knowledge. This was based, among others, on the elaborate books on "Tonpsychologie" by C. Stumpf and G. Révész, on O. Abraham's research into absolute hearing, E. Kurth's energetic theory of musical experience, and a lot of empirical work that had been directed to the perception of tone distances and intervals, melodies, timbre, as well as rhythmic structures. Much of this older literature is summarized in Wellek (1963). Although much of what was presented did depend on methods such as introspection and phenomenological description, it cannot be said that empirical work was scarce or even lacking at all. One may just point to Stumpf's many experiments on "Verschmelzung" and consonance, to Köhler's extensive experiments on timbre (Köhler, 1910a, 1910b, 1913, 1915), that led to the identification of formants, to Koffka's experiments on rhythm perception (Koffka, 1909), or to various experiments carried out by v.Hornbostel and Abraham on tonal distances and tonal brightness (von Hornbostel, 1926), as well as by Straub on tone height vs. tone chroma (Straub, 1929). Music psychology then wasn't just that *subjective* as perhaps is believed by many of our contemporaries. It is still worth while taking a look at these past achievements. For example, one could mention the empirical research on intonation practice in singing at Seashore's laboratory in Iowa, or similar research done on intonation of folk songs, and even on Habà's microtonal work in Berlin by Abraham and Kreichgauer. Also, the beginnings of empirical research into musical enjoyment, as was executed by Weld (1912), deserves some attention. Not all of this was undertaken in musicological institutes and seminars. In fact, much of the work was actually done in laboratories of psychology, but it was nevertheless evidently directed towards musical phenomena and therefore, it fitted into the systematic study of music.

At Berlin, Leipzig, Vienna, Graz, and other places, the close cooperation between musicologists and psychologists had culminated in the emergence of the Gestalt school. The cooperation, however, almost abruptly came to an end with the rise of Nazism in Germany and neighboring countries, when leading scholars in both fields, among others W. Köhler, K. Lewin, and E. v.Hornbostel emigrated (Schneider, 1993a).

After World War II, it was difficult to restore systematic musicology as well as Gestalt psychology in Europe, not only because of the emigration of key figures, but also because of a shift in scientific paradigm. In particular, the rise of *cybernetics* and *information theory* in the 1950ies had an immediate impact on many disciplines. American psychology, at the time dominated by behavioristic trends (Lundin, 1967), strongly influenced European thinking. With respect to methodology, testing of hypotheses according to, most of all, the Pearson-Neyman-paradigm became (and still is) almost obligatory (Wottawa, 1990). For the above mentioned reasons, in addition to a number of methodological problems that could not be solved by that time, such as testing of issues of representation and the lack of brain scanning technology, the Gestalt approach soon began to fall apart. At least, it lost much of its attractiveness and internationally acclaimed innovative position. Instead, it met severe criticisms especially from behavioristic and operationalistic quarters. There had been too many *Gestalt laws*, and perhaps not enough hardcore explanations to account for these, notwithstanding the great amount of experimental work that had been done over decades (Allport, 1967, Chap.5).

Both the impact of cybernetics and information theory, and the empirical research based on behavioristic psychology, made their way also into systematic musicology (Dahlhaus, 1971). At the same time, there was a lot of research done in other disciplines ranging from acoustics and physiology to sociology and aesthetics that was, and partly still is, relevant to systematic musicology (Elschek, 1992). The few researchers, representing systematic musicology as a discipline on its own, couldn't but to try to follow these new developments as well as contribute to at least some of the focal points, such as semiotics (Faltin & Reinecke, 1973; Broeckx, 1975; Stefani, 1976, 1987; Karbusicky, 1986). Yet, a handful of scholars would not be sufficient to obtain the critical mass that is necessary to yield an amplitude of research that could stand the comparison to the former Gestalt school.

What happened, instead, was that a generation of musicologists gradually shifted towards music sociology and the related field of psycho-sociology as a basis for understanding music (de la Motte-Haber, 1982). It was believed that music, at the highdays of the International Avant-Garde, could be better understood as a social phenomenon, rather than a merely perceptive or cognitive one. After all, music did no longer concern too much about pitches (traditionally the focus of systematic musicology), but instead included noises and environmental sounds in so-called *open structures* (Sabbe, 1977). Music sociology from the anti-bourgeois point of view, was the approach advocated by the Marxist-oriented musicology of the eastern European countries (Marothy, 1974).

3 The Gestalt Approach Revisited

It is well-know that music played a major role in the emergence of the Gestalt school. Although in fact different schools of Gestalt theory did exist, the approach in general can be characterized as a conviction that what happens to a part of an organized whole or *Gestalt*, is determined by intrinsic laws inherent in this whole. Physicalism was embodied in the hypothesis of *psychoneural isomorphism* which states that the dynamics of representations that are evidently present in psychological processes, are reflected at a neural level.

The Gestalt movement dates back to the late 19th century, and gained prominence by about 1920. By that time, different schools of Gestalt psychology in Würzburg, Berlin, Graz, Vienna, Frankfurt, and Leipzig were engaged in the study of auditory phenomena. They focused on rhythm and pitch perception, melody, consonance, timbre, structural organization of musical space, monotic vs. dichotic, as well as directional hearing (von Hornbostel, 1926; Wellek, 1934; Vernon, 1935; Sandig, 1939). The research topics, in short, covered almost every aspect that duly could have been covered by a relatively small group of very active and innovative researchers.

Though its sparkle by 1950 was gone, Gestalt psychology never really disappeared, and instead continued to produce works of prime importance to both general psychology and music psychology in particular (Fraisse, 1957, 1974; Frances, 1958; Wellek, 1963). The Gestalt approach that by some had been proclaimed *dead* already, lived up to enjoy *revivals*, on both sides of the Atlantic (Ertel, Kemmler, & Stadler, 1975). Gestalt thinking gradually gained a new impetus, and was found to be of particular importance in combination with then up-to-date trends in cybernetics and information science (Miller, Galanter, & Pribram, 1960). Another major event was the publication of U. Neisser's *Cognitive Psychology* (Neisser, 1967) that, in many respects, announced the emergence of the cognitive sciences. A lot of what has happened since, is summarized by H. Gardner, in his book on the so-called *cognitive revolution* (Gardner, 1987). Recently, new *Gestalt laws* have been proposed by Rock and Palmer (1990), namely the law of *enclosure*, and the law of *connectedness*. Seen from this point of view, the revival of Gestalt concepts was inevitable for various reasons:

- Gestalt perception is so predominantly present in our everyday life that it will be difficult, if not impossible, to omit what is subsumed under configurational approaches to perception, such as the notions of perceptual learning, and figure-ground and context-dependent perception. The descriptive power of Gestalt concepts has often been acknowledged for their plausibility in terms of human experience, and thus, closeness to the *real world*. Gestalt concepts play a central role in aesthetic theories that focus on the interplay between ratio and affect (Broeckx, 1981).
- There was already a wealth of experimental data on Gestalt perception available that had been adduced in formulating those many so-called *laws* (Helson, 1933). Of the many experimental findings, in particular those relating to *proximity*, *common fate*, and tendency towards *good forms* ("Prägnanz"), only a few hardly ever have been questioned. Today, these laws still provide a prominent heuristics for experimental psychology (Bregman, 1990).
- Behavioristic and operationalistic approaches in psychology and related fields soon were considered to be too mechanistic, and sometimes not even well founded as methodology (Plutchik, 1963). In particular, the work of S. Stevens on psychophysical measurement of pitch and loudness was almost completely refuted for methodological fallacies (Kristof, 1969). Experiments with pure tones, for example, may help to delineate particular functions of subjective pitch and loudness perception, but their relevance for complex tones in context may be questioned (Moore, 1989; Zwicker & Fastl, 1990).
- It has been shown (e.g. by J. Piaget) that behavior is guided by schemata, and that such concepts can be derived from perception and other cognitive processes. Schemata which are relevant to the recognition of patterns, can be considered as being relatively stable, on the one hand, and to hold for many subjects, on the other (Klix, 1971). The notion of schema is again a central concept in cognitive science and musicology (Seifert, 1991; Arbib, 1995; Leman, 1995).
- A shift of focus and methodology has also been effectuated by disciplines such as linguistics, anthropology and ethnomusicology. In these fields, the significant role of conceptualization, worldviews, cognitive categories,

memory, and mental evaluation of peripheral sensory input was demonstrated, often in a comparative way. In ethnomusicology too, there was this new and exciting evidence in favour of the Gestalt view, as shown in the performance practice of, for example, West African drumming, and Kiganda Amadinda music (Kubik, 1973, 1983; Wegner, 1993).

By about 1970, the tide had definitely turned to cognitive studies. This, in combination with the new tools offered by the information and engineering sciences, led to the emergence of an approach called *cognitive musicology*. This approach, which innovated systematic musicology from inside, was first heavily influenced by the formal methods introduced by mathematical logics and computational linguistics and in particular also by the powerful computer-based methods of the so-called artificial intelligence research (Kunst, 1976; Laske, 1977, 1986; Longuet-Higgins, 1987). Cognitive science itself was much the outcome of a post-neopositivistic reaction to behaviorism in the early sixties. It was above all based on the manipulation of information by means of sophisticated computational tools.

The development of cognitive musicology as a subfield of artificial intelligence research led to a particular conception on the nature of musical representations–a concept without which no computational account is possible. Influenced by formal description tools, musical representations were conceived of in terms of *symbols* that are manipulated by *rules*. In this so-called *symbol systems* approach (Newell, 1982), propositional representations of music were believed to be a proper starting point for the study of musical cognition. Later on (mid 1980ies), issues of musical representation were reconsidered and worked out in terms of a *subsymbolic* approach, starting from sounds and representations based on auditory images and neural networks (Leman, 1989, 1993; Seifert, 1993; Leman, 1995; Cosi, De Poli, & Lauzzana, 1994; Toiviainen, Kaipainen, & Louhivuori, 1995; Toiviainen, 1996). With this auditory-based approach, it became possible to connect a methodology based on computation with an epistemology rooted in the tradition of naturalism.

Meanwhile, even one of the ill-fated claims made by some of the Gestaltist's, that of psychoneural isomorphism, so vigorously defended by W.Köhler, has regained interest within the cognitive paradigma (Scheerer, 1994). The subsymbolic approach is now regarded as an appropriate tool to elevate cognitive science above the status of mechanized *folk psychology*. It offers a computational methodology that is in line with the naturalistic epistemology of traditional systematic musicology. It promises an integrated approach to psychoacoustics, auditory physiology, Gestalt perception, self-organization and cognition. In the search for universal psychological laws, methodology and empirical foundation are required to be as *hard* as in the physical sciences (Shepard, 1989).

4 Cognitive and Systematic Musicology in the Age of Computing

The forces that nowadays pull towards an integration of the music sciences are based on a computation oriented methodology. Apparently, different fields such as psychoacoustics, computer modelling, artificial intelligence research, psychology, semiotics, and anthropology got somehow involved with cognition and computation:

- In psychoacoustics and auditory physiology, research remained no longer restricted to the study of sensation. Nor is behaviorism the only prevailing doctrine for an experimental setup. Global auditory phenomena, such as rhythmical grouping (van Noorden, 1975), pitch fusion (Terhardt, 1974), and other forms of *context-dependent* perception (Plomp, 1976) have become genuine topics of research (Parncutt, 1989). Complementary to purely empirical research, computer modelling has become an important domain of research. Many of the results obtained in digital speech processing appear to have a direct impact on musicological research (Födermayr & Deutsch, 1992; Leman, 1994b, 1994a).
- The information processing approach to psychology (Lindsay & Norman, 1977) and computer modelling based on principles of artificial intelligence research became a popular paradigm for the study of music cognition (Laske, 1977, 1988). In turn, many so-called AI-labs were interested in music as a domain of application for new logical heuristics and formal representation (Balaban, Ebcioglu, & Laske, 1992). Recent tendencies now show a particular interest in interactive computer systems that often combine aspects of both symbolic and subsymbolic representation (Camurri, 1990; Camurri, Canepa, Frixione, & Zaccaria, 1992; Rowe, 1993; Camurri, Frixione, & Innocenti, 1994; Povall, 1995; Camurri & Leman, 1997). The concern with man-machine interaction led to a special interest in the role of musical expression and rhythm perception.
- Digital signal processing and its application to music entered the field in the early sixties (Mathews, 1969; Mathews & Pierce, 1991). It gave the researcher the total control over sound analysis and synthesis. Whereas in the past, researchers had to rely on resonance tubes, tuning forks, and electronic circuits, it now became possible to synthesize complex sound objects which, in turn can be used in experimental conditions. MusicV, a most popular sound compiler, was indeed developed as a means to explore the fascinating effects of ambiguous sound objects that lead to illusions of pitch and rhythm (Risset, 1978). In composition, however, this most powerful tool turned out to become a nightmare, especially when the subtile constraints of human perception and cognition are neglected. The control over sound, however, provides many new opportunities for a phenomenological approach to musical imagination (Schaeffer, 1966; Risset & Wessel, 1982; Risset, 1992). As a result, phenomenology stands

no longer in contrast with naturalist tendencies. The technology of sound computing (De Poli, Piccialli, & Roads, 1991; Roads, De Poli, & Pope, 1997) therefore should be seen as a main condition for the revival of systematic musicology.

- The seventies also introduced semiotics as a main paradigm for music research (Monelle, 1992). It was a primary conceptual framework of how to deal with music as basic entity of information. One approach adopted a formal linguistic point of view (Nattiez, 1975; Ruwet, 1975; Laske, 1977). Another approach sought the foundations of music in a Peircian theory of signs and cybernetics (Faltin & Reinecke, 1973; Posner & Reinecke, 1977; Karbusicky, 1986). Semiotics based on cybernetics is nowadays considered to be an integrated part of cognitive musicology.
- At the turn of the 1980s, studies in context dependent pitch perception (Krumhansl & Shepard, 1979; Krumhansl & Kessler, 1982; Krumhansl, 1990), timbre perception (Risset & Wessel, 1982; Grey, 1977) and segregation (McAdams & Bregman, 1979) marked the beginning of a vast amount of experimental research in Gestalt-based music psychology. In comparison with the former Gestalt school, experimental psychologists nowadays have some formal tools available for data analysis. Techniques based on multidimensional scaling (MDS) and clustering analysis now offer opportunities to map out the Gestalt-based organisation of internal memory (Shepard, 1982).

5 The Future of Systematic Musicology

The history of systematic musicology shows a tendency towards (i) a more powerful control of the acoustical basis of music (owing to an improved technology for sound manipulation), (ii) a growing interest in empirical foundation and validation of human information processing capabilities, and (iii) a growing interest in the role of the cultural environment, in particular the role of so-called cognito-social schemata or worldviews. What kind of developments may we expect for the future?

The success for a revival of systematic musicology may depend on (i) the ability to elaborate the appropriate methodology which would allow the factual integration of the above mentioned tendencies into models that explain and predict observed context-dependent phenomena in music, such as pitch, rhythm and timbre perception, expressivity, and emotionality related to musical perception, composition, performance, and this within the proper socio-cultural environment, and (ii) the ability to accumulate results from different fields and communicate with scientists from different disciplines dispersed over different areas all over the world.

Some of the developments that may become important in the future are related to the organisation of interdisciplinary research, the exploration of empirical research in relation to computer modelling, the use of advanced technology fully integrated within the musicology community, the accumulation of knowledge, the epistemological basis of naturalism, the integration of environmental approaches, and the application of musicological knowledge in technological applications:

- Interdisciplinarity. Music is a rich research domain which allows many different types of control and formalization: from sounds, to scores, to musical objects in multimodal environments. In addition, the topics of interest include problems of perception, performance, analysis, as well as composition of music. Laboratories with interest in music research will probably remain dispersed over many different academic departments including engineering, psychology, musicology, physics, neurology, sociology, communication, medicine. It is highly probable, therefore, that the success of systematic musicology will depend on the ability to form alliances over different disciplines and hence to make a new synthesis of obtained results. Theoretically speaking, computer modelling may turn out to become the cement of this segment of musicological universe because it allows accumulation of obtained results from many different fields. Practically speaking, a dispersion of laboratories in different fields and different places all over the world offers no longer a problem of communication in the information society.
- Empirical data and modelling. Empirical data will remain the most important source for musical theory building. A distinction can be made between data from behavioral research, brain research and computer simulation. Data from computer simulations can of course not be considered as pure empirical data but computer output can be put on *equal terms* with empirical data provided that the simulation is adequate and that both types of data can be compared in the context of a correlational analysis. Obviously, the output of computer simulation is considered to be valuable if and only if the computer model itself is based on an empirically approved theory of underlying mechanisms.
- Advanced technologies. Musicology, as a modern science, will have to become acquainted with advanced technologies that are nowadays used in brain science. We may restrict ourselves here to the observation that studies, which focus on the relationships between behavior and brain, will have to rely on advanced technologies such as electroencephalogram (EEG), functional magnetic resonance imaging (fMRI), magnetic encephalogram (MEG), positron emission tomography (PET) and other related techniques (Toga & Mazziotta, 1996; Sergent, 1996). In addition, super computing, multivariate statistical analysis, and other tools will be needed to put systematic musicology at the frontline of scientific research into perception and cognition. Given the demands of ecological modelling, adequate computer models heavily rely on the most advanced techniques for fast computing.
- Accumulation of knowledge. Another task is the development of sophisticated computational tools which build on results obtained in different

fields. The computational methodology allows, more than ever before, the accumulation of research results that are achieved in often very specialized research topics, such as numerical techniques, cochlear dynamics, physical modelling, object-oriented programming, etc.

- Naturalism. The observed correlations between the output of computer models and the psychological data forms at present a starting point for a reconsideration of what it means to rationally understand music. Research based on this idea is characterized by a tendency towards *naturalism*, i.e. the belief that the understanding of music should ultimately be based on methods which are paradigmatically exemplified in the natural sciences. Hence the interest in musical sounds as the starting point of musical modelling. It is believed that what happens in the human brain when listening to music is, apart from the specificities of the auditory system, highly dependent on the characteristics of the acoustical properties of the musical signals. Naturalism in musicology started with the work of H. v.Helmholtz (1863/1968) but it took more than one century before an appropriate methodology, i.e. computer modelling, would provide an opportunity to integrate different approaches for the study of context-dependent issues in music.
- Integration of environmental approaches. Ethnomusicology is much concerned with studies of people in their natural and cultural environment. One of the important conclusions to be drawn from this research is that musical cognition, though based on principles some of which can be considered *universal*, is more than just sound processing since it incorporates cultural aspects of multimodal categorization (Wegner, 1993; Moisala, 1995). Worldviews and categorical thinking are based on representations that involve active engagement of all senses in music making. The connection between the causal-based naturalist approaches in representation and action science remains a challenge for the future of systematic musicology.
- Application of musicology in technological applications. Hybrid architectures, in which low-level (sensorial and perceptive) information processing is combined with high-level (cognitive) planning of actions, offer a challenge for the integration of different representational systems within a global cognitive architecture (Camurri & Leman, 1997). The growing industrial interest in interactive multimedia systems, in which knowledge about human musical communication, such as pitch and rhythm perception, expression etc..., has to be incorporated with different sensorial modalities and perceptual systems, may provide an unique opportunity for musicology to prove its usefulness in technology oriented applications.

6 Conclusion

With the emigration of leading scholars both in musicology and psychology after 1933, the decline of both systematic musicology and Gestalt psychology in Europe was inevitable. After the second world war, the tradition of systematic musicology was only partially restored and research activities became subjected to centrifugal forces that led to disintegration. Loss of coherence can be attributed to a tendency towards specialization. A field such as psychoacoustics developed is own dynamics based on a behavioristic paradigm and independently from any Gestalt-based theory or application. On the other hand, some of the alleged sub-disciplines such as sociology of music and music theory developed into rather specialized fields and incorporated new topics such as the study of popular music and jazz, media research, and avant-garde music.

In the early 1970ies, under the influence of *post-neopositivism*, the need for an integration and unification of dispersed systematic approaches was felt (Apostel, 1972). Yet, a proper methodology, by which important new findings in empirical sciences and engineering could be integrated with earlier and contemporary theories of musical phenomena, was still missing. Although systematic musicology, at that time, produced some valuable empirical research (e.g. Hesse, 1972) as well as highly abstract theories of music (e.g. Boretz, 1969a, 1969b; Seeger, 1977), it became clear that the search for a new methodology was a central problem of concern.

During the past decades, several attempts have been made to re-establish a discipline of fundamental music research based on a methodology grounded, in the main, in computation. In order to achieve valid models of music perception and musical experience in general, stringent methods have been adopted that depend, most of all, on *formalization*, *experimentation* and, more recently, *computer simulation*. According to the approach chosen, reductionist explanation is preferred against qualitative description, and data gathering has been based on quantitative experimentation rather than on introspection. Theories have been elaborated by means of computational models that are more explicit than those metaphors that were introduced earlier this century. At different levels of research, from data analysis, sound synthesis, simulation of processes of perception, to the production of music, computers have become an important tool. These tools now form the methodological backbone for a naturalistic research programme. The programme, once started by H.v.Helmholtz and C.Stumpf in the 19th century, has been gradually transformed into a modern cognitive systematic musicology, an approach that can be fully integrated with the most advanced fields of modern science.

The netto effect seems to result in a transformation of musicology towards a powerful modern cognitive science. A main task of the new systematic musicology is therefore to combine the findings from different research domains, to show their real musical relevance and to work out, on this basis, the foundations for a rational understanding of musical phenomena.

References

Allport, F. (1967). *Theories of perception and the concept of structure.* New York, NY: Wiley.

Apostel, L. (Ed.). (1972). *De eenheid van de cultuur: Naar een algemene systeemtheorie als instrument van de eenheid van ons kennen en ons handelen.* Boom: Meppel.

Arbib, M. (1995). Schema theory. In M. Arbib (Ed.), *The handbook of brain theory and neural networks.* Cambridge, MA: The MIT Press.

Balaban, M., Ebcioglu, K., & Laske, O. (Eds.). (1992). *Understanding music with AI: Perspectives on music cognition.* Cambridge, MA: The MIT Press.

Boretz, B. (1969a). Meta-variations II: Sketch of a musical system. *Perspectives of New Music, 8*, 49–111.

Boretz, B. (1969b). Meta-variations I: Studies in the foundations of musical thought. *Perspectives of New Music, 8*, 1–74.

Bregman, A. (1990). *Auditory scene analysis: The perceptual organization of sound.* Cambridge, MA: The MIT Press.

Broeckx, J. (1975). *Contemporary values on musical style and aesthetics.* Antwerp: Metropolis.

Broeckx, J. (1981). *Muziek, ratio en affect: Over de wisselwerking van rationeel denken en affectief beleven bij voortbrengst en ontvangst van muziek.* Antwerp: Metropolis.

Camurri, A. (1990). On the role of artificial intelligence in music research. *Interface – Journal of New Music Research, 19*, 219–247.

Camurri, A., Canepa, C., Frixione, M., & Zaccaria, R. (1992). HARP: A framework and a system for intelligent composer's assistance. In D. Baggi (Ed.), *Readings in computer generated music.* Los Almitos, CA: IEEE Computer Society Press.

Camurri, A., Frixione, M., & Innocenti, C. (1994). A cognitive model and a knowledge representation system for music and multimedia. *Journal of New Music Research, 23*, 317–347.

Camurri, A., & Leman, M. (1997). AI-based music signal applications. In C. Roads, G. De Poli, S. Pope, & A. Piccialli (Eds.), *Musical signal processing.* Lisse, The Netherlands: Swets & Zeitlinger.

Chailley, J. (1967). *Expliquer l'harmonie.* Paris: Editions Rencontre.

Cosi, P., De Poli, G., & Lauzzana, G. (1994). Auditory modelling and self-organizing neural networks for timbre classification. *Journal of New Music Research, 23*, 71–98.

Dahlhaus, C. (Ed.). (1971). *Einführung in die systematische Musikwissenschaft.* Cologne: A. Volk.

de la Motte-Haber, H. (1982). Umfang, Methode und Ziel der Systematischen Musikwissenschaft. In C. Dahlhaus & H. de la Motte-Haber (Eds.),

Systematische Musikwissenschaft. Wiesbaden: Akademische Verlangsgesellschaft Athenaion.

De Poli, G., Piccialli, A., & Roads, C. (Eds.). (1991). *Representations of musical signals.* Cambridge, MA: The MIT Press.

Elschek, O. (1992). *Die Musikforschung der Gegenwart* (Vols. 1, 2). Vienna: Stiglmayr.

Ertel, S., Kemmler, L., & Stadler, M. (Eds.). (1975). *Gestalttheorie in der modernen Psychologie.* Darmstadt: Steinkopff.

Faltin, P., & Reinecke, H. (Eds.). (1973). *Musik und Verstehen: Aufsätze zur semiotischen Theorie, ästhetik und Soziologie der musikalischen Rezeption.* Cologne: A. Volk, H. Gerig.

Födermayr, F. (1983). Zum Konzept einer vergleichend-systematischen Musikwissenschaft. *Musikethnologische Sammelbände, 6,* 25–39.

Födermayr, F., & Deutsch, W. (1992). Musik als Geistes- und Naturwissenschaftliches Problem. In W. Gratzer & A. Lindmayr (Eds.), *De Editione Musices – Festschrift Gerhard Croll zum 65. Geburtstag.* Laaber: Laaber-Verlag.

Fraisse, P. (1957). *Psychologie du temps.* Paris: Presses Universitaires de France.

Fraisse, P. (1974). *Psychologie du rythme.* Paris: Presses Universitaires de France.

Frances, R. (1958). *La perception de la musique.* Paris: Vrin.

Gardner, H. (1987). *The mind's new science: A history of the cognitive revolution.* New York, NY: Basic Books.

Graf, W. (1980). *Vergleichende Musikwissenschaft: Ausgewählte Aufsätze.* Vienna, Föhrenau: Elisabeth Stiglmayr.

Grey, J. (1977). Multidimensional perceptual scaling of musical timbres. *The Journal of the Acoustical Society of America, 61,* 1270–1277.

Helson, H. (1933). The fundamental propositions of Gestalt psychology. *Psychological Review, 40,* 13–32.

Hesse, H. (1972). *Die Wahrnehmung von Tonhöhe und Klangfarbe als Problem der Hörtheorie.* Cologne: A. Volk.

Hostinský, O. (1879). *Die lehre von den musikalischen klängen.* Prag.

Karbusicky, V. (1986). *Grundriß der musikalischen Semantik.* Darmstadt: Wissenschaftliche Buchgesellschaft.

Klix, F. (1971). *Information und Verhalten.* Bern, Stuttgart: Huber.

Koffka, W. (1909). Experimental-Untersuchungen zur Lehre vom Rhythmus. *Zeitschrift für Psychologie, 52,* 1–109.

Köhler, W. (1910a). Akustische Untersuchungen 1. *Zeitschrift für Psychologie, 54,* 241–289.

Köhler, W. (1910b). Akustische Untersuchungen 2. *Zeitschrift für Psychologie, 58,* 59–140.

Köhler, W. (1913). Akustische Untersuchungen 3a. *Zeitschrift für Psychologie, 64,* 92–105.

Köhler, W. (1915). Akustische Untersuchungen 3b. *Zeitschrift für Psychologie, 72,* 1–192.

Kristof, W. (1969). *Untersuchungen zur Theorie psychologischen Messens.* Meisenheim am Glan: A. Hain.

Krumhansl, C. (1990). *Cognitive foundations of musical pitch.* New York, NY: Oxford University Press.

Krumhansl, C., & Kessler, E. (1982). Tracing the dynamic changes in perceived tonal organization in a spatial representation of musical keys. *Psychological Review, 89,* 334–368.

Krumhansl, C., & Shepard, R. N. (1979). Quantification of the hierarchy of tonal functions within a diatonic context. *Journal of Experimental Psychology: Human Perception and Performance, 5,* 579–594.

Kubik, G. (1973). Verstehen in afrikanischen Musikkulturen. In H. Reinecke & P. Faltin (Eds.), *Musik und Verstelten.* Cologne: A. Volk, H. Gerig.

Kubik, G. (1983). Kognitive Grundlagen afrikanischer Musik. In A. Simon (Ed.), *Musik in Afrika.* Berlin: Museum für Völkerkunde.

Kunst, J. (1976). Making sense in music I: The use of mathematical logic. *Interface – Journal of New Music Research, 5,* 3–68.

Laske, O. (1977). *Music, memory and thought.* Ann Arbor, MI: University of Microfilms International.

Laske, O. (Ed.). (1986). *Cognitive musicology.* Ghent. (Special issue of CCAI – Journal for the integrated study of artificial intelligence, cognitive science, and applied epistemology, Vol. 3, No. 3)

Laske, O. (1988). Can we formalize and program musical knowledge? An inquiry into the focus and scope of cognitive musicology. In M. Boroda (Ed.), *Musikometrika I.* Bochum: Studienverlag Dr. N. Brockmeyer.

Leman, M. (1989). Symbolic and subsymbolic information processing in models of musical communication and cognition. *Interface – Journal of New Music Research, 18,* 141–160.

Leman, M. (1993). Symbolic and subsymbolic description of music. In G. Haus (Ed.), *Music processing.* Madison, WI: Oxford University Press, A-R Editions.

Leman, M. (Ed.). (1994a). *Auditory models in music research. Part II.* Lisse, The Netherlands: Swets & Zeitlinger. (Special issue of Journal of New Music Research)

Leman, M. (Ed.). (1994b). *Auditory models in music research. Part I.* Lisse, The Netherlands: Swets & Zeitlinger. (Special issue of Journal of New Music Research)

Leman, M. (1995). *Music and schema theory: Cognitive foundations of systematic musicology.* Berlin, Heidelberg: Springer-Verlag.

Lindsay, P., & Norman, D. (1977). *Human information processing: An introduction to psychology.* London: Academic Press.

Longuet-Higgins, C. (1987). *Mental processes: Studies in cognitive science.* Cambridge, MA: The MIT Press.

Lundin, R. (1967). *An objective psychology of music.* New York, NY: The Ronald Press. (Originally published in 1953)

Marothy, J. (1974). *Music and the bourgeois, music and the proletarian.* Budapest: Akad. Kiado.

Mathews, M. (1969). *The technology of computer music.* Cambridge, MA: The MIT Press.

Mathews, M., & Pierce, J. (Eds.). (1991). *Current directions in computer music research.* Cambridge, MA: The MIT Press.

McAdams, S., & Bregman, A. S. (1979). Hearing musical streams. *Computer Music Journal, 3*, 26–43.

Miller, G., Galanter, E., & Pribram, K. (1960). *Plans and structure of behavior.* New York, NY: Holt, Rinehart & Winston.

Moisala, P. (1995). Cognitive study of music as culture: Basic premises for cognitive ethnomusicology. *Journal of New Music Research, 24*, 8–20.

Monelle, R. (1992). *Linguistics and semiotics in music.* Chur: Harwood Academic Publishers.

Moore, B. C. (1989). *An introduction to the psychology of hearing.* London: Academic Press.

Nattiez, J. (1975). *Fondements d'une sémiologie de la musique.* Paris: Union Générale d'Editions.

Neisser, U. (1967). *Cognitive psychology.* New York, NY: Appleton-Centuri-Crofts.

Newell, A. (1982). The knowledge level. *Artificial Intelligence, 18*, 87–127.

Parncutt, R. (1989). *Harmony: A psychoacoustical approach.* Berlin, Heidelberg: Springer-Verlag.

Plomp, R. (1976). *Aspects of tone sensation.* London: Academic Press.

Plutchik, R. (1963). Operationism as methodology. *Behavioral Sciences, 8*, 234–241.

Posner, R., & Reinecke, H. (Eds.). (1977). *Zeichenprozesse: Semiotische Forschung in der Einzelwissenschaften.* Wiesbaden: Akademische Verlagsgesellschaft Athenaion.

Povall, R. (1995). Compositional methods in interactive performance environments. *Journal of New Music Research*, 109–120.

Riemann, H. (1914/15). Ideen zu einer "Lehre von den Tonvorstellungen". In *Jahrbuch Peters* (Vol. 21/22). Peters.

Riemann, H. (1916). Ideen zu einer "Lehre von den Tonvorstellungen". In *Jahrbuch Peters* (Vol. 23). Peters.

Risset, J. (1978). *Hauteur et timbre des sons* (Tech. Rep.). Paris: Centre Georges Pompidou.

Risset, J. (1992). The computer as an interface: Interlacing instruments and computer sounds; real-time and delayed synthesis; digital synthesis and processing; composition and performance. *Interface – Journal of New Music Research, 21*, 9–20.

Risset, J., & Wessel, D. (1982). Exploration of timbre by analysis and synthesis. In D. Deutsch (Ed.), *The psychology of music.* New York, NY: Academic Press.

Roads, C., De Poli, G., & Pope, S. (1997). *Musical signal processing.* Lisse, The Netherlands: Swets & Zeitlinger.

Rock, I., & Palmer, S. (1990). The legacy of Gestalt psychology. *Scientific American, 263*, 48–62.

Rowe, R. (1993). *Interactive music systems.* Cambridge, MA: The MIT Press.

Ruwet, N. (1975). Théorie et méthodes dans les études musicales: quelques remarques rétrospectiveset prélimina. *Musique en Jeu, 17*, 11–36.

Sabbe, H. (1977). *Het muzikaal serialisme als techniek en als denkmethode.* Ghent: Rijksuniversiteit Gent, Faculteit Letteren en Wijsbegeerte.

Sandig, H. (1939). Beobachtungen an Zweiklängen in getrenntohriger und beidohriger Darbietung: Ein Beitrag zur Theorie der Konsonanz. In *Neue psychologische Studien* (Vol. XIV). Munich: Beck.

Schaeffer, P. (1966). *Traité des objets musicaux: Essai interdisciplines.* Paris: Editions du Seuil.

Scheerer, E. (1994). Psychoneural isomorphism: Historical background and current relevance. *Philosophical Psychology, 7*, 183–210.

Schneider, A. (1993a). Musikwissenschaft in der Emigration: Zur Vertreibung von Gelehrten und den Auswirkungen auf das Fach. In H. Heister, C. Maurer-Zenck, & P. Petersen (Eds.), *Musik und Musiker im Exil: Folgen des Nazismus für die internationale Musikkultur.* Frankfurt/M.: S. Fischer.

Schneider, A. (1993b). Systematische Musikwissenschaft: Traditionen, Ansätze, Aufgaben. *Systematische Musikwissenschaft, 1*, 145–180.

Schneider, A. (1993c). Zur Situation der Systematischen Musikwissenschaft. *Systematische Musikwissenschaft, 1*, 73–95.

Seeger, C. (1977). *Studies in musicology 1935–1975.* Los Angeles, Berkeley: University of California Press.

Seifert, U. (1991). The schema concept: A critical review of its development and current use in cognitive science and research on music perception. In A. Cammuri & C. Canepa (Eds.), *Proceedings of the IXth Colloquium on Musical Informatics.* Genova: AIMI/DIST.

Seifert, U. (1993). Systematische Musikwissenschaft als Grundlagenforschung der Musik. *Systematische Musikwissenschaft, 1*, 195–223.

Sergent, J. (1996). Human brain mapping. In R. Pratt & R. Spintge (Eds.), *Music medicine, Vol.2.* St. Louis: MMB.

Shepard, R. (1982). Structural representations of musical pitch. In D. Deutsch (Ed.), *The psychology of music.* New York, NY: Academic Press.

Shepard, R. (1989). Internal representations of universal regularities: A challenge for connectionism. In L. Nadel, L. Cooper, P. Culicover, & R. Harnish (Eds.), *Neural connections, mental computation.* Cambridge, MA: The MIT Press.

Stefani, G. (1976). Analisi, semiosi, semiotica. *Rivista Ital. di Musicologia, 11*, 106–125.

Stefani, G. (1987). A theory of musical competence. *Semiotica, 66*, 7–22.

Straub, W. (1929). Tonqualität und Tonhöhe. *Archiv für die gesamte Psychologie, 69*, 289–395.

Terhardt, E. (1974). Pitch, consonance, and harmony. *The Journal of the Acoustical Society of America, 55*, 1061–1069.

Toga, A., & Mazziotta, J. (Eds.). (1996). *Brain mapping.* London: Academic Press.

Toiviainen, P. (1996). Optimizing auditory images and distance metrics for self-organizing timbre maps. *Journal of New Music Research, 25*, 1–30.

Toiviainen, P., Kaipainen, M., & Louhivuori, J. (1995). Musical timbre: Similarity ratings correlate with computational feature space distances. *Journal of New Music Research, 24*, 282–298.

van Noorden, L. (1975). *Temporal coherence in the perception of tone sequences.* Unpublished doctoral dissertation, Technische Hogeschool, Eindhoven.

Vernon, P. (1935). Auditory perception. *British Journal of Psychology, 25*, 265–283.

von Helmholtz, H. (1863/1968). *Die Lehre von den Tonempfindungen als physiologische Grundlage für die Theorie der Musik.* Hildesheim: Georg Olms Verlagsbuchhandlung.

von Hornbostel, E. (1926). Psychologie der Gehörserscheinungen. In A. Bethe (Ed.), *Handbuch der normalen und pathologischen Physiologie* (Vols. XI, 1). Berlin, Heidelberg: Springer-Verlag.

Wegner, U. (1993). Cognitive aspects of Amadinda xylophone music from Buganda: Inherent patterns reonsidered. *Ethnomusicology, 37*, 201–241.

Weld, H. (1912). An experimental study of musical enjoyment. *American Journal of Psychology, 23*, 245–308.

Wellek, A. (1934). Der Raum in der Musik. *Archiv für die gesamte Psychologie, 91*, 395–443.

Wellek, A. (1963). *Musikpsychologie und Musikästhetik.* Frankfurt/M.: Akad. Verlagsanstalt.

Wottawa, H. (1990). Einige Überlegungen zu (Fehl)Entwicklungen der psychologischen Methodenlehre. *Psychologische Rundschau, 41*, 84–97.

Zwicker, E., & Fastl, H. (1990). *Psychoacoustics: Facts and models.* Berlin, Heidelberg: Springer-Verlag.

Systematic, Cognitive and Historical Approaches in Musicology

Jukka Louhivuori

Department of Musicology, University of Jyväskylä, PL 35 Jyväskylä, Finland

Abstract. The aim of this paper is to discuss the relationship between systematic, cognitive and historical musicology. This will be done by outlining the historical and epistemological backgrounds, and comparing the methods and objects. It is argued that systematic and cognitive musicology have a common methodological background with a focus on the systematic way of conducting research. However, quite a lot of studies in cognitive musicology apply nonsystematic methods as well; even a historical approach to cognition is needed. The main tenet of this paper is that the backgrounds of cognitive and systematic musicology are sufficiently similar for a close and fruitful cooperation. However, because the object of the research in cognitive musicology is the human mind, also soft methods such as field work, interviews and observations have to be applied as well. In that sense, cognitive musicology may broaden the research field of systematic musicology. Cognitive musicology, especially simulations with artificial neural networks, may give new possibilities to apply systematic methods in the study of some aspects of the history of music.

1 Introduction

The division of musicology into two separate branches, called *historical* and *systematic*, is often attributed to Adler (1919) and it has been used as an ideal classification of how to organize musicological activities at the institutional level. Academic professions are undertaken according to this classification and curriculum plans most often reflect the duality between the historical and systematic approach. The colleagues from both branches indeed have rather separate engagements. They organize separate international congresses and concentrate on very specialized topics. If somebody is studying the physical aspects of timbre, then a paper about the biographical details of Josquin Desprez is probably not at the focus of this person's research interest and vice versa, the person studying the life of Josquin may not be directly interested in sonological approaches to timbre analysis. Thus, although both work on music, the subject and the methodological approaches can be very different.

Next to historical and systematic musicology, cognitive musicology has emerged as a new branch of musicology. The cognitive approach originates from the seventies, at a time where musicologists got interested in artificial intelligence research and information psychology. These so-called cognitive sciences have always been somehow related to philosophical questions, but the methods have been based on those of the natural sciences. Especially the human brain has been an object of intensive research. The functioning of the brain has been studied using psychological experiments, mathematical models (in artificial intelligence and artificial neural networks), brain scanning and neurophysiological research based on single cell recordings of animal brains. In the last decade, a growing interest has been observed in research topics which aim to understand human expression, affect and emotion. The question has been raised whether it would be possible to explain emotional aspects of human behavior from the cognitive point of view. Other questions pertain to: What is the relationship between emotions and neural processing in the brain. Is it really possible to explain every aspect of human behavior in terms of physical facts (the relationship between mental states and the physical states of the brain)? The general tendency is that, in order to understand human cognitive processes, cultural and other softer aspects of human behavior should be taking into account more seriously (Krumhansl, 1995). In the field of cognitive musicology, a similar development has taken place. A growing interest in emotions and cultural aspects of cognition (cognitive ethnomusicology) is clearly to be seen (Castellano, Bharucha, & Krumhansl, 1984; Krumhansl, 1995).

The relationships between cognitive musicology, historical musicology and systematic musicology are somewhat unbalanced. Historical aspects of cognition, for example, have thus far not been widely discussed. Yet, we do have good reasons to assume that music history could indeed contribute to research topics where methods of cognitive science/musicology are applied. One such research topic is related to the history of tonality. Would it be possible to study the historical development of tonality using the cognitive approach? The connection between cognitive musicology and systematic musicology, on the other hand, has been more straightforward. The joint conference of *JIC96-Brugge* housed the Fourth International Symposium of Systematic Musicology and the Second Conference of Cognitive Musicology. It could be interpreted as an indication of a common scientific interest between the systematic and the cognitive approach (Leman, 1995; Schneider, 1993; Seifert, 1993). Would it be possible to find a similar common scientific interest between the systematic and the historical approach, the two main and most traditional branches of musicology?

During the the Third International Symposium of Systematic Musicology (at Zeillern, Austria in 1995) the question of the topics of the next conference was raised. One suggestion was to concentrate on the relationship between systematic and historical musicology from the point of view of cognitive musicology. It is well known that the approaches in systematic and historical musicology are thought to be quite separate from each other (Seeger, 1939; Wiora, 1948). Cognitive musicology might perhaps change this relationship in a way that may have far-reaching consequences. The background of this idea comes from the tradition of cognitive musicology, which in some respects differs from the tradition of systematic musicology. Although studies in history of music are quite rare in both branches, I claim that cognitive musicology provides an interesting opportunity to study the history of music with methods familiar to researchers working in the field of systematic musicology. Cognitive musicology could be a natural link between historical and systematic musicology. Especially the simulation of dynamic systems (e.g. using artificial neural networks) offers a systematic method for the study of at least some important aspects in music history.

2 Why Classification

The division into systematic and historical musicology has been criticized for being too vague and unsatisfactory. The study of the acoustics of historical instruments is an example of a systematic study of a historical topic. It can be classified as either systematic or historical (Fig.1). I believe that cognitive musicology is able to broaden the set of studies in music history even in a much more significant way.

However, the fact that the terms *systematic*, *cognitive* and *historical* do not belong to the same class of concepts, makes comparison more difficult. The word *systematic* refers to methodology, the term *cognitive* refers to the object of the study, while *history* refers to the period of time under discussion. Using these terms in an ideal classification schema is therefore a risky enterprise. Therefore, the question can be asked if we need any classification schema at all.

In some cases, classifications prevent or inhibit the natural development of science. International and national research groups often have their own (often interdisciplinary) research traditions, which may make it difficult to classify them as either systematic, historical, ethnomusicological or cognitive. What is, for example, the natural reference group of the researchers who are studying mental processes from the anthropological point of view?

In a similar way, narrow classifications may even prevent research groups to approach new research paradigms in a flexible way. Now and then a new paradigm arises, such as ethnomusicology, which changed the content and methodology of existing paradigms and which developed independent new subfields. If the tradition is very strong and the contents of the subfields are too rigidly defined, new paradigms may not find the structures to support the development of new ideas. Classifications may hamper the development of science, they often contribute to the fact that people working in a certain field of music get into a closed system.

The power of systematic, cognitive and historical musicology also largely relies on the volume and quality of researchers working together. The future of these fields will depend on the ability of researchers and research communities to renew and adapt to the significant changes occurring in science, and on the power of attraction of the field to allure the most talented researchers.

3 Towards a New Classification Schema

The need for a new classification schema, in which cognitive musicology is added as a subfield next to systematic and historical musicology, can be argued in terms of intrinsic and extrinsic reasons. The intrinsic reasons come from within the research society, the extrinsic reasons come from outside, for example from the needs and values of society.

3.1 Intrinsic Reasons

Sharing Insight Concerning Objects and Methods A main reason for establishing a classificatory connection between the subfields of musicology, is concerned with insights about common objects and methods. To what extend can we say that these subfields still share a common methodology and object of study?

Starting from the physical carrier of music, we notice already some differences in interest. For example, in order to understand the processing of auditory information, we indeed need to be aware of the physical properties of sound, but cognitive science is more interested in processes occurring at higher levels of our cognitive system (Gardner, 1985) and therefore it seems to go beyond the traditional focus on acoustics. In addition, musical acoustics might be prerequisite for historical studies of musical instrument but physical properties of sound have not at all been at the core of historical musicology.

In methodology, both systematic and cognitive musicology do have a number of common focus points which differ with historical musicology. The word systematic indeed implies that the main methodological demand is to study music in a systematic way. This has been a central demand also in the field of cognitive musicology.

Yet cognitive musicology has also a softer side, which is one of the arguments to classify it differently from systematic musicology. Because the object of the research is the human mind, also soft methods have been applied. The anthropology of music cognition, which aims to study the human mind in its cultural context, is of great importance for the understanding of musical thinking. Research in this field has been conducted using methods, such us field work, interviews and observations (Baily, 1988; Blacking, 1973; Cole, 1975; Herndon & McLeod, 1981; Kippen, 1987; Moisala, 1991, 1995). An understanding of the musical concepts of people representing different cultures requires qualitative methods, which are often based on field work lasting sometimes for several years.

At first glance, the history of music and cognitive musicology seem to exist far away from each other. Although it is often impossible to study the history of musical thinking with systematic methods, this kind of study is still needed in order to understand human cognition. Some questions treat how listeners, composers and musicians think about music, the kind of beliefs they have, the concepts they have used and the theories they have had about music. Cognitive science offers an interesting opportunity to study the history of music in a systematic way. It is possible to use for example artificial brain models in studying the development of musical styles, scales, modes, tonal systems etc. Artificial brain models could be placed into different historical musical environments to study how the artificial brain organizes musical inputs and how the output of the system depends on the changing environment. Cognitive musicology, therefore, is in many respects the bridge between systematic and historical musicology.

Volume of Research New classifications require much more than someone offering a new method and common research object. The volume of one certain type of research may have a great impact on the emergence of a new classification. Every discipline tolerates a certain amount of research which is methodologically quite far away from the main stream. For example, ethnomusicological studies have been conducted hundreds of years in Europe without any special need to name it as ethnomusicology. It was only when the volume of these types of studies grew and the society became conscious of the importance of this point of view that there was a pressure to call it a new branch of musicology. In the last decades, the volume of research in cognitive musicology has grown considerably as well. The need for new classifications comes when the content of old classifications does not correspond well to the existing scientific practice.

Dissatisfaction with the Traditional Paradigms The need for a new classification is often motivated by a feeling of dissatisfaction with the present methods, research objects and results. This feeling can be based on the idea that the prevailing paradigm is inadequate to solve the scientific problems that the scientific community considers crucial. A good example of this is a change of paradigm within cognitive science itself. A decade ago there was a strong belief that artificial intelligence based on rule based systems could be a meaningful method to study the mental processes of human beings. Later it became obvious that it is not possible to understand or explain human thinking, especially musical thinking, solely by using rule based systems.

The main argument against traditional musicology was directed towards methods, but many authors have criticized other aspects as well. It was argued, for example, that music should be studied in an authentic context, which implies a demand for ecological validity, and that psychological tests about timbre should be conducted using natural sounds in natural musical contexts. The brain has indeed been developed during millions of years. In order to survive, it had to process sounds and to adapt itself to a sound environment. Therefore, it is not at all clear that the brain processes unnatural sounds in the same way as natural sounds. This basic fact has often been neglected in traditional research and it is a main argument to extend paradigms in music research towards cognitive issues.

One of the most important arguments against traditional musicology, both systematic and historical, is the lack of studies about musical processes, such as composition (Laske, 1988). Dynamical aspects of music have been neglected, owing, perhaps, to the fact that western music is analytically studied from scores. This may give the wrong impression that music is independent of time. Cognitive musicology, on the other hand, has stressed the fact that music results from complex and dynamic processes of the brain. Although we may have the impression that an orchestra is playing far away on the stage, actually everything during the act of attending the concert and listening to music, happens inside the brain. The impression of distance and space around us is merely a consequence of mental processes.

Epistemology Other reasons in favor of a new classification schema pertain to epistemological and ontological questions. As long as we are studying physical aspects of sound, it appears that questions about knowledge acquisition, storing, and processing, are not in the centre of interest. Yet when the main focus is on cognitive aspects of timbre, for example, it is no longer possible to avoid epistemological and ontological questions. What actually is essential in musical thinking? What, for example, is the role of the human body in perception? Musical thinking should not be separated from action through bodily movements (Maturana & Varela, 1987; Varela, Thompson, & Rosch, 1991). The work of a composer depends heavily on physical activities needed in compositional processes and the way our mind adapts itself to a musical environment depends on the instruments through which the mind communicates with the external world. Similar questions can be raised about the role of cultural contexts. In that respect, one may ask whether it is possible at all to understand cognition by studying only the brain, without taking into consideration social and cultural context?

Cognitive musicology therefore can not avoid taking a stand on epistemological and ontological questions. The fact that the philosophical background of cognitive science is based on physicalism offers a good starting point for a fruitful co-operation with systematic and cognitive musicology.

History Musicology has had a long lasting tendency to concentrate on the history of music and, within that context, also on music theory. The latter is strongly based on musical practice. In fact, music theory has been written on the musicians' terms as theoretical concepts arose from the need to serve composers and musicians.

The scientific basis of music theory, therefore, has been weak and the main method of testing theories has been based on intuition. This pragmatic approach is now considered to be insufficient because it is not adapted to practical applications. A good example of the lack of theories in central topics of music theory is tonality. The lack of a theory of tonality is curious if we take into consideration the central role of the concept in Western music.

The history of systematic musicology is strongly rooted in the development of the natural sciences. Serious attempts have been made to study music from a scientific point of view. Yet the lack of appropriate methods and tools has prevented these studies from reaching their goals.

The background of cognitive science is similar in that cognitive science has always been based on the idea of studying music using scientific methods, as understood in natural sciences. Two different paths can be seen. Cognitive musicology was strongly influenced by the development of computer technology, brain research, informatics and linguistics (Chomsky, 1957). Also the history of psychology has had a strong influence. The attempt of psychology at the end of the 19th century of gaining a position as a science among other sciences led to the adoption of the methods of natural sciences. The consequence was an increase in laboratory tests and animal experiments. An attempt was made to explain human learning by using test with rats and other mammals. This trend is known as behaviorism.

Behaviorism and cognitive science share a belief in the methods of natural sciences. However it should not be forgotten that cognitive psychology was a reaction against behaviorism. The main criticism was that human learning and other mental processes should be studied in ecologically valid circumstances, which means that instead of unnatural test material (such as meaningless series of numbers) human behavior should be studied using test material the human mind is accustomed to processing (Piaget, 1952). It was soon understood that human learning is qualitatively different from the learning processes of rats. Learning of language is something more than conditioning. Behaviorists argued that the study of human mental processes should be conducted by studying overt behavior. Cognitivists argued that it is possible to study cognition using more direct ways, for example by means of brain research and computer simulations of brain processes.

Systematic and cognitive musicology have a common background, especially if we look at methodology. It is also clear that the multidisciplinary nature of cognitive science has had a strong influence on the development of methods in cognitive musicology. The backgrounds of cognitive and systematic musicology are sufficiently similar to allow for a close and fruitful co-operation. It is not at all clear whether this is also the case with historical musicology.

Convincing Power of Scientific Results New disciplines should be able to show much more convincing scientific results than is expected from the studies of existing paradigms. The latter have had time to show their scientific power. The results that cognitive musicology is able to present are without any doubt significant. The study of music cognition has given a deeper insight into the concepts of musical pitch, tonality, perception, learning and the production of music (composition, improvisation, variation etc.) as well as the psychomotor aspects of singing and playing. Models and theories about musical concepts are now on a much firmer basis than was the case a decade ago. Studies in the field of cognitive musicology have not only given a deeper understanding of various concepts but they have also directed attention to new research topics. Simulations of brain processes have shown how essential it is to take into consideration the bodily aspects of music. Also it has become clear how and in what sense musical thinking is different from or similar to thinking of other human activities. The importance of the dynamic aspects of music should be mentioned as well (Kaipainen, 1994; Kangas, 1991). Linguistics has up to now studied human thinking and mental processes as symbol manipulation. The significant role of emotion, intuition and nonsymbolic information processing in music might be relevant for other disciplines as well (Valentine, 1995).

Results in computer simulations, psychological experiments and brain research give good reasons to believe that the strategy chosen by cognitive musicologists has been at least in the right direction. Whether these results are convincing enough is another question. As a representative of cognitive musicology I believe that these results have shown the power of this paradigm.

Clarity of Scientific Argumentation The scientific community should consider differences between the old and new paradigms to be significant. One can ask for example if cybernetics has had sufficient independence to be able to be accepted as a new paradigm. Although cybernetics has had significant influence on the development of cognitive science, this discipline is only seldom mentioned.

The goal of cognitive musicology is relatively clear: to help us to understand and explain the complexity of musical experience, thinking and behavior. The question should be asked whether there are any aspects of music which could not be classified as cognitive study? Is cognitive musicology a replica of the concept of musicology? In my opinion the answer is no. We still need studies about the lives of composers, of musical styles and instruments. While biographical studies of composers could focus more on cognitive processes taking place during musical activities, all the facts we get from the biographies of composers give us a better understanding of the compositional processes.

3.2 Extrinsic reasons

In calling for a modification of the traditional classification schema, we should not forget to mention the role and relevance of the politics of science. Cognitive musicology as a multidisciplinary science is particularly sensitive to political structures and institutions but the problem is that it applies the methods of natural sciences onto soft research problems, such as musical thinking. Institutions that grant financial support are typically organized according to the traditional classifications of disciplines and hence, interdisciplinary fields are in danger of staying in no-man's land.

In order to carry out research we should get the acceptance of society. This acceptance is reflected into the politics of science. We therefore should ask ourselves what kind of studies is society (National Funds, European Union etc.) willing to support and what kind of institutions should be approached.

Systematic musicology has an established position on the scientific map. This is not the case with cognitive science or cognitive musicology. The methods and objects are close to the rest of the research community, such as psychology, computer science or philosophy. As such, it is not easy to convince the scientific community of the independent nature of cognitive studies. The field is still labile and in practice spreads into many different scientific environments. We have some excellent psychologists studying music cognition in the departments of psychology, physicists working in the departments of music, musicologists working in the departments of information science and so on. The difficulty of classifying cognitive science makes it difficult to find the financial support which is always the basic precondition for research.

4 Conclusion

Systematic and cognitive musicology have much in common, such as the belief in a systematic way of conducting research based on methods adopted from the natural sciences. Until now, most of the studies in cognitive musicology have been systematically conducted and in this respect, one could easily argue in favour of a classification of cognitive musicology as a subfield of systematic musicology. The problem with this point of view, however, is that cognitive musicology includes a much broader field, oriented towards a humanistic approach, such as cognitive ethnomusicology (cognitive anthropology). In addition to the methods of natural science, it applies softer methods than is usual in the field of systematic musicology.

The history of music is a field which has had a less prominent role in systematic or cognitive musicology. Despite of the lack of tradition in applying systematic methods in the field of music history I believe that this approach is able to offer new and fascinating research topics for researchers of both branches. Especially methods applied in cognitive musicology are appropriate in the field of historical musicology, such as modelling of cognitive processes. Other soft research topics exist where systematic methods could be applied such as the study of emotion and feeling, affects, intentions, perhaps even consciousness. These are research topics that cognitive science, too, has neglected, but which have come into the focus of attention of the research community mainly due to studies done in the arts, especially within music.

Figure 1 shows the three branches of musicology that have been discussed in this paper. They are related to each other and a few examples of typical research topics have been mentioned in order to help comparison. Cognitive musicology may significantly broaden the research field of systematic musicology towards topics which, from a musical point of view, are of great interest

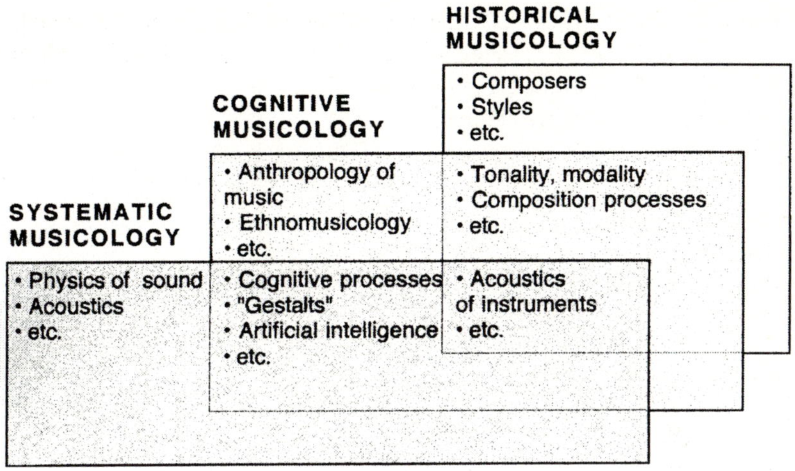

Fig. 1. Examples of some research topics that link systematic, cognitive, and historical musicology together

and should be in the main focus of musicologists, but which have been neglected so far, partly because of the limitations of systematic research methods. Cooperation with people working in the field of historical musicology may provide an opportunity to apply the new methodology onto questions that have a historical interest. The cooperation with cognitive musicologists who have developed methods for simulating human cognitive processes may be of particular interest.

I see two possible tendencies for the future. One is that both systematic and cognitive musicology continue to live their independent lives in close connection with each other. The other is a gradual integration of systematic and cognitive musicology, separate from historical musicology. This kind of intrinsic change of the paradigm might be quite beneficial for both the systematic and cognitive field. The methodological power and flexibility, as well as the significance of research topics of cognitive musicology, would join with the volume of research and researchers that systematic musicology can offer. This combination is strong enough to cause rapid and significant improvement in both fields. This kind of musicology is attractive enough to allure the best forces of the next generation of young researchers. The most appealing scenery for the future is a closer contact of systematic and cognitive field with historical musicology. This cooperation would have the most interesting and challenging consequences for the future of musicology and might change radically classifications in musicology and lower borders between different fields of music research.

References

Adler, G. (1919). *Methode der Musikgeschichte.* Leipzig: Breitkopf & Härtel.

Baily, J. (1988). Anthropological and psychological approaches to the study of music theory and musical cognition. *Yearbook for Traditional Music, 20*, 114–125.

Blacking, J. (1973). *How musical is man?* Seattle: University of Washington Press.

Castellano, A., Bharucha, J., & Krumhansl, C. (1984). Tonal hierarchies in the music of North India. *Journal of Experimental Psychology: General, 113*, 394–412.

Chomsky, N. (1957). *Syntactic structures.* The Hague: Mouton.

Cole, M. (1975). An ethnographic psychology of cognition. In R. Brislin, S. Bochner, & W. Lonner (Eds.), *Cross-cultural perspectives on learning.* New York, NY: John Wiley & Sons.

Gardner, H. (1985). *The mind's new science: A history of the cognitive revolution.* New York, NY: Basic Books.

Herndon, M., & McLeod, N. (1981). *Music as culture.* Darby: Norwood Editions.

Kaipainen, M. (1994). Dynamics of musical knowledge ecology: Knowing-what and knowing-how in the world of sounds. *Acta Musicologica Fennica, Helsinki, 19.*

Kangas, J. (1991). Time-dependent self-organizing maps for speech recognition. In *Artificial neural networks: Proceedings of the ICANN-91, Espoo, Finland.* Amsterdam: Elsevier Science Publishers.

Kippen, J. (1987). An ethnomusicological approach to the analysis of musical cognition. *Music Perception, 5*, 173–196.

Krumhansl, C. (1995). Music psychology and music theory: Problems and prospects. *Music Theory Spectrum, 17*, 53–90.

Laske, O. (1988). Introduction to cognitive musicology. *Computer and Music, 12*, 43–57.

Leman, M. (1995). *Music and schema theory: Cognitive foundations of systematic musicology.* Berlin, Heidelberg: Springer-Verlag.

Maturana, H., & Varela, F. (1987). *The tree of knowledge: The biological roots of human understanding.* Boston, London: New Science Library Shambhala.

Moisala, P. (1991). *Cultural cognition in music: Continuity and change in the Gurung music of Nepal.* Jyväskylä: Gummerus.

Moisala, P. (1995). Cognitive study of music as culture: Basic premises for cognitive ethnomusicology. *Journal of New Music Research, 24*, 8–20.

Piaget, J. (1952). *The origins of intelligence in children.* New York, NY: International University Press.

Schneider, A. (1993). Systematische Musikwissenschaft: Traditionen, Ansätze, Aufgaben. *Systematische Musikwissenschaft, 1*, 145–180.

Seeger, C. (1939). Systematic and historical orientation in musicology. *Acta Musicologica, XI.*

Seifert, U. (1993). Systematische Musikwissenschaft als Grundlagenforschung der Musik. *Systematische Musikwissenschaft, 1*, 195–223.

Valentine, E. (1995). Deconstructing cognition: Towards a framework for exploring non-conceptualised experience. In P. Pylkkänen & P. Pylkkö (Eds.), *New directions in cognitive science: Proceedings of the International Symposium, Saariselkä, 1995* (pp. 1–10). Helsinki: Publications of the Finnish Artificial Intelligence Society.

Varela, F., Thompson, E., & Rosch, E. (1991). *Embodied mind: Cognitive science and human experience.* Cambridge, MA: The MIT Press.

Wiora, W. (1948). Historische und systematische Musikforschung. *Die Musikforschung, I*, 171–191.

Empiricism, Gestalt Qualities, and Determination of Style:

Some Remarks Concerning the Relationship of Guido Adler to Richard Wallaschek, Alexius Meinong, Christian von Ehrenfels, and Robert Lach

Michael Weber

Institute of Musicology of the University of Vienna, Universitätsstrasse 7, A-1010 Vienna, Austria

Abstract. G. Adlers acquaintance and friendship with philosophers such as A. Meinong, C. von Ehrenfels – the founder of the theory of Gestalt qualities –, and E. Mach, and Adler's general interest in philosophy were conducive to his familiarity to the contemporary discussion about empiricism and the challenge of the humanities by science. Therefore, the search for empirical substantiation of speculative ideas caused great influence on Adler's concept of music history research. Though Adler shared this basic concern with the representatives of comparative musicology his academic career did not bring forth a prolific co-operation with his colleagues at the University of Vienna, R. Wallaschek and R. Lach.

1 Introduction

Looking back a century to the first decades of academic musicology, it can be stated that a close relationship to other disciplines such as philosophy, history, psychology, anthropology, and physics seemed to be more or less a matter of course. Especially in the center of Europe, at the Universities of Berlin, Prague, and Vienna, the ongoing developments in the connected fields of philosophy and psychology, attracted the increased attention of musicologists.

Guido Adler (1855–1941), Professor at the German University of Prague and at the University of Vienna, was typical of such scholars with broad fields of interest (Adler, 1935). His publications concerning the theory and method of musicological research illustrate the exceedingly strong influence of the contemporary discussion of humanities challenged by the exact sciences at that time. Adler witnessed such debate in his dealings with philosophers and physicists such as Alexius Meinong (1853–1920), Friedrich Jodl (1849–1914), Ernst Mach (1838–1916), Carl Stumpf (1848–1936), Christian von Ehrenfels (1859–1932), and Ludwig Boltzmann (1844–1906) (Wessely, 1986).

The following attempts to summarize some of the influences on Adler's intellectual world by Meinong, Professor of Philosophy at the University of Graz, by von Ehrenfels, Professor of Philosophy at the German University of Prague and founder of the theory of Gestalt qualities, and by Mach, Professor of Physics at the University of Graz and at the German University of Prague, and Professor of Philosophy at the University of Vienna, and by a few other scholars from the fields of philosophy and psychology. Adler's attitude towards the representatives of comparative musicology at the University of Vienna, namely Richard Wallaschek (1860–1917), and Robert Lach (1874–1958), will be elucidated briefly.

2 Adler and the Representatives of Comparative Musicology

It is appropriate to refer to Adler as a key figure in the development of Austrian musicology (Antonicek, 1986a). In addition to its general importance for musicology Adler's famous essay on *Extent, Methods, and Purpose of Musicology* (Adler, 1885) from 1885 is regarded as one of the most significant contributions concerning the discipline's development at the University of Vienna. On the one hand, this text – the first in a series of further publications – bears eloquent witness to the tendency towards empiricism at that time. On the other hand, the division of musicology into a historic and a systematic branch was definitely conducive to the academic rise of comparative musicology.

Wallaschek was able to habilitate in 1896 for *Music Psychology* and *Aesthetics of Musical Art* at the University of Vienna (Graf, 1980). According to W. Graf (1974), this fact "may be seen as the beginning of Comparative Musicology in Austria" (p.16). Apparently Adler, who was Ordinarius for *History and Theory of Musical Art* at the University of Vienna from 1898, was not entirely convinced of his colleague's qualifications for a musicological professorship. In a confidential letter (Eder, 1995, p.179f.) Adler described Wallaschek as follows:

> "[...] he is actually an aesthetician, having published a (very feeble) *Aesthetic of Musical Art* in one volume (Wallaschek, 1886), and then an ethnomusicological book (Wallaschek, 1903)[...]. He may not know particularly much about the psychology of music, he does not seem to be sufficiently trained in philosophy or in an empirical-experimental direction" [1] (Wallaschek, 1905, 1930)

Nevertheless, in 1908, Adler finally gave up his original opposition to Wallaschek's appointment as Extraordinarius (Antonicek, 1986a).

In 1920 Lach, having habilitated five years earlier, succeeded as Associate Professor for Comparative Musicology, Music Psychology, and Aesthetics of Musical Art at the University of Vienna (Graf, Jancik, Meister, Nowak, &

---

[1] G. Adler to A. Meinong, Vienna, June 24, 1899 (Eder, 1995, p.179f.): "[...] er ist eigentlich Ästhetiker, hat eine (sehr schwache) *Ästhetik der Tonkunst* in einem Band veröffentlicht und dann ein musikethnologisches Werk [...]. Allzuviel wird er auch nicht von Musikpsychologie verstehen, er scheint nicht genügend philosophisch vorgeschult, nicht in empirisch-experimenteller Richtung".

Schenk, 1954). Adler had been a member of the application commission and had tried to prevent the appointment (Antonicek, 1986a) referring to "the requirements of the subject of Musicology in its entirety (historical and systematic branch)" (p.188).

One year earlier, on the occasion of a request for a musicological professorship at the University of Graz, Adler (Eder, 1995) had actually announced to his friend Meinong:

> "One must dispense with aesthetics [...]. I do not know any scientifically well-founded music aesthetician, probably this is the case with former Music Aesthetics on the whole. Comparative Musicology ([Music] Ethnology) is a side- or a secondary track, where the publications of the Academy [of Sciences] may be parked. For the German-Austrian universities, only competent music historians may be considered, perhaps those, with [music] theory as well"[2] (p.268f.).

Considering these reservations, a prolific co-operation between Adler, who was always closely connected with the inductive method, and the Viennese representatives of the related branch of comparative musicology could not transpire.

3 Meinong's Influences on Adler

Adler's close connection to the leading contemporary exponents of the empirical trends in philosophy (Wessely, 1986) is reflected in many ways in his own work. Impressions of Franz Brentano's (1838–1917) lectures at the University of Vienna (Adler, 1935; Eder, 1995) may be considered to have been of crucial significance, as well as the lifelong friendship to Meinong (Eder, 1995), the founder of the theory of objects and of the first Austrian experimental-psychological laboratory established at the University of Graz in 1894 (Meinong, 1978b).

Adler himself reported (Eder, 1995), for example, having read Brentano's *Psychology from an Empirical Point of View* (Brentano, 1924, 1925, 1968)[3] and, of having "been absorbed" by Meinong's philosophical papers, again and again having gained insight, "that I can learn various things for my subject"[4] (p.261) from them.

[2] G. Adler to A. Meinong, Vienna, August 15, 1919 (Eder, 1995): "Auf die Ästhetik ist zu verzichten [...]. Ich kenne keinen wissenschaftlich fundierten Musikästhetiker, wohl ist überhaupt dies bei der bisherigen Musikästhetik der Fall. Die vergleichende Musikwissenschaft (Ethnologie) ist ein Neben- oder Seitengeleise, auf das die Akademien [der Wissenschaften] ihre Druckprodukte schieben können. Für die deutsch-österreichischen Universitäten kommen nur tüchtige Musikhistoriker eventuell mit der Beigabe der Theorie in Betracht" (p.268f.).

[3] G. Adler to A. Meinong, Hinterbruehl, July 31, 1878 (Eder, 1995, p.55)

[4] G. Adler to A. Meinong, Vienna, August 31, 1917 (Eder, 1995): "[...] Vertiefung [...], daß ich mancherlei für mein Fach lernen kann" (p.261).

Distinct influence (Eder, 1995)[5] was left by Meinong in that well-known section of Adler's essay on methods (Adler, 1885) from 1885, putting the inductive method (Kalisch, 1988) in the center of interest:

> "The *method* of musicological research conforms to the nature of the subject being researched: [...] for the examination of the laws of art in different eras and their organic associations and developments, the art historian will use the very same methods as the natural-scientist: preferably the *inductive* method. He will give prominence to the common facts of various examples, separate the differences, and make use of abstraction by neglecting particular components of tangible conceptions and by preferring others. Also, the assertion of hypotheses is not ruled out"[6] (p.15).

During the composition of his inaugural lecture for Vienna, Adler (Eder, 1995) "was often close to you [Meinong] in mind, for your consorting marked me indelibly. [...] How I would love to work at your side again!"[7] (p.177), Adler confessed to Meinong in 1899.

Adler's disagreement (Eder, 1995)[8] with Meinong's review of Stumpf's book *Music Psychology (Tonpsychologie)* for the *Vierteljahrsschrift für Musikwissenschaft (Musicological Quarterly)* (Meinong, 1885/1978a), as well as their different views on the validity of the law of causality (Adler, 1919; Eder, 1995) [9] could not do any harm to their mutual esteem. Nevertheless, their continual desire – lasting for decades – to resume lecturing at the same University, as they had done in the early 1880's in Vienna, remained unfulfilled.

---

[5] G. Adler to A. Meinong, Vienna, July 2, 1884; A. Meinong to G. Adler, Graz, July 4, 1884 (Eder, 1995) (p.86f.).

[6] G. Adler (Adler, 1885): "Die *Methode* der musikwissenschaftlichen Forschung richtet sich nach der Art des zu Erforschenden: [...] zur Erforschung der Kunstgesetze verschiedener Zeiten und ihrer organischen Verbindung und Entwicklung wird sich der Kunsthistoriker der gleichen Methode bedienen wie der Naturforscher: vorzugsweise der *inductiven* Methode. Er wird aus mehreren Beispielen das Gemeinsame abheben, das Verschiedene absondern und sich auch der Abstraction bedienen, indem von concret gegebenen Vorstellungen einzelne Theile vernachlässigt und andere bevorzugt werden. Auch die Aufstellung von Hypothesen ist nicht ausgeschlossen" (p.15).

[7] G. Adler to A. Meinong, Vienna, June 24, 1899 (Eder, 1995): "[...] verkehrte ich im Geiste oft mit Ihnen, denn Ihr Verkehr ließ untilgbare Spuren in mir zurück. [...] Wie gern arbeitete ich wieder an Ihrer Seite!" (p.177).

[8] E. g. A. Meinong to G. Adler, Graz, October 18, 1884 (Eder, 1995, p.94f.)

[9] A.Meinong to G. Adler, Graz, August 11, 1919; G. Adler to A. Meinong, Vienna, August 15, 1919 (Eder, 1995) (p.267f.).

4 Von Ehrenfels' Foundation of the Theory of Gestalt Qualities

Brentano's lectures were also of decisive significance for von Ehrenfels' many-sided work. Like Adler, von Ehrenfels was attached by lifelong friendship to his teacher Meinong (Fabian, 1986; Weinhandl, 1960; Kindinger, 1965). Among his philosophical – on theory of cognition, theory of value, and social philosophy, among others – and his psychological papers von Ehrenfels' later fame was established by his article *On "Gestalt Qualities"* (von Ehrenfels, 1890/1988a) from 1890, which he wrote as lecturer at the University of Vienna (Simons, 1988), inspired by Mach's reflections (Mulligan & Smith, 1988). More than forty years later, von Ehrenfels (von Ehrenfels, 1890/1988c) summarized his considerations as follows:

> "The starting point of the theory of Gestalt qualities was the attempt to answer a question: what is melody? The most obvious answer: the sum of the individual tones which make up the melody. But opposed to this is the fact that the same melody may be made up of quite different groups of tones, as happens when the same melody is transposed into different keys. If the melody were nothing other than the sum of the tones, then we would have to have different melodies, since different groups of tones are involved. [...] The decisive step in the founding of the theory of Gestalt qualities was now the assertion on my part, that if the memory images of successive tones are present as a simultaneous consciousness-complex, then a presentation of a new category can arise in consciousness, a unitary presentation, which is connected in a peculiar manner with the presentations of the relevant complex of tones. The presentation of this whole belongs to a new category for which the name 'founded content' came into use [on the proposal of Meinong 1969; Kindinger 1965 [10]; Simons, 1988]. Not all founded contents are intuitive in nature and related to the presentation of a melody. There are also non-intuitive founded contents, as for example relations. What is essential to the relation between the founded content and its fundament is the one-sided determination (Bedingtheit) of the former by the latter. Every founded content necessarily requires a fundament. A given complex of fundamental presentations is able to support only a quite specific founded content. But not every fundament must as it were be crowned and held together by a founded content"[11] (p.121f.).

[10] C. von Ehrenfels to A. Meinong, Vienna, June 3, 1891 (Kindinger, 1965, p.74f.)

[11] C. von Ehrenfels (von Ehrenfels, 1890/1988c): "Der Ausgangspunkt von der Lehre über Gestaltqualitäten war der Versuch der Beantwortung einer Frage: Was ist Melodie? Nächstliegende Antwort: Die Summe der einzelnen Töne, welche die Melodie bilden. Dem steht aber gegenüber die Tatsache, daß dieselbe Melodie aus ganz verschiedenen Tongruppen gebildet werden kann, wie es beim Transponieren derselben Melodie in verschiedene Tonarten erfolgt. Wäre die Melodie nichts anderes als die Summe der Töne, so müßten, weil hier verschiedenen Tongruppen vorliegen, auch verschiedene Melodien gegeben sein. [...] Der entscheidende Schritt für die Begründung der Lehre von der Gestaltqualität war nun

Von Ehrenfels' discovery of an additional element in the perception of complex sensual stimuli was accompanied by the almost simultaneous description of *figural elements* ("figurale Momente") (p.203–212) in the *Philosophy of Arithmetics* by Edmund Husserl (1859–1938) (Husserl, 1992b; Simons, 1988; Smith, 1988). Still von Ehrenfels' transfer of the concept of *Gestalt qualities* to non-psychic areas, like his cosmogony, aesthetics, and theory on prime numbers, was also conducive to the spreading of the term Gestalt qualities, so that Husserl's (Husserl, 1992a) "Einheitsmoment" (p. 237) as well as Meinong's (Meinong, 1971) *founded object* ("fundierter Gegenstand") (p.399f.) could not prevail (Simons, 1988).

5 Von Ehrenfels' Acquaintance with Adler

Adler and von Ehrenfels had known each other at least since the beginning of the 1880's. Thus, in the spring of 1883, Adler reported (Eder, 1995) to Meinong, who had been an Extraordinarius for Philosophy at the University of Graz since the previous term (Meinong, 1978b): "Sometimes I see Ehrenfels in concerts"[12] (p.69). In addition, they both were involved in the first series of performances of Richard Wagner's (1813–1883) *Parsifal* one year earlier in Bayreuth (Adler, 1935), during the course of which von Ehrenfels even made a *pilgrimage* on foot (Fabian, 1986). However, Meinong had declined Adler's invitation to accompany him (Eder, 1995).

Von Ehrenfels' enthusiasm about Wagner, which he shared with Anton Bruckner (1824–1896) (Brod, 1979), who had taught him musical composition from 1880 to 1882 (Fabian, 1986), is demonstrated by a dozen essays and in some of his dramatic operas (Winkler, 1986). He let Adler have a copy of his two articles *On the Clarification of the Dispute about Wagner ("Zur Klärung der Wagner-Controverse")* (von Ehrenfels, 1896/1986b) and *The Musical Architectonics* ("Die musikalische Architektonik in Wagner's

die Behauptung meinerseits: Wenn die Erinnerungsbilder der aufeinanderfolgenden Töne als ein gleichzeitiger Bewußtseinskomplex vorliegen, so kann im Bewußtsein eine Vorstellung neuer Kategorie auftauchen, und zwar eine einheitliche Vorstellung, welche auf eine eigentümliche Weise mit den Vorstellungen des betreffenden Tonkomplexes verbunden ist. Die Vorstellung dieses Ganzen gehört einer neuen Kategorie an, für welche der Name 'fundierte Inhalte' üblich wurde. Nicht alle fundierten Inhalte sind anschaulicher Natur und der Melodievorstellung verwandt. Es gibt auch unanschauliche fundierte Inhalte, wie z. B. die Relation. Das Wesentliche des Verhältnisses zwischen dem fundierten Inhalt und seinem Fundament ist die einseitige Bedingtheit jenes durch dieses. Jeder fundierte Inhalt bedarf notwendig eines Fundaments. Ein bestimmter Komplex von Fundamentalvorstellungen vermag nur einen ganz bestimmten fundierten Inhalt zu tragen. Aber nicht jedes Fundament muß von einem fundierten Inhalt gleichsam gekrönt und zusammengehalten werden" (p.168).

[12] G. Adler to A. Meinong, Vienna, March 31, 1883 (Eder, 1995): "Ehrenfels sehe ich manchmal in Konzerten" (p.69).

Ring des Nibelungen") (von Ehrenfels, 1896/1986a) from 1896 "for kind inspection"[13]. The first article was completed with hand-written, not always approving comments (Antonicek, 1986b) by his fellow-professor at the German University of Prague. In accordance, Adler's report (Eder, 1995) to his friend Meinong from a stay in Karlsbad was ambiguous:

> "Now I do have the welcome opportunity to talk about you more often, namely with Ehrenfels [...] – lately he gave a lecture on the musical motif with Wagner in the German Craftsman Association [in Prague] [...]. I was not present – but had the opportunity to become acquainted with his views in person and in writing (in two articles) [...]. Since talking or writing about matters of business and profession is not suitable for a spa, I reserve the review for a later date"[14] (p.161).

Aside from the discourse about common preferences in music and shared experiences with lessons by Bruckner (Adler, 1935), a philosophical exchange of ideas may have taken place, similar to the ones Adler used to have with Meinong or Jodl (Eder, 1995) among others. Von Ehrenfels presented Adler with his habilitation paper *On Feeling and Will* ("Über Fühlen und Wollen") (von Ehrenfels, 1887/1988b; Eder, 1995)[15] and his famous essay *On "Gestalt Qualities"* (von Ehrenfels, 1890/1988a) "with best regards"[16].

6 Mach's Influences on Adler's Determination of Style

Adler's interest in science, which went beyond the boundaries of music history research, was reflected only to a minor extent in his editing of the *Vierteljahrsschrift für Musikwissenschaft*. Approximately a mere fifth of the articles and a third of the book reviews contained in the ten annual volumes were from the field of systematic musicology (Gebesmair, 1996). In retrospect, Adler proudly called Mach, Meinong, the physicist Max Plank (1858–1947), and Stumpf his collaborators, but regretted the rejection of the philosopher and psychologist Wilhelm Wundt (1832–1920) very much (Adler, 1935).

[13] Both presented copies are stored in the Library of the Institute of Musicology of the University of Vienna with the numbers B 5727/14 and /17.

[14] G. Adler to A. Meinong, Karlsbad, April 8, 1897 (Eder, 1995): "Habe ich ja jetzt willkommene Gelegenheit, auch über Sie öfter zu sprechen, und zwar mit Ehrenfels [...] – letzthin hielt er im Deutschen Handwerkerverein [in Prag] einen Vortrag über das musikalische Motiv bei Wagner [...]. Ich war nicht dabei – hatte aber Gelegenheit seine Ansichten persönlich und schriftlich (in zwei Aufsätzen) [...] kennenzulernen. Da es nicht kurgemäß ist, über Geschäfte und Berufsangelegenheiten zu sprechen oder zu schreiben, behalte ich die Besprechung für später mir vor" (p.161).

[15] G. Adler to A. Meinong, Prague-Weinberge, November 17, 1888 (Eder, 1995, p.124)

[16] The presented copy of the essay is stored in the Library of the Institute of Musicology of the University of Vienna with the number B 5739/17.

Mach – once with Stumpf and once with Jodl – had decisively contributed to Adler's appointment at the German University of Prague and at the University of Vienna (Antonicek, 1986a; Eder, 1995). In addition, Mach, who was appointed Professor for *Philosophy, especially History and Theory of Inductive Sciences* (Eder, 1995), exercised considerable influence on the music historian in other ways, too (Blaukopf, 1995). For example, in an attempt to "exact substantiation" of style characteristics of music historical epochs (Blaukopf, 1991), Adler (Adler, 1930/1981) proposed the "important methodical instrument of statistic specifications" (p.793) in his *Handbook of Music History*.

For Adler's procedure of *Determination of Style* (Adler, 1911, 1919) as well as the above, equivalent intentions were not unimportant, for

> "the stylistic execution of a work of art may [...] be perceived and comprehended scientifically as well. Concerning this [...] the following factors may be considered: the course of recognition is either inductive or deductive; in this case, it is followed twice"[17] (p.11),

as he wrote in his book *Method of Music History*. Accordingly, the "analogy of the method of artistic science to the one of natural-science" (p.15), called for by Adler (1885), gained significant importance in his basic concept of music history research (Adler, 1919; Heinz, 1968; Kalisch, 1988):

> "Summing up, it may be pointed out that the responsibility of music history is not the exploration of the Beauty in musical art, but the recognition of the course of development of music in works and creators"[18]. (p.13).

In Adler's eyes (Adler, 1919), the outcome of a "critical analysis determinating style" should – with reference to Mach, Meinong's *Objects of Higher Order* ("Gegenstände höherer Ordnung") (Meinong, 1971), and von Ehrenfels' *Gestalt Qualities* – ultimately consist of the

> "accentuating and clarifying in the genetic succession of art phenomenons the Common, the Special, and the Individual within particular periods of art, epochs, and schools"[19] (p.112).

[17] G. Adler (Adler, 1911): "Die stilistische Ausführung eines Kunstwerkes kann [...] auch wissenschaftlich erkannt und erfaßt werden. Hierzu kommen [...] folgende Momente in Betracht: der Weg der Erkenntnis ist entweder induktiv oder deduktiv, hier wird er doppelt beschritten" (p.11).

[18] G. Adler (Adler, 1919): "Zusammenfassend sei hervorgehoben, daß die Aufgabe der Musikgeschichte nicht die Erkundung des Kunstschönen in der Tonkunst, sondern die Erkenntnis des Entwicklungsganges der Musik in Werken und Schaffenden ist" (p.13).

[19] G. Adler (Adler, 1919): "[...] in der genetischen Folge der Kunsterscheinungen das Allgemeine, Besondere und Individuelle innerhalb der einzelnen Kunstperioden, Zeitabschnitte, Schulen herauszuheben, klarzustellen" (p.112).

7 Accordance and Disaccordance between Adler and Lach

With his *style determinating* concept of music history research, Adler also came close to views represented by Lach as well, who was – against Adler's will (Adler, 1935; Antonicek, 1986a) – in 1927 appointed as his successor as Ordinarius for Musicology at the University of Vienna.

In the *Studies on the Developmental History of Ornamental Melopoei* – published in 1913 with a dedication to Adler – the importance of which was pointed out by the bearer of that dedication himself (Adler, 1911), Lach (Lach, 1913) drew the conclusion that:

> "the same developmental stages, phases, and forms underlying respectively appearing in the purely formal-genetic order of evolution, may be demonstrated purely historically, in the history of music, in exactly the same order"[20] (p.524).

Adler's twofold idea of music and its history as *organism* e.g. (Adler, 1935; Angerer, 1988) and his assumption of a general processuality (Schneider, 1984; Adler, 1919) were definitely related to that. In the application of the typological procedure in determinating style (Heinz, 1969; Kalisch, 1988; Schneider, 1984), Adler was of the same opinion as other representatives of comparative musicology, too (Schneider, 1976).

Adler agreed less with Lach's definition (Lach, 1924) of the position of comparative musicology which should, after all, "integrate any musical and music historical events in the continuum of a process of *nature*, a process of natural-scientifically conditioned inner necessity" (p.12). Lach's assignment (Lach, 1924) of music history research to a comparative musicology aiming at systematic goals, "wherein any musical and music historical disciplines (as music theory, music history, music psychology, music aesthetics, and their auxiliary disciplines) are just different sides of the application of this principle, just different methodic means in the service of the one, *natural-scientific* approach"[21] (p.12), went far beyond Adler's empirical attempts. Adler agreed with Lach (1921), however, who finally admitted in completion of a study by Alois Höfler (1853–1921), a former student of Meinong, following von Ehrenfels' *Gestalt Qualities*, namely that

[20] R. Lach (Lach, 1913): "[...] daß [...] dieselben Entwicklungsstufen, -phasen und -formen, die der rein formal-genetischen Evolutionsreihe zugrunde liegen, bzw. in ihr auftreten, in genau derselben Reihenfolge auch rein historisch, in der Geschichte der Musik, sich nachweisen lassen" (p.524).

[21] R. Lach (Lach, 1924): "[...] sich ihr also alle musikalischen und musikhistorischen Erscheinungen in das Kontinuum eines *Natur*prozesses, eines Prozesses von naturwissenschaftlich bedingter innerlicher Notwendigkeit, eingliedern [...]. [...] in der alle musikalischen und musikhistorischen Disziplinen (also Musiktheorie, -geschichte, -psychologie, -ästhetik und deren Hilfsdisziplinen) nur verschiedene Seiten der Anwendung dieses Prinzips, nur verschiedene methodische Mittel im Dienste dieser einen, *naturwissenschaftlichen* Betrachtungsweise sind" (p.12).

> "what is corresponding always and everywhere in the complete consideration of any process or phenomenon of natural science as well as of science of art and of humanities: that the Real and the Phenomenal is just an allotropy, a heteronomy of the Ideal, the Mental, and the Psychical"[22] (p.149).

8 Conclusion

It becomes apparent, that Adler did in fact share some opinions with the representatives of comparative musicology. A logical co-operation, however, did not take place which might have been decisively due to Adler (Antonicek, 1986a). He was irritated mainly by Wallaschek's non-experimental orientation, and probably, by his tendency towards feuilletonistic writing. In the case of Lach, the reasons for disapproval can be seen to a greater extent in the ideologic-political sphere. Adler and Lach agreed above all concerning the at least partial replacement of the traditional concept of music research by a new type of scientific attitude oriented towards the ideal model of methods of the natural-sciences.

Adler's acquaintance and friendship with a number of philosophers (Wessely, 1986) went well beyond the exchange of ideas that had been customary between music-educated and literary fellow professors and chamber music partners at that time. Adler was associated with Meinong, von Ehrenfels, and Stumpf, who came from the philosophical school of Brentano, and who were closely tied to the descriptive psychology during their entire lives, and with Mach and Boltzman, who compared to the first group took the view of a rather positivistic biased empiricism, but also with Jodl, who can clearly be incorporated in the positivistic tradition (Haller, 1993). Although direct adoptions can only rarely be proven, Adler's references to the corresponding passages in the philosophical papers of his friends and the frequent reference to the inductive method in Adler's own articles and books are doubtlessly based upon this exchange of ideas.

That Adler's concept of an integrated musicology, inspired by his philosophical considerations and his foresight concerning the discipline's future, could not be realized appropriately was evidently caused, to an inconsiderable portion, by the attitude which Adler had about his possible partners on the side of comparative musicology. Accordingly – and this tradition followed by his numerous students – Adler was able to promote a situation which, viewed in its entirety, still seems to be characteristic to a greater or lesser extent of musicology today.

[22] R. Lach (Lach, 1921): "[...] was auch sonst immer und überall die gesamte Betrachtung aller natur- wie kunst- und geisteswissenschaftlichen Prozesse und Phänomene übereinstimmend aufweist: daß das Reale und Phänomenale nur eine Allotropie, eine Heteronomie des Idealen, des Geistigen, des Psychischen ist" (p.149).

Acknowledgement

I gratefully acknowledge Ms. Sabine Pawikovsky, Mrs. Sharon Gredler, and Ms. Jennifer Gredler for their help with the preparation of the English version including the translation of the quotes.

References

Adler, G. (1885). Umfang, Methode und Ziel der Musikwissenschaft. *Vierteljahrsschrift für Musikwissenschaft, 1*, 5–20.

Adler, G. (1911). *Der Stil in der Musik 1: Prinzipien und Arten des musikalischen Stils.* Leipzig: Breitkopf & Härtel.

Adler, G. (1919). *Methode der Musikgeschichte.* Leipzig: Breitkopf & Härtel.

Adler, G. (1935). *Wollen und Wirken: Aus dem Leben eines Musikhistorikers.* Vienna, Leipzig: Universal-Edition.

Adler, G. (1981). Die Wiener klassische Schule. In G. Adler (Ed.), *Handbuch der Musikgeschichte 3: Dritte Stilperiode 2* (2nd ed.). Munich: Deutscher Taschenbuch Verlag. (Original work published 1930)

Angerer, M. (1988). Methodenprobleme der musikalischen Stilgeschichte: Guido Adler und Ernst Kurth. *Schweizer Jahrbuch für Musikwissenschaft, N.S. 6/7*, 43–59.

Antonicek, T. (1986a). Musikwissenschaft in Wien zur Zeit Guido Adlers. *Studien zur Musikwissenschaft, 37*, 165–193.

Antonicek, T. (1986b). Wagner, Bruckner und die Wiener Musikwissenschaft. In O. Wessely (Ed.), *Bruckner Symposion 1984: Bruckner, Wagner und die Neudeutschen in Österreich, Bericht.* Linz: Anton Bruckner-Institute.

Blaukopf, K. (1991). Natur- und Kulturwissenschaft: Ernst Machs Einfluß auf die Wiener Kunstsoziologie. In K. Behne, E. Jost, E. Kötter, & H. de la Motte-Haber (Eds.), *Musikwissenschaft als Kulturwissenschaft: Festschrift zum 65. Geburtstag von Hans-Peter Reinecke.* Regensburg: Gustav Bosse Verlag.

Blaukopf, K. (1995). *Pioniere empiristischer Musikforschung: Österreich und Böhmen als Wiege der modernen Kunstsoziologie.* Vienna, Leipzig: Verlag Hölder-Pichler-Tempsky.

Brentano, F. (1924). *Psychologie vom empirischen Standpunkt 1.* Leipzig: Felix Meiner Verlag.

Brentano, F. (1925). *Psychologie vom empirischen Standpunkt 2: Von der Klassifikation der psychischen Phänomene, mit neuen Abhandlungen aus dem Nachlaß.* Leipzig: Felix Meiner Verlag.

Brentano, F. (1968). *Psychologie vom empirischen Standpunkt 3: Vom sinnlichen und noetischen Bewußtsein. Äußere und innere Wahrnehmung, Begriffe* (Rev. 2. ed.). Hamburg: Felix Meiner Verlag.

Brod, M. (1979). *Streitbares Leben: Autobiographie 1884-1968.* Frankfurt/M.: Insel Verlag.

Eder, G. (1995). *Guido Adler und Alexius Meinong: Eine Freundschaft in Briefen.* Amsterdam, Atlanta: Editions Rodopi.

Fabian, R. (1986). Leben und Wirken von Christian von Ehrenfels: Ein Beitrag zur intellektuellen Biographie. In R. Fabian (Ed.), *Christian von Ehrenfels: Leben und Werk.* Amsterdam: Editions Rodopi.

Gebesmair, A. (1996). Interdisziplinarität und Empirismus in der Musikforschung. Guido Adlers Naheverhältnis zum naturwissenschaftlichen und soziologischen Denken am Beispiel der "Vierteljahresschrift für Musikwissenschaft" (1885–1894). In I. Bontinck (Ed.), *Wege zu einer Wiener Schule der Musiksoziologie: Konvergenz der Disziplinen und empiristische Tradition.* Vienna, Mühlheim a. d. Ruhr: Guthmann-Peterson.

Graf, W. (1974). Die vergleichende Musikwissenschaft in Österreich seit 1896. In *Yearbook of the International Folk Music Council* (Vol. 6). International Folk Music Council, and International Music Council.

Graf, W. (1980). *Vergleichende Musikwissenschaft: Ausgewählte Aufsätze.* Vienna, Föhrenau: Elisabeth Stiglmayr.

Graf, W., Jancik, H., Meister, R., Nowak, L., & Schenk, E. (Eds.). (1954). *Robert Lach: Persönlichkeit und Werk – Zum 80. Geburtstag.* Vienna: Musicological Institute of the University of Vienna.

Haller, R. (1993). *Neopositivismus: Eine historische Einführung in die Philosophie des Wiener Kreises.* Darmstadt: Wissenschaftliche Buchgesellschaft.

Heinz, R. (1968). *Geschichtsbegriff und Wissenschaftscharakter der Musikwissenschaft in der zweiten Hälfte des 19. Jahrhunderts: Philosophische Aspekte einer Wissenschaftsentwicklung.* Regensburg: Gustav Bosse Verlag.

Heinz, R. (1969). Guido Adlers Musikhistorik als historisches Dokument. In W. Wiora (Ed.), *Die Ausbreitung des Historismus über die Musik: Aufsätze und Diskussionen.* Regensburg: Gustav Bosse Verlag.

Husserl, E. (1992a). *Logische Untersuchungen II/1: Untersuchungen zur Phänomenologie und Theorie der Erkenntnis (Text nach Husserliana XIX/1).* Hamburg: Felix Meiner Verlag.

Husserl, E. (1992b). *Philosophie der Arithmetik (Text nach Husserliana XII).* Hamburg: Felix Meiner Verlag.

Kalisch, V. (1988). *Entwurf einer Wissenschaft von der Musik: Guido Adler.* Baden-Baden: Verlag Valentin Koerner.

Kindinger, R. (Ed.). (1965). *Philosophenbriefe: Aus der wissenschaftlichen Korrespondenz von Alexius Meinong, 1876–1920.* Graz: Akademische Druck- u. Verlagsanstalt.

Lach, R. (1913). *Studien zur Entwicklungsgeschichte der ornamentalen Melopöie: Beiträge zur Geschichte der Melodie.* Leipzig: C. F. Kahnt Nachfolger.

Lach, R. (1921). Gestaltunbestimmtheit und Gestaltmehrdeutigkeit in der Musik: Bei- und Nachträge zu Höflers Abhandlung "Tongestalten und lebende Gestalten". In A. Höfler (Ed.), *Naturwissenschaft und Philosophie: Vier Studien zum Gestaltungsgesetz – 2. Tongestalten und lebende Gestalten.* Vienna: Alfred Hölder.

Lach, R. (1924). *Die vergleichende Musikwissenschaft, ihre Methoden und Probleme.* Vienna, Leipzig: Hölder-Pichler-Tempsky.

Meinong, A. (1969). Zur Psychologie der Komplexionen und Relationen. In A. Meinong, R. Kindinger, & R. Haller (Eds.), *Abhandlungen zur Psychologie.* Graz: Akademische Druck- u. Verlagsanstalt.

Meinong, A. (1971). Über Gegenstände höherer Ordnung und deren Verhältnis zur inneren Wahrnehmung. In A. Meinong & R. Haller (Eds.), *Abhandlungen zur Erkenntnistheorie und Gegenstandstheorie.* Graz: Akademische Druck- u. Verlagsanstalt.

Meinong, A. (1978a). Rezension von C. Stumpf "Tonpsychologie 1". In A. Meinong & R. Haller (Eds.), *Selbstdarstellung, Vermischte Schriften.* Graz: Akademische Druck- u. Verlagsanstalt. (Original work published 1885)

Meinong, A. (1978b). A. Meinong (Selbstdarstellung). In A. Meinong & R. Haller (Eds.), *Selbstdarstellung, Vermischte Schriften.* Graz: Akademische Druck- u. Verlagsanstalt.

Mulligan, K., & Smith, B. (1988). Mach and Ehrenfels: The Foundations of Gestalt Theory. In B. Smith (Ed.), *Foundations of Gestalt theory.* Munich, Vienna: Philosophia Verlag.

Schneider, A. (1976). *Musikwissenschaft und Kulturkreislehre: Zur Methodik und Geschichte der Vergleichenden Musikwissenschaft.* Bonn: Verlag für systematische Musikwissenschaft.

Schneider, A. (1984). *Analogie und Rekonstruktion: Studien zur Methodologie der Musikgeschichtsschreibung und zur Frühgeschichte der Musik.* Bonn: Verlag für systematische Musikwissenschaft.

Simons, P. (1988). Einleitung. In R. Fabian (Ed.), *C. von Ehrenfels, Philosophische Schriften 3: Psychologie, Ethik, Erkenntnistheorie.* Munich, Vienna: Philosophia Verlag.

Smith, B. (1988). Gestalt theory: An essay in philosophy. In B. Smith (Ed.), *Foundations of Gestalt theory.* Munich, Vienna: Philosophia Verlag.

von Ehrenfels, C. (1986a). Die musikalische Architektonik. In R. Fabian (Ed.), *C. von Ehrenfels, Philosophische Schriften 2: Ästhetik.* Munich, Vienna: Philosophia Verlag. (Original work published 1896)

von Ehrenfels, C. (1986b). Zur Klärung der Wagner-Controverse: Ein Vortrag. In R. Fabian (Ed.), *C. von Ehrenfels, Philosophische Schriften 2: Ästhetik.* Munich, Vienna: Philosophia Verlag. (Original work published 1896)

von Ehrenfels, C. (1988a). Über "Gestaltqualitäten". In R. Fabian (Ed.), *C. von Ehrenfels, Philosophische Schriften 3: Psychologie, Ethik, Erkenntnistheorie.* Munich, Vienna: Philosophia Verlag. (Original work published 1890; English translation in B. Smith (Ed.), *Foundations of Gestalt Theory.* Munich, Vienna: Philosophia Verlag)

von Ehrenfels, C. (1988b). Über Fühlen und Wollen. Eine psychologische Studie. In R. Fabian (Ed.), *C. von Ehrenfels, Philosophische Schriften 3: Psychologie, Ethik, Erkenntnistheorie.* Munich, Vienna: Philosophia Verlag. (Original work published 1887) von Ehrenfels, C. (1988c). Über Gestaltqualitäten. In R. Fabian (Ed.), *C. von Ehrenfels, Philosophische Schriften 3: Psychologie, Ethik, Erkenntnistheorie.* Munich, Vienna: Philosophia Verlag. (Original work published 1890; English translation in B. Smith (Ed.), *Foundations of Gestalt Theory.* Munich, Vienna: Philosophia Verlag)

Wallaschek, R. (1886). *Ästhetik der Tonkunst.* Stuttgart: W. Kohlhammer.

Wallaschek, R. (1903). *Anfänge der Tonkunst.* Leipzig: Johann Ambrosius Barth.

Wallaschek, R. (1905). *Psychologie und Pathologie der Vorstellung: Beiträge zur Grundlegung der Aesthetik.* Leipzig: Johann Ambrosius Barth.

Wallaschek, R. (1930). *Psychologische Aesthetik.* Vienna: Rikola Verlag.

Weinhandl, F. (1960). Christian von Ehrenfels, sein philosophisches Werk. In F. Weinhandl (Ed.), *Gestalthaftes Sehen: Ergebnisse und Aufgaben der Morphologie – Zum hundertjährigen Geburtstag von Christian von Ehrenfels.* Darmstadt: Wissenschaftliche Buchgesellschaft.

Wessely, O. (1986). Vom wissenschaftlichen Denken Guido Adlers. In J. Lederer (Ed.), *Gedenkschrift Guido Adler.* Föhrenau: Elisabeth Stiglmayr.

Winkler, G. J. (1986). Christian von Ehrenfels als Wagnerianer. In R. Fabian (Ed.), *Christian von Ehrenfels: Leben und Werk.* Amsterdam: Editions Rodopi.

Gestalt Concepts and Music: Limitations and Possibilities

Mark Reybrouck

University of Leuven, Blijde inkomststraat 21, B-3000 Leuven, Belgium

Abstract. This paper is concerned with possible applications of Gestalt concepts in relation to music. The concepts are examined as to their limitations and possibilities. We propose a reappraisal of some older insights and an enlargement of the Gestalt concepts from a rather intuitive to an operational approach, arguing strongly for using the methodology of interdisciplinary research, and stressing the importance of the cognitive view of elaboration and processing of musical structure by the listener's mind. This involves a shift from structural description to a functional approach, with special emphasis on musical information processing. Furthermore attention is directed to the specificity of music as a temporal and sounding art, stressing the role of memory and imagination, and the tension between actuality and virtuality in the construction of musical Gestalts.

1 Introduction

Gestalt-theoretical insights belong to the common equipment of psychological research. They strongly influenced learning theories, in stressing the importance of the progressive and meaningful organization of the perceptual field, and are now enjoying a renewed level of interest. To quote Hofstadter: "In any case, it will be good when AI people are finally driven back to looking at the insights of people working in the 1920's, such as Wittgenstein and his *games*, Köhler and Koffka and Wertheimer and their *Gestalts*, and Terman and Binet, and their IQ-problems" (Hofstadter, 1986, p.638). Modern technological means and the refinement of measuring devices and methodologies of research, however, suggest orientations other than the rather intuitive formulations of the early pioneers. The proposal of this paper is to explore this new impetus in outlining a possible enlargement of Gestalt concepts with special attention to the interdisciplinary approach and the role of the listener in the experience of Gestalts.

2 Gestalt Theory and Music: Limitations and Possibilities

The application of Gestalt-theoretical concepts to the domain of music is not without risks. The intuitive character of their general claims and the highly speculative character of extrapolation of experiences from the visual domain into other fields of experience are waiting for empirical support. Much groundwork still has to be done, not only in the empirical domain but also at the conceptual level. It may be interesting therefore to trace some general outlines of the foundation of Gestalt theory.

2.1 Philosophical and Biological Foundations

Gestalt theory (Köhler, 1929; Wertheimer, 1958; Koffka, 1935) was originally a psychological theory, that widened its claims to general philosophical conceptions about biological and physical facts . The theory therefore partly fits with the organismic conception of von Bertalanffy (1949), a view that transcends the mechanistic and vitalistic views of science by stressing the idea of totality. The starting points of this conception of the world are three fundamental principles of modern biology: the principle of totality (not only the parts and the processes, but also their relationships), the organic principle (the organization leans upon the hierarchical order of the organism), and the principle of dynamics (living beings not merely exist, but they *occur*). The organismic conception, therefore, is opposed to the three traditional paradigms of scientific methodology: the analytic-summative conception that analyses first, to put the parts together thereafter, the machine-theoretical conception that reduces the order of living processes to the conditions that are inherent in the structure of the organisms, and the reaction-theoretical conception of the organism as an automaton, that reacts in a pre-established way on stimulation of the senses. Organismic theory stresses the importance of the system as a whole, besides its dynamical and essentially active aspects. The problem stated sounds quite modern and is illustrative of this crucial point: it is the biological way of thinking that grounded Gestalt theory (von Bertalanffy, 1949).

The early Gestalt concepts, however, are greatly intuitive and philosophical in essence. This is especially true in the case of C. von Ehrenfels, who was the first to introduce the Gestalt concept as an overall quality of a content of consciousness, that transcends its component parts. This global concept has two major qualities (the Ehrenfels-criteria): the whole is more than the sum ("Übersummenhaftigkeit") and the Gestalt can be transposed ("Transponierbarkeit") (von Ehrenfels, 1890). The philosophical and psychological theories that were the outcome of these early conceptions, however, did not always resist the challenges of empirical evidence. So Gestalt theory claims that in perception one can grasp *immediately* a configuration that is already organized. The act of perception should recognize the configuration on the basis of a fundamental isomorphism between the *structure of the object* and the *psychophysiological structure* (Gestalt-like patterns of excitation in the brain) of the perceiver (Guillaume, 1937). This premise however has been largely contested by post-Gestaltist authors (Piéron, 1955).

2.2 Gestalt Concepts Revisited: an Operational Approach

Skilled perception and schooled listening are not concerned with isolated entities, but with their organization in structures and configurations. Structuring, however, is an active affair that leans heavily upon the higher functions of the brain (Reybrouck, 1989). The grouping phenomena of Gestalt psychology therefore can be formulated in terms of organizational principles, knowledge acquisition, categorization, schematization and abstraction, and this in a dynamic rather than a static way, since music is a dynamic system characterized by totality, organization and teleology (Stoianova, 1975). Music, in that sense, can be defined as a sound producing organism (Coker, 1972; Reti, 1961; de Sélincourt, 1820), and the musical experience should be the outcome of an interaction between the listener and the musical organism. Music, thus defined, is an organic structure and music analysis has to be broadened from a structural description to a description in terms of processes (Rosenthal, 1989). To quote von Bertalanffy: "Organic structures are themselves expression of an ordered process; and they are only maintained in and by this process. Therefore, the primary order of organic processes must be sought in the processes themselves, not in the pre-established structures." (von Bertalanffy, 1967, p.73).

An operational description of this idea is possible by substituting a system for the organism. This system should be an open system (von Bertalanffy, 1949, 1950), i.e. an organized whole that maintains a state of equilibrium under continuous import and export of material. What matters here is the conception of a dynamic morphology (Thom, 1972, 1980; Petitot, 1983, 1985) that stresses the importance of processes of growth. The structure of the organism therefore has to be conceived as a quasi-stationary state, that maintains itself for a while and then changes and disappears.

The listening process should mirror this development in a kind of morphodynamic analogy between the sonorous articulation and its image in the listener's mind. Schooled listening therefore not only involves the extraction of a morphological lexicon (Petitot, 1989) as a set of discrete entities, but also the construction of a relational network between them that allows a process-like description in terms of continuation and growth. The description of the combinations of the individual elements in a discrete-digital way, therefore, has to be supplemented with an analog description of the development of the whole, as a kind of envelope placed over the individual elements.

What matters here is an operational description of structures that group temporal courses in a broader organic chain. For this reason there is need for a conceptual apparatus that allows a gentle transition from a mechanical concept of form to an organic one, and that stresses the importance of structuring by the listener as well.

In this context we once again refer to the concept of *Gestalt* of von Ehrenfels as an overall quality that exists besides and above the components. The conceptualization concerning these qualities is closely related to Husserl's phenomenological theory of time (Husserl, 1928) (both Husserl and Ehrenfels were inspired by Brentano) and especially his conception of the synthesizing function of time. The latter leans upon Brentano's idea that the grouping of a sequence of representations is not possible without being the simultaneous object of the knowing mind, that combines them in one singular act of consciousness. One has to make a distinction, therefore, between the unchangeable but divisible experience of real time and the essentially indivisible and continuous flow of the inner experience of time.

The importance of these earlier insights must be emphasized. Their intuitive formulation, however, is waiting for an operational approach, especially with respect to the definition of the *time objects* as structural Gestalt-units and the transformations that can be applied to them.

One must further bear in mind that the *perceived* structures are not necessarily isomorphic with the *sounding* structures as such, because of the role of schemes and knowledge mediating between stimulus and response. The operational approach, therefore, not only applies to the structural description of music as an independent variable (the argument), but also to the perception by the listener as a dependent variable (the image) and the mediation between them (the function that maps the argument to its image).

What really matters here is not the structural description of the music as an artefact, but the whole set of transformations on these structures by the perceiving mind, from mere pattern recognition and heuristic procedures to more elaborate processes of information processing. And the formal approach par excellence for describing this, uses a formal language as suggested by semiotics (Morris, 1937/1975) and by neopositivism (Carnap, 1937/1971). In this approach the structures and patterns are considered as signs that can be studied on a syntactical (the formal relations between the signs), a semantical (the meaning of the signs) and a pragmatical level (the effects of the signs on the user).

Semiotics, besides, has essentially an interdisciplinary base, but leans heavily upon symbolic logics. The categorization processes, e.g., that are at the core of schooled listening, can be formalized as predication processes, that apply to individual elements but also to complex configurations. In either case, the result is a set of propositions that are reducible to subject-predicate relationships. A more promising approach, however, is the enlargement of these syllogistic categories to a description in terms of *predicate and arguments*, *operator and operandum*, or *functor and arguments*. The profits of this approach are the possibilities of introducing algebraic methodologies in defining predicates and arguments as variables, that can be filled in without restrictions. From here, of course, one can skip gently to set theoretical, group theoretical and topological methodology. A Gestalt (structure), then, can be defined as a geometrical figure (a set of points), and the recognition of that Gestalt as performing an identification algorithm on transformations in geometrical sense (mere recognitions should be identical transformations, the other transformations should be classified as similarity transformations).

3 Epistemological Implications or Gestalt Theory between Nature and Nurture

Gestalt perception of music is not merely a matter of musical structure, but is essentially dependent on structuring by the listener, since dealing with music is an active and constructive process. The construction, however, is not an arbitrary one, but can be directed as a result of learning and instruction and Gestalt principles may be helpful here.

3.1 Structural Description and Functional Approach

A starting point is the assumption that objects of knowledge and of perception have a system character. This is closely related to the holistic and biological way of thinking in stating that global structures resist destruction of some of their parts. Individual elements, in fact, are better perceived within a context (Palmer, 1977) than in isolation. This underlines the greater stability of global patterns that are stronger the more abstract they are. McNeill (1971) e.g. suggested an analogy between the characteristics of a sentence and a biological system. Every thought or every conscious perception should be correlated with a biolinguistic system with a very short term of life. The application of these insights to other domains of perception is probable.

The system character of objects of knowledge, however, is not sufficient ground for its perception by the listener. The structural description of the objects, therefore, has to be supplied with a functional approach. Piaget (1968) has stressed the necessity of adding a subject to the structures as a centre of functioning. But the functional approach has been emphasized by other scholars as well. J. von Uexküll e.g. developed his theory of the "Umwelt" (world around us) as the combination of a perceptual and effector world, that are linked together by a functional cycle (von Uexküll, 1934/1957). This cycle operates by means of trigger mechanisms. Objects that are selected on account of their importance act as perceptual cue bearers, while other objects pass unnoticed. Objects with an operational meaning on the other hand act as functional cue bearers. Both are related in the sense that the functional qualities affect the perceptual ones. They transform the object of perception by giving it a functional tone. Perception and effectuation are linked together, causing a closed loop. This is the "Umwelt" of the animal, as the sum total of perceiving and performing.

"Umwelt"-research is highly informative and is directed primarily to the determination of the perceptual triggers of the existing stimuli. Every subject builds up relations with the external environment, selecting some of them to give them special meanings, and to construct his specific world or "Umwelt".

3.2 In Search of Universals of Knowing and Perception

The functional approach provides an interesting extension of traditional perception theory, in linking sensorium and motorium. The objects of perception and knowledge receive an operational meaning and can be influenced by processes of learning and selection (Edelman, 1989). Anyhow, the "Umwelt" of most animals is a rather closed universe that does not change spontaneously. It can be interesting therefore, to investigate the organization principles of perception and knowledge that are operating on a certain level of biological functioning (Edelman, Gall, & Cowan, 1988; Wallin, 1991).

At the perceptual level there are the Gestalt laws, as formulated by Wertheimer (1922, 1958). The epistemological quality, however, of these laws is quite ambiguous, since there is some tension between what is innate (nature) and what is learned (nurture). Future research has to be directed to the physiological and psychobiological bases of perception, to uncover the innate equipment of the human species. Bruner (1957) postulated some perceptual primitives or identities that are not the outcome of a learning process. They concern innate categories as motion, causation, intention, identity, equivalence, time and space. And probably these epistemological primitives lean upon capacities that are still more primitive.

The problem that arises here is the question of the existence of universals of knowing and perception. The answer is affirmative if one considers them as being the formal objects of a primary code (Bystrina, 1983). This concerns all information without sign character (e.g. the rule-systems that are reducible to the genetic code, perception codes and intra-organismic codes). What matters here, once again, is the neurological and psychobiological substructure of perception. A domain of special interest in this connection is the study of ecological variables in perception (Martindale & Moore, 1989; Neisser, 1987) and musical universals in man and animal (Hulse & Page, 1988).

Rule-systems that are the outcome of learning processes, on the other hand, are classified as secondary code. They link production and reception of signs and sign-complexes, as elements of a broader context. What is concerned here are symbolic processes that are removed some distance from the sensory material.

3.3 Cognitive Abstraction: Categorization, Schematization and Prototypicality

The first order of cognitive abstraction is symbolization, since symbols are not real things, but things that stand instead of them, and this in several kinds of relationships. One common point, however, is their lack of concrete and idiosyncratic qualities. The process of abstraction, therefore, is characterized by the extraction of invariants, so that differences in sensory modalities or modalities of perception evoke the same perceptual image. Thus one is classifying the input in categories on the basis of common features, that are stored in a more or less schematic way, so combining the methodologies of categorization (Bruner, 1957; Rosch, 1973; Rosch, Mervis, Gray, Johnson, & Boyes-Braem, 1976; Neisser, 1987; Edelman, 1989; Gjerdingen, 1990), schematization (Smolensky, 1988; Bharucha, 1987), priming (Martindale & Moore, 1989) and perception theory.

Categorization, however, is not a passive registration of ready-made stimuli, but a constructive process, that integrates sensation, perception and cognition. Listening, therefore, is a matter of information processing, that integrates several levels of processing in one global process (Camilleri, 1989; Neisser, 1967; Seifert, 1993). At the lowest level there are systems for recognizing patterns whose information is contained in the external world. Assigning these patterns to music, however, cannot be done on a priori-grounds, for the delimitation of musical patterns is greatly influenced by the selection processes and the decoding mechanisms of the listener. The feature theory or ecological theory of perception (Gibson & Gibson, 1957) states that the perceiver learns to discriminate a number of distinctive features. At the core of this theory is an increase in differentiation and discrimination as the result of a perceptual learning process. Schema theory, on the contrary, starts from the assumption that a schema or model is built up in the listener's mind. Both hypotheses are not opposed but are complementary in the sense that the construction of a standard as a reference leans upon learning to discriminate more and more features to integrate them in a schema that acts as a prototype for discrimination between categories and for classification of patterns in well-defined categories.

Perception, thus, is dependent on the search requirements of the perceiver. These can be minimal in an attempt to structure the stimuli in the most economical way, but can be very significant as well. Yet, some configurations impose their structures in a more compelling way. The organizing principles at work here are almost unequivocally referring to the imposition of stability on external stimuli, as has been stated in the law of pregnancy of Koffka (1935) and the law of good form of Wertheimer (1922). They claim that good form is the most simple, the most regular and the most symmetrical of the possible forms in the actual circumstances.

3.4 Perception as an Active and Transactional Process

Perception, as an active and transactional process between perceiver and environment, is an organization of the perceptual input in search of meaning. The central question is the role of learning and instruction, since the expert sees patterns of organized information where the novice sees only unrelated details (Davidson & Welsh, 1988). At the core of this problem is the everlasting debate of nativism versus empiricism: is perception the outcome of construction as a result of interaction between sensory and motor behavior (nurture) or are there innate receptor mechanisms that function in an autonomous way (nature) (Tighe & Dowling, 1993; Hargreaves, 1986)? The traditional dichotomy, however, has been weakened since perception and learning theories are gradually coming together, redefining the nativistic conception as a kind of information extraction and the empiricistic one as an important moment in perception theory (Hargreaves, 1986).

In connection with this, perception can be defined as a kind of learning with former experiences, this being as important as actual learning in the perceiving act. Perception, thus, cannot ignore processes of acculturation, enculturation, and conditioning. And this emphasizes the importance of the reader over the text, as has been stressed by reception aesthetics (Iser, 1970; Jauss, 1975, 1977). Much will depend, however, on the construction of an internal model of the external environment that can function as a trigger mechanism for Gestalt-like recognition of stimulus configurations (Sloboda, 1985). For these schemas are directing our attention, in establishing correspondences between type figures of the external stimuli and those that are stored in memory. And this, in fact, is a simple interpretation of schema theory: new structures are matched against existing ones.

4 Applying Gestalt Concepts on Music Analysis

One has to be cautious in applying Gestalt concepts on music analysis. Gestalt-theoretical concepts and their experimental research focused primarily on visual experiences. The translation of insights from the visual to the auditive domain is highly speculative, since music, as a temporal art, is essentially discursive. In contrast with a geometrical figure, that is described as a whole when looking at it, a musical figure needs a successive presentation. The grasping of its meaning, then, is *polythetic* in that it can be grasped only by hearing the music as it unfolds step by step (Wright, 1995). This hampers the holding of a musical Gestalt as an immediate and directly experienced whole. Music, however, can be grasped in a monothetic way, if the discursive processes are coded as discrete things as in the case of conceptualization (McAdams, 1989). The monothetic grasping of temporal Gestalts can then be defined as a combination of actual and virtual impressions of sounds. The epistemological nature of these Gestalts, however, has still to be clarified.

A second problem is the complexity of the musical texture, that is mostly composed of multiple layers. In order that Gestalt perception can be directed to one of these layers (figure/ground) or to the sum of these, there is, in both cases, some focusing and direction of attention by the listener (Sloboda, 1985; Dowling & Harwood, 1986). Auditive perception, therefore, is a result of heuristic processes that analyse the auditive field in component parts (Bregman, 1978, 1981, 1990). The extraction of meaningful constituents out of a complex acoustic input is not an arbitrary process, but is constrained by principles of grouping and dividing, as described by Gestalt psychology.

Dividing, however, is an analytical process, and there is a danger of reducing the biological way of thinking to a structural description. The problem has been formulated by Souris (1976) who stated that traditional music theory dismisses the internal reactions of the elements. Real music, therefore, only exists when it is sounding, with every sound being responsible for the overall texture. There is much sense, then, in speaking of auditory events or scenes rather than of isolated structures (Bregman, 1990; Krumhansl, 1992; McAdams, 1989).

A third problem is the need for simultaneous and successive decoding of the music, since musical structure is composed of simultaneous and successive Gestalts (Volkelt's, 1959, "Simultangestalten" and "Verlaufsgestalten"). The grasping of successive Gestalts leans upon temporal decoding mechanisms, in an attempt to segregate the sonorous articulation in temporal Gestalt-units, as distinct spans of time that are both internally cohesive and externally segregated from comparable time-spans immediately preceding and following them (Tenney & Polańsky, 1980; Jones, 1982, 1976; Jones & Bolz, 1989; Jones & Holleran, 1992; McAdams, 1989). The operational approach of simultaneous decoding, on the other hand, is more difficult, because simultaneity can be actual or virtual. Actual simultaneity leans upon real sonorous articulation, and can be the object of empirical research. Virtual simultaneity leans upon memory and imagination, and cannot be studied in a direct way.

There is, finally, the problem of mental involvement with the music. Two possibilities of tapping this moment-to-moment history are open here: the listening process mirrors the actual sonorous articulation in inexorable time or the process involves mental operations that are independent of inexorable time. This distinction is important, because it places constraints on using one Gestalt concept for describing both groupings in real time (real sonorous groupings) and in the imagination. The latter are dependent on the synthetic function of consciousness, substituting virtual images for actual perception. This can be done in an *a priori* (synthesis by anticipation) or an *a posteriori* way (synthesis from memory). The solution, however, of this problem is not easy and is waiting for further semiotic research (Reybrouck, 1995).

5 Conclusion

In this paper we presented an overview of some Gestalt concepts as applied to music. We have proposed an enlargement of the traditional concepts focusing primarily on an operational and interdisciplinary approach. Nevertheless we are deeply impressed by the modernity of some of the early insights as e.g. the organismic and the functional approach. These have to be placed, however, in the framework of modern methodology with special emphasis on the neuropsychological and psychobiological approach. Besides, there is the definition of music as a temporal art, as music is essentially a function of time. Listening, therefore, is a discursive process, that leans upon memory and imagination. Another point finally, is the role of mediation by the listener's mind, because there is no linear relation between the musical structure and the listening behavior. Music analysis, therefore, has to be broadened from a structural to a functional approach, stressing the importance of learning processes and modulating factors of attention and motivation.

References

Bharucha, J. (1987). Music cognition and perceptual facilitation: A connectionist framework. *Music Perception*, *5*, 1–30.

Bregman, A. (1978). The formation of auditory streams. In J. Requin (Ed.), *Attention and performance VII.* Hillsdale, NJ: Erlbaum.

Bregman, A. (1981). Asking the "what for" question in auditory perception. In M. Kubovy & J. Pomerantz (Eds.), *Perceptual organization.* Hillsdale, NJ: Erlbaum.

Bregman, A. (1990). *Auditory scene analysis: The perceptual organization of sound.* Cambridge, MA: The MIT Press.

Bruner, J. (1957). On perceptual readiness. *Psychological Review*, *64*, 123–152.

Bystrina, I. (1983). Kodes und Kodewandel. *Zeitschrift für Semiotik*, *5*, 1–22.

Camilleri, L. (1989). A modular approach to music cognition. *Interface – Journal of New Music Research*, *18*, 33–44.

Carnap, R. (1971). *The logical syntax of language.* London: Routledge and Kegan Paul. (Original work published 1937)

Coker, W. (1972). *Music and meaning: A theoretical introduction to musical aesthetics.* London: Free Press / Collier - Macmillan.

Davidson, L., & Welsh, P. (1988). From collections to structure: The developmental path of tonal thinking. In J. Sloboda (Ed.), *Generative processes in music: The psychology of performance, improvization and composition.* Oxford: Clarendon Press.

de Sélincourt, B. (1820). Music and duration. *Music and Letters*, *1*, 286–293.

Dowling, W., & Harwood, D. (1986). *Music cognition.* New York, NY: Academic Press.

Edelman, G. (1989). *Neural darwinism: The theory of neuronal group selection.* New York, NY: Oxford University Press.

Edelman, G., Gall, W., & Cowan, W. (1988). *Auditory function: Neurobiological bases of hearing.* New York, NY: John Wiley and Sons.

Gibson, E., & Gibson, J. (1957). *Principles of perceptual learning and development.* Englewood Cliffs, NY: Prentice Hall.

Gjerdingen, R. (1990). Categorization of musical patterns by self-organizing neuronlike networks. *Music Perception*, *7*, 339–370.

Guillaume, P. (1937). *La psychologie de la forme.* Paris: Flammarion.

Hargreaves, D. (1986). *The developmental psychology of music.* Cambridge: Cambridge University Press.

Hofstadter, D. (1986). *Metamagical themas: Questing for the essence of mind and pattern.* Harmondsworth: Penguin Books.

Hulse, S., & Page, S. (1988). Toward a comparative psychology of music perception. *Music Perception*, *5*, 427–452.

Husserl, E. (1928). Vorlesungen zur Phänomenologie des inneren Zeitbewusstseins. *Jahrbuch für Philosophie und Phänomenologische Forschung, 9*, 367–489.

Iser, W. (1970). *Die Appelstruktur der Texte: Unbestimmtheit als Wirkungsbedingung literarischen Prosa.* Konstanz: Universitätsverlag.

Jauss, H. (1975). Der Leser als Instanz einer neuen Geschichte der Literatur. *Poetica, 7*, 325–344.

Jauss, H. (1977). Aesthetische Erfahrung und literarische Hermeneutik. In *Bd. I, Versuche im Feld der ästhetischen Erfahrung.* Munich: Fink.

Jones, M. (1976). Time, our lost dimension: Toward a new theory of perception, attention, and memory. *Psychological Review, 83*, 323–355.

Jones, M. (1982). Music as a stimulus for psychological motion II: An expectancy model. *Psychomusicology, 2*, 1–13.

Jones, M., & Bolz, M. (1989). Dynamic attending and responding to time. *Psychological Review, 96*, 459–491.

Jones, M., & Holleran, S. (1992). *Cognitive bases of musical communication.* Washington, DC: American Psychological Association.

Koffka, K. (1935). *Principles of Gestalt psychology.* New York, NY: Harcourt, Brace, and Company.

Köhler, W. (1929). *Gestaltpyschology.* New York, NY: Liveright.

Krumhansl, C. (1992). Internal representations for music perception and performance. In M. Jones & S. Holleran (Eds.), *Cognitive bases of musical communication.* Washington: American Psychological Association.

Martindale, C., & Moore, K. (1989). Relationship of musical preference to collative, ecological, and psychophysical variables. *Music Perception, 6*, 431–446.

McAdams, S. (1989). Contraintes psychologiques sur les dimensions porteuses de forme en musique. In S. McAdams & I. Deliège (Eds.), *La musique et les sciences cognitives.* Brussels: Mardaga.

McNeill, D. (1971). Sentences as biological systems. In P. Weiss (Ed.), *Hierarchically organized systems in theory and practice.* New York, NY: Hafner.

Morris, C. (1975). *Foundations of the theory of signs, Vol. 1, Nr. 2.* Chicago, IL: University of Chicago Press. (Original work published 1937)

Neisser, U. (1967). *Cognitive psychology.* New York, NY: Appleton-Centuri-Crofts.

Neisser, U. (1987). *Concepts and conceptual development: Ecological and intellectual factors in categorization.* New York, NY: Cambridge University Press.

Petitot, J. (1983). Paradigme catastrophique et perception catégorielle. *Recherches sémiotiques (RS/SI), 3*, 207–245.

Petitot, J. (1985). *Les catastrophes de la parole: De Roman Jakobson à René Thom.* Paris: Maline.

Petitot, J. (1989). Perception, cognition et objectivité morphologique. In S. McAdams & I. Deliège (Eds.), *La musique et les sciences cognitives.* Brussels: Mardaga.

Piaget, J. (1968). *Le structuralisme.* Paris: Presses Universitaires de France.

Piéron, H. (1955). *Rapport au symposium "la perception".* Paris: Presses Universitaires de France.

Reti, R. (1961). *The thematic process in music.* London: Faber and Faber.

Reybrouck, M. (1989). Music and the higher functions of the brain. *Interface – Journal of New Music Research, 18*, 73–88.

Reybrouck, M. (1995). *Spanning en ontspanning in de muziek: Een semiotische benadering van de omgang met muziek.* Unpublished doctoral dissertation, Katholieke Universiteit Leuven, Louvain.

Rosch, E. (1973). On the internal structure of perceptual and semantic categories. In T. Moore (Ed.), *Cognitive development and the acquisition of language.* New York, NY: Academic Press.

Rosch, E., Mervis, C., Gray, W., Johnson, D., & Boyes-Braem, P. (1976). Basic objects in natural categories. *Cognitive Psychology, 8*, 382–439.

Rosenthal, D. (1989). A model of the process of listening to simple rhythms. *Music Perception, 6*, 315–328.

Seifert, U. (1993). *Systematische Musiktheorie und Kognitionswissenschaft – Zur Grundlegung der kognitiven Musikwissenschaft.* Bonn: Verlag für systematische Musikwissenschaft.

Sloboda, J. (1985). *The musical mind: The cognitive psychology of music.* London: Clarendon Press.

Smolensky, P. (1988). On the proper treatment of connectionism. *Behavioral and Brain Sciences, 11*, 1–74.

Souris, A. (1976). *Conditions de la musique et autres écrits.* Brussels: Editions de l'Université de Bruxelles.

Stoianova, I. (1975). L'énoncé musical. *Musique en jeu, 19*, 23–57.

Tenney, J., & Polansky, L. (1980). Temporal Gestalt perception in music. *Journal of Music Theory, 24*, 205–241.

Thom, R. (1972). *Stabilité structurelle et morphogenèse.* New York, NY: Benjamin.

Thom, R. (1980). *Modèles mathématiques de la morphogenèse.* Paris: Bourgois.

Tighe, J., & Dowling, W. (1993). *Psychology and music: The understanding of melody and rhythm.* Hillsdale, NJ: Lawrence Erlbaum Associates.

Volkelt, H. (1959). Simultangestalten, Verlaufsgestalten und Einfülung. *Zeitschrift für experimentelle und angewandte Psychologie, 6*, 357–371.

von Bertalanffy, L. (1949). *Das biologische Weltbild.* Bern: Francke Verlag.

von Bertalanffy, L. (1950). The theory of open systems in physics and biology. *Science, 111*, 23–29.

von Bertalanffy, L. (1967). *Robots, men and minds: Psychology in the modern world.* New York, NY: Braziller.

von Ehrenfels, C. (1890). Über "Gestaltqualitäten". *Vierteljahrschrift für Wissenschaft, 14.*

von Uexküll, J. (1957). A stroll through the worlds of animals and men, A: Picture book of invisible worlds. In *Instinctive behavior: The development of a modern concept.* New York, NY: International Universities Press. (Original work published 1934)

Wallin, N. (1991). *Biomusicology: Neurophysiological, neuropsychological and evolutionary perspectives on the origins und purposes of music.* New York, NY: Pendragon Press.

Wertheimer, M. (1922). Untersuchungen über die Lehre von der Gestalt. *Psychologische Forschung.*

Wertheimer, M. (1958). Principles of perceptual organization. In D. Beardslee & M. Wertheimer (Eds.), *Readings in perception.* Princeton, NJ: Van Nostrand.

Logic, Gestalt Theory, and Neural Computation in Research on Auditory Perceptual Organization

Randolph Eichert, Lüder Schmidt, and Uwe Seifert

Institute of Musicology, University of Hamburg, Neue Rabenstr. 13,
D-20354 Hamburg, Germany

Abstract. In our paper we discuss the Berlin Gestalt theory, its relation to modelling with artificial neural networks and ask if Gestalt theory may serve as a paradigm for future research on music perception in the realm of (computational) auditory perceptual organization. To answer this question, we made a logical analysis of important Gestaltist concepts from the point of view of philosophy of science, discuss a particular model of brain functioning and perception which is often assumed to be a modified follow up of Gestalt theory, and study Gestalt theory in relation to modelling perceptual organization with artificial neural networks using some well known facts from automata theory. The paper is divided into three parts: (i) Gestalt theoretical concepts such as "Gestaltqualität", *functional whole* ("Wirkungszusammenhang") and *emergence* are studied in discussing prior logical interpretations of Gestalt theory, (ii) K. Pribram's holographic brain metaphor is discussed in relation to W. Köhler's concept of a *functional whole* and related to artificial neural network research, (iii) the relevance of Gestalt theory – especially the principle of "Prägnanz" and the concept of a *functional whole* – in connection to artificial neural networks for research on perceptual organization is analysed. The idea of *emergent* computations is discussed with regard to concepts from automata theory. In our conclusion we present an answer to our main question and give some hints for future research.

1 Introduction

The Gestalt movement, which started with the work of C. von Ehrenfels in the late 19th century, had a profound influence on psychological research in perception, reasoning, and social relations. Wertheimer (1974) gives a brief historical account from the point of view of perceptual structure. Three main schools can be distinguished:

- the Berlin school, called the *Gestaltists*, with W. Köhler, M. Wertheimer, K. Koffka and W. Metzger as main proponents.
- the Leipzig school, or the so called "Ganzheitspsychologen", with F. Krüger and A. Wellek, and
- the Graz school, proposing the notion of "Komplexionen", as represented by A. Meinong and his followers.

The main characteristics and differences of these schools are that the Berlin school assumes an *isomorphism* between the mental and the physical. So at least there is no mind-body-problem and the explanation of mental phenomena can be achieved in terms of the natural sciences, e. g. by neuropsychological and physical theories. In contrast to this, the Leipzig school proposes a dualism between mind and brain. A special entity like the soul and special methods are needed to explain mental functions; mind cannot be reduced to or explained by the natural sciences. The Graz school assumes that one needs a special mental process to explain the phenomena described by Gestaltists. Today, many scientists (e.g. E. Terhardt, H. Bruhn, A. Bregman, F. Lerdahl and R. Jackendoff) use Gestalt concepts and principles in their research on music perception and cognition to explicate musical phenomena. Artificial neural network models are used as tools to simulate music cognition (Griffith & Todd, 1994; Leman, 1995; Todd & Loy, 1991) and researchers in this area often refer to Gestalt concepts. But neither the logical structure of Gestalt concepts nor the status of connectionist modelling is theoretically clear. They are used metaphorically. So one may question the status of Gestalt theory as a paradigm (Gregory, 1974; Kuhn, 1976) for research on perceptual organization, which is attempted in this paper.

2 The Logical Explication of Gestalt Concepts: Ehrenfels' "Gestaltqualtität" and Köhler's "Wirkungssystem"

Gestalt concepts like *dependence system*, *part-whole-relation*, *emergence* etc. have been analyzed by many philosophers of science (Bergmann, 1944; Grelling & Oppenheim, 1937/1938/1988a, 1937/1938/1988b; Henle, 1942; Leinfellner, 1966; Rescher, 1953; Rescher & Oppenheim, 1955). Three serious problems have often been mentioned in connection with the writings using concepts from Gestalt theory: (i) the expression *Gestalt* is used in different manners by different authors, (ii) often many different expressions are used for one single fact without further explanations, (iii) in many cases there is not even one clear concept. Therefore, the primary objective of logical analysis is to suggest definitions, that allow for the following:

> "When the concepts thus determined are appropriately inserted into sentences which appear characteristic of the Gestalt theorists, these sentences turn out neither trivial nor empty of sense" (Grelling & Oppenheim 1937/1938/1988a, p.192) [1]

The logical analysis revealed that one has to distinguish carefully between the Gestalt concepts of "Gestaltqualität" which is based on the writings of C. von Ehrenfels, and of a "Wirkungssystem" developed by the Berlin

[1] "Setzt man die so festgelegten Begriffe in jeweils passender Weise in charakteristisch erscheinende Sätze der Gestalttheoretiker ein, so werden diese Sätze weder trivial noch sinnleer" (Grelling & Oppenheim, 1937/1938a, p.211).

school, especially by W. Köhler. The analysis of Grelling and Oppenheim (1937/1938a) often makes use of logical concepts introduced by Carnap (1929). They start with the idea of transposition of complexes. Hence let us suppose that there is an operation which permits to transpose a complex A into a complex B, such that the two complexes stand in correspondence relation. Grelling and Oppenheim call this operation a *transposition* of complexes. On this basis the authors propose now the following explication of the *Gestalt*:

> "the Gestalt (of a complex with respect to a correspondence) is the invariant of transpositions (of the complex with respect to the transpositions)" (Grelling & Oppenheim, 1937/1938/1988a, p.196) [2]

We may say that a Gestalt is a class of complexes, such that all members of the class stand in pairs in correspondence with respect to the values of a particular value classificator. Therefore, a melody-Gestalt is simply a class of sequences of tones (complexes) which stand in correspondence to each other, i.e. the value classificators *interval* of each complex show in pairs the same values. Furthermore, according to Grelling and Oppenheim this concept must be distinguished from the concept of Gestalt as a classificator. Gestalt as a classificator means a function whose arguments are single complexes and whose results are Gestalt-individuals. Examples for Gestalts as classificators are the smell, the taste, or the color of objects, which can be noticed at a body like the melody at a sequence of tones. Thus, Gestalt as a classificator coincides with Ehrenfels' "Gestaltqualität".

Grelling and Oppenheim showed that the Gestalt concept "Wirkungssystem" used by Köhler and others is quite different from that of Ehrenfels' "Gestaltqualität". In contrast to Ehrenfels' "Gestaltqualität", Köhler's concept is called *dependence system* ("Wirkungssystem") by Grelling and Oppenheim. Unfortunately, the authors provide only brief remarks about dependence systems. First, they describe a whole w as something they call "a system with respect to a particular relation" (Grelling and Oppenheim, 1937/1938a, p.220). The part-of-relation R is defined in relation to a certain operation called *decomposition* (Sect.3). An example of such a system according to Grelling and Oppenheim is a telephone net, where connections hold among all members. In order to define their concept of dependence system, Grelling and Oppenheim further refer to Carnap's idea of determination and suggest that a system w is a dependence system if R is the ternary relation of determination. They skip the formal problem that their relation R is binary and Carnap's determination is ternary. So, on this basis it would be exaggerated to speak of an explication of the concept of dependence system by Grelling and Oppenheim. Rather, their remarks give a vague hint of what kind of reflections underlay their account. Seventeen years later Oppenheim

[2] "Gestalt (eines Komplexes mit Bezug auf eine gewisse Korrespondenz) ist die Invariante von Transformationen (des Komplexes mit Bezug auf die Korrespondenz) [...]" (Grelling & Oppenheim, 1937/1938a, p.216).

together with N. Rescher (Rescher & Oppenheim, 1955) give an improved explication of Gestalt concepts and the concept of dependence systems by avoiding the difficulty that stems from the use of Carnap's concepts. But in contrast to Grelling and Oppenheim, Rescher and Oppenheim do not want to construe any particular Gestalt concepts. Rather, they are interested in various types of Gestalt concepts that "often have applications to one and the same object of scientific enquiry" (Rescher & Oppenheim, 1955, p.106). Hence, dependent attributes and structural properties can be regarded as possible aspects of one whole and what they try to construe are rather "conditions of adequacy which underlie talk of wholes" (Rescher & Oppenheim, 1955, p.90). Nevertheless, Rescher and Oppenheim provide a more formal definition of Gestalt as complex and dependence system than Grelling and Oppenheim did. A further result of their explication is to have shown the importance of attributes of wholes and their distinction between *shared*, *unshared*, *derivable* and *underivable* attributes of wholes, as the logical properties of attributes play a significant role for the so-called *emergence* of features of wholes.

3 Emergence: Part, Wholes, Decompositions and Theories

Since some Gestaltists laid claim to the explanatory value of the concept of *emergence*, there was an increasing interest in a philosophical or logical clarification of this vague concept. Partly the examination took place together with other philosophical problems, as in the paper of Garnett (1942), who asserts that "a theory of emergence is not consistent with the strict demands of scientific method". Other philosophers focussed more on the question what the expression *emergence* actually means. A first paper of this kind was Henle's article about the status of emergence and its relation to novelty, predictability and simplicity (Henle, 1942). Two years later Henle's ideas were taken up and continued (Bergmann, 1944). Although the philosophical papers mentioned above dealt with the problem of emergence in a sophisticated manner, a real step towards a logical explanation of emergence was primarily due to Hempel and Oppenheim's famous essay *Studies in the Logic of Explanation* (Hempel & Oppenheim, 1948). The authors show that emergence is a relative, not an absolute concept: "emergence of a characteristic is not an ontological trait inherent in some phenomena; rather it is indicative of the scope of our knowledge at a given time; thus it has no absolute, but a relative character; and what is emergent with respect to the theories available today may lose its emergent status tomorrow" (Hempel & Oppenheim, 1948, p.150).

Three important observations lead to this opinion:

- To ask whether or not a property of a whole is emergent presupposes that the whole is decomposed into parts: "Before we can significantly ask whether a characteristic W of an object w is emergent, we shall therefore have to state the intended meaning of the term *part of*. This can be done by defining a specific relation *Pt* and stipulating that those and only those objects which stand in *Pt* to *w* count as parts or constituents of *w*" (Hempel & Oppenheim, 1948, p.148).

- In order to infer a property of a whole from the knowledge about its parts, it is necessary to have a complete characterization of the parts belonging to the whole, i.e. there must be a certain class of attributes *G* and a sentence, that states for every part which attribute it possesses. Hence, "the occurrence of a characteristic may be emergent with respect to one class of attributes and not emergent with respect to another" (Hempel & Oppenheim, 1948, p.148).
- The predictability of a property of a whole depends obviously on the available theories and their laws. As mentioned above, there is no guarantee that if a characteristic is emergent relative to a theory *T* it is also emergent with respect to another theory *T*'. Therefore, emergence is a thrice relativized concept: "The occurrence of a characteristic *W* in an object *w* is emergent relatively to a theory *T*, a part relation *Pt*, and a class *G* of attributes if that occurrence cannot be deduced by means of *T* from a characterization of the *Pt*-parts of *w* with respect to all the attributes in *G* " (Hempel & Oppenheim, 1948, p.151).

A few years later, Rescher and Oppenheim (1955) took up the ideas of Hempel and Oppenheim and gave a systematic, step by step development on the basis of the primitives *part* and *whole*. But they avoid mentioning *emergent* properties and rather speak of a *D-G-T-underivable attribute* of a whole. The symbol *D* denotes here a *decomposition*, i.e. any possible (but not necessary physical) dissection of a whole into parts. In order to speak about parts of a whole, it is necessary to decompose a whole into parts. This holds not only for *analytical* researchers, but also for Gestaltists, as Leinfellner (1966, p.219) pointed out:

> "... one can only speak of a whole in contrast to parts. So, in order to be able to make statements about the whole, one must be able to identify parts at least within this whole, since it must be structured = possess parts as emphasized by all Gestalt psychologists. One must break up the whole into parts, analyze it, to be able to establish that it is a whole, i. e. that it possesses holistic properties which differ from the partial properties and are not shared by the parts." [3].

[3] "...von einem Ganzen kann man nur im Gegensatz zu Teilen sprechen. Man muss also, um über das Ganze Aussagen machen zu können, zumindest innerhalb dieses Ganzen Teile feststellen können, weil es ja, wie alle Gestaltpsychologen betonen, und wie noch behandelt werden wird, strukturiert sein muss = Teile haben muss. Man muss das Ganze in Teile zerlegen, analysieren, um überhaupt feststellen zu können, dass es ein Ganzes ist, d.h. dass es holistische Eigenschaften hat, die sich von den partiellen Eigenschaften unterscheiden und die die Teile nicht haben"

In most cases, there are many possibilities to decompose a whole into parts. A very interesting question in this context is, how many possibilities there are to decompose a whole into parts. One example: Suppose we have two sequences of tones in different keys but with the same melody, and a decomposition of the sequences of tones provides two sets of tones which have no tone in common, i.e. the intersection of the sets is the null set. In this case, the same melody is an emergent property of the whole, and a sentence that ascribes the same melody to the different sets does not follow from a set of sentences that describe the properties of the parts. But there is no law, axiom, or rule that forbids us to decompose the sequences of tones into ordered sets of intervals. Now, the melody is no longer an emergent property of the whole, because it follows from a characterization of the parts of the whole.

In the history of science, there are many examples for the advance of theories, the consequences of which show that a property can lose its status as an emergent property. Some people call this scientific progress. However, even if emergence were a particular characteristic of properties of wholes, there would be no reason to believe, that this is the case only for certain theories like Gestalt theory. This aspect was emphasized in Nagel (1971, p.372): "But the logical point constituting the core of the doctrine of emergence is applicable to all areas of inquiry and is as relevant to the analysis of explanations within mechanics and physics generally as it is to discussions of the laws of other sciences". For that reason, emergence as a concept is not restricted to any specific theory; moreover, it is not even a concept that is characteristic for any particular branch of science.

4 Preliminary Conclusion

Let us take a brief look at the results of our above considerations for current research.

- There are at least two different meanings of *Gestalt*: Ehrenfels' "Gestaltqualität" as equivalence class, and Köhler's functional whole or dependence system.
- Empirical scientists cannot trust that the philosophical analysis of Gestalt concepts provides clear explications or definitions for their theories. To avoid the danger of being regarded as metaphysicists, they themselves must make a great effort to use precise concepts in connection with their research.
- We agree with Nagel's (1971, p.372) statement, that the doctrine of emergence is nothing more than "a thesis concerning the logical relation between certain statements". Therefore, it seems to us that the claim of the existence of mysterious emergent or holistic attributes is a metaphysical rather than an empirical statement. Already in 1952, Nagel states that "the mere fact that a system is a structure of dynamically interrelated parts does not suffice, by itself, to prove that the laws of such a system

cannot be reduced to some theory developed initially for certain assumed constituents of the system" (Nagel, 1952, p.30).

- Emergence is an important scientific concept only in so far as it has a heuristic value for scientific research, but it possesses no explanatory value.

It is important to notice that the explication of *emergence* by Hempel, Oppenheim, Nagel and others is a matter of logic, not of facts. The result of their explication, i.e. that *emergence* is a relative, not an absolute concept, will hold for any theory using this concept. Therefore, *emergence* will remain a concept without explanatory value until its supporters propose a new adequate explication or definition for an absolute concept of emergence. But to take up the challenge and solve the logical problems is very difficult, as the recent philosophical debates about the prospects of nonreductive physicalism and supervenience show (Beckermann, Flohr, & Kim, 1992; Hoyningen-Hüne, 1996; Stephan, 1994).

5 Gestalt Theory and the Brain

An important field of research for Gestalt psychologists is the relation between brain and neural function, and perception. It is here that part of the concepts explicated in the previous sections were developed. As we will not take up the details of the logical analysis reported in Sect.2-3, but rather present an informal account, there will be some slight deviations in terminology: instead of the term *dependence system*, which will be reserved for the logical construct, we will apply the term *functional whole*, coined and used by Köhler in his own writings (Köhler, 1947, p.136), or alternatively *physical system* as a literal translation of Köhler's term "physikalisches System" – one of the key concepts in his approach to the Gestalt problem. Regarding the concept of emergence, our remarks will reveal that Köhler indeed intended to show the existence of absolutely non-derivable "Gestalten", taking a position criticized by philosophers (see above) and not necessarily shared by other researchers in the field. We feel justified to focus on the work of Köhler as he was the one to set forth the notion of psychophysical isomorphism and to most vigorously explore the neurophysiological implications (Scheerer, 1994). Isomorphism itself is not discussed, as it "[...] was never formally defined but introduced by way of examples" (Scheerer, 1994, p.184); to us, the discussion of the notion of isomorphism seems to require some formal explication. Gestalt psychological research into the functioning of the brain provides for a connection to neuropsychology on the one hand, here exemplified by some remarks on the holonomic theory of perception advocated by K.Pribram; on the other hand, links to attempts at modelling (aspects of) perception on computers can be explored. This latter aspect will be taken up in detail in the concluding sections of this paper.

5.1 Functional Wholes, Emergence and Field Theory

Two of the notions discussed in the previous section, namely the concepts of *functional whole* ("Wirkungssystem") and of *emergence* are vital to the approach to Gestalt theory developed by W. Köhler and his students. This is illustrated in the introduction to his book *Die physischen Gestalten in Ruhe und im stationären Zustand*, the opening of which reads "Gestalten are called after v.Ehrenfels those psychic states and events, whose characteristic properties and effects cannot be composed from properties and effects of the same sort of their so-called parts." (Köhler, 1920, p.IX). That the term *composed* is to be understood as *deduced* can be seen from some remarks about chemical compounds. According to Köhler, properties which cannot be explained by properties of the constituent chemical elements may be understood as "Gestalten" in the sense cited. But Köhler continues his argumentation by hinting at the possibility that further research in the field of (physical) chemistry might reveal ways to reduce the properties of compounds to basic physical principles (Köhler, 1920, p.XI). He turns to physics to find firm ground to show the existence of "Gestalten". This resort to physics seems typical of the belief commonly held at the end of the 19th and the beginning 20th century, that physics was essentially complete with only minor adjustments to be worked out.

The paradigm introduced by Köhler in this context is the *physical system* as opposed to the *and-sum* ("Und-Summe") (Köhler, 1920, p.41). A *pure sum* ("reine Summe") is characterized by the following definition (Köhler, 1920, p.42): "A *together* is a pure *sum* of *parts* or *pieces* if it can be produced from them, taking one after the other without changing any of the *parts* by the composition. And conversely: A *together* is a *pure sum* if by removal of *parts* or *pieces* neither the *remaining together* nor the *parts* are changed". Physical systems are descriptively characterized by the fact, that by a change at one point of a configuration the whole distribution is affected, illustrated e.g. by the distribution of electrical charge on a conductor surface. Logical explication of the so defined concepts of *additivity* ("Summativität") and *non-additivity* ("Nicht-Summativität") leads to a vast system of different notions of additivity (Rausch (1937/1967) distinguishes 1488 different notions). The opposition of *physical systems* and *pure sums* is problematic, as pointed out by Keiler (1980, p.90): "There is no precise criterion to determine when a *together* is to be addressed as a *physical system* or as a *sum*; rather, we have to regard the transition from *sum* to *system* as a gradual one, depending on the relative strengths of external and internal interactions of *parts*." This conception of Gestalt forms the basis of Köhler's theory of psychophysical Gestalten, according to which all psychological facts are intimately connected to processes in the brain, these brain processes possessing a Gestalt character.

The connection between brain and psychological events is described by the assumption of psychophysical isomorphism (Köhler, 1969, p.66): "Psychological facts and the underlying events in the brain resemble each other in all their structural characteristics." Köhler took the underlying events in the brain to be DC currents which were not restricted to neural fibers but resulted from slow graded potentials produced at the synaptic junctions. He tried to produce empirical evidence in favour of his assumption by measuring brain potentials. The theory of psychophysical Gestalten was criticized from various points of view, experimental (Pribram, 1975, 1991, p.XXVI) as well as theoretical (Keiler, 1980, 1981). As Keiler (1981) argues, the experiments cited to disprove Köhler's theory cannot be regarded as conclusive due to problems in the experimental procedure and in the interpretation of the results. Just as problematic are Köhler's own experiments, the application of electrodes to the scalp or the brain entails a change in the physical system under consideration. This should have produced changes in the perception of the experimental situation. However, no such changes were recorded (Keiler, 1981, p.112).

5.2 Field Theory and Holonomic Theory of Perception

Although Köhler's neuro-electrical interpretation of psychophysical isomorphism is widely held to be disproved (Pribram, 1991, p.XXVI), there are features that remained influential for other veins of research. The idea of *physical systems* was taken up by Wygotski, Luria and others and brought together with results of Pavlov. This synthesis resulted in a new conception of *functional system* (for details see again Keiler 1981, p.113). This conception differs from Köhler's Gestalt approach mainly in two ways (Keiler, 1981, p.112):

- Köhler's reduction of *Gestalt* to *structure* is avoided (the meaning of structure in this context remaining to be discussed).
- The dogma of non-derivability (emergence) of Gestalt is rejected; rather, the *functional system* is taken to be the result of a process. It is – apart from the general theme of the relation of brain function to perception – in taking up these (indirect) influences and in describing brain processes in terms of fields of activation that are not confined to neural cells that Stadler (1981) sees Pribram's holonomic theory of perception as a modern theory of perception obeying fundamental postulates of Köhler's theory of psychophysical Gestalten.

5.3 Some Fundamental Aspects of the Holonomic Theory

Led by evidence showing that even large amounts of damage to neural systems did not necessarily entail deterioration in the performance of these systems, neuroscientists concluded, that "the neural elements necessary to the recognition and recall processes must be distributed throughout the brain systems involved" (Pribram, Nuwer, & Baron, 1974, p.417). As these requirements seemed to be fulfilled by optical holograms, which allow for reconstruction of the whole picture, even when most of the photographic material is destroyed, the analogy was taken up and explored extensively. Especially, a network model of Fourier holography was developed on the basis of "elements and junctional characteristics plausibly like those in neural networks" (p.454). Other processing models were discussed, arguing that the Fourier process yielded the strongest response on recognition (p.442). Thus, the holographic hypothesis seemed to be sufficiently founded to yield a workable paradigm for the exploration of brain function.

Nevertheless, this hypothesis was challenged by Willshaw (1981), who showed that the essential features of a holographic store can be achieved by a simpler correlographic model. Kohonen (1989, p.212) points out several problems arising from the holographic hypothesis. The holographic hypothesis was modified and incorporated by Pribram into his holonomic theory of brain and perception (Pribram, 1991, 1975). In this theory he attempts to present "a neural systems analysis of the brain-behavior relationship, which takes into account processing levels, allows the perceptual experience to be analyzed into basic functional modules that are at the same time separable and interpenetrating" (Pribram, 1991, p.1). These functional modules are arranged in hierarchical layers. This hierarchical layering, however, does not entail purely bottom-up processing in a classical sense. Rather, "each level entails both feedforward and feedback operations: thus, the paradox of the separable yet unitary nature of the perceptual experience can be accounted for" (p.4).

Thus, the holistic nature of Pribram's theory is explicated in terms of the connections and interactions of different brain systems. Taking into account Pribram's intention to come up with mathematical transformations to describe these interactions (p.2) and his assumption of these transformations being *paralinear*, i.e. essentially linear with slight nonlinearities affording an improvement in coupling, the term *holonomic* may be interpreted as "holistic, with lawful (*paralinear*) connections and interactions of the parts". In dropping the postulate of unidirectional, bottom-up information processing in favour of the concept roughly outlined here, brain function in perception can possibly be described in terms of a *dependence system* in the sense explicated (or rather shown not to be explicated) above. Pribram's theory differs in yet another way from more classical neural theory by taking into account the fact that neural processing to a large extent is achieved by localized (dendritic) micro processes (Schmitt, Dev, & Smith, 1976; Shepherd & Koch, 1990; Shepherd, 1994). He thus denies the all-importance of the all-or-nothing paradigm in neural processing, which still seems to dominate a considerable part of textbook literature on computer modelling (Parberry, 1994; Rojas, 1993), and auditory perception (Luce, 1993). For fundamentals and critique, see Shepard (1994, 1990). Even in the holonomic theory, perceptual experience such as the perception of object forms or the separation of figure from ground is to be regarded as the result of a process involving various kinds of interactions of different neural systems, again refuting Köhler's doctrine of absolute non-derivability of Gestalt. As we have seen, Köhler's approach to the Gestalt problem has been stimulating for current neuropsychological research, partly supported by neurophysiological data. Nevertheless, some of his interpretations and findings have to be rejected or revised. The idea of neural systems analysis in connection with the above mentioned features may indicate the applicability of the Gestalt concept of dependence system to a formal description of the brain-behavior relationship.

6 Perceptual Organization: Gestalt Principles, Laws and Connectionism

In the last decade there has been a revival of neural network modelling in brain research (Anderson & Hinton, 1981) and of Gestalt psychology in the psychology of perception now called *perceptual organization* (Pomerantz & Kubovy, 1986). This research is strongly related to research in artificial perception, especially in research on vision and visually guided behavior in robotics as well as pattern recognition and scene analysis (Zucker, 1995; Nelson, 1976; Edelman & Reeke, 1990). Artificial perception, perceptual organization and brain modelling furthermore converge with measurement theory and psychophysics (Mausfeld, 1994b, 1994a) .

In psychology of music, K.Pribram's holographic model of memory and perception (Pribram et al., 1974; Pribram, 1991) is often referred to in connection with artificial neural networks because of its holistic and Gestaltist properties (Bruhn, 1995). Naturally, the question arises of how these research areas – artificial perception, perceptual organization, brain modelling, psychophysics and research on music perception – are related, what the benefits and dangers of their convergence are, and what may be their common paradigm for research on perception (Gregory, 1974; Pomerantz & Kubovy, 1986; Mausfeld, 1994b, 1994a). Most important to perceptual organization in connection with artificial perception and computer simulation of perceptual processes are Köhler's conception of functional whole and the principle of "Prägnanz", another key concept of Gestalt theory (Zucker, 1995; Pomerantz & Kubovy, 1986). This principle is assumed to be the main organizational principle of (conscious) mental phenomena and serves as a key to demonstrate the priority of holistic mental organization over the proximal physical stimulus. It is assumed that there is enough evidence for the "Prägnanz" principle to be well supported empirically by many Gestalt laws. But this evidence has been questioned by Pomerantz and Kubovy. After discussing several Gestalt laws such as the law of proximity, good continuation, closure, symmetry etc. in detail, they (Pomerantz & Kubovy, 1986, 36-31-6.1) conclude: "The evidence for favoring the "Prägnanz" principle is somewhat thin. Whereas the dominant view of perceptual organization held by psychologists in general is most often centered around the Gestalt approach (as witnessed by most textbook treatment of perception), the Gestalt explanations of Gestalt phenomena are often inadequate, vague, or simply wrong. The various Gestalt laws of organization are on a fairly safe ground when considered as descriptions of organizational tendencies, although even here they often fail to give us predictions of everyday perception or (in the absence of a formal model) to tell us when one law will prevail over another." Further, the only possible empirically testable explanation of the "Prägnanz" principle which has been given by the Gestaltists is the assumption of isomorphism. The assumption of isomorphism between brain states and mental states was intended to give the basis for an explanation of perceptual processes governed by the law of "Prägnanz" in terms of a physical theory, the theory of fields and so assure that principles and laws of Gestalt theory – especially the principle of "Prägnanz" – will be empirically testable by brain research. But as Pomerantz and Kubovy (1986, 36-12-3.1) remarked: "The only specific brain mechanisms proposed by the Gestaltists were readily falsified (...). It is now clear although the brain is a volume conductor of electric currents, these currents are irrelevant to perception and so cannot be plausible physiological mechanisms for achieving "Prägnanz"."

Although this view, which is based on experimental studies made by Lashley, is widely accepted, these results have been questioned too (Sect.5.1). Nevertheless, today it is assumed that artificial neural networks provide a good framework for Gestalt theoretical research in general (Palmer, 1990; Scheerer, 1994) and especially for research on music perception (Bruhn, 1995; Leman, 1995), because these systems are models of the brain which allow to study Gestalt concepts as they may be embodied by the brain. The main idea to translate the core concepts of Gestalt theory such as the principle of "Prägnanz" and the functional whole to neural networks (Scheerer, 1994, p.72) is based on the conceptual distinction between function and structure. The function of the system is metaphorically described by the "Prägnanz" principle. The structure of the system is captured by the concept *functional whole* ("Wirkungszusammenhang"). This structure could be interpreted in terms of neural networks as its topological structure represented by a graph, and the "Prägnanz" principle may be interpreted as an optimization problem, i.e. as a search for a local minimum in a value space of a mathematical function. Although artificial neural networks seem to be a powerful tool for exploring the basic principles which govern perception and the functioning of the brain, there is a tendency – especially in some parts of the scientific research communities of psychology and computer science – to ascribe a mysterious property called *emergent* computation to the computation done by (artificial) neural networks. Often their behavior is interpreted in an ontologically unjustified way in claiming that these systems do have more or another – perhaps not analyzable – computational power than any other physically described system. In order to see that there is no special or mysterious computational aspect connected with e.g. holographic models or any other models implemented in artificial neural networks to study brain functions and perception even in connection with Gestalt concepts, we must have a glance at (artificial) neural networks from the point of view of automata theory.

7 Neural Nets and Automata

Connectionism, neural engineering and computational neuroscience originated with the research on (artificial) neural nets done by McCulloch and Pitts (1943/1965). Although as actually interpreted, they gave a logical model of the transmission of the electrical activity of interconnected nerve cells, one should bear in mind that it was not their main goal to explicate the factual functioning of nerve nets, but to show that nerve nets can realize the same logical functions as those studied in propositional and predicate logic and that no mysterious entity like mind or soul is needed to explicate this kind of reasoning (McCulloch & Pitts, 1943/1965, pp.21-22,37-39). They related their results on the computational capabilities of neural nets obtained in applying the axiomatic method to empirical research on recursion theory and Turing machines (McCulloch & Pitts, 1943/1965, pp.22,35). In general, a neuron or a whole neural network may be viewed as an (finite) automaton, i.e. as a state system (Arbib, 1964; Parberry, 1994; Rojas, 1993; Sontag, 1995; Suppes & Rottmayer, 1974, for details on automata theory and perception). An automaton can be interpreted as a symbol processing device operating on strings. Different classes of automata and their computational power are studied in automata theory. The class of automata with the highest computational power are Turing machines; the least computational complexity in the hierarchy is exhibited by the finite automata. Computational power is restricted by memory size. Finite automata have finite memory; Turing machines on the other hand are provided with potentially infinite (external) memory and are called infinite automata. The class of recurrent nets is a modern version of artificial neural nets. They can be interpreted as either finite or infinite automata. The capacities of these neural networks to simulate infinite automata are discussed by Sontag (1995). The author emphasizes the fact that two biological interpretations of infinite memory must be distinguished. In the first interpretation, an animal's (infinite) long term memory is associated with the environment (Arbib, 1964, pp.1-30), as nicely illustrated in Braitenberg (1986, pp.26-28,106). In the second interpretation, no distinction is made between external and internal memory. Short term memory and long term memory are located in the same system: the brain. In this latter interpretation, activation values or connection strengths represent long term memory. It can be shown that a recurrent neural net with roughly 1000 neurons is capable to simulate a universal Turing machine.

Thus, neural networks do not exhibit any mysterious emergent computations, and there is a powerful framework and reference point provided by automata theory for the analysis of the computations exhibited by neural networks as well as for theoretical analysis of other (physical) computational devices. In principle, neural networks and computers under appropriate idealization fall into the same class of automata, they are universal Turing machines. Turing machines represent an abstract boundary for the study of information processing systems in general.

8 General Conclusion

To answer our question posed at the end of Sect.1 concerning the status of Gestalt theory as a paradigm for research in perceptual organization, we want to add a brief remark on the use of the term paradigm. By *paradigm* we mean a set of beliefs with some of them as accepted knowledge or – often implicit - unquestioned theoretical concepts. The concepts relate, select, and let us see facts in a certain light, thereby directing the lines of research. For the following reasons, we do *not* regard Gestalt theory as an appropriate paradigm:

- We suggest that a careful distinction must be made between Gestalt theory and Gestalt psychology. Gestalt theory cannot function as an explanatory theory.
- Main concepts like functional whole, emergence, the principle of "Prägnanz" and the Gestalt laws point to problems to be solved, but give no solution to the indicated problems. So at best they are of heuristic value for perceptual psychology and the study of perceptual organization. To refer to these problems, it may be convenient to resort to Gestalt psychology as a phenomenalistic accumulation of observed facts requiring further explanation (Haubensack, 1985, pp.32-33).
- The explanatory value of the concept of emergence is denied, if its use is intended to be theoretical: logical analysis reveals the need for a threefold relativation.
- Neuropsychological and computational modelling may possibly be interpreted within the logical framework provided by the concepts of functional whole and dependence system.
- Neural computation is distinguished only from a pragmatic point of view, affording special tools for implementation. Theoretically, it doesn't matter that computation is performed in a parallel and distributed way. Rather, the underlying principles have to be specified.

Proposals for future research:

- Auditory computation and perceptual organization (Bregman, 1990; Ellis, 1995; Hawkins, McMullen, Popper, & Fay, 1996; Kubovy & Pomerantz, 1981; Leman, 1995, 1997; Pomerantz & Kubovy, 1986; Shamma, 1995) should be explored as the interface where research on hearing from neuroscience, psychophysics, measurement theory and perceptual psychology meets music theoretical cognitive modelling from musicology to study together perceptual organization in audition.

- Some main problems to be solved in research on auditory perceptual organization are related to the questions: What is an auditory perceptual code? How can we define the stimuli of complex auditory perceptions? How are percepts organized? What does the cognitive architecture of the auditory system looks like?
- To answer these question one has to determine the functions of the system first. One has carefully to define the organization of percepts to test them by computational models.
- Therefore, and because in multidisciplinary empirical research one needs a common language to report, discuss, compare and evaluate theories and concepts, we propose to use some tools provided by metalogic, i.e. model – and recursion theory, as a general framework and reference point for further research (Lehmann, 1985).

References

Anderson, J., & Hinton, G. (1981). Models of information processing in the brain. In G. Hinton & J. Anderson (Eds.), *Parallel models of associative memory.* Hillsdale, NJ: Erlbaum.

Arbib, M. (1964). *Brains, machines and mathematics.* New York, NY: McGraw-Hill.

Beckermann, A., Flohr, H., & Kim, J. (1992). *Emergence or reduction? Essays on the prospective of nonreductive physicalism.* Berlin, New York: Walter de Gruyter.

Bergmann, G. (1944). Holism, historicism, and emergence. *Philosophy of Science, 11,* 209–221.

Braitenberg, V. (1986). *Künstliche Wesen: Verhalten kybernetischer Vehikel.* Braunschweig: Vieweg.

Bregman, A. (1990). *Auditory scene analysis: The perceptual organization of sound.* Cambridge, MA: The MIT Press.

Bruhn, H. (1995). Gehör. In L. Finscher (Ed.), *Die Musik in Geschichte und Gegenwart, Vol. 3.* Kassel: Bärenreiter.

Carnap, R. (1929). *Abriss der Logistik mit besonderer Berücksichtigung der Relationstheorie und ihrer Anwendungen.* Vienna: Julius Springer.

Edelman, G., & Reeke, G. (1990). Is it possible to construct a perception machine? *Proceedings of the American Philosophical Society, 134,* 36–73.

Ellis, D. (1995). *Hard problems in computational auditory scene analysis.* (http://sound.media.mit.edu/~dpwe/writing/hard-probs.html)

Garnett, A. (1942). Scientific method and the concept of emergence. *The Journal of Philosophy, 39,* 477–486.

Gregory, R. (1974). Choosing a paradigm for perception. In E. Carterette & M. Friedman (Eds.), *Handbook of perception, Vol. I: Historical and philosophical roots of perception.* New York, NY: Academic Press.

Grelling, K., & Oppenheim, P. (1937/1938a). Der Gestaltbegriff im Lichte der neuen Logik. *Erkenntnis, 7,* 211–225.

Grelling, K., & Oppenheim, P. (1937/1938b). Supplementary remarks on the concept of Gestalt. *Erkenntnis, 7,* 357–359.

Grelling, K., & Oppenheim, P. (1988a). The concept of Gestalt in the light of modern logic. In B. Smith (Ed.), *Foundations of Gestalt theory.* Munich, Vienna: Philosophia Verlag. (Original work published 1937/1938)

Grelling, K., & Oppenheim, P. (1988b). Supplementary remarks on the concept of Gestalt. In B. Smith (Ed.), *Foundations of Gestalt theory.* Munich, Vienna: Philosophia Verlag. (Original work published 1937/1938)

Griffith, N., & Todd, P. (Eds.). (1994). *Music and creativity.* (Special issue of Connection Science: Journal of Neural computing, Artificial Intelligence, and Cognitive Research 6(2-3))

Haubensack, G. (1985). *Absolutes und vergleichendes Urteil. Eine Einführung in die Theorie psychischer Bezugssysteme.* Berlin, Heidelberg: Springer-Verlag.

Hawkins, H., McMullen, T., Popper, A., & Fay, R. (1996). *Auditory computation.* New York, NY: Springer-Verlag.

Hempel, C., & Oppenheim, P. (1948). Studies in the logic of explanation. *Philosophy of Science, 15*, 135–175.

Henle, P. (1942). The status of emergence. *The Journal of Philosophy, 39*, 486–493.

Hoyningen-Hüne, P. (1996). Supervenient/Supervenienz. In J. Mittelstraß (Ed.), *Enzyklopädie Philosophie und Wissenschaftstheorie, Bd. 4.* Stuttgart, Weimar: Verlag J.B. Metzler.

Keiler, P. (1980). Isomorphie-Konzept und Wertheimer-Problem. Beiträge zu einer historisch-methodologischen Analyse des Köhlerschen Gestaltansatzes, Teil I. *Gestalt Theory, 2*, 78–112.

Keiler, P. (1981). Isomorphie-Konzept und Wertheimer-Problem. Beiträge zu einer historisch-methodologischen Analyse des Köhlerschen Gestaltansatzes, Teil II. *Gestalt Theory, 3*, 93–128.

Köhler, W. (1920). *Die physischen Gestalten in Ruhe und im stationären Zustand. Eine naturphilosophische Untersuchung.* Braunschweig: Vieweg.

Köhler, W. (1947). *Gestalt psychology.* New York, NY: Liveright.

Köhler, W. (1969). *The task of Gestalt psychology.* Princeton, NJ: Princeton University Press.

Kohonen, T. (1989). *Self-organization and associative memory.* Berlin, Heidelberg: Springer-Verlag.

Kubovy, M., & Pomerantz, J. (1981). *Perceptual organization.* Hillsdale, NJ: Erlbaum.

Kuhn, T. (1976). *Die Struktur wissenschaftlicher Revolutionen.* Frankfurt/M.: Suhrkamp.

Lehmann, G. (1985). *Modell- und rekursionstheoretische Grundlagen psychologischer Theorienbildung.* Berlin, Heidelberg: Springer-Verlag.

Leinfellner, W. (1966). Logische Analyse der Gestalt: Logik und Gestaltpsychologie. *Studium Generale, 19*, 219–235.

Leman, M. (1995). *Music and schema theory: Cognitive foundations of systematic musicology.* Berlin, Heidelberg: Springer-Verlag.

Leman, M. (Ed.). (1997). *Foundations of pitch and timbre perception.* Lisse, The Netherlands: Swets & Zeitlinger. (Special issue of Journal of New Music Research, Vol. 26, nr. 2)

Luce, R. (1993). *Sound and hearing: A conceptual introduction.* Hillsdale, NJ: Erlbaum.

Mausfeld, R. (1994a). Hermann v.Helmholtz: Die Untersuchung der Funktionsweise des Geistes als Gegenstand einer wissenschaftlichen Psychologie. *Psychologische Rundschau, 45*, 133–147.

Mausfeld, R. (1994b). Methodologische Grundlagen und Probleme der Psychophysik. In T. Herrmann & W. Tack (Eds.), *Methodologische Grundlagen der Psychologie.* Göttingen: Hogrefe.

McCulloch, W., & Pitts, W. (1965). A logical calculus of ideas immanent in nervous activity. In *Embodiments of mind.* Cambridge, MA: The MIT Press. (Original work published 1943)

Nagel, E. (1952). Wholes, sums, and organic unities. *Philosophical Studies, 3*, 17–32.

Nagel, E. (1971). *The structure of science: Problems in the logic of scientific explanation.* London: Routledge and Kegan Paul.

Nelson, R. (1976). On mechanical recognition. *Philosophy of Science, 43*, 24–52.

Palmer, S. (1990). Modern theories of Gestalt perception. *Mind and Language, 5*, 289–321.

Parberry, I. (1994). *Circuit complexitiy and neural networks.* Cambridge, MA: The MIT Press.

Pomerantz, J., & Kubovy, M. (1986). Theoretical approaches to perceptual organization. In K. Boff, L. Kaufman, & J. Thomas (Eds.), *Handbook of perception and human performance. Vol. II: Cognitive processes and performance.* New York, NY: Wiley.

Pribram, K. (1975). Towards holonomic theory of perception. In S. Ertel, L. Kemmer, & M. Stadler (Eds.), *Gestalttheorie in der modernen Psychologie.* Darmstadt: Steinkopff.

Pribram, K. (1991). *Brain and perception: Holonomy and structure in figural processing.* Hillsdale, NJ: Erlbaum.

Pribram, K., Nuwer, M., & Baron, R. (1974). The holographic hypothesis of memory structure in brain function and perception. In D. Krantz, R. Atkinson, R. Luce, & P. Suppes (Eds.), *Contemporary developments in mathematical psychology. Vol. II: Measurement, psychophysics and neural information processing.* San Francisco, CA: Freeman.

Rausch, E. (1967). Über Summativität und Nichtsummativität. In *Psychologische Forschung. Zeitschrift für Psychologie und ihre Grenzwissenchaften.* Darmstadt: Reprint of Wissenschaftliche Buchgesellschaft. (Original work published 1937)

Rescher, N. (1953). Mr. Madden on Gestalt theory. *Philosophy of Science, 20*, 327–328.

Rescher, N., & Oppenheim, P. (1955). Logical analysis of Gestalt theory. *The British Journal for the Philosophy of Science, 6*, 89–106.

Rojas, R. (1993). *Theorie der neuronalen Netze. Eine systematische Einführung.* Berlin, Heidelberg: Springer-Verlag.

Scheerer, E. (1994). Psychoneural isomorphism: Historical background and current relevance. *Philosophical Psychology, 7*, 183–210.

Schmitt, F., Dev, P., & Smith, B. (1976). Electronic processing of information by brain cells. *Science, 193*, 114–120.

Shamma, S. (1995). Auditory cortex. In M. Arbib (Ed.), *The handbook of brain theory and neural networks.* Cambridge, MA: The MIT Press.

Shepherd, G. (1990). The significance of real neuron architectures for neural network simualitions. In E. Schwartz (Ed.), *Computational neuroscience.* Cambridge, MA: The MIT Press.

Shepherd, G. (1994). *Neurobiology* (3rd ed.). New York, NY: Oxford University Press.

Shepherd, G., & Koch, C. (1990). Introduction to synaptic circuits. In G. Shepherd (Ed.), *The synaptic organization of the brain.* New York, NY: Oxford University Press.

Sontag, E. (1995). Automata theory and neural networks. In M. Arbib (Ed.), *The handbook of brain theory and neural networks.* Cambridge, MA: The MIT Press.

Stadler, M. (1981). Feldtheorie heute – von Wolfgang Köhler zu Karl Pribram. *Gestalt Theory, 3*, 185–199.

Stephan, A. (1994). Theorien der Emergenz – Metaphysik oder? *Grazer Philosophische Studien, 48*, 105–115.

Suppes, P., & Rottmayer, W. (1974). Automata. In E. Carterette & M. Friedman (Eds.), *Handbook of perception, Vol. I: Historical and philosophical roots of perception.* New York, NY: Academic Press.

Todd, P., & Loy, D. (1991). *Music and connectionism.* Cambridge, MA: The MIT Press.

Wertheimer, M. (1974). The problem of perceptual structure. In E. Carterette & M. Friedman (Eds.), *Handbook of perception, Vol. I: Historical and philosophical roots of perception.* New York, NY: Academic Press.

Willshaw, D. (1981). Holography, associative memory, and inductive generalization. In G. Hinton & J. Anderson (Eds.), *Parallel models of associative memory.* Hillsdale, NJ: Erlbaum.

Zucker, S. (1995). Perceptual grouping. In M. Arbib (Ed.), *The handbook of brain theory and neural networks.* Cambridge, MA: The MIT Press.

Knowledge in Music Theory by Shapes Musical Objects and Sound-Producing A

Rolf Inge Godøy

Section for Musicology, University of Oslo, P.O.Box 1017, Blindern N-0315, Norway

Abstract. Music theory must try to deal with *emergent qualities*, such as contour, texture, timbre and tone semantics, and this necessitates recognizing musical objects as holistic entities. Representations by *shapes* can be useful here, as shapes are inherently holistic. The idea of shapes is seen as applicable to several aspects and modalities at work in musical imagery, providing images at various levels of resolution of both the unfolding sounds and the sound-producing actions. The paradigm of shapes is seen as well supported by several contemporary domains of thought, but in need of extensive development as an alternative to more abstract approaches in music theory.

1 Introduction

The aim of this paper is to present some ideas on how the thinking of musical objects as *shapes* can enhance knowledge in music theory. *Musical objects* are here understood as delimited segments of actual sonic unfolding (i.e. single or composite sounds, tones, chords, rhythmic patterns, phrases or even more extended sections) having various perceived *emergent qualities* such as timbre, texture and contour, qualities which can not be represented locally at any single point in time, but only as holistic images of the entire musical object. This paper is based on the conviction that thinking of these perceived emergent qualities as shapes is a privileged mode of representation in our minds by producing *images of dynamic unfolding*, providing more or less stable images of an otherwise ephemeral sonorous flux.

2 Notions of Shape

Images of shapes, produced by reiterated *drawings* (mentally, or on paper, or on the computer screen), can make explicit otherwise tacit or ineffable knowledge of musical objects by increasing our awareness of as many features as possible, i.e. mapping out unmapped territory. This is a matter of directing our attention towards various features in musical sound by active efforts of imagination and representation, or by what is called an intentional focus in the phenomenological domain of thought. We may for instance consider the act of drawing the shape of the perceived loudness trajectory of a sound, i.e.

of what could be called its overall envelope. This will probably for many people have the effect of directing attention towards this feature, clarifying and producing knowledge about the dynamic unfolding of the sound. By listening to the sound again and modifying the drawing of the envelope and reiterating several times this act of listening and drawing, the effect will probably be that of producing a progressively refined image and hence progressively more knowledge about this particular sound. This process of oscillation between sound and various shape-images can be similarly applied to several other features of the musical object (e.g. timbral, textural and contoural unfolding), as well as at any level of resolution from the minuscule inflections and transients of single tones, what R. Barthes aptly termed "the grain of the voice" (Barthes, 1977), to the more macroscopic features of phrases and even whole sections of musical works.

Actually, such cognition by shapes is now well known and indispensable in the domain of digital synthesis and signal processing, where the oscillation between considering the perceived qualities of sounds and the corresponding shape-images (such as the shapes of overall dynamic envelopes, shapes of stationary spectra and shapes of spectral evolution, shapes of frequency fluctuations, etc.) is quite simply the everyday working method (Risset, 1991; Cogan, 1984; Wessel, 1985; Moore, 1990). However, this idea of gaining knowledge by establishing correlations between features of musical sound and various shape-images is also based on fairly extensive material from the domains of Gestalt, both classical (von Ehrenfels, 1890/1988; Stumpf, 1883/90; Koffka, 1963; Köhler, 1947; Wertheimer, 1967) and its contemporary applications to music (Bregman, 1990) and cognitive linguistics (Johnson, 1987; Lakoff, 1987), as well as from the domain of phenomenology (Husserl, 1980; Schaeffer, 1966; Chion, 1983; Miller, 1982) and various domains of the cognitive sciences, in particular those of mental imagery (Denis, 1989; Finke, 1989; Kosslyn, 1994), and morphodynamical theory (Petitot, 1985b; Thom, 1983).

I believe this last mentioned domain provides the most thorough and systematic foundation for applications of notions of shape in the human sciences, combining ideas from phenomenology on the emergence of the relatively stable or solid out of the flux of lived experience with ideas from mathematics in a general *hermeneutics by geometrization*. In the words of R.Thom: "......the first objective is to characterize a phenomenon as a form, as a 'spatial' form. To understand means then first of all to geometrisize" (Thom, 1983, p.6). Very briefly, it can be said that cognition by shape is here seen as deeply rooted in the human cognitive faculties, being the essential element of all understanding as well as the very condition for x to arise in the first place. The argument for this is that if we were continuously submerged in an uninterrupted stream of impressions, there would be no way that meaning could arise as there would be no qualitative discontinuities, no chunking of information, and thus that our impressions of the outside world would be totally amorphous.

Husserl and other phenomenologists, such as Ricoeur, have similar arguments for the necessity of *stepping out of*, or *interrupting*, the continuous flux and *lumping* together impressions to form chunks or Gestalts, or what I here call shapes. As Ricoeur has pointed out, distantiation to the immediate experience is for this reason necessary for any meaning to arise in the first place (Ricoeur, 1981). In this sense, we could speak of a seemingly contradictory and profoundly enigmatic relationship between the continuous flux and the discontinuous solid in human experience in general and in musical cognition in particular. Yet on the other hand, probably nobody would deny the transformation of continuous audio-acoustic flux into more solid or stable chunks with some kind of *meaning* in our cognitive apparatus, as well as in the cognitive apparatus of other living animals, such as for instance research on the categorization of sounds in animals (Harnad, 1987). We should thus here in our context quite simply suppose that this incessant transformation from flux to solid is probably endowed by evolution, and rather shift our attention to what the consequences of this will be for knowledge in music theory. Epistemologically, the crucial question for us is how well the dynamical qualities of musical sound are accounted for in our representations. Considering the various features of musical sound as bundles of concurrent trajectories as mentioned above (i.e. trajectories for overall loudness, for spectral evolution, for perceived pitch, for textural evolution, etc.), representations should then be capable of preserving the dynamical qualities of musical sound by representing these various trajectories as shapes. In sum, the *shape-paradigm* in music theory I am discussing here is precisely an effort to represent knowledge in a way which conserves the dynamics of musical sound, perhaps paradoxically so, by drawing solid images of continuous flux.

In this enterprise of drawing shapes, we could speak of two poles, one based primarily on *analogical* representations and another based on *signal representations* of musical sound. As for the latter, it is well known that although signal based representations of musical sound such as wave forms and spectrograms can be informative and useful for many purposes, there is still a considerable distance between the complexity and richness of what can be heard and what can be seen, even though there are always choices to be made (such as *window* size and shape, how much or what to account for in terms of psycho-acoustic filterings, etc.) which may enhance the information in the images. However much work is now going for improving the possibilities here, amongst other things with wavelets or multi-resolution images which can improve the time/frequency resolution, as well as several intelligent algorithms for estimating perceived qualities like pitch, loudness and various timbral/textural patterns. As for what I call *analogical* shape representations (with the required *technology* of paper and pencil or even just imagination and a comfortable arm chair) it is well founded in various cognitive literature (Black, 1962; Holyoak & Thagard, 1995, as well as the above mentioned references to cognitive linguistics and morphodynamical thought), but most of all well implemented in the work of P. Schaeffer (Schaeffer, 1966; Chion,

1983), summarized in the various shapes designated in his morphological matrix (Schaeffer, 1966, pp. 584-587). Ideally we should have a convergence of these two approaches in the future so that the signal based shapes correlate more directly with analogical shapes drawn on the basis of phenomenological investigations of musical objects. For the moment, the analogical approach is of course readily available, only limited by our imagination.

However, advocating such an uninhibited use of analogical representations and visual metaphors does of course raise difficult questions as to the relationship between the visual and auditive as well as motoric modalities in musical cognition. It should be possible to have an open eye for relevant research on the confluence of modalities (Damasio, 1989), yet at the same time to have a basically *pragmatic* approach of considering the *productivity* of thinking all the emergent qualities of musical objects as shapes. This pragmatic aspect may be considered common to all applications of metaphor. As stated by M. Black, there is in the use of metaphors initially a transfer of words by *catachresis*, which is the *improper* use of a word in a new context out of lack for a proper term. When the transferred or metaphorical use has become more established, it will no longer have this character of catachresis (Black, 1962, pp. 32-33). Of particular interest for us here, it is possible to document the ubiquity of visual metaphors in the discourse and epistemological attitudes of our culture in general (de Man, 1978; Sweetser, 1990), as well as in musical discourse in particular, with visual metaphors such as *rough*, *smooth*, *bright*, *dull*, *curved*, *flat*, etc. and more abstract but indispensable spatial metaphors such as *high*, *low*, *short*, *long*, *thick*, *thin*, etc. (Deutsch, 1984; von Ehrenfels, 1890/1988; Godøy, 1993). Although we of course could not claim that these visual metaphors are universal, they are after all frequently used, and are probably even indispensable in our work in music theory and analysis, as they fulfil a need to point out features in the musical sound. What I am proposing here is an extension of this pragmatic use of visual metaphors to include shape-images of as many as possible of the features of musical sound, thus enhancing our knowledge and discriminatory capabilities.

3 Geometrization of Musical Objects

If we accept the idea that musical experience is based on the perception of a continuous audio-acoustic flux which somehow is chunked into more or less stable entities in our cognition, we could also claim that musical cognition is based on *retrospective images* of flux, and hence, that we are in fact dealing with *musical imagery* when talking about musical cognition. Although this point may seem obvious, research into musical and auditory imagery seems still to be in its beginnings (Reisberg, 1992), compared with the more extensively researched field of what has commonly been called *mental imagery*, a field which has primarily been concerned with visual imagery (Denis, 1989; Finke, 1989; Kosslyn, 1994). Whereas this visual imagery research has developed quite sophisticated methods and models for investigating the content and transformation of images, research in music cognition from the last couple of decades has largely focused on what could be called symbol based systems in the sense of mapping out, or proposing explanations, models or algorithms for the cognition of pitch, tonality and tonal harmony as applicable to large categories of musical practice, something which is clearly consistent with western preoccupations with abstractions, discretizations and symbols. With the exception of research in digital synthesis and signal processing as well as some auditory research, what may with P.Schaeffer be called the *morphology of musical objects* has received considerably less attention.

However, we do now see tendencies of more ecological (i.e. less based on abstract, ideal (platonic) systems, more based on evolutionary conditions) and subsymbolic orientations (i.e. considering the complex substrate from which more stable and *simple* entities or symbols emerge), as well as a renewed interest in shapes or forms in general, a kind of "Naturphilosophie" inspired by investigations into self-organization in various sciences. It must be remembered that the point of departure for Schaeffer was always the *sonic event* in the sense of a segment of actual sonorous unfolding, investigated by the method of "sillon fermé" (meaning the numerously reiterated listening to the segment), gradually mapping out as many as possible of the features of the sound, and not by starting out with a set of abstractions based on neat, clean discretizations of idealized entities. Schaeffer was very clear about the distinction between the *concrete* and the *abstract* in music theory, and has also demonstrated lucidly why and how abstract systems of pitch have been given such privileged status in western musical thought.

As stated by Schaeffer, and more recently extensively documented by auditory research, musical objects are multidimensional entities, usually comprising several concurrent tendencies of change or fluctuations in harmonic content, frequency and amplitude, which again can be seen as combined into various higher order features like the timbral or textural qualities of a sound. Schaeffer made a multidimensional matrix for several of these higher level dimensions (with sub-dimensions), using various metaphors to give these dimensions names like *mass* (overall harmonic content), *grain* (fast and small fluctuations in frequency, amplitude and/or harmonic content) and *allure* (slower fluctuations in the sound), all of which were seen as time-dependent evolving elements, hence as what I here call shapes, and all of which would have sub-shapes at higher levels of resolution, i.e. when zooming into the sub-dimensions of the main dimensions (see sound examples #1 and #2 in the Appendix for illustrations of *mass*, *grain* and *allure*). Although Schaeffer's multidimensional (and multi-resolution) morphological ordering of musical objects may be seen as a kind of system, his point was all the time that of characterizing the *surface qualities* of the musical object, preserving the actual spatio-temporal layout of the musical object at various levels of resolution, and not to reduce the musical object to a collection of abstract and spatio-temporally collapsed symbolic entities.

To preserve the actual spatio-temporal nature of musical objects, *geometry* must be the primary element in representation, primordial to other more abstract, symbolic and/or numerical representations. Geometric representations have other advantages as well, such as being highly efficient in transmitting large amounts of information *at a glance* or in a broad-band manner (Tufte, 1983, 1990), as may be indicated by briefly confronting some attributes of geometric and non-geometric representations: *broad-band* vs. *serial*, *quantal* vs. *sequential*, *figure* vs. *points*, *analog* vs. *discrete*, as well as *quality* vs. *quantity*. Also, geometric representations of qualities as shapes have the advantage of being more in accordance with the approximate nature of perception and cognition, compared with exact or bit-mapped kinds of representation, by being *fault-tolerant*, i.e. that deviations from one shape to repetitions of that same shape are acceptable within certain limits. In this respect, cognition by shapes is well in accordance with connectionist principles, as well as with categorical perception, i.e. having quite narrow spaces of inter-categorical boundaries but relatively large areas for intra-categorical variation (Petitot, 1985a), something which has lead S. Harnad to suggest the term *approximationism* for categorical perception (Harnad, 1987).

4 Shapes and Variants

The acquisition of knowledge by shapes in music theory may then be seen as a two-stage process:

1. Drawing images of the various features of the musical object, and
2. Imagining or actually generating sonorous variants of the musical object where the shapes of the various features may be varied.

In this way, we could speak of both an internal and an external geometry of the musical object in music theory. The internal geometry is that which encompasses an image of the various features of the singular musical object, such as the envelope of its overall perceived loudness. Should we wish to investigate the effect of the envelope shape on the perceived quality of a tone, we could produce a series of variants where the shape of for instance the attack portion of the envelope was incrementally varied from gradual to steep, and as we know, this would result in a series of tones incrementally varied from a *bowed* to a *struck* kind of attack (see sound example #3 in the Appendix for an illustration of this). In this sense, we have gained knowledge about the attack feature of the tone, both in terms of its internal geometry by drawing the envelope shape, and in terms of its external geometry by seeing it synoptically in relation to a number of other possible but rejected variant shapes of the envelope. The same principle of shape and variant shapes could of course be applied to several other features of a musical object, such as various subdimensions of its harmonic and/or textural content. Again, this is in fact what most people working with digital synthesis and signal processing do all the time: Incrementally varying the shape of some feature or set of features and listen to the resultant sound. More precisely, this is what J.-C. Risset has termed the *analysis by synthesis* strategy of investigating musical sound. Knowledge is here then a matter of knowing the relative position of a certain musical object in relation to other possible, but rejected, variants. Epistemologically, this means that knowledge is not based on absolute values, but on the relative position of a particular musical object in relation to a large number of other musical objects.

In principle, I see nothing wrong with extending this analysis by synthesis through shape variation to the domain of note-based music. In fact, I believe it could give us valuable knowledge about several emergent qualities in note-based music, in particular textural features. Music theory and analysis has mostly avoided emergent qualities like texture and timbre because these can not be easily reduced to abstract symbolic entities, but on the contrary have to be treated on the global level as shapes, precisely preserving the spatio-temporal layout of the music. It is also possible to investigate the effects of changes in modality by this same procedure, i.e. by preserving durations, dynamics, octave placement, as well as performance characteristics such as phrasing, articulation and tempo variations, but incrementally varying the pitches from one modality to another by each variant. In this way, it is possible to investigate the effects of different modal schemata in a fairly realistic context, thus to evaluate modality in a truly global manner. An example of a set of variant modalities applied to the C-major prelude by Chopin is given in the Appendix, concluding with a version of this prelude which has preserved the original C-major modality but which has been *spatially collapsed*, i.e. has had the octave placement eradicated by a modulo 12 operation so that all the tones are squeezed into one octave (see sound example #4 in the Appendix). This should give us some indication of the effects of such a drastic spatial deformation compared with the more modest spatial deformations of the modality variants.

I see this process of mapping out the internal and the external geometry of the musical object by variants (imagined or real) as producing quite useful knowledge, notably knowledge which has not been accessible to music theory before because of its preoccupation with abstract and frequently spatio-temporally collapsed symbolic structures. For a critique of music theory, see Godøy (1993). In addition to this, thinking of various qualities of musical objects as shapes with certain limits to variation or deformation may be seen as an *alternative to a rule-based approach to musical cognition*, with the idea that well-formedness may be represented as shapes with certain acceptable deviations, precisely as is the case in synthesis and signal processing. Also, thinking shapes could be seen as a *generative* principle in the sense of enabling *mapping from known to unknown* (projection) as well as *gradual transformation from one shape to another* (deformation), preserving some features while discarding some and acquiring new ones. Again, this is well known from synthesis and signal processing, e.g. by projecting the spectral shape (formants) from one sound on to another, or by gradually deforming the spectral shape (gradual change of formantic peaks) of a sound.

5 Shapes and Sound-Producing Actions

It must be noted that generativity by shapes as an alternative to rules is advocated by cognitive linguistics with the ideas of metaphoric projections and image schemata. Moreover, regarding the *ecological idea* in cognitive linguistics of bodily based image schemata (Johnson, 1987; Lakoff, 1987), it could be stated that there are in fact bodily correlates to sounds as shapes, for instance with timbre as correlated both to the stationary shape of the vocal tract for vowels (formants), as well as to the shape of the changing shape of the vocal apparatus (tongue, lips, etc.) for transient vocal articulations. As we know, the early (and contested) version of categorical perception was based on a motor theory (Stevens, 1972; Harnad, 1987). In this connection, we should remember the various everyday language shape-metaphors for sound qualities, common in our culture but not necessarily universal, such as *hollow*, *narrow*, *open*, *closed*, etc. Also, it would seem reasonable to assume that the shapes of the sound-producing actions or gestures contribute to the drawing of shapes in the auditory image, such as the percussionists mallet hitting the drum in one stroke, the pianists hand sliding up the keyboard in a scale run, the shape of the pianists hand when playing a certain chord, etc.

For this reason, I believe it could be fruitful to recognize musical objects in our imagination as *composite with regards to different modalities*, in most cases comprising images of the sound-producing actions as well as of *the sound itself*. Research in auditory imagery seems to suggest such a cross-modal, composite makeup of the musical object in our imagination (Reisberg, 1992), just as it has been documented that visual imagery tend to mobilize more *primary* perceptive and motor faculties in our cognitive apparatus (Kosslyn, 1994). In this way, we could think of the musical object, *auditory image* (McAdams, 1984) or *auditory stream* (Bregman, 1990) as in most cases comprising both action and sound as in the triangular model of Fig1.

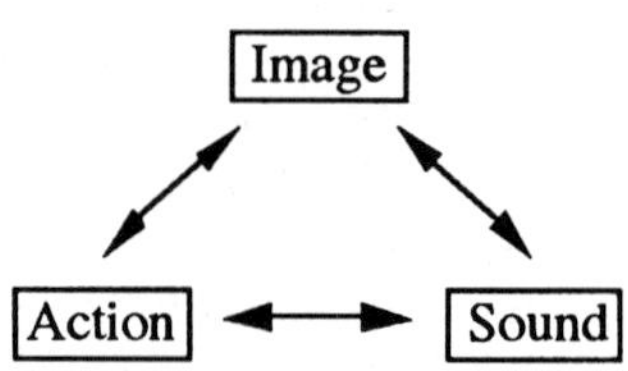

Fig. 1. Triangular model of image, action, and sound

Considering the shape of the sound-producing actions as equally important for our images of musical objects as the “purely” auditive signal, leads to some other attractive points as well, some already thematized in relevant literature, some more a matter of enlightened guesswork at the moment:

- The *coherence* of the sound is in many cases dependent upon identification of source, in other words on the sound-producing action or set of actions, as is frequently pointed out by Bregman (1990). On the basis of learned associations between source and sound, disparities or incoherences in the signal by itself can be overridden in our perception because we perceive the coherence of the action or actions causing the sound. Thus, imagining the sound-producing action in the musical object is inherently holistic in the sense that a complex sound may be subsumed by a *simple* gesture (such as the mallet hitting the tamtam) or sets of gestures (such as the sticks repeatedly hitting the snare drum in a roll).
- Images of musical objects based not only on the *pure* signal but on the shape of the sound-producing actions as well, have the advantage of being inherently *hierarchical* in the sense that actions are hierarchical (such as the picking up an object with my hand being hierarchical from the *executive* intention downwards to the fine and coordinated movements of every joint in my arm, hands and fingers), thus shedding light on questions of grouping or parsing as well as *central* vs. *peripheral* or ornamental elements in musical objects. This may be clearly seen in the case of rhythmic grouping (as is the case of alternations between the 3/4 and 6/8 versions of a rhythmic pattern of one quarter note, two eight notes and one quarter note, with a Gestaltist exclusive allocation to the one or the other), where it is difficult to see how groupings may be arrived at in a bottom-up manner without recourse to some top-down schematic inclusion by action hierarchies. Such action hierarchies are in fact well known in the actions of musical performance and in the actions of dance.
- Also, actions are eminently *transferrable* and *deformable*, allowing for a generativity by reusing the shape of the sound-producing action in another context (transfer), with or without some altering of shape (deformation).
- Finally, action is eminently *scalable* in the sense of allowing both fast and slow motion re-play in our imagination, hence being indispensable for musical imagery by enabling rapid *scanning* of musical passages as well as slow, protracted or reiterated contemplation of sonic events. Regardless the speed of unfolding, imagined actions will intrinsically conserve a spatio-temporal layout, and this will help preserving the spatio-temporal layout of the musical object as a whole, i.e. be an antidote against unfortunate abstractions. Preserving the image of the sound-producing gestures will in most cases also preserve the textural and contoural qualities of the musical object in our imagination.

To my knowledge, the action aspect of musical objects is mostly an unexplored territory, with some notable exceptions such as D. Sudnow's work *Ways of the Hand* (Sudnow, 1978), a remarkable introspective account of musical cognition in jazz improvisation by the shapes of the hands (in chords and passages) and the shapes of movement of hands (in passages, progressions and textures). We should welcome more research in this direction, as I believe it is now necessary to depart from the purely *signal-based* kind of musical cognition towards a more holistic perspective, seeing musical objects as primordially *events*, as concrete in the sense of occurring in time-space (both the *external* and the *internal* of our imagination), almost always complex in the sense of being *messy*, never being *neat* and cleanly discretized, yet also as *simple* and unitary in the sense of action-intentions. The point with my triangular model of image, action and sound is then to integrate the main aspects of musical imagery in the conceptual scheme of thinking shapes.

6 Conclusion

The reason for proposing this *shape-paradigm* in music cognition, and in particular in musical imagery, is epistemological. Listening to musical sound will in most cases (or always) produce an holistic image with various kinds of emergent qualities, not just the mentioned qualities of contour, texture and timbre, but equally well tonality, global harmonic quality, or what M. Leman has denoted *tone semantics* (Leman, 1995), as no group of tones within a certain time span will be innocent of such emergent qualities. We could in a sense say that there is an *emergent quality superiority effect* in musical listening, similar to the documented *word superiority effect* in reading, meaning that words are recognized faster than single letters (Haberlandt, 1994).

It seems that neural networks can simulate some aspects of this process of forming cumulative memory images of musical objects, demonstrating how certain profiles of tone centers and tone semantics can emerge from the internal context of musical objects. However, I believe there is still a challenge to develop more explicit shape representations for various modal qualities in order to capture the audibly distinctive qualities of otherwise quite close variants, such as the mentioned variants of the Chopin prelude (sound example #4 in the Appendix). I am not sure of how such modal variants could be represented by musical schema theory, as I see this as not really a question of tone center images (the tone centers could be considered more or less the same in all the variants) but rather as a question variant modal *flavors*, to use a metaphor of Persichetti (1962). This would be analogous to timbral variants of a chord in some kind of instrumental setting (orchestra, string quartet, etc.) where the chord could be judged as *the same* but where there would be changes in the spectral envelope between the variants. In such a case, we would speak of a series of variants shapes of the spectrum. I would guess that in the case of the modal variants of the Chopin prelude, we could then by analogy speak of a similar series of variant shapes of the *spectrum*, here understood as the cumulative retrospective image of intervallic qualities over a given time window, hence transforming Persichettis *flavors* to shapes.

This remains of course to be investigated further, but it is interesting to see that an *holistic* approach to intervallic qualities in particular and to musical imagery in general was in fact proposed at the very beginning of Gestalt theory by von Ehrenfels (1890/1988, in particular pp.91-92). It is quite clear from this article by von Ehrenfels that the effect of context was crucial for all music cognition, and that the cumulative retrospective images of musical objects with their global emergent qualities was in fact the point of departure for many ideas in early Gestalt theory.

In this paper, I have tried to give a sketch of how the *shape-paradigm* could be a practical and theoretically reasonably well founded strategy for representing knowledge about emergent qualities. The epistemological foundation for this is the phenomenological idea of musical objects seen *at a glance* as a *retentional image* in memory. Although the relationship between the continuous flux of musical sound and the more stable retentional images of this flux may be seen as profoundly enigmatic, I have tried to argue that it should be possible to conserve the dynamical nature of musical sound in shape representations, provided that we conserve the various features as trajectories. This goes equally well for the sound-producing actions (moving hands, fingers, arms, feet, etc.) as for the audio-acoustic flux. In this sense I see no difference between retentional images of musical sound and retentional images of other events (such as walking, running, dancing, swimming, etc.).

With the two mentioned poles of this shape-paradigm, i.e. analogical representations and signal based representations, it is of course my hope that signal based representations will develop to much higher levels of sophistication in the coming years, whereas the analogical representations will remain more a matter of diligent observation and structuring of the pertinent features of musical objects, a process which as was in fact initiated with the work of P. Schaeffer. Also, we should work towards means for *dynamic* and *multi-modal* shape-representations in music theory, allowing *panning* between different vantage points, *zooming* between different levels of resolution, as well as *rotating* between different emergent qualities of musical objects. For the moment however, I think there is a convergence of material from several domains of thought which supports a working hypothesis of the privileged role of shapes in music cognition in general and in musical imagery in particular.

Appendix

Sound Examples (CD-tracks 1-4)

- Sound Example #1: A scale of 7 variants of a synthesized bass like sound, gradually increasing its *mass* (overall harmonic content) and *grain* (textural fluctuation) density. This sound was produced by modal synthesis of a bowed string, i.e. by physical modelling, by progressively adding vibrating modes in the model using the Modalys software from IRCAM.

- Sound Example #2: The fourth variant sound of the previous example is here first presented unmodified, then with the application of a stationary and lastly a fluctuating formantic filtering, using the formantic envelope of a trombone wah-wah mute as the filter. This illustrates the projection of formantic shape on to a sound as well as the *allure* (slower fluctuation) of a sound in the last sound of this example. The signal processing was done with the Audiosculpt software from IRCAM.
- Sound Example #3: A scale of 7 variant attacks on the same sound as in the previous example, proceeding from gradual to steep and illustrating the effect of the attack shape on the quality of the sound.
- Sound Example #4: A series of variants of the C-major prelude by Chopin, starting out with the original, then continuing with variants in minor, phrygian, lydian and finally again in C-major but octave compressed by a modulo 12 operation. This allows for a comparison of the effect of changing the modality of a musical object while keeping the other features constant, as well as the effect of retaining the original modality but deforming the spatial layout in the last variant. The performance is by a MIDI sequencer playing a sampled piano.

References

Barthes, R. (1977). *Image-music-text.* Glasgow: Fontana/Collins.

Black, M. (1962). *Models and metaphors.* Ithaca, London: Cornell University Press.

Bregman, A. (1990). *Auditory scene analysis: The perceptual organization of sound.* Cambridge, MA: The MIT Press.

Chion, M. (1983). *Guide des objets sonores.* Paris: Editions Buchet/Chastel.

Cogan, R. (1984). *New images of musical sound.* Cambridge, MA: Harvard University Press.

Damasio, A. (1989). Time-locked multiregional retroactivation: A systems-level proposal for the neural substrates of recall and recognition. *Cognition, 33*, 25–62.

de Man, P. (1978). The epistemology of metaphor. *Critical Inquiry, 5*, 13–20.

Denis, M. (1989). *Image et cognition.* Paris: Presses Universitaires de France.

Deutsch, D. (1984). Musical space. In W. Crozier & A. Chapman (Eds.), *Cognitive processes in the perception of art.* Amsterdam: Elsevier Science Publishers.

Finke, R. A. (1989). *Principles of mental imagery.* Cambridge, MA: The MIT Press.

Godøy, R. (1993). *Formalization and epistemology.* Oslo: Det historisk-filosofiske fakultet.

Haberlandt, K. (1994). *Cognitive psychology.* Needham Heights, MA: Allyn and Bacon.

Harnad, S. (Ed.). (1987). *Categorical perception.* Cambridge: Cambridge University Press.

Holyoak, K., & Thagard, P. (1995). *Mental leaps: Analogy in creative thought.* Cambridge, MA: The MIT Press.

Husserl, E. (1980). *Vorlesungen zur Phänomenologie des inneren Zeitbewusstseins.* Tübingen: Max Nimeyer Verlag.

Johnson, M. (1987). *The body in the mind.* Chicago, IL: The University of Chicago Press.

Koffka, K. (1963). *Principles of Gestalt psychology.* New York, NY: Harcourt, Brace, and World.

Köhler, W. (1947). *Gestalt psychology.* New York, NY: Liveright.

Kosslyn, S. (1994). *Image and brain.* Cambridge, MA: The MIT Press.

Lakoff, G. (1987). *Women, fire and dangerous things. What categories reveal about the mind.* Chicago, IL: The University of Chicago Press.

Leman, M. (1995). *Music and schema theory: Cognitive foundations of systematic musicology.* Berlin, Heidelberg: Springer-Verlag.

McAdams, S. (1984). The auditory image: A metaphor for musical and psychological research on auditory organization. In W. Crozier & A. Chapman (Eds.), *Cognitive processes in the perception of art.* Amsterdam: Elsevier Science Publishers.

Miller, I. (1982). Husserl's account of our temporal awareness. In H. Dreyfus (Ed.), *Husserl, intentionality, and cognitive science.* Cambridge, MA: The MIT Press.

Moore, F. (1990). *Elements of computer music.* Englewood Cliffs, NJ: Prentice Hall.

Persichetti, V. (1962). *Twentieth century harmony.* London: Faber and Faber.

Petitot, J. (1985a). *Les catastrophes de la parole de Roman Jacobson á René Thom.* Paris: Maline.

Petitot, J. (1985b). *Morphogenèse du sens I.* Paris: Presses Universitaires de France.

Reisberg, D. (Ed.). (1992). *Auditory imagery.* Hillsdale, NJ: Lawrence Erlbaum Associates.

Ricoeur, P. (1981). *Hermeneutics & the human sciences.* Cambridge, Paris: Cambridge University Press, Editions de la Maison des Sciences de l'Homme.

Risset, J. (1991). Timbre analysis by synthesis: Representations, imitations and variants for musical composition. In G. De Poli, A. Piccialli, & C. Roads (Eds.), *Representations of musical signals.* Cambridge, MA: The MIT Press.

Schaeffer, P. (1966). *Traité des objets musicaux: Essai interdisciplines.* Paris: Editions du Seuil.

Stevens, K. (1972). The quantal nature of speech: Evidence from articulatory-acoustic data. In E. David & P. Denes (Eds.), *Human communication: A unified view.* New York, NY: McGraw-Hill.

Stumpf, C. (1883/90). *Tonpsychologie.* Leipzig: Hirzel. (two volumes)

Sudnow, D. (1978). *Ways of the hand.* Cambridge, MA: Harvard University Press.

Sweetser, E. (1990). *From etymology to pragmatics.* Cambridge: Cambridge University Press.

Thom, R. (1983). *Paraboles et catastrophes.* Paris: Flammarion.

Tufte, E. (1983). *The visual display of quantitative information.* Cheshire, CT: Graphics Press.

Tufte, E. (1990). *Envisioning information.* Cheshire, CT: Graphics Press.

von Ehrenfels, C. (1988). On "Gestalt Qualities". In B. Smith (Ed.), *Foundations of Gestalt theory.* Munich, Vienna: Philosophia Verlag. (Original work published 1890)

Wertheimer, M. (1967). Laws of organization in perceptual forms. In W. Ellis (Ed.), *A source book of Gestalt psychology.* London: Routledge & Kegan Paul.

Wessel, D. (1985). Timbre space as a musical control structure. In C. Roads & J. Strawn (Eds.), *Foundations of computer music.* Cambridge, MA: The MIT Press.

Statistical Gestalts – Perceptible Features in Serial Music

Elena Ungeheuer

Robert-Schumann-Hochschule für Musik, Fisherstrasse 110, D-40476 Düsseldorf, Germany

Abstract. Serial composing in the electronic studio opened new fields of musical perception. In this regard the discovery of statistics plays an important role. Parallels can be seen with the development in physics, where statistics helped to integrate the probabilistic behavior of quants into the causalistic concept of the physical world. Over and above that, Stockhausen's serial masterpiece *Gesang der Jünglinge* shows how serial composing of mass phenomena was orientated towards the creation of dynamic Gestalts on a statistical basis, being referential figures for the composer as well as for the listener.

1 Introduction

In the middle of our century, after the cultural disaster of the second world war, composers of avantgarde music got interested in the mechanisms and non-linearities of auditory perception. There are several reasons for that:

- First of all, there was an artistic orientation towards aesthetical renewal in which composers wanted to be deliberated from outweared and abused clichés of musical expression. They created a new music and a new way of listening to music.
- A second aspect is that the composers who searched for new musical solutions in the electronical studios were confronted with problems of auditive perception in many respects. In order to synthesize sounds it was important to analyse sounds as well as to analyse the conditions of perceiving sounds both physiologically and psychologically. A simple additive synthesis of arbitrary sine waves indeed does not fuse necessarily into one unified whole. Depending on specific dynamical repartitions, frequency ratios near to the overtone scale and other particularities, such as layerings of sine waves, are interpreted as polyphonic chords. In this context, P. Boulez objected to K. Stockhausen's *Studie I* in 1953 that it contained much more chords than new sounds (Boulez, 1966, p.196). H. Pousseur tried to solve the same problem in his first electronic piece *Seismogramme* with the help of individual envelopes which he designed for every sine wave before overlaying them in order to build up complex sounds (Pousseur, 1955, p.44).

– A third reason is that composers working in the electronical studio were interested in technical as well as in psychoacoustical advises. At the studios and festivals of avantgarde music there were always scientists visiting. One of the famous acousticians and phoneticians of that time was W.Meyer-Eppler, teaching at Bonn University (Ungeheuer, 1992), whose actual researches in the acoustical domain influenced musical composing. Owing to this interaction with scientists, composers were encouraged to widen simplistic physical and acoustical concepts.

2 Serial Composing and Statistics

Statistical phenomena entered the world of serial composition through the use of noise, that is to say after the first electronic compositions have been built up by superimposing sine waves, in particular in the works produced in 1953-54. The use of electronically generated noise as a category of sound had strong musical consequences. One is related to the premise of *serialism* which, at the time, demanded a correspondence between the characteristics of the material and the rules of form-building, between micro- and macro-organization (Stockhausen, 1963, p.141). The kinds of the noise used can be examined by studying how noise was handled at that time. First of all, noise was considered as a complex phenomenon which could be treated by filtering. By using filters, third or octave filters, precisely determined parts could be cut out of white noise, and a colored noise could be obtained. The apparatus of telecommunication engineering, with which the studios for electronic music were equipped, allowed noise to be defined through its bandwidth specification. Interestingly, in Stockhausen's *Gesang der Jünglinge*, bandwidth and filtering treatment were also applied to musical figures consisting out of sine waves, impulses and vocal sounds (Decroupet & Ungeheuer, 1993).

The inner structure of noise can be characterized as being irregular. In the serial view, noise is situated in opposition to a sine wave, firstly because a sine wave has the most regular Gestalt in the acoustical domain. Secondly with regard to the serial scale *single Gestalt* and *mass phenomenon*, because a sine wave is a single frequency – noise is always formed out of a mass of frequencies.

Regularity is a question of order and, following the definitions of information theory, order is a state lacking indetermination. However, determined forms in nature are less probable than indeterminate ones, nevertheless humans are constantly engaged in forming ordered and determined Gestalts out of the mass of non-specific phenomena. The theorem of entropy states that all natural phenomena develop towards an increasing chaotic situation, distributing all particles equally in the space. The most balanced situation is the situation of the biggest chaos and this situation has the highest degree of probability (Wiener, 1964). As such, noise can be considered as a chaotic phenomenon. Electronically generated noise does not allow the composer to control all partial frequencies. In order to reach an equal distribution – not only on the level of frequencies but in all kinds of sonic dimensions – and nevertheless controlling them in detail, Stockhausen invented a method of row permutation, which he called "statistische Reihenpermutation" (Stockhausen, 1963, p.54). The given row of departure was permuted following a principle that avoided the repetition of yet existing configurations.

Statistical tools were fascinating composers at the electronic studios of the fifties in an exceptional manner. Nevertheless the relationships between statistics and serial composing have thus far not yet been sufficiently investigated. Asking for the role of statistics in compositional processes, we are immediately pointed to the works of I. Xenakis who is known to have used stochastical formulas in order to generate mass phenomena of sounds. But Xenakis' interest in problems of indetermination went far beyond that. Heisenberg's "Unbestimmtheitsgleichung" was one of the most suggestive starting points for his musical thinking (Hoffmann, 1994, p.87).

Xenakis studied carefully those developments in physical thinking, having in mind that the extension of our concept of the world through the principle of indetermination – even if it is limited to the micro-dimension – has to take into account the validity of given rules. Without rules there is no way of doing scientific research and without rules there is no way of generating music. He also experienced that statistics can help. Nature follows a statistical causality when we accept that the determinative principle of prediction is no longer related to single events (Hoffmann, 1994, p.88), but to the statistics of many events.

3 From Statistics to Gestalts in Electronic Serial Music

The serial pole-pair *single Gestalt* vs. *mass phenomenon* is particularly suited to be exploited in dialectical movements. Working with electronical techniques both phenomena can mutually be exchanged through the simple means of changing the playing speed of the tape recorder. Playing a sequence of tones in a slow speed produces the impression of single sound events, of Gestalts, whereas the fast playing of the sequence may fuse the sound events together. Changing the playing speed of the tape recorder can be conceived of as the auditory equivalent of changing the distance of looking in the visual level. Mass phenomena can thus be analysed by changing the scope from near to far. In order to determine the contours of the mass as a whole, its bandwidths, its average and its tendency to developing, the researcher takes a distance from the mass phenomenon that allows him to observe it as a Gestalt.

How can we account for the fact that *Gesang der Jünglinge*, a classical example of serial composition in the electronic domain, displays such an abundance of distinguishable Gestalts? Is the scene somehow staged in the foreground, regardless of the *serial framework*, or does serial construction suddenly become tangible?

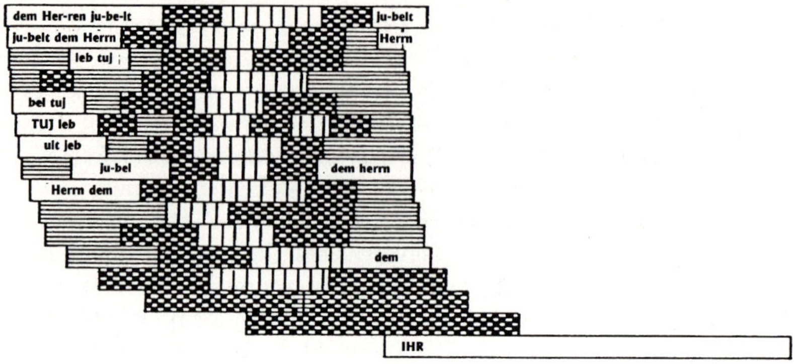

Fig. 1. Symmetrical repartition of timbres in sound complex no. 5 from the beginning of part F in Stockhausen's *Gesang der Jünglinge*. The picture shows vocal sounds with syllables, horizontal lines standing for sine waves, squares standing for impulses, vertical lines standing for noise. The sketch transcription is taken from Stockhausen's edition of the sketches of *Gesang der Jünglinge*, Kürten: Stockhausen-Verlag, 1983, p. III F/2 und III F/25 (pagination of all sketches follows the catalogue by Decroupet, 1993)

Obviously, it is not the multitude of sonic interrelationships that characterizes the serial music of the fifties and sixties as opposed to non-serial music, rather it is the expansion of the actual categories of musical relationships, a development, observed from one piece to the next one, towards new and previously unknown dimensions. This is already inherent in the basic serial idea, which embodies a gradual connection – a reconciliation – of extremes: a concept extendable both horizontally and vertically. That is why after the so-called period of *punctualism*, the first stage of serial music in which multi-layered definitions of the parameters yielded a succession of *punctual* events, Gestalts made their appearance. In a well-known comment on his *Klavierstück I* (Stockhausen, 1963, p.63,64), originally conceived for broadcasting, Stockhausen focused his listeners' attention on the specific directions in which intervals and pitch relations of larger groups of sounds evolved: rising-falling, rising-rising, falling-rising-falling, etc. The relations between adjacent punctual events yielded groups recognizable as Gestalts.

After a first stage characterized by the organization of static sounds such as in Stockhausen's *Studie I* and K. Goeyvaerts' *Komposition Nummer 5*, composers became increasingly interested in the transitions between sonic phenomena. This necessarily involved the organization of dynamic sounds. In his comments on serial music G.-M. Koenig emphasizes the importance of a principal "aversion to unresolved contrasts" (Koenig, 1991, p.292) to account for the emergence of the idea of a timbre continuum in which the three types of timbres produced by different kinds of generators – sine wave, impulse, noise – were auditorily related.

Auditory changes in timbre were hard to achieve in the fifties by the cumbersome technique of cutting and pasting tapes, because each sound particle had to be realized separately before being integrated into one of the layers to be superimposed and finally synchronized in a complex result. On the other hand, this method gave composers maximum control on the detail level. Stockhausen opted for this procedure when he started with the realization of *Gesang der Jünglinge*. One result of the repartition of timbres was based on the Gestalt shown in Fig.1. In other parameters, too, this complex sound exhibits an evolutionary form characterized by a double movement of divergence and convergence, e.g. in the organization of the pitches.

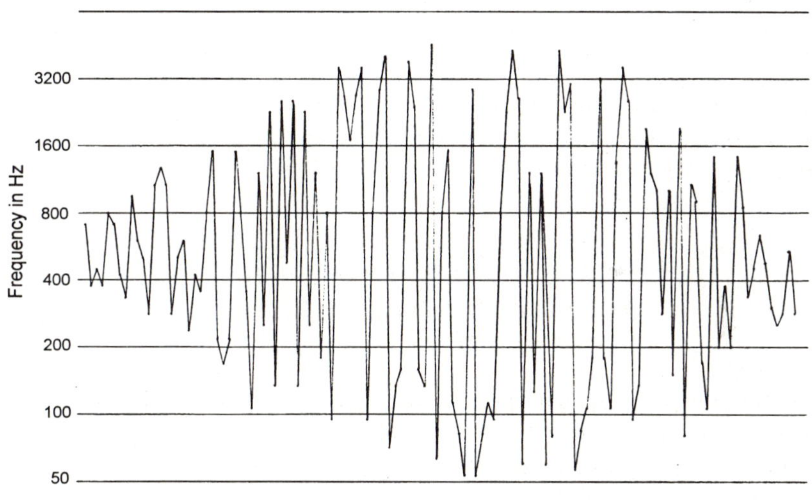

Fig. 2. Frequency evolution of sound complex no. 5 in Stockhausen, *Gesang der Jünglinge*, x-axis: time, y-axis: frequency in Hz; sketch transcription *Gesang der Jünglinge* (Decroupet, 1993, p.III F/3)

At first these Gestalts were only accessible to the composer – and also, perhaps, to the assistant or technician who helped to realize them. They emerged against a background represented by the totality of possible sound values, and in making use of simple geometrical forms (ellipse, straight line, parabola, etc.) they obeyed such classical rules of Gestalt-forming as proximity, identity and a tendency towards *good Gestalt*, as demonstrated by Wertheimer, Köhler and Koffka for visual objects [1].

The passage discussed here represents the first 30 seconds of the final section of *Gesang der Jünglinge* (beginning at 8'39") and it consists exclusively of mosaic sounds played in succession or partly overlapping. These complex sounds either emerge from silence or provide a context in which the foreground-background relationships change in accordance with the listener's focus of attention. The context, which exceeds the sum of individual sensations always results from the concentration of sounds generated. In that concentration, each mosaic sound acts independently as a condensed smallest particle displaying a multi-layered change of texture.

4 The Audibility of Statistical Gestalts

The basic thesis of Gestalt theory, i.e. that configurations of sensations are perceived primarily and their details in the second place, is often achieved in a work of art by composing a mass structure:

> "The extremely short average distances between sounds entries (and also their durations) indicate the enormous density of this *statistical* structure, in which individual sound events merge into massive complexes. " (Stockhausen, 1983, p.V/4) [2]

Producing a composition by working with cutting and pasting lengths of tape and using recording machines capable of operating at various speeds, focused the composer's interest on the change between these aural perspectives:

> "The transference of results from communication theory and experimental phonetics, and the new considerations of nature this prompted, enabled me to discover compositions which I classify as *statistical forms* [...] In statistical forms I employed groups and points in an attempt to mediate between collectives organized in accordance with the rules of large numbers. The problem is that in certain circumstances the same elements act as collectives (statistically defined massive complexes), while in others they are perceptible as groups and as points."(Stockhausen, 1963, pp.229,235).

For *Gesang der Jünglinge*, the category of statistical Gestalts classifies those complexes which display a clearly perceptible contour but whose *filling* cannot be analysed on hearing because of the vast number of sound particles.

[1] In the terms of Gestalt theory, a Gestalt is characterized by its "Übersummenhaftigkeit" (a Gestalt is more than its sum) (von Ehrenfels, 1890/1960, p.19).

[2] Stockhausen's comment on the beginning of the first *rhythmic inset*, where the durations tend – as in the mosaic sounds – towards the limit value of 1/20 of a second in Stockhausen (1983, p.V/4).

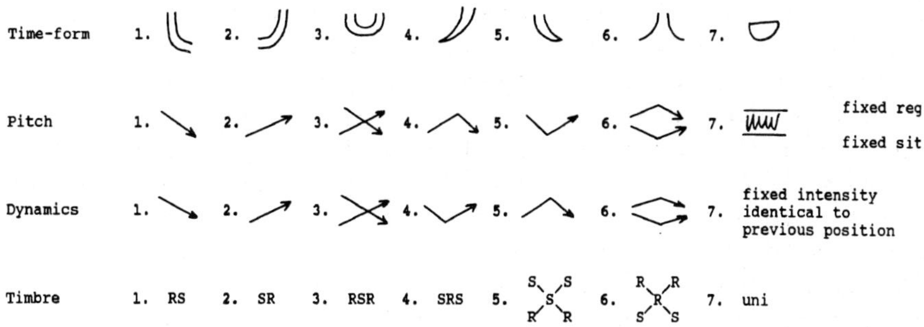

Fig. 3. Overview of evolutionary forms for the sound complexes at the beginning of part F in Stockhausen, *Gesang der Jünglinge* valable for the sound dimensions time-form (part cut out of the basic rhythmical structure), pitch, dynamics and timbre; sketch transcription *Gesang der Jünglinge* (Stockhausen, 1983, p.I/5)

On the detail level, the degree of determination by the composer may vary; while every single element of the mosaic sounds is predetermined in every dimension, the inner events of the multi-layered vocal and impulse complexes of sections A and B were realized by approximation. In the case of the impulse complexes, Stockhausen and two assistants modulated each layer separately by interpreting zigzag lines drawn by the composer directly. This was done by setting the various machines to influence the duration of the impulses, their pitch (by selecting filter positions and varying these settings during the process) and their intensity.

> "We agreed to meet at a certain point, regardless of how we would get there (because what happened in the meantime could only be notated statistically); as a result, a form layer is constituted through a process – a process which the man who taught me information theory, Meyer-Eppler, called *aleatoric*". (Stockhausen, 1989, p.337sq.)[3]

The choirs in the two first sections of *Gesang der Jünglinge* were realized in a similarly approximate fashion: the boy, whose unmodulated voice was the source of all the vocal sounds in the composition, could only approximately comply with Stockhausen's instructions (which were very precise, at least in stipulating that all the layers of one and the same complex should be of the same length). Confronted with the results, Stockhausen adopted a more pragmatic attitude, synchronizing the different layers the best he could, all of which having different total durations in the end.

How, though, did the composer devise the specific treatment of these Gestalts? An early compositional sketch related to the first mosaic sounds

[3] "Wille zur Form und Wille zum Abenteuer", Interview of R. Frisius with K. Stockhausen, 8.1.1978.

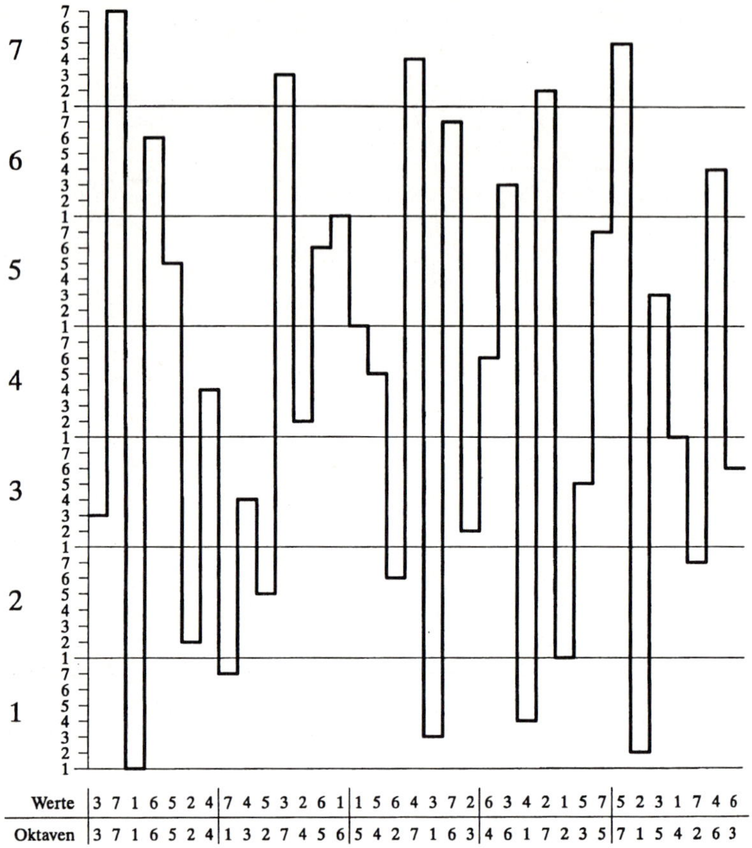

Fig. 4. Draft of a basic sequence of durations in Stockhausens *Gesang der Jünglinge*. The x-axis represents time, the y-axis shows values of basic durations. Sketch transcription from *Gesang der Jünglinge* (Stockhausen, 1983, p.I/1)

(Fig. 3) shows a preliminary graphic notation for the timbre evolution, already revealing the entire range of *evolutionary forms* ("Bewegungsformen"), as Stockhausen called moving statistical phenomena since 1954 [4].

For *Gesang der Jünglinge*, Stockhausen developed evolutionary forms as a compositional tool that enabled him to mediate between processes in each sound dimension as well as the parameters of sound modulation (i.e. the realization mode). These evolutionary forms are notated as zigzags and may be applied to both the micro and the macro level, relating to a single sound or to the whole sound complexes. They are perception-oriented Gestalts: they describe the progressive change from one value to another in one parameter,

[4] "Von Webern zu Debussy. Bemerkungen zur statistischen Form", (Stockhausen, 1963, p.77).

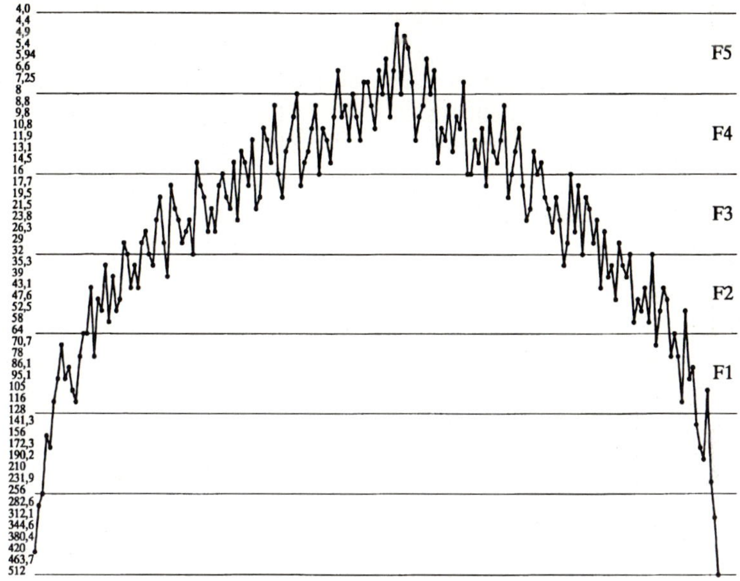

Fig. 5. Statistical Gestalts: evolutionary forms of the basic durations in Stockhausens *Gesang der Jünglinge*. The x-axis represents time, the y-axis shows durations in centimeters. Sketch transcription from *Gesang der Jünglinge* (Stockhausen, 1983, p.III F/14-15)

the ear having enough time to appreciate the initial situation and consciously follow the change of state that occurs in this dimension.

Evolutionary forms are certainly involved in the composition of mass phenomena. Such phenomena can be observed from a distance and described as a moving Gestalt by applying statistical methods. A comparison of two early drafts, showing the distribution of the basic durations at the initial beginning of the piece, shows how in the development of the composition, Stockhausen moved away from the idea of *equal* distribution. The latter was obtained by means of statistical permutation in order to reach an overall tendency which includes equal distribution of the brief fragments inside of the overall form. The basic duration is established by the entry delay between two successive sounds regardless of either real duration, which depends on further definition. The first draft looks like the one in Fig. 4. The scale of duration values (on the right) is based on octave ratios and also on an equal subdivision into seven steps per octave. The permutations of the row (below) indicate that the time values are selected on the basis of equally distributed octaves and steps per octave. The result is a constant, rapid alternation of long and short durations from event to event. The pre-programmed statistical equal distribution gives the ear no chance to adapt to a sounds true duration. Bounced back and forth between short sounds and sound complexes composed of a multitude of short particles, the aural impression becomes diffuse and qualitative differences are reconciled.

Stockhausen duly decided on the distribution of durations, characterized by a distinctly oriented movement (from long to short and back again to long events), the values always distributed equally over the respective octaves. Perception is, if you like, *taken by the hand* – guided towards the various limits of its qualitative levels, as shown in Fig. 5.

5 Conclusion

Serial music of the fifties found one of its bases in Schoenbeg's twelve-tone technique, generalized for all musical dimensions. The period of *punctualism*, which is partly characterized by non-oriented distributions of the sound-events and the resulting impression of continual dispersal of the sonorous information, was transgressed as early as spring 1952, when Boulez, followed by Stockhausen, began to compose in different ways with grouping devices. The result, both at a technical as well as perceptual level, was the emergence of musical sense-units constituted by a variable number of sounds related to one another by common features (e.g. same register, same dynamics, and so on).

As a consequence of his first experiments in electronic music, Stockhausen designed for *Gesang der Jünglinge*, parallel to the evolution in time acoustically inherent to each sound, the idea of evolutionary forms in all parameters. These evolutionary forms led to the conception of immediately perceptible musical Gestalts, which were from the point of view of the composer controlled either step by step through precise value-changes in a given parameter or on the basis of their global behavior (the detailed realization being left to approximate interpretations of graphic prescriptions). For the listener, the result could in both cases be appreciated only qualitatively, that is to say statistically, following the global shape of the generated Gestalts.

References

Boulez, P. (1966). Auprès et au loin. In *Relevés d'apprenti.* Paris: Seuil.

Decroupet, P. (Ed.). (1993). *Sammlung Karlheinz Stockhausen, Musikmanuskripte.* Basel: Paul-Sacher-Stiftung.

Decroupet, P., & Ungeheuer, E. (1993). Son pur – bruit – médiations: Matières, matériaux et formes dans "Gesang der Jünglinge" de Karlheinz Stockhausen. In *Genesis. Revue internationale de critique génétique* (Vol. 4). Paris: Jean Michel Place Archivos.

Hoffmann, P. (1994). *Amalgam aus Kunst und Wissenschaft. Naturwissenschaftliches Denken im Werk von Iannis Xenakis.* Frankfurt/M.: P. Lang.

Koenig, G. (1991). *Ästhetische Praxis. Texte zur Musik* (Vol. 1). Saarbrücken: Pfau.

Pousseur, H. (1955). Strukturen des Baustoffs. In *Die Reihe 1.* Vienna: Universal Edition.

Stockhausen, K. (1963). *Texte 1.* Cologne: Dumont.

Stockhausen, K. (1983). *Gesang der Jünglinge.* Stockhausen edition.

Stockhausen, K. (1989). *Texte 6.* Cologne: Dumont.

Ungeheuer, E. (1992). *Wie die elektronische Musik "erfunden" wurde... Quellenstudie zu Werner Meyer-Epplers Entwurf zwischen 1949 und 1953.* Mainz: Schott.

von Ehrenfels, C. (1960). Über "Gestaltqualitäten". In *Gestalthaftes Sehen. Ergebnisse und Aufgaben der Morphologie (Zum hundertjährigen Geburttag von Christian von Ehrenfels.* Darmstadt: Wissenschaftliche Buchgesellschaft. (Original work published 1890)

Wiener, N. (1964). *Mensch und Menschmaschine.* Frankfurt/M.: Athenäum.

II

From Pitch to Harmony

"Verschmelzung", Tonal Fusion, and Consonance: Carl Stumpf Revisited

Albrecht Schneider

Institute of Musicology, University of Hamburg, Neue Rabenstr. 13,
D-20354 Hamburg, Germany

Abstract. This article discusses C. Stumpf's concept of "Verschmelzung", as part of his philosophical and psychological theories including consonance since these are not generally known today. In particular, it is shown how Stumpf had elaborated his concept in certain respects from the *Tonpsychologie* (Stumpf, 1883, 1890) to his final writings. To allow a more adequate understanding of Stumpf, also his ideas on *concordance* and *discordance* are briefly summarized. Further, it is argued that "Verschmelzung" is related to, yet not identical with modern views of *tonal fusion*. In connection with Stumpf's criteria for integral hearing ("einheitliches Hören"), issues and experimental findings on tonal fusion are surveyed, and are interpreted also with respect to neurophysiological mechanisms that are possibly relevant to perception of fusion. It can be hypothesized that fusion – as being prerequisite to the experience of consonance – is a *natural* rather than a *cultural* phenomenon. Finally, Stumpf's assertion according to which the sensational attribute of *roughness*, and the perceptual quality of *dissonance*, can be separated from each other, is reexamined on the basis of experimental data.

1 Introduction

The theory of consonance played a central role in writings on psychology of music since H. v.Helmholtz (1870) had outlined a concept that was based on, in the main, acoustical and physiological observations. Different from this approach, C. Stumpf (1883, 1890, 1898, 1911) developed his own view which emphasizes the notion of "Verschmelzung". According to Stumpf (1890, 64ff., 127ff.), stimuli that are suited to bring about tonal fusion, can also explain the perceptual quality of consonance.

In certain respects, consonance and dissonance are taken as categories relevant to apperception rather than to mere sensation of stimuli. Apperception – a notion derived from the medieval Latin word "adpercipere" – is found in philosophical works on epistemology from Descartes, Leibniz and Kant, to modern psychological treatises on perception and understanding provided by Wundt, v.Helmholtz, Brentano, Stumpf, and Husserl. It denotes, first of all, a principle that defines conditions prerequisite to allow coherent human experience by way of continuous synthesis (cf. Kant's *Critique of Pure Reason*, B 131ff.). Further, apperception has been regarded as a specific mode of *inner* or mental perception that, different from *normal* perception, requires vigorous attention as well as intentional reference towards objects and circumstances

(Brentano, 1924/28; Wundt, 1896/1928, 244ff.). Whatever is perceived in this way, also will be evaluated and categorized in terms of previous experience; this process of "valde clare et distincte percipere"–as Descartes and Leibniz had put it–will lead to full apperception of phenomena of which we gain conscious awareness.

2 Historical and Theoretical Background

2.1 Stumpf's Concept of "Verschmelzung"

As is well-known, C. Stumpf as a philosopher and psychologist devoted much of his work to problems in perception. Though he also had studied vision, it was research on music perception and apperception, condensed in his book *Tonpsychologie*, and in many other seminal publications, that was acknowledged as original, apt, and stimulating with respect to cognitive foundations of musicology. In a number of ways, Stumpf can be related to phenomenology, on the one hand, and to Gestalt psychology, on the other. Not only was he the teacher of E. Husserl, W. Köhler and K. Koffka, he also was in close contact with M. Wertheimer and collaborated with E. von Hornbostel. Though Stumpf's ideas and considerations are close to what became Gestalt psychology after 1910 (Boring, 1950, Chaps.18, 23; Pongratz, 1967, p.301ff), he remained skeptical of several of the Gestaltist's propositions (e.g., the idea of psychophysical isomorphism as was advanced by W. Köhler), mainly for epistemic reasons (Stumpf, 1939, §15).

In his *Tonpsychologie* as well as in other publications, Stumpf discussed the notion of "Verschmelzung" that, however, was not always correctly understood by his contemporaries (Stumpf, 1907, p.22f). Since Stumpf's *Tonpsychologie* was written very much from a philosophical point of view (as was the "Psychologie vom empirischen Standpunkt", elaborated by his teacher and friend, F. Brentano (1924/28)), the concept of "Verschmelzung" has to be understood as being a part of his epistemological theory, in the first place, though it is of course related to experience and psychological observation. Stumpf's *Tonpsychologie* – a book that, at a first glance, is devoted to technical details of music perception – in many chapters addresses fundamental philosophical and psychological issues, namely: how are coherent objects constituted, in our perception and thinking? What are the relations between parts and wholes? How is synthesis of (earlier, recent, actual) experience achieved in apperception? What are the cognitive functions that make such apperception possible? (Stumpf, 1939, §§11-24).

To recapitulate the essentials of Stumpf's concept: "Verschmelzung" comprises two aspects in that it refers (i) to a structural property already inherent in certain given formations ("Gebilde"), and (ii) to a specific cognitive function. Basically, it can be addressed as one of four types of relations established, cognitively, between objects, elements, and features that are perceived.

These types are labelled, in Stumpf's (1883, p.96ff) terminology, (1) *plurality* ("Mehrheit"), (2) *comparison* ("Steigerung"); though Stumpf speaks of "Vergleichung", the idea is that by comparison one recognizes degrees in quality or intensity inherent in phenomena that are perceived, thus, a term that perhaps expresses this idea better would be: *comparative*, (3) *similarity* ("Ähnlichkeit"), (4) *fusion* ("Verschmelzung"). Obviously, these four types by themselves make up an ordinal scale in which plurality indicates weak relations between classes of objects, percepts, etc., while perfect fusion constitutes structural identity. The notion of "Verschmelzung" thus can be illustrated, as in fact it was, by geometrical configurations. Husserl (1976, p.77) argued, in this respect, that perfectly congruent shapes are identical so as "to repeat each other".

Formations (shapes, chords, etc.) that are prone to *triggering*, in subjects, the perceptual process of "Verschmelzung" have to be conceived of as being non-additive yet rather topological constructs so as to allow certain transformations to retain features as being invariant. These make up configurations that are of relevance as a "Verhältnisganzes" (Stumpf, 1873, p.106ff.; Stumpf, 1883, 1890, 1939, pp.229ff; Husserl, 1913/1980, part III), that is, as a highly integrated, *complex whole* (K. Koffka) that by attentive subjects is identified by the relations between parts rather than by the parts itself. For example, in music, a specific chord is identified by the intervals between notes rather than by the notes.

Though the term "Verschmelzung" can be literally translated by *melting*, what is meant by Stumpf is *not* that this process will bring about coalescence, in the sense that, for example, several notes played simultaneously, will become, in perception, one, thereby loosing their identity or quality. Of course, configurations can be so prominent (compared to recognition of single parts), and so tight, that analysis of the whole into parts for subjects is more difficult. This aspect indeed has been stressed by Stumpf especially in his *Tonpsychologie* (1890, p.64f., p.128). In that configurations (e.g. complex chords such as the well-known *Tristan*-chord) are recognized instantly, in one act of perceiving, the parts of such a configuration are of relevance as contributing to a "Empfindungsganzes" (the coherent percept of a *Gestalt* that is based on certain relations of parts rather than on a mere sum of such parts).

However, Stumpf did *not* mean that there is no mental act of discrimination with respect to such formations (Stumpf, 1907, p.23). Quite the contrary, for the expert listener, any set of notes (pairs, triples, more complex chords) that are presented simultaneously, will be conceived of as forming a certain spatial arrangement, and thus, as a more or less distinct configuration. Since the relations between the elements are of prime importance, such a configuration can become a "Tonganzes", a *sonority* in which the elements or parts are fused into the whole though they still can be distinguished, and identified, by the attentive listener.

Notes being parts of a "Tonganzes" tend to fuse to various degrees, and this holds true even with respect to harsh dissonances. Fusion thereby is not totally absent from dissonant intervals; rather, these are perceived to fuse only to a low degree, as "Verschmelzung" can be scaled (Stumpf, 1890, p.135ff; Stumpf, 1919, p.110). Accordingly, *complete dissonance* thus is seen as the *lowest degree of fusion* (Stumpf, 1939, p.261). Or, put in another way, pairs or triples of simultaneously given dissonant notes will be recognized to belong to the pole of plurality much more than to that of fusion (Stumpf, 1883, p.101). What has been said so far can be summarized in a scheme as follows:

overall perceptual relation	A	B	C	D
between n objects, features:	plurality →	comparative →	similarity →	fusion

While plurality points to formations that can be mentally analyzed as comprising of a number of elements that are well separable, the comparative relation basically constitutes an ordinal scale [<, >] for judgements: of two simultaneous or successive notes, one typically can be judged to be higher than or louder than the other. As to judgements of similarity, these can be based on recognition of parts or of relations between parts being similar. Similarity of relations has been defined, since antiquity, as *analogy*. Also, judgements can consider similarity of basic qualities, or other criteria. Finally, extreme similarity could lead to congruence and thus, to fusion. With respect to perceptual analysis and apperception, the main theme in the *Tonpsychologie*, however, fusion is different from the three other types in that even the sharpest analysis of, for example, triads, results in the plain fact that *three* notes constitute but *one* chord. It was this specific case, in which parts can be perceived as parts and at the same time as a *whole* (that itself is a configuration made up of parts, and of the relations between them, to result in one coherent "Verhältnisganzes") that Stumpf (1890, p.67ff., p.193ff.) considered to be a major problem to theories of both tone (fundamental plus harmonics) and chord perception, including the theory of consonance.

The more notes tend to fuse, the more they *penetrate* each other. "Verschmelzung" thereby is equivalent to "Durchdringung" (Stumpf, 1911, p.324; Stumpf, 1939, p.261). This does not imply, however, that the elements that make up a configuration, or the relations between them, will no longer be discernible. At least subjects who have some musical experience will be able to identify the notes that make up even the most consonant chord. A chord, therefore, by musical subjects is *not* heard as *one*, as has been claimed recently by Bregman (1990, p.508ff.) and Huron (1991, p.153); it is always analyzed into its constituents.

Stumpf, on several occasions (Stumpf, 1897, p.290; Stumpf, 1898, p.43; Stumpf, 1911, p.323), rejected views that–in misreading his own statements–held that chords cause the experience of *oneness*. This is only the case with unmusical persons who in Stumpf's famous experiments had (con-)fused two tones for being one (Stumpf, 1890, p.142ff.; Stumpf, 1897). Similar results have been reported in Faist (1897), and recently in DeWitt and Crowder (1987). For the musical mind, however, a triad does *not* result in apperception of *oneness* ("Einheit") yet in apperception of coherence and integrity, that is, "Einheitlichkeit" (Stumpf, 1911, p.323).

This explanation, given by Stumpf to clarify matters, indicates that "Verschmelzung" is a concept that has to be understood from a Gestalt perception background: if we are listening to a completely harmonious chord (Fig.1), we can focus on one or several constituents (notes and/or their respective partials) for a moment, then return to the chord as a complex whole, then turn to another constituent realizing its relation to other parts as well as to the overall configuration, etc. Thereby, our attention kind of *swings* between parts and whole, however, we are attentive to but one stimulus configuration that is *looked at* (listened to) from different angles.

What Stumpf found a little perplexing, though, is the fact that the experience of "Verschmelzung" persists even after the perceptual analysis of a given chord, the apperception of the individual notes that form its parts, recognition of the relations between the notes and of the respective chord structure, has been accomplished. Thus, "Verschmelzung" somehow must be attributed to integration of sensory input and to the synthesis of single percepts into configurations or "Gestalten" on a cortical level (Stumpf, 1890, p.184). These are stored in memory so that conscious experience of "Verschmelzung", in principle, would even be possible by way of mental imagery (Stumpf, 1919, p.19; Stumpf, 1939, §14).

Not to be misunderstood as being *subjective* in his approach (as some might suspect he was), Stumpf (1890, p.127ff.; 1939, p.244) insisted that "Verschmelzung" is a structural property already inherent in a number of simultaneous notes that thus would have to be recognized, rather than established, by the attentive listener. He/she can only experience "Verschmelzung" because stimulus features, on the one hand, and the structural organization of the auditory system and the brain, on the other, provide a material basis to do so. One can find similar strains of thinking in, among others, Piaget and Chomsky.

It should be mentioned, finally, that "Verschmelzung", besides "Verknüpfung" (the mental act of *connecting* perceptual objects as well as thoughts, see Külpe, 1893, T. II; Stumpf, 1907, p.23f.), has been regarded as a fundamental mode of cognitive activity that aims at constituting coherent empirical objects (Brentano, 1924/28; Stumpf, 1919, p.103f.), and is by no means restricted to the auditory modality, see also (Gurvitch, 1975, p.66ff.).

2.2 "Verschmelzung" and Consonance, Consonance and Concordance

Stumpf (1898, p.42ff.) believed "Verschmelzung" to be the major reason also for consonance of, basically, pairs of tones. In distinguishing two tones that make up a consonant pair (e.g., *c* - *g*; *c* - *e*), we at the same time realize that these form a coherent whole. The reason for "Verschmelzung" thus is that two sensations are firmly connected, and thereby constitute a peculiar type of experience that can be distinguished from that of *similarity*. For example, the two notes *c* and *c♯* (*cis*) are quite similar, however, they do not fuse at all. Relating the two principles of similarity and fusion, Stumpf (1898, p.50) proposed a scheme according to which the *similarity* of any two tones depends on their linear frequency difference ($f2 - f1$), while *fusion* is dependent on frequency ratios of pairs of tones ($f2 : f1 = n_i/n_j$; n = natural numbers 1, 2, 3, ..., n) . This scheme accounts for, on the one hand, the two-dimensional theories of pitch that comprise of tone height, and tone chroma, respectively. On the other, it distinguishes between arbitrary tone distances ($f1 \leftrightarrow f2$), and distinct musical intervals such as the fifth, fourth, etc. The two principles, according to Stumpf (1898, p.60), interact in such a way that those intervals which bear the same degree of fusion ("Verschmelzungsgrad"), are still different with respect to tonal distance (e.g., major and minor third, major and minor sixth).

It should be mentioned that Stumpf, for theoretical reasons, had refused two-dimensional theories of pitch even in his *Tonpsychologie*, and had maintained that pitch in fact has to be understood as a one-dimensional quasi-continuum. He started to revise his view in later publications (Stumpf, 1898), and had switched to the two-dimensional concept of pitch as proposed by Brentano and Révész by 1914 (Stumpf, 1914/15, p.309ff.). Without being able to go into details here (Schneider, 1991, 1997b, part 3; Schneider & Beurmann, 1991, p.181ff.), this issue–that caused severe problems in Stumpf's approach to pitch perception, and also in his concept of consonance–again has to do with relations of parts to wholes, and with psychic functions of connecting parts as well as with the postulate of the *integrity* ("Einheitlichkeit") of apperception that had been elaborated by Kant. Good questions, in this respect, are: will musicians and/or listeners organize tones/pitches perceived along one, two, or even m dimensions under *real life* conditions? How are these dimensions interrelated? What role do geometrical models of m-dimensional pitch play in the musical mind of performers and listeners in actual musical situations?

Stumpf's observations supported the hypothesis that consonant intervals are based on simple integer frequency ratios, notwithstanding small deviations from precise ratios that may occur in actual music performances. His findings were corroborated by later experiments that established a hierarchy of intervals whereby these can be regarded as phenomenally distinct *anchor points* on the quasi-continuum of pitch or, rather, on a quasi-continuum of frequencies on which many pairs of pitches may be realized, each having a different size with respect to $f1 \leftrightarrow f2$, (Reinecke, 1964; Hall & Hess, 1984).

In that, Stumpf took "Verschmelzung" to be the reason also for our experience of consonance, he still had to find a reason why "Verschmelzung" can occur. The explanation he tentatively gave pointed to neurophysiological functions of the auditory pathways (Stumpf, 1898, p.50ff.). By the end of his life, he was not so sure that "Verschmelzung" can be taken as a full *explanation* to consonance (in the sense of cause and effect, cf. Stumpf, 1924, p.246; Stumpf, 1926, p.281). Rather than being the cause of specific relationships experienced in the perception of pairs of tones, "Verschmelzung" and consonance were regarded as a consequence to such relationships that are already inherent in certain stimuli, and which thus must be detected by some neurophysiological mechanism of pattern recognition. He discussed several of them, thereby pointing to, among others, the approach known as *peak picker* theories of pitch that are closely related to the detection of recurrent amplitude maxima in periodic waveshapes (Schneider, 1997b, p.123ff.).

It is interesting to see that Stumpf, who had advanced his concept of "Verschmelzung" in opposition to Helmholtz' physiologically based theory of consonance, finally had to resort to neurophysiological explanations. He found another field, though, where his original ideas on psychic judgement functions proved to be most useful and apt: that of the perception of triads and more complex chord structures as well as that of the relationship of musical chords with respect to harmony. Stumpf thereby introduced his principle of *concordance* that he (Stumpf, 1911, p.332) defined as follows: in arbitrary chords, we shall look for the largest possible number of notes within each octave that are consonant in relation to each other, starting with the note being lowest in pitch, and progressing among the consonant notes from the stronger degrees of consonance [as defined in the *Tonpsychologie*] to the weaker ones.

The main reason for introducing this principle, of course was to allow comparison of such chords as 4:5:6 (natural major) and 10:12:15 (natural minor, both are derived, in this fashion, from the series of harmonics). Another reason, however, was that Stumpf theoretically argued from the point of view that hearing is based on just intonation. His experiments on tonal fusion in fact had employed tones played in this tuning system. Modern piano tuning, however, is based on equal temperament, to the effect that two chords *c* - *e♭* - *g* and *c* - *d♯* - *g* cannot be distinguished anymore by their respective intonation, which is the same notwithstanding actual musical notation and context. Therefore, to comprehend chord progressions the way they should be according to musical syntax, and regardless of ambiguous (or even wrong) intonation, we have to emphasize the role of relational thinking ("beziehendes Denken"). Mainly to account for problems arising from *thinking in just intonation* versus *performing in equal temperament*, Stumpf elaborated his concept of concordant and discordant chords that also is an attempt to deal with more complex configurations than just pairs of consonant tones. He claimed, in effect, emancipation of apprehension from mere perception. At about the same time, H. Riemann came up with similar ideas.

Distinguishing between consonance and concordance, Stumpf (Stumpf, 1911, p.341) stated that while *consonance* is a matter of immediate sensation, *concordance* is a matter of apprehension and relational thinking. In Stumpf's concept, concords are such chords which are based on the diatonic scale steps of each key. For example, the triad *e - gis - h (e - g♯ - b)* is a concord if e can be regarded as the tonic, but will be conceived of as a discord in a C-major context notwithstanding we are listening to the same sound stimulus. Further, even the d-minor chord in C-major would have to be considered as a discord, according to the principle of concordance, if played on a piano with equal temperament because of the impurity of intonation, resulting in a mistuned major third *f* - a, and an (acoustically) imperfect fifth *d* - a. The d-minor chord is apprehended, however, as a concord because of its syntactic relations within a C-major context.

Stumpf thus argued that, in apperception, elements not precisely fitting each other in terms of small integer ratios or whatever formal criterion of relationship, may still be assessed to make up *conclusive* configurations. Thereby, configurations are seen as constituted, in a final stage, in apperception, and as being *relatively* independent from the actual sensory input. Since we know, in music, of such constructs as "Scheinkonsonanzen" and "Trugschlüsse" that, to be sure, make sense only with respect to relational thinking, Stumpf's concepts of concordance and discordance can be put into a more general framework of musical apprehension.

3 Fusion/Consonance vs. Roughness/Dissonance: Some Issues and Findings

Though "Verschmelzung" should be regarded, in Stumpf's view, primarily as a matter of conscious awareness ("Bewusstseinstatsache"), it is dependent on perceptual parameters, and thus, on characteristics of sound stimuli as well as the sensory input to be processed in the auditory pathways. Stumpf, in contrasting what he (Stumpf, 1926, 280ff.) labelled *integral* vs. *plural* hearing, discussed certain conditions that are helpful for *integral* hearing (including the experience of "Verschmelzung"). These are (i) harmonicity of partials of complex tones, (ii) just intonation of intervals comprising several complex tones, (iii) spatial identity of partials, that is, partials of complex tones will be best *integrated* by listeners if they appear to come from one source and/or direction, (iv) synchronicity of onset of all partials (which thus should be phase-locked), (v) spectral envelope and intensity relations among partials (which should conform to certain patterns), (vi) neurophysiological integration mechanism, (vii) psychological factors such as training, adaption, attention, etc. From this concept it becomes quite obvious that *tonal fusion* is dependent on physical as well as psychoacoustic features of actual sound stimuli, and that tonal fusion in turn is regarded as prerequisite to the experience of "Verschmelzung" (and consonance) by Stumpf.

In a number of experiments, conditions relevant to determine perceptual parameters related to the experience of tonal fusion and consonance, on the one hand, and of roughness and dissonance, on the other, have been explored (Plomp & Levelt, 1965; Terhardt, 1974; Plomp, 1976; Coninx, 1977; Mathews & Pierce, 1980; McAdams, 1982; Aures, 1985; Moore, Peters, & Glasberg, 1985; Patterson, 1990; Houtsma & Smurzynski, 1990; Sethares, 1993; Bregman, 1990, Chaps.2-3; Schmidt, 1996). Fusion, in this respect, can be observed in (i) the sensation of single complex tones made up of n partials, as well as in (ii) two simple tones, each having but one sine component, and in (iii) two or more complex tones, of n partials each, sounded simultaneously. It was found that fusion in especially single complex tones (but also in several complex tones played at a time) is largely dependent on (i) spectral energy distribution, (ii) harmonicity vs. inharmonicity of spectral content, (iii) familiarity with the spectral envelope of sounds, as well as on (iv) modulation that is affecting one, several, or all, of the spectral components. Thereby, tonal fusion as perceived rests on both temporal, and spectral, features of sounds.

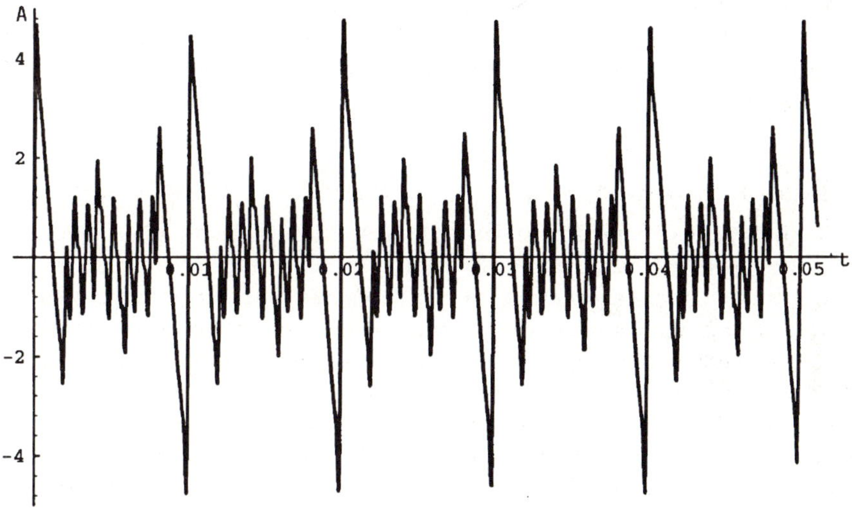

Fig. 1. Waveshape, harmonic major chord. Reproducible with Mathematica using the formula: Plot[Sin[2Pi 300t] + 0.5Sin[2Pi 600t] + 0.33Sin[2Pi 900t] + 0.25Sin[2Pi 1200t] + 0.2Sin[2Pi 1500t] + 0.166Sin[2Pi 1800t] + 0.143Sin[2Pi 2100t] + Sin[2Pi 400t] + 0.5Sin[2Pi 800t] + 0.33Sin[2Pi 1200t] + 0.25Sin[2Pi 1600t] + 0.2Sin[2Pi 2000t] + 0.166Sin[2Pi 2400t] + 0.143Sin[2Pi 2800t] + Sin[2Pi 500t] + 0.5Sin[2Pi lOOOt] + 0.33Sin[2Pi 1500t] + 0.25Sin[2Pi 2000t] + 0.2Sin[2Pi 2500t] + 0.166Sin[2Pi 3000t] + 0.143Sin[2Pi 3500t], {t,0,0.0511}, PlotRange→All] (= 51 ms of sound)

To illustrate the concept of fusion, we may generate a sound that is a major chord made up of the constituents 300, 400, 500 Hz. If we now add partials to these, to the effect that we have three complex tones (each of which has seven harmonics, the amplitudes of which are arranged to conform to 1/N, a reasonable scaling factor, see also Hutchinson & Knopoff, 1978), the result is the waveshape shown in Fig.1[1]. As can be easily seen, this is a very regular, highly symmetrical and periodic structure with strong peaks at the beginning of each period because all the components are coupled in phase. Listening to this sound, we not only can detect the three constituents of the chord yet also some of the partials, notwithstanding that these blend perfectly. The experience of the attentive listener thus is likely that he/she will perceive the chord both as a multiplicity (Parncutt, 1989, p.92) *and* as a unitary, highly coherent structure. In recognizing this double nature of fusion as triggered by specific sounds (Stumpf had explored fusion making use of pipe organ mixture stops), we can shift attention between constituents, and the overall blend, of the sound in question, that is, between parts and whole. This is, to be sure, exactly the phenomenon that was described by Stumpf as "Verschmelzung". Experiencing fusion thus includes recognition of (a) the constituents, of (b) a certain pattern of constituents as they fit to each other, and (c) of a configuration that can be regarded as being stable, and kind of self-contained.

The experience of roughness, in contrast, is quite different. In case one presents two simple tones, having the frequency ratio 16 : 15 (a minor second), and the frequencies 160 and 150 Hz, respectively, beats at the frequency $f_s = | f_1 - f_2 |$ will become clearly audible as the beats are complete, that is, amplitude of the waveshape resulting from the two components comes to zero once per modulation cycle because of total interference at $\tau/2$ if the two components have equal amplitude and phase (Schneider, 1997b, p.95ff.). The sensational phenomena known as beats (f_s is relatively low so as to allow, at least in many cases, counting of the number of beats per time unit) and roughness (f_s typically is too high to allow such counting anymore, and the beats constitute a sequence of fast pulses) has been known for centuries, and has been used to tune up instruments such as organs and cembali. It was Helmholtz (1870, Chaps.8,10-11) who discussed the influence of – as he called them – *disturbances* on interval sensation, and the perception of consonance and dissonance, in particular. His observations on beats and roughness, along with remarks on patterns of component spacing in complex tones, and on critical distances between such components, have been confirmed, in almost every aspect, by more recent research (Plomp & Levelt, 1965; Terhardt, 1976). Especially the well-known V-curve of dissonance derived from estimation of maximum roughness (equivalent to minimum consonance) as a function of frequency separation of two sine components (Kameoka & Kuriyagawa, 1969), stems from considerations Helmholtz had made one century earlier. Since the sensation of beats, or of marked roughness, often is equated

[1] All sound examples referred to in this article have been generated on a NeXT running *Mathematica*©.

with the perceptual quality of dissonance, it follows that consonance can be defined, as Helmholtz indeed did, and what Stumpf (1898) criticized, *ex negatione*, namely that the experience of consonance depends on the absence of roughness and beats in sound stimuli.

Helmholtz offered, however, also one *positive* principle, the coincidence of partials of two (or more) complex tones, as an explanation to the perception of consonance. This aspect has been emphasized also in more recent theories of consonance, and even in some that are based on neuronal processes, rather than just on inner ear functions (Hesse, 1972).

Another aspect related to consonance is that of musical harmony. Harmony, to be sure, points to tonal relations that can be expressed, in a consistent and satisfactory manner, by ratios of small integers (Vogel, 1993). Besides, such small integer ratios have been found valid in judgements on the optimum size of tonal intervals (Reinecke, 1964), and of course with respect to both the experience of tonal fusion as elaborated by Stumpf (1890, 1897, 1898), and the principle of coincidence of partials (which are also aptly called *harmonics*). Typically, pairs of simple or complex harmonic tones, the frequencies of which have simple ratios, fuse very well, and reduce roughness and beats to a minimum. As can be easily checked by experiment, the resulting waveshape becomes as simple, and as clear with respect to periodicity, as can be achieved for a given musical interval or chord. Consider the *natural* seventh 7:4 as compared to that of 16:9, and both to the seventh 1:1.7818 in equal temperament, the major seventh chord 4:5:6:7 as compared to its tempered counterpart, a major chord in a Vallotti and Young tuning as compared to Pythagorean tuning. For examples, see Schneider (1994, p.235f., Figs.6,7; 1997a). By contrast, pairs of complex tones with *stretched* or otherwise non-coinciding partials are prone to cause so-called *Helmholtz-beats* (Rasch, 1984) even if their fundamental frequencies form just intervals (like 3:2, 4:3, etc.). Further, it has been argued that pitch and fusion are *isomorphic* (Cohen, 1984, p.384) in that ambiguity of pitch increases if partials in a harmonic complex sound are mistuned (Moore et al., 1985). Finally, inharmonic sounds made up of n components of irregular frequency spacing, and of equal amplitudes, do not blend yet tend to dissolve into parts, have no clear periodicity, and often will produce quite severe modulation effects (Schneider, 1996, 1997b).

Harmony also has been based on pitch perception, and explained in terms of the theory of so-called virtual pitch (Terhardt, 1976, p.130ff.). This view does not contradict or preclude what has been argued here so far, since the pitches (or roots) derived from a virtual pitch model (and, similarly, from the Houtsma/Goldstein *optimum processor* as well as from Wightman's *pattern recognition model* (Lyon & Shamma, 1996, p.242f.) are the less ambiguous the more a number of sine components assumed to have been resolved in the cochlear filter bank, and to be processed thereafter on higher stages of the auditory pathway, have frequencies that conform to small integer ratios. For example, if we generate a sound comprising of the harmonics no. 2-15

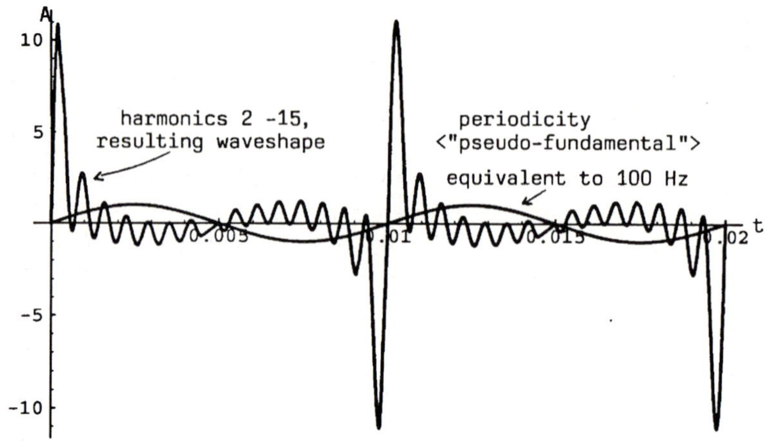

Fig. 2. Harmonics no. 2-15: resulting waveshape; "pseudo-fundamental". Reproducible with Mathematica using the formula: Plot[Evaluatel[{Sin[2Pi 200t] + Sin[2Pi 300t] + Sin[2Pi 400t] + Sin[2Pi 500t] + Sin[2Pi 600t] + Sin[2Pi 700t] + Sin[2Pi 800t] + Sin[2Pi 900t] + Sin[2Pi lOOOt] + Sin[2Pi 1100 t] + Sin[2Pi 1200t] + Sin[2Pi 1300t] + Sin[2Pi 1400t] + Sin[2Pi 1500t], Sin[2Pi lOOt]}, {t, 0.00, 0.02}, PlotRang→All]]

(Fig.2), the resulting waveshape will exhibit a very clear periodicity that, in turn, gives rise to unambiguous pitch defined by the period of the signal that exactly equals, in this case, the frequency of the *missing fundamental* (i.e., 100 Hz, because of $t = 1/f$, $\tau = 10ms$). There is some evidence that among vertebrates, *mammalia* and also the class of *aves* possess of neural mechanisms that are suited to detect periodicity from spike trains and interspike intervals (see below).

It is possible, however, to create waveshapes of complex sounds that, though having a totally harmonic spectrum made up of sine components all coupled in phase as well as the same clear-cut periodicity of our first example, will not be sensed as being perfectly fused in terms of sensory "Wohlklang" (Aures, 1985). This is the case, for example, with the waveshape shown in Fig.3 as here the higher harmonics have been given increasingly stronger amplitudes while the phase relations remain unchanged. Thereby, the spectral envelope (Fig.4) is not just unfamiliar, the sound generated with this waveshape also is perceived as being sharp or even *piercing* in quality – though the sound fairly resembles that of a tenor sax. According to judgements of 35 musically trained subjects who were confronted with this stimulus plus musical examples played with this sound, it lacks the pleasantness usually connected with tonal fusion (Berlyne, 1971, p.239ff.). In one experiment, 14 subjects were asked to judge both fusion and roughness of this sound on

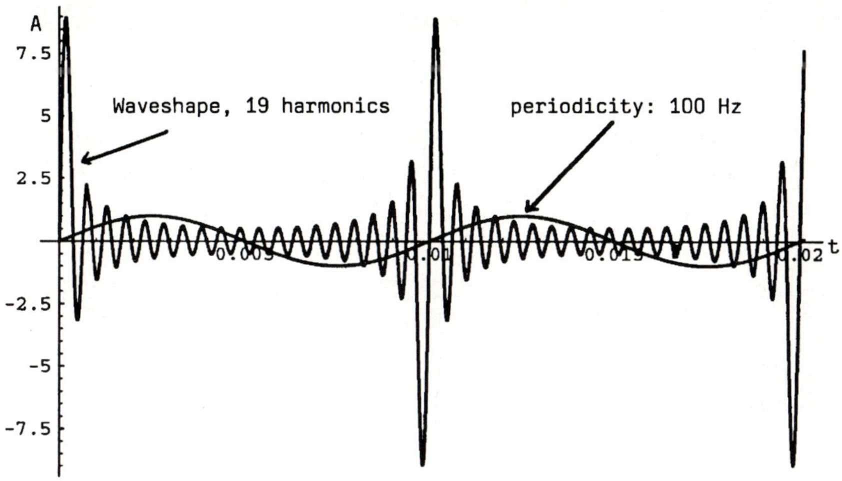

Fig. 3. Waveshape, "Sax" sound; periodicity: 100 Hz. Reproducible with Mathematica using the formula: Plot[{O.1Sin[2Pi 100t] + 0.15Sin[2Pi 2OOt] + O.2Sin[2Pi 300t] + 0.25Sin[2Pi 400t] + 0.3Sin[2Pi 500t] + 0.35Sin[2Pi 600t] + 0.4Sin[2Pi 700t] + 0.45Sin[2Pi 800t] + 0.5Sin[2Pi 900t] + 0.55Sin[2Pi 1000t] + 0.6Sin[2Pi 1100t] + 0.65Sin[2Pi 1200t] + 0.7Sin[2Pi 1300t] + 0.75Sin[2Pi 1400t] + 0.8Sin[2Pi 1500t] + 0.85Sin[2Pi 1600t] + 0.9Sin[2Pi 1700t] + 0.95Sin[2Pi 1800t] + Sin[2Pi 1900t], Sin[2Pi 100t]}, {t, 0, 0.02010991}, PlotRange→All]

scales of seven degrees each. The results are as shown in Table 1. With respect to variance, both data files are homogeneous; F[2.57, p <0.05] = 1.28; t-test yields t[2.056, p <0.05] = 2.158.

Similar effects can be observed in a sound created from combining a cosine pulse with a function $1/sin(2\pi xt)$ that produces a symmetrical waveshape (Fig.5) and a spectrum comprising very many harmonics (Fig.6). The frequency values given in the spectra due to sampling rate conversion (48 kHz to 44.1 kHz) have to be multiplied by 1.088435 to yield the original frequency data of the sounds as were played to the subjects from a DAT. The cosine pulse component, by the way, is of minor importance in this case, and has been added only to reinforce the altogether very clear periodicity of the sound that thereby gives rise to the sensation of two strong pitch components (≈

Table 1. Sound *Sax*, Scales of fusion and roughness are from 1 to 7 (1 = minimum, 7 = maximum), number of subjects = 14

	aM	median	SD	variation coefficient
fusion	2.96	2.75	1.8	60.8%
roughness	4.54	5	2.04	45.03%

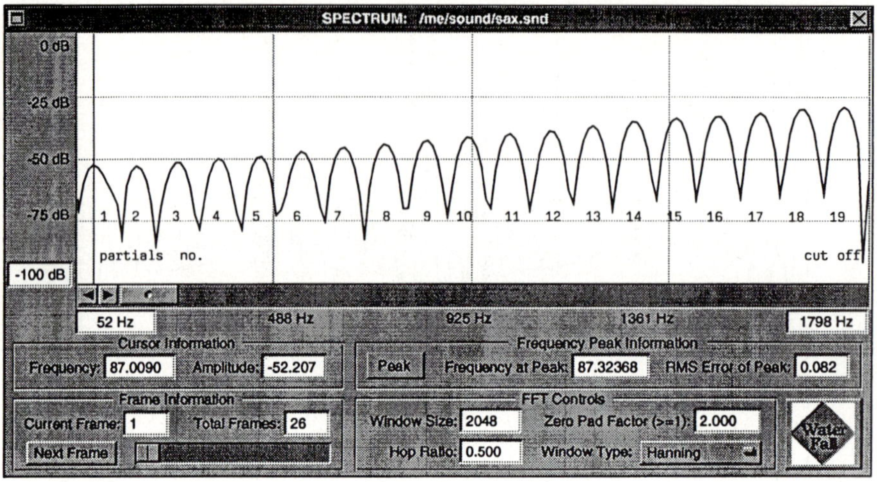

Fig. 4. Spectrum, "Sax" sound, 19 partials, rising slope of envelope

G2, G3) plus a number of partials that can also be identified by ear (Table 2). Variances in these two data files are not homogeneous, $F(3.91, p < 0.01) = 4.505$. Therefore, t-test is problematic (though it would be highly significant at $t[3.707, p < 0.001] = 5.617$).

It can be inferred from these results, as well as from many other findings, that low ratings for fusion basically correspond to high ratings for roughness. Thus, these two sensory qualities can be considered as being opposed to each other.

The reason why fusion has been rated quite low in both cases reported above, while roughness gained relatively high ratings in these sounds, could be attributed, most of all, to spectral energy distribution. Compared to the normal additive synthesis of n harmonics (Fig.1), there is a shift of energy in the *sax*-sound towards higher harmonics, with the cut-off frequency of the spectrum set, in this case, at 1.9 kHz (Fig.4). Of course, with the given spacing of components (the frequency increment per partial is 100 Hz), interaction of partials is likely due to the concept of critical bands. Further aspects perhaps

Table 2. Sound *CosPulse*, Scales of fusion and roughness are from 1 to 7 (1 = minimum, 7 = maximum), number of subjects = 14

	aM	median	SD	variation coefficient
fusion	2.36	2	1.55	65.74%
roughness	4.93	5	0.73	14.81%

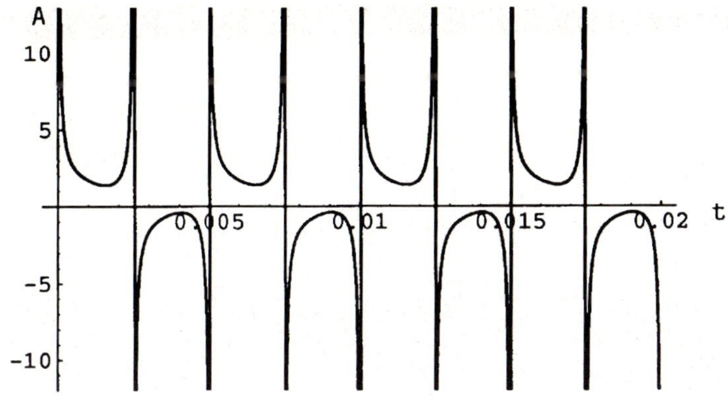

Fig. 5. Pulse-type sound, waveshape for 20 ms. Reproducible with Mathematica using the formula: Plot[Evaluate[Cos[2Pi 100t]^2 + 1/Sin[2Pi 200t], {t, 0, 0.02}]]ar

worth to be considered, especially with respect to the second sound, are spectral density and the flow of energy caused by this type of signal.

- First, as to critical bands, this factor of course is of relevance to the perception of roughness – as is known from very many experiments involving amplitude modulated sounds to produce first-order beats, etc. (Plomp, 1976, Chap. 3; Coninx, 1977). The concept of critical bands (CB) – that has been used to explain a variety of phenomena in psychoacoustics from loudness and masking to perception of pitch and consonance, and in man and animal alike (for an overview, see Scharf 1970) – however bears many assumptions as to the nature of the periphery of the auditory system, and of the performance of the cochlear *filter bank*, in particular.
  Though the concept is a useful one, more recent findings have demonstrated that "none of these assumptions is strictly correct" (Moore, 1993, p.58). For example, the band pass filters assumed to be linear, symmetrical, and overlapping each other, are in fact neither linear nor symmetrical as soon as SPL is above a (relatively low) threshold. Further, the exactitude of pitch discrimination as has been found to be 2-5 cent in expert listeners (Meyer, 1979), obviously is much better than predicted by the concept of critical band. Moreover, the ability to hear out even very high spectral components in complex sounds (see, however, Duifhuis, 1970, 1971), is completely at odds with the critical band approach. Of course, also sensitivity to second-order beats produced by slightly mistuned harmonic intervals presented in dichotic condition, as well as other auditory phenomena cannot be satisfactorily explained by *place* mechanisms involving the concept of critical band (Hesse, 1972; de Boer, 1976). Thus, in addition to peripheral *filtering* that, in recent research (Keidel, 1992),

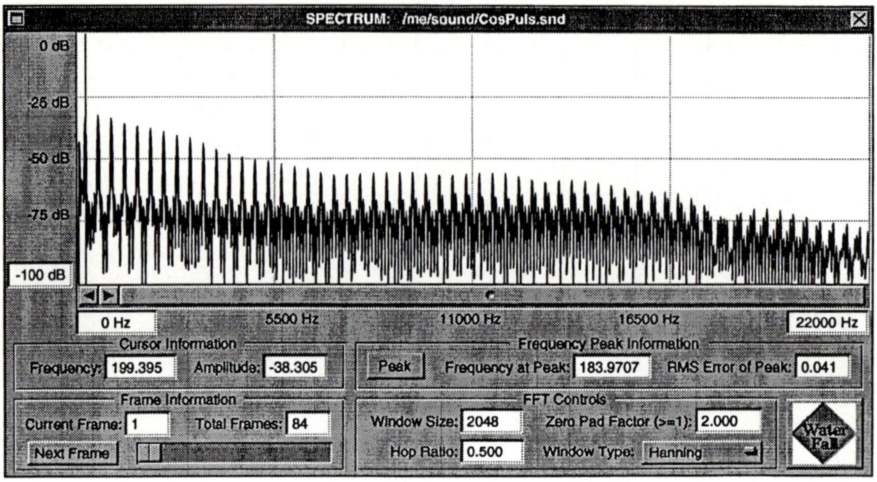

Fig. 6. Resulting spectrum, Pulse-type sound (Fig.6), Spectral Energy 0 - 22 kHz

has been found to be more precise than predicted by conventional critical band estimates, one has to consider neural mechanisms of the auditory pathway where critical bands have a quite different meaning (Ehret, 1988).

Neural mechanisms of pitch processing that are suited to improve the performance of the cochlear filter bank are especially of significance with respect to multi-harmonic tone complexes since higher partials which presumably cannot be resolved by the filter bank (Duifhuis, 1970, 1971) may be processed separately, and as groups, by use of temporal information (Houtsma & Smurzynski, 1990; Kohlrausch & Houtsma, 1992). As has been emphasized just recently, such mechanisms are well possible "because our ability to detect a change in the frequency of a single tone is an order of magnitude better than that implied by the critical band..." (Lyon & Shamma, 1996, p.243). This increase in performance is ascribed to temporal cues inherent in the signal that will be encoded in the temporal fine structure of spike trains, and seemingly can be processed in several nuclei of the auditory pathway. There is some empirical evidence available for such coding and processing of temporal patterns (Langner, Schreiner, & Albert, 1992; for temporal aspects, see also Delgutte, 1996, p.190ff.).

As far as *fusion* and *consonance* (and not just pitch) of sounds is concerned, one may also point to neural information processing on higher levels. Keidel and co-workers (Keidel, 1975) found evidence in the midbrain (corpus geniculatum mediale) of the cat for networks synchronized by clocks that respond to specific stimulus frequencies f_n. Maxima in the

PST-histograms extracted from demodulated spike trains were found, for each cell unit, to conform to f_n, $2f_n$, $3f_n$, etc. This type of network was therefore regarded as kind of a *neural harmonic analyzer* (Keidel, 1992, p.349ff.). A model capable of calculating what has been labelled the central spectrum from interspike intervals has been proposed by Srulovicz and Goldstein (1983). This, again, selects frequencies that are close to characteristic frequencies of nerve fibers plus its harmonics.

Furthermore, one has to consider both the ear and the auditory pathway not just from the afferent but also from the efferent side, as well as to take into account active elements – known as *Kemp's echo* or *cochlear amplifier* (Keidel, 1992, p.304ff.; Zenner, 1993) – that are suited, on the one hand, to significantly improve performance in frequency discrimination and other analytical tasks as well as the dynamic response. On the other, *active* elements in the transduction network may change the actual input signal characteristic: that's why there are nonlinearities. Most interesting findings, in this respect, have been reported on the performance of outer hair cells, as well as Hensen's cells, in the organ of Corti. As could be demonstrated by experiment, these cells, if mechanically driven to resonance by simple sine components, will respond by generating a number of harmonics not found in the input signal (Khanna, Ulfendahl, & Flock, 1989a, 1989b). Consequently, compared to the input signal, the output to the neural pathway is both much more complex, and heavily nonlinear as the vibratory response of these cells gains a waveform that contains up to ten odd and even harmonics, with a maximum if the stimulus frequencies are set at about 180-220 Hz. By analogy to electronic devices known from studio equipment, one could address these cells, that also play an essential role in the cochlear amplifier circuit, to function as *harmonic enhancers*.

– Second, as to spectral density and energy flow (that can be calculated by integration methods), it seems possible that sounds such as that shown in Figs.5-6, if presented with only medium SPL of about 55-60 dB to subjects, may bring about saturation in the auditory pathway. Because of the very many harmonics, and spectral energy available up to 22 kHz, saturation could be expected according to aspects known as spread of excitation of nerve fibers, and also according to both rate-place and temporal-place approaches of intensity coding (Delgutte, 1996, p.173ff.). As here in fact we are not just dealing with intensity levels, but rather with spectral density and spectral energy (flow), it is likely that these parameters interact in some way, to be accounted for by a combined spatio-temporal model (Evans, 1992, p.298f.; Lyon & Shamma, 1996, p.252ff.).

In any event, the roughness assigned to the sound in question 18is felt even in a totally harmonic context such as a C-major cadence; this waveshape and spectrum, according to the ratings made by subjects (n = 14), are not regarded as pleasant (Table 3).

Table 3. Cadence, C-major, Sound: CosPulse, Scales [fusion; roughness] 1 - 7

	aM	median	SD	variation coefficient
fusion:	2.714	3	1.204	44.37%
roughness:	4.821	5	1.613	33.445%

Though the aM for fusion here is a little higher than for the sound outside a musical context, and the aM for roughness a little less (what has to be attributed to the harmonic cadence), the situation remains unchanged: this sound, deliberately created to be as periodic and harmonious as is possible, and to allow for utmost fusion, rather is perceived as dissonant. As the pitch of this sound is no problem at all, we may hypothesize that the sensation of roughness occurs consequent to an overshoot of spectral energy, and especially in higher frequency registers notwithstanding the more conventional (i.e., downwards slope) spectral envelope.

4 Stumpf's Claim on Possible Separation of Dissonance and Roughness: Experimental Findings

The concept of consonance as outlined by Helmholtz (1870) and other scientists, according to Stumpf (1898) was insufficient, most of all, in that dissonance had been attributed to the sensation of roughness or beats, and consonance to the absence of these *disturbances*. Stumpf not only objected to this *negative* definition, he even claimed that there was not necessarily a strong link between the perception of *dissonance*, on the one hand, and the sensation of *roughness* or *beats*, on the other. In fact, one might find quite complex chords in music that will be judged to be totally dissonant in apperception, while the actual sound is free from any audible roughness. To illustrate the case, Stumpf (1898, p.6ff.) gave a number of examples in musical notation, and generated the chords (Fig.7) by means of a set of tuning forks (as their sound spectra come close to pure tones which Stumpf found to be suited best for the experiment as there will be no partials to interact which each other).

For these chords that for our experiments have been re-synthesized by means of a computer, ratings on scales directed to the basic qualities of consonance and roughness should be low. If the first scale refers to consonance, and consonance is conceived of as being the opposite of dissonance, it can be outlined as to range from minimum consonance ($\equiv$ maximum dissonance) to maximum consonance. All the chords can be expected to yield low ratings on this scale that ranges from 1 to 9 because of the high degree of dissonance ($\equiv$ low consonance) inherent in the chord structure.

However, also ratings for roughness ($1 =$ minimum, $9 =$ maximum roughness) should be low since, according to Stumpf, dissonance and roughness can

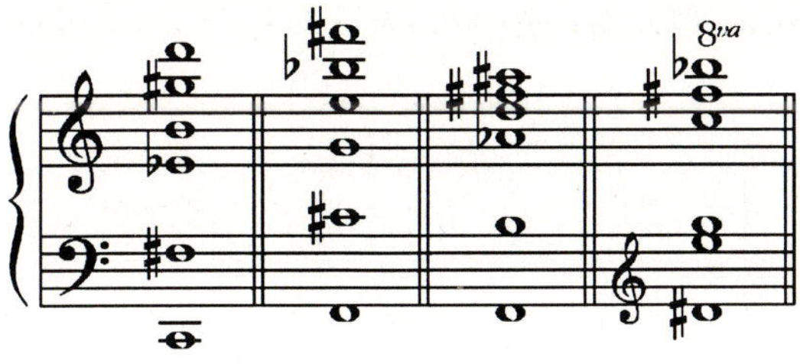

Fig. 7. Four chords designed by Carl Stumpf (1898)

be separated while Helmholtz and his followers would held that whenever a chord is perceived as being very dissonant, it is at the same time perceived as a sound with marked roughness or beats.

To test Stumpf's hypothesis, his four chords composed of six notes each were generated in just intonation, as Stumpf argued from the point of view of this system then, and also in order to minimize roughness caused by deficient tuning. The sine components used for additive synthesis are shown in Table 4.

Table 4.

	Chord 1[Hz]	Chord 2[Hz]	Chord 3[Hz]	Chord 4[Hz]
Component No1	66	88	88	309.38
"" 2	183.3	275	247.5	396
"" 3	316.8	396	422.4	495
"" 4	495	660	594	1056
"" 5	825	950.4	733.34	1466.67
"" 6	1188	1466.67	916.67	1900.8

The sounds were generated using a sampling rate of 44.1 kHz, each sound lasting for 2 seconds. To avoid any click or crack at the onset, initial amplitude of each sound starts from the negative[2]. The sounds generated were played to 36 musically trained subjects, all students in musicology at the University of Hamburg, who were asked, in a first trial, to judge the chords with respect to consonance (1-9), and in a second trial (performed two weeks after the first) with respect to roughness (again, on a scale running from 1 to 9). Thus, we

[2] In the syntax of Mathematica, synthesis of, e.g., the first chord thus is achieved by the computation command: Play[-3 (Sin[2Pi 66t]+Sin[2Pi 183.3t]+Sin[2Pi 316.8t]+Sin[2Pi 495t]+Sin[2Pi 825t]+Sin[2Pi 1188t]), {t, 0, 2}, SampleRate→44100, PlayRange→All].

Table 5. Descriptive statistics, 4 chords, 2 trials

	Degree of consonance (1-9)				Roughness (1-9)			
Chord number	1	2	3	4	1	2	3	4
Mean [arith.]	4.88	3.88	4.44	3.26	4.11	5.31	4.78	6.2
Median	5	3.5	4.5	3	4	5	4.5	7
Standard Dev.	1.3	1.63	1.54	1.75	1.82	1.75	1.82	1.92
Range	1-8	1-7	1-7	1-7	1-8	2-8	2-9	2-9

have 8 data files (4 chords, 2 trials) (Table 5). Correlation of the 4 + 4 files (consonance, chord 1, roughness, chord 1, consonance, chord 2, roughness, chord 2; etc.) gave low coefficients for r_{xy} between 0.053 and 0.171, and in no case was found to be significant (t-test of r_{xy}: n.s.).

Bartlett-tests performed on the data showed all variances to be homogeneous. Ratings for the chords (each chord taken as a *treatment*) however differ considerably since ANOVA yields F[5.793], $p < 0.001 = 6.934$ for the chords 1 to 4 with respect to *consonance*, and F[5.772], $p < 0.001 = 8.316$ for chords 1 to 4 with respect to *roughness*. Finally, a strict multiple-mean-test (Scheffé-test) with respect to *consonance* (files 1-4) yields D[1.071], $p < 0.05 = 1.176$ for contrast 3/4 (=chords nos. 3 and 4), D[1.071], $p < 0.05 = 1.029$ for contrast 1/2 (=chords nos. 1 and 2), and D[1.576], $p < 0.001 = 1.618$ for contrast 1/4 (chords nos. 1 and 4). With respect to *roughness*, contrast 1/2 is at D[1.221], $p < 0.05 = 1.194$, contrast 3/4 is at D[1.483], $p < 0.01 = 1.417$, and contrast 1/4 yields D[1.795], $p < 0.001 = 2.083$.

As can be seen from the basic statistics, especially chord no. 4 yields quite low ratings for consonance, and high ratings for roughness, with chord no. 2 not doing much better. Therefore, at least data of these two chords clearly indicate that low ratings for consonance are paralleled by high ratings for roughness. As we did in fact equate low degrees of consonance with its complementary perceptual quality, namely dissonance, it can be inferred that dissonance indeed rather goes along with roughness than to be a separable quality or *dimension* (as to the distinction of *integral* vs. *separable* as well as to so-called *interacting dimensions*, cf. Schneider & Beurmann, 1994, p.116). This interpretation is corroborated also by the negative results obtained in the correlational analysis (see above): if the chords employed in our experiment should have led to the perception of dissonance with yet no perception of roughness, low ratings on the *consonance*-scale should match low ratings on the *roughness*-scale, and both ratings then would have to correlate in a significant way for each of the chords 1-4 (what in fact they didn't).

Though Stumpf cautiously choose the notes (and their frequency distances) to be spaced wide enough to avoid the critical band (as was discussed already by Helmholtz; Stumpf 1898, p.6f. addressed the same concept under the label of "Merklichkeitsgrenze"), his claim according to which the perceptional quality of dissonance can be separated from the sensation of roughness is not supported by our data. Apparently, subjects still assign the attribute of roughness to sounds in spite of the fact that the frequency distances between pairs of components do not violate critical bands. Roughness as is obviously *perceived* from, in particular, chord no. 4, can be a attributed to (i) the very unregular, unsteady waveshape of this sound (the first 43 ms of which are shown in Fig.8), and (ii) the shift of energy towards higher frequencies what increases perceived dissonance (see above, Tables 1, 2; and also model calculations in Hutchinson & Knopoff, 1978, p.13ff.). After all, we may still consider dissonance to be closely associated with roughness (even though perception of roughness is not necessarily restricted to phenomena dependent on critical band mechanisms).

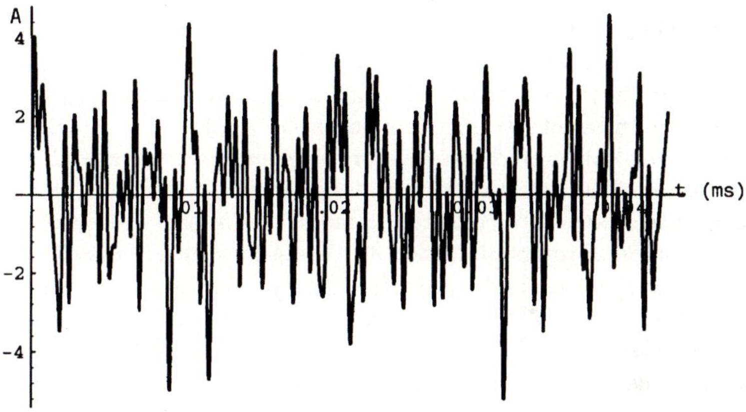

Fig. 8. "Stumpf chord" no. 4, resulting waveshape, 43 ms of sound

5 Conclusion

In his attempt to establish a coherent theory of music perception, labelled "Tonpsychologie", C. Stumpf had to discuss, among other issues, the concepts of consonance and dissonance. Regarding the mental activity understood as *apperception*, Stumpf reasoned that physical attributes of stimuli, and the sensations they may cause, are just not enough to explain the specific experience of consonance. In order to emphasize the singular perceptual quality of highly consonant intervals, he referred to "Verschmelzung" that with respect to perception of intervals had been considered as early as antiquity

(Stumpf, 1898, p.41). The notion of "Verschmelzung" had been elaborated, in a different though related context, in philosophical and psychological treatises on epistemics and cognition from Kant to Lotze (the latter, besides Brentano, being Stumpf's mentor), and then by Stumpf himself as well as by Wundt and Külpe. While Stumpf first had taken the cognitive principle of "Verschmelzung" as an explanation to consonance, he later thought that the ultimate cause for this perceptual phenomenon has to be found on a neurophysiological level. One of the reasons was that the experience of consonance (and, along with it, also that of cognitive "Verschmelzung") worked best with intervals of tones having small integer frequency ratios such as 3:2, 4:3, 5:4, etc. The structure that is decisive as to bringing about the specific effect thereby is regarded as being inherent in the stimulus configuration, on the one hand, and in the organization of the auditory pathway including its central stages, on the other. Thus, Stumpf, in this issue as well as in others, adopted a *nativistic* point of view though he always also underpinned the importance of factors of learning and adaptation.

The present article has surveyed some later findings and concepts that, at least in essence, are suited to confirm the view according to which principles of harmonic generation and analysis are of prime importance also to auditory perception. The same aspect that appears as *templates* of harmonics in various pitch extraction models, again is addressed in neurophysiological investigations of interspike interval histograms where maxima often are found to conform to small integer ratios (Keidel, 1989, p.173ff). The fact that these occur so prominently both as regards the structure of musical sounds, and thus, auditory stimuli, and with respect to perceptual mechanisms, has nothing to do with number mysticism yet with periodic functions as well as harmonic principles frequently found in nature (animated as well as inanimated). In an attempt apparently to avoid any nativistic explanation to pitch and interval perception (as well as consonance), and instead emphasizing learning (the old *nature* vs. *nurture* controversy), it was claimed that "the sense of musical intervals is acquired by processing speech" (Terhardt, 1974, p.1066). Strange enough, however, neural modelling of speech perception, and in particular of vowels, because of the periodicity of the signals basically exhibits the same response pattern known from the *neural harmonic analyzer* mentioned above. That is, there are clear maxima at harmonics of the fundamental frequency, of the respective signal that is processed with this realistic rate-place model using a human cochlear map (Delgutte, 1996, p.202f.). Musical intervals, moreover, are found in different contexts of animal communication prior to our systems of speech (Tembrock, 1978). It is not very likely, then, that speech was needed first to "shape" auditory perception, and thereby bring about also sensations of consonance. What was left for human experience, as Stumpf did elaborate, is to develop and improve cognitive functions necessary to fully apprehend musical structure as well as to understand our own processes relevant to listening to, and enjoying, music.

Appendix

Sound Examples (CD-tracks 5-10)

- Sound Example #1: Harmonic major chord, each of the components comprising of seven harmonics.
- Sound Example #2: Additive synthesis of nineteen harmonics. The amplitudes of these partials increase with their respective number. The sounds resembles that of a tenor sax.
- Sound Example #3: Pulse type high energy sound. Labeled for convenience cosine pulse.
- Sound Example #4a: Some brief musical examples, played with a sampled synthesized sax sound
- Sound Example #4b: Some brief musical examples, played with a sampled synthesized cosine pulse sound.
- Sound Example #5: Stumpf's chords number 1, 2, 3, and 4 (each repeated three times).

References

Aures, W. (1985). Der sensorische Wohlklang als Funktion psychoakustischer Empfindungsgrössen. *Acustica, 58*, 282-290.

Berlyne, D. (1971). *Aesthetics and psychobiology.* New York: Appleton-Century-Crofts.

Boring, E. (1950). *A history of experimental psychology* (2nd ed.). Englewood Cliffs, NJ: Prentice-Hall.

Bregman, A. (1990). *Auditory scene analysis.* Cambridge, MA: The MIT Press.

Brentano, F. (1924/28). *Psychologie vom empirischen Standpunkt, Vols.1-3.* Leipzig: F. Meiner. (Vol.1 was first published in 1874)

Cohen, E. (1984). Some effects of inharmonic partials on interval perception. *Music Perception, 1*, 323-349.

Coninx, F. (1977). *On the perception of changes in amplitude and frequency of acoustic signals.*

de Boer, E. (1976). On the "residue" and auditory pitch perception. In W. Keidel & W. Neff (Eds.), *Handbook of sensory physiology, Vol. V,3.* Berlin, Heidelberg: Springer-Verlag.

Delgutte, B. (1996). Physiological models for basic auditory percepts. In H. Hawkins, T. McMullen, A. Popper, & R. Fay (Eds.), *Auditory computation.* New York, Berlin, Heidelberg: Springer-Verlag.

DeWitt, L., & Crowder, R. (1987). Tonal fusion of consonant musical intervals: The oomph in Stumpf. *Perception and Psychophysics, 41*, 73-84.

Duifhuis, H. (1970). Audibility of high harmonics in a periodic pulse. *The Journal of the Acoustical Society of America, 48*, 888-893.

Duifhuis, H. (1971). Audibility of high harmonics in a periodic pulse II. *The Journal of the Acoustical Society of America, 49*, 1155-1162.

Ehret, G. (1988). Frequency resolution, spectral filtering, and integration on the neuronal level. In G. Edelman, W. Gall, & W. Maxwell (Eds.), *Auditory function: Neurobiological bases of hearing.* New York, Chichester: Wiley.

Evans, E. (1992). Auditory processing of complex sounds: An overview. *Phil. Trans. R. Soc. London Series B, 336*, 295-306.

Faist, A. (1897). Versuche über Tonverschmelzung. *Zeitschrift für Psychologie, 15*, 102-131.

Gurvitch, A. (1975). *Das Bewußtseinsfeld.* Berlin, New York: de Gruyter. (Ed. from the original typescript by W. Fröhlich)

Hall, D., & Hess, J. (1984). Perception of musical interval tuning. *Music Perception, 2*, 166-195.

Hesse, H. (1972). *Die Wahrnehmung von Tonhöhe und Klangfarbe als Problem der Hörtheorie.* Köln: A. Volk.

Houtsma, A., & Smurzynski, J. (1990). Pitch identification and discrimination for complex tones with many harmonics. *The Journal of the Acoustical Society of America, 87*, 304-310.

Huron, D. (1991). Tonal consonance versus tonal fusion in polyphonic sonorities. *Music Perception, 9*, 135-154.

Husserl, E. (Ed.). (1913/1980). *Logische Untersuchungen, Vol.II,1.* Halle: M. Niemeyer. (1913, reprinted as 6th edition in Tübingen: M. Niemeyer 1980)

Husserl, E. (1976). *Erfahrung und Urteil. Untersuchungen zur Genealogie der Logik.* Hamburg: F. Meiner. (Ed. by L. Landgrebe, 4th Edition)

Hutchinson, W., & Knopoff, L. (1978). The acoustic component of western consonance. *Interface, 7*, 1-29.

Kameoka, A., & Kuriyagawa, M. (1969). Consonance theory part I: Consonance of dyads. *The Journal of the Acoustical Society of America, 45*, 1451-1459.

Keidel, W. (1975). *Physiologie des Gehörs. Akustische Informationsverarbeitung.* Stuttgart: G. Thieme.

Keidel, W. (1989). *Biokybernetik des Menschen.* Darmstadt: Wiss. Buchges.

Keidel, W. (1992). Das Phänomen des Hörens: Ein interdisziplinärer Diskurs. *Naturwissenschaften, 79*, 300-310; 347-357.

Khanna, S., Ulfendahl, M., & Flock, Å. (1989a). Modes of cellular vibration in the organ of corti. *Acta otolaryng. Stockholm, Suppl. 467*, 183-188.

Khanna, S., Ulfendahl, M., & Flock, Å. (1989b). Waveforms and spectra of cellular vibrations in the organ of corti. *Acta otolaryngol. Stockholm, Suppl. 467*, 189-193.

Kohlrausch, A., & Houtsma, A. (1992). Pitch related to spectral edges of broadbent signals. *Phil. Trans. R. Soc. London B, 336*, 375-382.

Külpe, O. (1893). *Grundriss der Psychologie.* Leipzig: Engelmann.

Langner, G., Schreiner, C., & Albert, M. (1992). Tonotopy and periodotopy in the auditory midbrain of cat and guinea fowl. *Advances in the Biosciences, 83*, 241-247.

Lyon, R., & Shamma, S. (1996). Auditory representations of timbre and pitch. In H. Hawkins, T. McMullen, A. Popper, & R. Fay (Eds.), *Auditory computation.* New York, Berlin, Heidelberg: Springer-Verlag.

Mathews, M., & Pierce, J. (1980). Harmony and nonharmonic partials. *The Journal of the Acoustical Society of America, 68*, 1252-1257.

McAdams, S. (1982). Spectral fusion and the creation of auditory images. In M. Clynes (Ed.), *Music, mind, and brain.* New York: Plenum Press.

Meyer, J. (1979). Zur Tonhöhenempfindung bei musikalischen Klängen in Abhängigkeit vom Grad der Gehörschulung. *Acustica, 42*, 189-204.

Moore, B. (1993). Frequency analysis and pitch perception. In W. Yost, A. Popper, & R. Fay (Eds.), *Human psychophysics.* New York, Berlin, Heidelberg: Springer-Verlag.

Moore, B., Peters, R., & Glasberg, B. (1985). Thresholds for the detection of inharmonicity in complex tones. *The Journal of the Acoustical Society of America, 77*, 1861-1867.

Parncutt, R. (1989). *Harmony: A psychoacoustic approach.* Berlin, Heidelberg: Springer-Verlag.

Patterson, R. (1990). The tone height of multi-harmonic tones. *Music Perception, 8*, 203-214.

Plomp, R. (1976). *Aspects of tone sensation.* London, New York: Academic Press.

Plomp, R., & Levelt, W. (1965). Tonal consonance and critical bandwidth. *The Journal of the Acoustical Society of America, 38*, 548-560.

Pongratz, L. (1967). *Problemgeschichte der Psychologie.* Bern, München: Francke.

Rasch, R. (1984). Theory of Helmholtz-beat frequencies. *Music Perception, 1*, 308-322.

Reinecke, H. (1964). *Experimentelle Beiträge zur Psychologie des musikalischen Hörens.* Hamburg: Sikorski.

Scharf, B. (1970). Critical bands. In J. Tobias (Ed.), *Foundations of modern auditory theory, Vol. 1.* New York, London: Academic Press.

Schmidt, L. (1996). Natürliche und unnatürliche Skalen? *Systematische Musikwissenschaft, 3*, 59-70.

Schneider, A. (1991). On concepts of "tonal space" and the dimensions of sound. In R. Spintge & R. Droh (Eds.), *Musicmedicine.* St. Louis: MMB.

Schneider, A. (1994). Tone system, intonation, aesthetic experience: Theoretical norms and empirical findings. *Systematische Musikwissenschaft, 2*, 221-254.

Schneider, A. (1996). Perception of inharmonic sounds: Theoretical implications, empirical findings. In R. Pratt & R. Spintge (Eds.), *Musicmedicine 2.* St. Louis: MMB.

Schneider, A. (1997a). On categorical perception of pitch and the recognition of intonation variants. In P. Pylkkänen (Ed.), *Mind, brain and physics.* Amsterdam: I.O.S. Press.

Schneider, A. (1997b). *Skala - Tonhöhe - Sound.* Bonn: Verlag für Systematische Musikwissenschaft. (Forthcoming)

Schneider, A., & Beurmann, A. (1991). Tonsystem, Frequenzdistanz, Klangformen und die Bedeutung experimenteller Forschung für die Vergleichende Musikwissenschaft. *Hamburger Jahrbuch der Musikwissenschaft, 11*, 179-223.

Schneider, A., & Beurmann, A. (1994). Tonhöhe - Intervall - Distanz: Zur Wahrnehmung von Klängen mit inharmonischen Spektren. *Systematische Musikwissenschaft, 2*, 113-143.

Sethares, W. (1993). Local consonance and the relationship between timbre and scale. *The Journal of the Acoustical Society of America, 94*, 1218-1228.

Srulovicz, P., & Goldstein, J. (1983). A central spectrum model: A synthesis of auditory nerve timing and place cues in monaural communication of frequency spectrum. *The Journal of the Acoustical Society of America, 73*, 1266-1276.

Stumpf, C. (1873). *Über den psychologischen Ursprung der Raumvorstellung.* Leipzig: S. Hirzel. (Reprinted in Amsterdam by Bonset 1965)

Stumpf, C. (1883). *Tonpsychologie, Vol. I.* Leipzig: S. Hirzel.

Stumpf, C. (1890). *Tonpsychologie, Vol. II.* Leipzig: S. Hirzel.

Stumpf, C. (1897). Neueres über Tonverschmelzung. *Zeitschrift für Psychologie, 15*, 280-303.

Stumpf, C. (1898). *Konsonanz und Dissonanz.* Leipzig: J.A. Barth. (Beiträge zur Akustik und Musikwissenschaft, Heft 1)

Stumpf, C. (1907). *Erscheinungen und psychische Funktionen.* Berlin: Akad. d. Wiss. and D. Reimer. (Abhandl. Königl. Preuss. Akad. der Wiss., Jg. 1906 4. Abhdl.)

Stumpf, C. (1911). Konsonanz und Konkordanz. *Zeitschrift für Psychologie, 58*, 321-355.

Stumpf, C. (1914/15). Über neuere Untersuchungen zur Tonlehre. In *Bericht über den 6. Kongress für exp. Psychol. 1914* (p. 305-344). Leipzig: J.A. Barth. (Reprinted in Beiträge zur Akustik und Musikwissenschaft, Heft 8, 1915, pages 17-56)

Stumpf, C. (1919). *Empfindung und Vorstellung.* Berlin: Akad. d. Wiss./d. Reimer. (Abhandl. Preuss. Akad. der Wiss., Jg. 1918, Phil.-hist. Kl.)

Stumpf, C. (1924). Carl Stumpf. In R. Schmidt (Ed.), *Philosophie in Selbstdarstellungen, Vol. 5.* Leipzig: F. Meiner.

Stumpf, C. (1926). *Die Sprachlaute.* Berlin: J. Springer.

Stumpf, C. (1939). *Erkenntnislehre* (Vol. 1). Leipzig: J.A. Barth.

Tembrock, G. (1978). *Bioakustik, Musik und Sprache.* Berlin: Akademie-Verlag. (Sitzungsber. Akad. Wiss. DDR, Mathemathik - Naturwiss. - Technik, Jg. 1978, 1/N)

Terhardt, E. (1974). Pitch, consonance, and harmony. *The Journal of the Acoustical Society of America, 55*, 1061-1069.

Terhardt, E. (1976). Ein psychoakustisch begründetes Konzept der musikalischen Konsonanz. *Acustica, 36*, 121-137.

v.Helmholtz, H. (1870). *Die Lehre von den Tonempfindungen als physiologische Grundlage.* Braunschweig: Fr. Vieweg. (Original work published in 1863)

Vogel, M. (1993). *On the relations of tone.* Bonn: Verlag für System. Musikwiss.

Wundt, W. (1896/1928). *Grundriss der Psychologie.* Leipzig: Engelmann. (1st ed. 1896, 15th ed. 1928)

Zenner, H. (1993). Hören. In R. Schmidt (Ed.), *Neuro- und Sinnesphysiologie.* Berlin, Heidelberg: Springer-Verlag.

Schema and Gestalt: Testing the Hypothesis of Psychoneural Isomorphism by Computer Simulation

Marc Leman[1] and Francesco Carreras[2]

[1] IPEM, University of Ghent, Blandijnberg 2, B-9000 Ghent, Belgium
[2] CNUCE/CNR, Via S. Maria 36, I-56126 Pisa, Italy

Abstract. The hypothesis of *psychoneural isomorphism* assumes a correspondence between mental representations of perceived phenomena and brain-states or cortical representations. This hypothesis is tested by means of a computer simulation of perceptual learning of context-dependent pitch. The method is based on an ecological modelling technique which is related to the Gestalt notion of *molar behavior*. A large scale computer simulation has been set up in which a physiological model is confronted with the strains and stresses of a realistic musical environment, in particular a sound recording of J.S.Bach's *Das Wohltemperierte Klavier*. The simulation suggests that the hypothesis of psychoneural isomorphism is a valid one. It predicts a particular schema or functional organization of neurons specialized in context-dependent pitch processing at the cortical level.

1 Introduction

The correlative study of behavioral data, physiological data and data from computer simulation may highly contribute to a better understanding of the processes that underlay musical cognition (Leman, 1995). Accordingly, it has been argued that computer simulation, which is the topic of this paper, may provide a working hypothesis for the development, interpretation and explanation of psychological and physiological observations. We discuss a way in which computer simulation may contribute to the study of what Koffka (1936, p.67) called *the central task of Gestalt psychology*: "the study of behavior in its causal connection with the psychophysical field". In particular, it will be shown that ecological modelling provides a way to test the hypothesis of psychoneural isomorphism.

Previous work has been directed towards (i) the development of an ecological approach to computer simulation of perceptual learning of music, and (ii) the comparison of the output of computer simulation with behavioral data. Both the ecological approach and the comparative approach have been developed in a series of studies that focus on context-dependent pitch perception (Leman, 1995; Leman & Carreras, 1996). In these studies, a schema theory of perceptual learning has been based on a *physiological* model of the human auditory system (Sect.3.2). The model consists of a sensorial, perceptive, and cognitive layer. When exposed to acoustical signals of context-dependent pitch configurations, the model keeps a trace of these configurations in a long-term memory store of the cognitive layer. The trace, embedded in its (artificial) neural carrier, is called a *schema* or *cognitive map*. When exposed to tonal music, most interestingly, the schema develops a salient functional response structure whose global organization is represented in terms of circles of fifths.

In this context, it is instructive to conceive of a schema in terms of its *analogical* and *topographical* properties. A representation is called *analogical* when the relationships between the physical objects are reflecting the relationships between the perceived objects. A representation is called *topographical* when the similarity relationships between the perceived objects are given in terms of a *distance metric*: similar configurations are then represented close to each other on the surface of the neural network carrier, while dissimilar configurations are represented at a distance from each other. The above mentioned studies, provide theoretical evidence that the geometric organization of the schema for context-dependent pitch perception in the human brain of listeners educated in Western tonal music tradition, may be characterized by a topographical layout which corresponds to the circles of fifths.

The studies furthermore show that the distance metric, which is embedded in the topographical layout of the schema, is an excellent starting point for a comparative analysis with data from psychological studies (Krumhansl, 1990; Bruhn, 1988). These studies have been based on the probe-tone technique which allows the characterization of context-dependent pitch configurations by means of a *probe*. In this technique, a diachronical pitch (or chord) configuration, called a context, is followed by a probe tone (or probe chord) and the listener has to rate the degree of fit between the probe and the context. By varying the probe, one obtains a profile which describes the context. Such profiles, as shown later on in this paper, form the basis of a comparative analysis with profiles deduced from the computer simulated schema.

A central methodological aspect of our approach concerns the realistic tailoring of the input (Leman, 1997). Due to intensive computations needed for the simulation of perceptual learning, the first studies were based on a limited data set, typically a set of chords, or relatively short excerpts of tonal music (Leman, 1995). More recently, the model has been implemented on a super-computer (Carreras & Leman, 1996) and tested in a more realistic musical environment (Leman & Carreras, 1996). In addition, a probe technique for the computer model has been worked out and the comparison with psychological and physiological data has been initially explored (Leman & Carreras, 1996).

The present paper will give an account of a large scale computer simulation of perceptual learning, together with a description of the way in which comparative tests have been realized. Attention is thereby focused on the nature of the involved methodology (ecological modelling for the simulation of molar behavior) as well as on its epistemological implications (testing the hypothesis of psychoneural isomorphism). The organization of the paper is such that we first introduce our notion of ecological modelling and schema theory in connection to the Gestalt concepts of *molar behavior* and *psychoneural isomorphism*. In a second part, we introduce the framework for ecological modelling. Next we give an overview of a large scale simulation of perceptual learning and an analysis of the neural structures that carry the schema for context-dependent pitch perception. In the third part, we compare the behavioral output of the physiological model with data from experimental psychology.

2 Relationship with Gestalt theory

Schema theory and related computer modelling deal with the study of how schemata get organized and how they constrain perception. The present study is therefore related to the study of behavior in its causal connection with the auditory acoustical environment.

2.1 Molar Behavior and Ecological Modelling

In a recent study which builds upon *Music and Schema Theory* (Leman, 1995), a computer simulation of perceptual learning of context-dependent pitch configurations was set up under realistic conditions, using real music played by a real musician on a real musical instrument (Leman & Carreras, 1996). The basis of this simulation is a recording of Book I of J.S. Bach's *Das Wohltemperierte Klavier* by Kenneth Gilbert [1]. The music is composed of 24 preludes and fugues in all tonalities. The total duration is 1 hour 50 minutes and 22 seconds. The instrument used in the recording is a harpsichord built by Jan Couchet, in Antwerpen, dated 1671. It is tuned in the old pitch of LA=392 Hz. Bach's music is processed by a physiological model of the auditory system which implements context-dependent pitch processing. The reduction of the psychophysical field to pitch is justified by the fact that Western tonal music has a rich syntax for synchronical as well as diachronical pitch configurations. Pitch configurations thereby operate as an independent category next to timbre and rhythm configurations.

Given this context of simulation, it is straightforward to relate Koffka's (1936) notion of *molar behavior* to the ecological modelling approach adopted here. Molar behavior is understood as the behavior which takes place in an environment endowed with forces, strains and stresses that determine what an organism will do in that environment. It is distinguished from molecular behavior: "the process which starts with an excitation on the sensory surface of an animal, is conducted by nerve fibers to nerve centers, switched over to new, efferent nerves, and ends in a muscle contraction or a gland secretion" (p. 26).

[1] Available on CD ARCHIV 413439-2

In ecological modelling, the aim is to expose a physiological model of cognitive information processing to the cultural constrained strains and stresses of a physical environment. In our particular example, the environment is Bach's performed music. The implied strains and stresses have an effect on the behavioral output of the physiological model. In this paper, we show that a probe technique, similar to the one used with human subjects, can be used to map out the behavioral constraints of the model.

From a methodological point of view, it is important to consider the role of the modeler. In ecological modelling, this role is limited to the specification of the type of interactions between (i) the physical representation of the environment and (ii) the physiological model of auditory information processing. In accord with the idea that (*molar*) behavior is determined by the forces and strains that are inherent in the environment, the model is assumed to be able to pick up aspects of the so-called *molar field* (here: the auditory environment) by *self-organization*.

2.2 Psychoneural Isomorphism and Schemata

Schema theory is nowadays a central notion in cognitive science. A critical analysis of the schema concept has been presented by Seifert (1991). Bregman (1990, p.40) defines a schema as a control structure in the human brain that is sensitive to some frequently occurring pattern, either in the environment, in ourselves, or in how the two interact. Arbib (1995) defines a schema both as a store of knowledge and the description of a process for applying that knowledge. In Leman (1995), a distinction has been made between long-term data-driven perceptual *learning* of a schema and short-term schema-driven *control*. The first is effectuated by dynamic principles of self-organization (and is the topic of the present paper), the second (not treated in this paper) is realized by associative processes that pertain to an attractor dynamics. The attractor dynamics implements the Gestalt principle of *Prägnanz* or tendency towards a good form. The representation inherent in a schema is thereby a result of its dynamic properties. Köhler too stressed the role of dynamic interactions between the parts of the self-regulating representational structure. Some aspects of the neural implementation of representation may have been misconceived by Köhler, but the general idea of a self-regulating brain map in which "the whole is more than the sum of its parts", is very clearly associated with the concept of schema.

W. Köhler (1924, 1971) has proposed that a connection between the physical level, the physiological level, and the behavioral level can be conceived of in terms of isomorphism relationships. Köhler's concept of *psychoneural isomorphism* defines the connection between the physiological and the behavioral levels. Its relevance to modern cognitive science has been discussed by Scheerer (1994). The type of isomorphism involved is of the second order (Shepard & Chipman, 1970) in that the representational structures resemble each other in terms of the relationships among the objects involved, not so much in terms of the object representations themselves. Köhler's concept of psychoneural isomorphism was defined in terms of system properties such as the idea that time-independent states are maintained as a whole and that they are conditioned by a total topography which has a dynamic constitution with localized reactions to specific stimuli. One region is thereby more strongly affected by near than by distant regions.

An interesting parallel can now be drawn between the notion of psychoneural isomorphism and schema theory, as it has been applied to the study of context-dependent pitch perception (Leman, 1995). Schemata are memories endowed with forces and strains that are formed by absorbing the forces and strains embedded in the physical environment. In our study, the response structure of the schema defines an isomorphism between the neural representation on the one hand, and the behavioral responses on the other hand.

In addition to the psychoneural type of isomorphism, an obvious connection exists between the acoustical properties of the sounds and their representation at the neural level. The latters are indeed generated by physiological mechanisms specialized in the processing of musical sound signals (Sect.3.2). We assume that the processing of periodicity pitch is neurologically predefined, while the processing of context-dependent pitch makes use of learned knowledge. The connection between what is learned and what is physiologically based, obviously, is an intricate one.

The hypothesis of psychoneural isomorphism can be put at test in computer simulations that subsume the ecological modelling approach. The response structure of a schema, which is carried by an artificial neural network that built up this schema by self-organization (by mere exposure to tonal music), can be mapped out and compared with response structures of human behavioral responses (Sect.5). The good degree of fit, between the behavioral output of the computer model and the behavioral responses of human listeners, gives rise to the hypothesis that a similar correspondence exists between the neural layer of the computer model and the neural representation in human listener. This implies that somewhere in the human cortex, probably close to the auditory sensory cortex, an area exists where neurons react in a way which recalls the topographical representation of the circle of fifths as it is found in the computer model. This prediction is furthermore nourished by the assumption that the underlying ecological methodology gives thus far the best chances for a realistic modelling of these processes. The underlying auditory model and principles of self-organization simulate the most essential aspects of human auditory processing and cortical self-organization. Neurophysiological studies will have to figure out if the hypothesis, now confirmed with computer simulations, is indeed a valid one (Janata, 1995).

3 The Framework for Ecological Modelling

The approach of ecological modelling defines the framework in which the Gestalt principle of psychoneural isomorphism will be tested. It consists of different so-called *environments*: the physical environment, the environment of human information processing, and the musical environment. An implementation environment pertains to the factual implementation of the model.

3.1 The Physical Environment

The physical environment in which music is situated is here reduced to its pure acoustical aspect. No visual or tactile effects are involved. The acoustical representation is based on a CD recording, from which we take a digitized waveform using a sampling rate of 20000 sa/s and a resolution of 16 bits.

3.2 The Human Information Processing Environment

The environment of human information processing system is reduced to processes of perceptual learning of context-dependent pitch information. The computer model of perceptual learning contains two parts:

- A perception module: It consists of a sensorial layer based on an auditory model for speech processing (Van Immerseel & Martens, 1992) and a pitch extraction part based on a periodicity analysis of neural discharge patterns in the auditory channels of that model (Leman, 1995). The acoustical front end implements outer and inner ear filtering and cochlear hydro-dynamical filtering by means of a bank of 20 asymmetric overlapping bandpass filters. The latters, also called *auditory channels* transform sounds into neural firing patterns, using half-wave rectification and dynamic range compression, short-term adaptation and synchrony reduction. For each neural firing pattern that comes out of one of the 20 auditory channels, a periodicity analysis is done by means of a short-term-autocorrelation analysis using delays of 0.4 ms. The analysis is applied to frames of 30 ms. The periodicity analysis is repeated every 10 ms. The *completion image*, which is similar to the so-called virtual pitch pattern (Terhardt, Stoll, & Seewann, 1982), is obtained by adding the short-term-autocorrelation results of a frame over the 20 auditory channels. The patterns which are obtained have 75 data points, from which we take a section of 56 data points.
  Figure 1 shows the completion images of 7.65 seconds of the first prelude of Bach's *Das Wohltemperierte Klavier* (first two measures). Each data point along the vertical axis represents a delay of the autocorrelation analysis from which we can calculate the frequencies in Hz-values. Yet, we are not interested in the precise pitch values but in the global Gestalt (pattern) that expresses the periodicity analysis. As the model is based

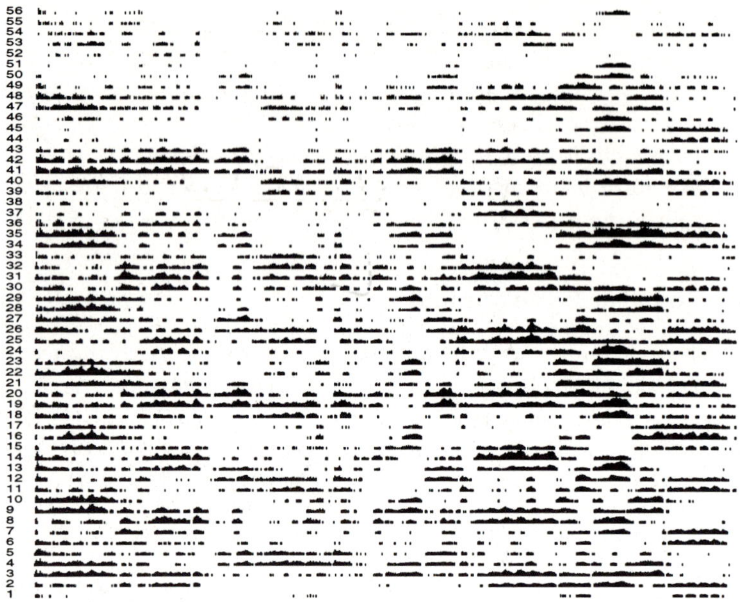

Fig. 1. Tone completion images formed by the processing of the first 7.65 seconds of Bach's *Das Wohltemperierte Klavier* (first two measures). The vertical axis represents the frequency content in terms of autocorrelation delays, the horizontal axis represents time

on *envelope* patterns of neural firing rates, it does not take into account the fine-temporal aspects of periodicity analysis either. As a consequence, aspects of pitch shift, as explained by the theory of Langner (1997), are not taken into account. The leaky-integrated results of completion images over time are called *context images*. The images give an account of the context-dependencies among perceived musical pitch patterns over a time period of about 3 seconds (Fig.2).

– A cognition module: It accounts for *data-driven* perceptual learning and it is based on the Self-Organizing Map (SOM)-architecture (Kohonen, 1995) using a two-dimensional array of neurons of 100 x 100 (=10000) neurons in a torus architecture. The latter means that the borders of the network are connected to each other. The learning of the network is data-driven, which means that it is not guided by any previous knowledge, rather, all information is coming from the environment. Learning by self-organization is realized at long-term and it yields a schema of context-dependent pitch perception. The network maps the 56-dimensional tone context images onto a two-dimensional grid of artificial neurons but it keeps the topographical order of the implied relationships that are imbedded in the auditory images. The various tone context images are thereby mapped onto a set of neurons so that neurons compete for representation

Fig. 2. Tone context images formed by leaky integration of the images in Fig.1. The vertical axis represents the frequency content in terms of autocorrelation delays, the horizontal axis represents time

and adapt to the invariant information in the images. The ability to generalize is a central property of perceptual learning and it is exactly this effect which is exploited in the computer simulations.

3.3 The Musical Environment

The musical environment is conceived of in terms of the set of cultural constraints imposed upon the acoustical environment. The cultural properties present in the recording of Book I of J.S.Bach's *Das Wohltemperierte Klavier* are representative for the context dependent pitch relationships used in the Western tonal music of the Baroque period. Bach's piece explores the tonal possibilities in all keys and therefore it provides a microcosmos of the entire tonal musical environment.

3.4 The Implementation Environment

The simulation has been performed by separating the perceptual and the cognitive module. First, the sounds have been processed with the perceptual module. The resulting tone context images are then stored in a file. Secondly, the file with tone context images is processed by the neural network. When physiological models are to be tested in a more realistic environment, then the amount of data is huge and the computational efficiency of the model becomes a critical factor in research (Carreras & Leman, 1996). The algorithm, which implements this cognitive layer, has been parallellized in an efficient way on a fast super-computer (IBM SP2) which is available at CNUCE/CNR Pisa.

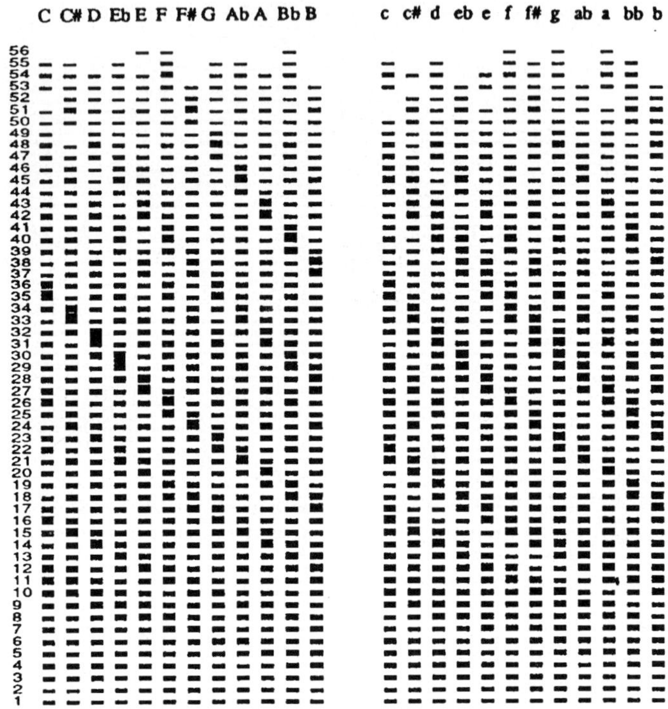

Fig. 3. *Tone center images* have been extracted from chord progressions that represent *cadences*, i.e. typical tonal stable configurations. The tone center images are based on *context images* which are produced by the perception layer. They have then been reduced to 24 images: 12 tone center images standing for 12 context-dependent pitch configurations of the major tonalities and 12 tone center images standing for 12 context-dependent pitch configurations of the minor tonalities. The patterns are here presented next to each other. The vertical axis represents the pitch content in terms of time-delays of the autocorrelation analysis

4 A Large Scale Simulation of Perceptual Learning

In this section, we discuss the results of a large scale simulation of perceptual learning with a particular focus on the analysis of the neural carrier of the emerged schema. The analysis provides a first step in the testing of the hypothesis of psychoneural isomorphism. (The second step will consist

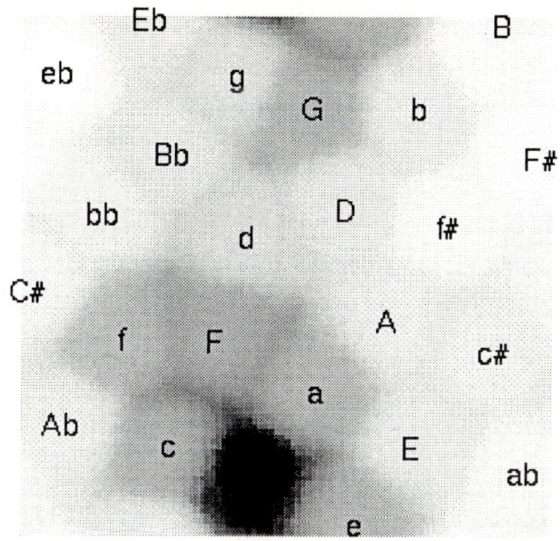

Fig. 4. Schema response to the tone center representing the context-dependent tone configuration of C major. The neural network which carries this schema contains 10000 neurons in a two-dimensional array of 100 x 100. The network has first been trained with the patterns shown in Fig.3. Afterwards, the tone center image of C major has been presented. The figure shows the activation of all the neurons in the network. Black means high activation, white means low activation

in showing that the response structure obtained corresponds with behavioral data.)

4.1 Parameters of Schema Learning

The SOM-algorithm (see Appendix A) allows the parametrization of several functional components (Kohonen, 1995). The main parameters are:

- The *number of epochs*, which defines the number of times the whole set of tone context images is processed by the neural network. It is necessary to randomize the data set for each epoch.
- The *learning rate*, which influences the degree of adaptation of the network to the tone context images. The learning rate can be held constant during the simulation, or one can gradually decrease it.
- The *radius of adaptation*, which selects the neurons that need to be adapted in response to input. From a geometrical point of view, the neurons adapted are those which lay in the neighborhood of the neuron which gives the highest response to a particular tone context image. The radius of adaptation should be wide in the beginning and gradually decrease to zero.

- The *similarity measure*, which defines the type of distance measure to be employed in calculating the response of the neurons to a particular tone context image given as input. In most of our simulations, Euclidian distances have been calculated. Minkowski metrics of higher order did not yield good results.

A number of simulations have been performed with different initial values of the above parameters. The trained map of each simulation was then analysed with various techniques, in order to assess the impact of the selected parameter configuration.

4.2 Analysis of the Emerging Schema

The processing of the musical input data by the model yields a schema by self-organization. Afterwards, the schema has to be analysed in order to obtain an overview of its implied structural organization. In this section, we restrict ourselves to two techniques. In a first technique, the schema is tested with images representing typical context-dependent pitch configurations. In a second technique, the schema is analysed by means of a data visualization method. A third technique is based on the use of probes. It is employed in the comparative analysis and will therefore be discussed in the next section.

- The first technique uses *tone center images* as test patterns. Tone center images have been extracted from chord progressions that represent *cadences*, i.e. typical tonal stable configurations. The chord progressions were synthesized sounds, using tone spectra that contain frequency components at octave intervals with an emphasis of the spectral components between 500 Hz and 1000 Hz (Shepard, 1964). Using a technique described in Leman (1995), 72 cadences of different types have been processed by the perception module. The resulting context images have then been reduced to 24 tone center images: 12 tone center images standing for 12 context-dependent pitch configurations of the major tonalities and 12 tone center images standing for 12 context-dependent pitch configurations of the minor tonalities (Fig.3).
  An interesting *theoretical* result has been obtained by using these 24 images as input patterns for a simulation of perceptual learning. Thus, instead of using Bach's music, we used here only those 24 images as input to the neural network. After learning we tested the network with the same 24 images. The result is shown in Fig.4. The figure gives the output of a two-dimensional array (a grid or map) of neurons (100 x 100). The labels c, F, f, a, e, etc... are put on the map at those places where the neurons give the highest response to an input of that type. The gray values show how the artificial neurons respond to an image representing the context-dependent pitch configuration of C major. The black spot shows the region with the highest response for C. There are several grey regions representing medium activation in the centers that correspond

to c, F, a, e, G, and other regions which are almost completely white (e♭, F♯, ...). These regions show little response to the tone center image. Given the torus architecture of the network, the global picture has a good resemblance with the topographical representations that have been obtained by Krumhansl (1990) in psychological experiments.

Fig. 5. (a) Three-dimensional representation of the schema response to the tone center representing the context-dependent tone configuration of C major. (b) Three-dimensional representation of the schema response to the tone center representing the context-dependent tone configuration of C♯ major

In the large scale computer simulation with Bach's music, we were curious to know whether the schema would (i) respond to these 24 patterns, and (ii) display some structure which is comparable to the theoretical result shown in Fig.4. Thus, in the large scale simulation, the network was first trained with Bach's music and afterwards tested with the 24 tone center images of Fig.3. Figure 5a-b shows that the network, after training, can differentiate between the tone center images of C (Fig.5a) and C♯ (Fig.5b). The network neurons are here represented by the horizontal plane, and the activation is represented by a landscape of peaks and valleys. Figure 6 shows the labels of the tone centers that stand for context-dependent tone configurations of the major tonalities. The grey values represent the activation response of the neurons to the tone center image of C (similar to Fig.5a but here presented as a density map). The analysis shows that the response of the network is differentiated, although the outcome is not so clearly ordered as in the theoretical case. Recall, however, that the tone center images are based on sounds with an entirely different spectral content than the sounds used for training the network. The fact that the training was done in the old tuning of

LA=392 Hz did not provide any problem with the test patterns, which are based on LA=440 Hz, using so-called *Shepard-spectra*.

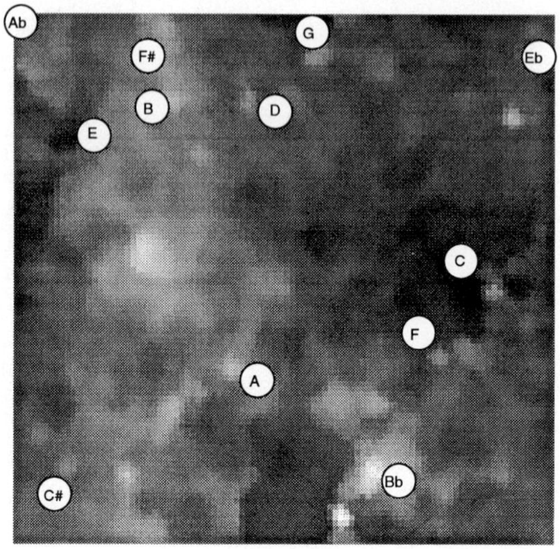

Fig. 6. Schema response to the tone center image of C-major. The labels show the response positions to the context-dependent tone configurations of the major tonalities

– A second technique, based on a visualization method developed by Kraaijveld et al. (1992) yields a so-called *boundary map*. In this map, the interpoint distances of the schema are graphically displayed to show the underlying structure of the data. Figure 7a displays the result of applying the visualization method to the idealized case (where only 24 tone center images are processed), while Fig.7b displays the results when visualization is applied to the schema that was trained with Bach's data. Obviously, Fig.7a shows nice boundaries of the clusters inherent in the schema. Such a clear boundary map could not be observed when the map was trained with Bach's music. The reason may be due to properties of both the visualization technique, and the self-organizing algorithm. First of all, the visualization technique is based on a local analysis of distances between neighboring neurons in the map. When classes overlap it is more difficult to apply this technique of visualization in a successful way. Secondly, the SOM-algorithm is sensitive to the statistical properties of the data. The music of Bach contains a lot of modulations and passages in which the tone center is less clearly expressed. On the map, such passages may cause a smooth transition from one tone center to another such that the boundaries become invisible with a local technique.

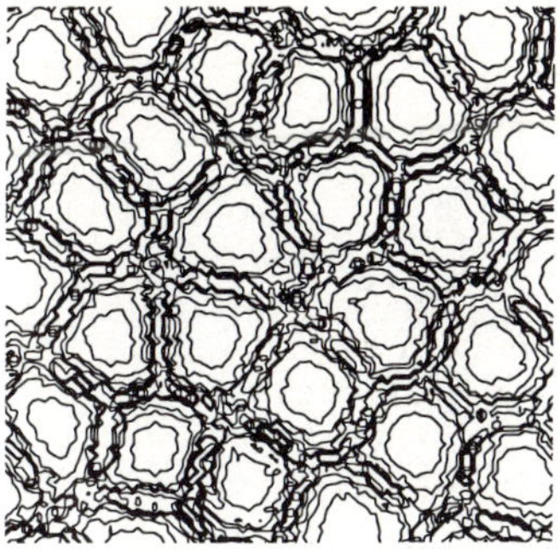

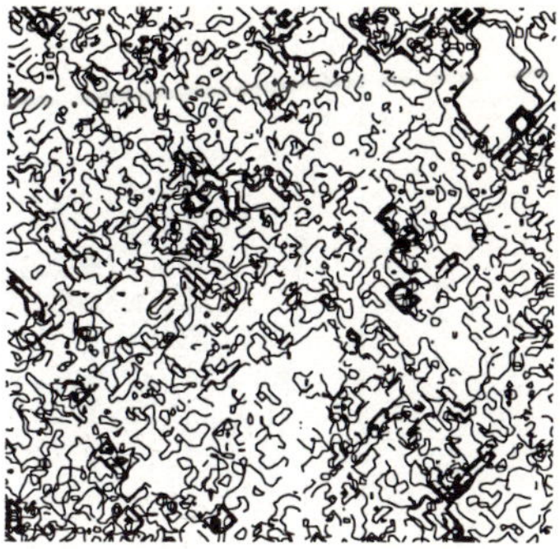

Fig. 7. (a) Boundary map of the schema shown in Fig.4. (b) Boundary map of the schema shown in Fig.6

In most of the simulations, apart from those which used non-Euclidean metrics, the network showed a tendency to organize itself into a global order. The simulations that use a variable learning rate and a repeated input of the data set (e.g. up to three times) have produced maps that have a better global ordering than the ones obtained by using a fixed learning rate of, e.g. 0.01. The use of a weight matrix trained with the ideal 24 tone center images as an initial matrix produced the most satisfactory map. Non randomization of the initial weight matrix introduced strong biases in the final results. In general, however, learning has been quite robust: small changes in parameters did not affect the output in any significant way.

5 A Comparison of Behavioral Output of the Model with Human Subjects' Responses

The analysis in this section aims to provide evidence for the second part of the hypothesis testing of psychoneural isomorphism. In particular, it is shown that a behavioral analysis of the model fits with a behavioral analysis of human responses. The analysis technique used in this section has much in common with the above mentioned probe tone technique used in Krumhansl's (1990) studies. The technique is based on taking two snapshots of the schema response during the processing of the musical sequence. One snapshot is taken after presentation of the context which, in this case is a I-IV-V sequence of

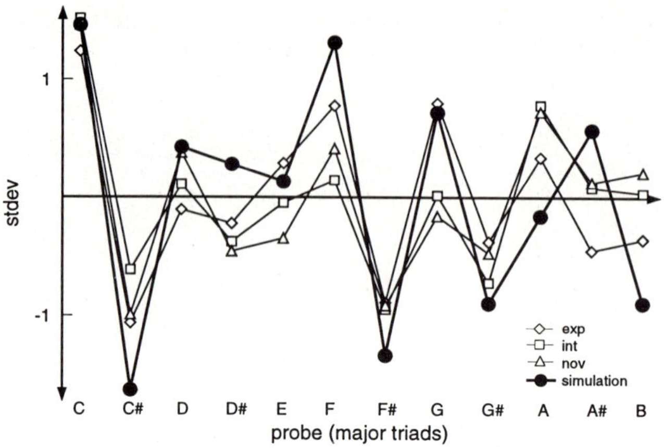

Fig. 8. Profile from experimental tests with musical experts (exp), novices (nov), subjects with initial training in music (int) and the computer (simulation). The context is C-major and major triads were used as probes

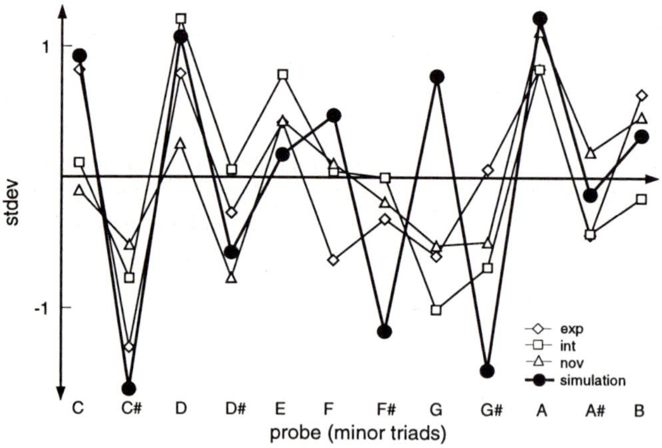

Fig. 9. Profile from experimental tests with musical experts (exp), novices (nov), subjects with initial training in music (int) and the computer (simulation). The context is C-major and minor triads were used as probes

Table 1. Correlations between the profiles obtained from expert, intermediate and novice subjects, computer simulation, data by Krumhansl (1990) and data by Bruhn (1988), (C-major context with major triads as probes)

	simulation	Bruhn (1988)	Krumhansl (1990)
exp	0.85**	0.91**	0.76**
int	0.70*	0.55	0.69*
nov	0.72**	0.59*	0.67*
mean	0.80**	0.72**	0.75**
simulation	1	0.87**	0.63*
Bruhn (1988)	0.87**	1	0.70*
Krumhansl (1990)	0.63*	0.70*	1

* Significant at $p < .05$
* Significant at $p < .01$

chords in the tonality of C-major, and another snapshot is taken after the presentation of the complete sequence which is a I-IV-V-X sequence of chords, where X is a variable that stands for one triad out of the 12 major and 12 minor triads (see Appendix B for audio examples). A snapshot of the schema response is represented by a vector of 10000 (100 x 100) data points. By calculating the similarity between the two vectors representing the snapshot of the schema response at two different time points it was possible to obtain a profile which stands for the network's reaction to context-dependent pitches. The similarity value obtained is the schema's degree of fit between the last chord of the sequence and the context. This is taken to be the equivalent of human behavioral responses when the degree of fit between the last chord heard and the context has to be rated. A profile is then obtained by putting next to each other the similarity results obtained with the different triad chords. Once such a profile has been obtained, it can be compared with profiles from human listeners. An experiment, which repeats the experiments done by Krumhansl (1990) has been set up with 20 subjects, students and personnel from the university of Ghent (Herreman, 1996). They had to listen to the sound sequences with which the computer model had been tested, and rate the degree of fit between the last chord and the context. The subjects could click an icon on a computer screen to hear a chord sequence and move the icon along a horizontal axis to express the degree of fit between the last heard chord and the context. The subjects were divided into three groups: musical novices, experts and an intermediate group. In Fig.8, a profile is shown in which the mean values for novices, experts and intermediate group are displayed next to the profile of the schema responses (called *simulation*). The values represent the degree of fit between the last chord of the sequences and the context, and this for the major triad chords based on the pitches c, c♯, d, d♯ etc.... Figure 9 shows the profiles for sequences with minor triads as probes.

Table 1 compares the profiles obtained from expert, intermediate and novice subjects with the results of the computer simulation and data provided by Krumhansl (1990) and Bruhn (1988). The table compares profiles which are based on C-major contexts with major triads as probes. The correlation between these profiles is significant. Table 2 compares the profiles which are based on C-major contexts with minor triads as probes. Here we notice that the correlation is less significant.

Table 2. Correlations between the profiles obtained from expert, intermediate and novice subjects, computer simulation, data by Krumhansl (1990) and data by Bruhn (1988), (C-major context with minor triads as probes)

	simulation	Bruhn (1988)	Krumhansl (1990)
exp	0.62*	0.49	0.55
int	0.54	0.63*	0.40
nov	0.61*	0.67*	0.55
mean	0.66*	0.67*	0.56
simulation	1	0.79**	0.73**
Bruhn (1988)	0.79**	1	0.63*
Krumhansl (1990)	0.73**	0.63*	1

* Significant at $p < .05$
* Significant at $p < .01$

6 Discussion of the Ecological Methodology Approach

In order to have a better understanding of the relevance of Gestalt-based computer simulations, it is instructive to have a look at the main characteristics of the simulated processes and the associated limitations of the ecological modelling approach. It is of outmost importance to consider these aspects in order to appreciate the strength of the obtained results.

- First of all, one should be aware of some specific properties of the perceptual simulated domain. A closer look reveals that it is characterized by (i) context-dependency, (ii) non-denotational semantics, (iii) unconscious information processing, and (iv) cultural factors embedded in sounds. Some of these aspects are favorable for ecological modelling but the particular implementation entails limitations as well.
  - *Context-dependency* means that the perception of a particular tone depends on previous tones. Context-dependency is obviously time constrained and temporal integration at a relatively large scale (3 seconds) has been taken into account. Integration seems to work quite

well when auditory representations are conceived of as images (vectors). Yet the type of integration is rather brute force: no account is given of grouping, rhythmical or motoric aspects involved in listening.

- Pitch (but also timbre) configurations may have a so-called signal function when they point to a particular cause of the sounds or events in the outside world. In most cases, however, sounds in music occur as *non-denotational* signs. Individual tones do not point to anything else, their semantics is self-referential. Within a context of other tones, however, tones may become the subject of a tone semantics. For example, they may become hierarchically related to other tones. A pitch may develop a certain function, for example, as tonic, sub-dominant, leading tone or other. The tonal system is based on a non-denotational semantics which allows a straightforward ecological modelling approach. A denotational semantics would be much more difficult to implement.
- *Unconscious* information processing means that the information processing is direct and automatic and that it does not involve any seemingly conscious thinking, effort or mental calculation. The semantics of tonal music indeed appears as something very natural and effortless. We often speak about the *sense of tonality* or the *tonal feeling* to express this observation. It is a favorable condition for ecological modelling.
- *Cultural factors embedded in sounds* is related to what has been called the musical environment. The context-dependent pitch configurations studied in tonal music are the outcome of a cultural development. So, when perceptual learning is studied, one may ask what is learned and what is physiologically determined. This is an intricate question of fundamental importance for the Gestalt approach. The physiological model provides a nuanced answer. The perception of the low pitch or virtual pitch, for example, is a fundamental aspect of context-dependent perception. There is now evidence that virtual pitch perception is a phenomenon which can be explained at the physiological level (Langner, 1997). In connection to this, the preference for certain harmonic intervals, in particular those that naturally result from periodicity analysis seems to be physiologically based as well. Yet at a higher level of information processing, in particular when tone configurations are involved, it can be argued that learning plays a prominent role. Learning may come down to an extraction of the statistical features inherent in the musical environment (and thus determined by cultural factors). However, during the cultural development of a society, learned knowledge may have an impact on the choice of musical instruments and the particular setup of pitch configurations. Over many generations of members of a cultural society, the feedback of the human cognitive system onto the acoustical environment may foster a tendency towards certain stable human-

environmental interactions. In this view, the Western tonal system is assumed to be the stable outcome of a complex feedback dynamics that took several centuries. Given the physiological constraints of the auditory system, which is assumed to be the same for different cultures, such a feedback may account for a wide variety of musical environments. It is known from studies in complex dynamics that very small changes at the beginning of a process can lead to very different stable outcomes at the end. That type of dynamics could explain the large differences in the musical environment of different cultures. A more detailed discussion of this dynamics, however, is out of the scope of this paper. We restrict ourselves here to the assumption that context-dependent pitch perception is learned and that that type of perceptual learning can be modelled in an ecological way. Cultural factors outside sound, however, are not taken to be part of the perceptual field in this model.

- Important aspects of music perception such as grouping, segregation, segmentation, are not taken into account either. In some cases, such as monodic pieces, these aspects may probably have a profound effect on the perception of the tonal center of a particular configuration of tones. Tone center perception is a most general aspect of context-dependent pitch perception. The fact that we do not take into account structural aspects is a limitation of the present model. Moreover, aspects that deal with musical structure, expression or emotion are typically not taken into account either. On the other hand, it can be argued that tone center perception is a basic and rather independent phenomenon of perception in which structural aspects play a role at a secondary level. Some aspects of expression, however, are indeed indirectly taken into account because their sonological patterns are somehow preserved in the auditory images of the pitches.
- Furthermore, perceptual representations are conceived of in terms of *auditory images* rather than *symbols*. Unlike the former assumption, this is not a choice or deliberate tracing but rather an essential principle of ecological modelling. One may conceive of many types of auditory images but they all are characterized by the fact that they *reflect* to some extend properties of the musical environment. Auditory images are ultimately causally shaped by the physical properties inherent in the sounds of the musical environment. Many computer models of music cognition,however, have thus far been based on symbol representations. They *point* to the objects in the world without reflecting any of the physical properties of the object. It can be argued that connectionist approaches to music are often in the spirit of symbolic information processing, thereby neglecting important aspects of the acoustical environment.
- Limiting ourselves to only one aspect of the acoustical environment has as a consequence that the molar behavior may be incompletely simulated. Ultimately, this restriction may lead to the observation that ecological

modelling cannot deal with the extra-musical sociological and cultural constraints unless we are able to simulate human perception and cognition in all its details, including visual and motoric activities, thinking, planning and so on. Hence, one could utter that ecological modelling is a priori leading to a dead end. We are quite aware of this argumentation: the impact of worldviews on one's musical conception and perception may be difficult to model in this paradigm (Moisala, 1995). Nevertheless, detailed studies may open many new insights. In particular, the type of low-level music perception simulated is assumed to be less influenced by worldviews than, say, musical composition. It is furthermore evident that techniques and concepts developed in ecological modelling may provide important suggestions for systematic musicologists to focus on the relationships between human information processing and the musical/sounding environment.

- Finally, as suggested in the introduction, there is one important technical aspect associated with ecological modelling. Using an artificial neural network of 10000 neurons as the cognitive layer of the physiological model, super computing power is needed in order to process about one hour of music in a way that is flexible enough for cognitive science research (Carreras & Leman, 1996). Such technical aspects are often neglected, or rather, they often force researchers to reduce the musical information with the risk of oversimplification. Special care has therefore to be taken that methodological problems do not interfere too much with epistemological assumptions of the ecological approach.

These observations show that the study of context-dependent pitch perception is constrained but rather well-defined. Ecological modelling is a powerful method, although it is connected to a particular aspect of the domain that is simulated.

7 Conclusion

The hypothesis of psychoneural isomorphism has been introduced by W. Köhler to explain the connection between mental representations of perceived phenomena and brain-states or cortical representations. In practice this hypothesis comes down to what Shepard and Chipman (1970) call a *second order isomorphism*: the statement that percepts are related to each other in the perceptual field as their cortical correlates are related to one another in the cortex, so that the structural relations of one field are reproduced in the other. Mental representations can be mapped out using methods of experimental psychology, while brain representations can be mapped out using different techniques related to the recording of cortical activity (such as EEG, MEG, fMRI, PET). In recent years, the accuracy of brain mapping techniques has been increased, yet the techniques are often limited to specific test paradigms in which the generation of a natural perceptual field or a genuine context is critical. At this point, computer modelling of brain physiology may provide an additional opportunity for hypothesis testing.

In this paper, the Gestalt hypothesis of psychoneural isomorphism has been tested by means of an ecological modelling approach applied to the perceptual learning of context-dependent pitch configurations. In particular, a physiological model of context-dependent pitch perception has been exposed to the physical representation of a recording of J.S.Bach's *Das Wohltemperierte Klavier*. The cognitive layer of the model developed a schema–or structured perceptual map–which emerged by self-organization. The schema was then tested: (a) by making an analysis of the neural carrier of the schema, (b) by making an analysis of the behavioral output. The neurons of the physiological model show a strong tendency towards a functional topography in terms of the circle of fifths. The behavioral output of the physiological model shows a convincing correlation with the behavioral responses of human subjects when confronted with the same stimuli. The correspondence between the behavioral results and the results obtained at the neural level of the model seems to confirm the hypothesis of psychoneural isomorphism.

The results of the computer simulations are in line with the hypothesis: (i) that the brain is capable, within the neurophysiological biased preference for consonant pitch intervals (Langner, 1997), of extracting an invariant statistics from the tonal music and keep it in internal memory, (ii) that the brain responds to context-dependent pitch configuration in an ordered and highly systematic way (reflecting this invariant statistics) (Krumhansl, 1990). Most importantly, the computer output shows a very high correlation with studies in experimental psychology in which this latter assumption has been tested on the basis of behavioral studies. The study shows that computer simulation may offer a bridge between the theory of tonal music, physiology of pitch perception (Schreiner & Langner, 1988; Langner, 1992, 1997), and experimental psychology (Bruhn, 1988; Krumhansl, 1990). The main result of this computer simulation was a particular hypothesis (or prediction) of psychoneural isomorphism: it states that the neuronal responses of the learned tonal schemata have a topographical organization in terms of the circle of fifths, and that this is the response structure which is also reflected in the behavioral studies.

Acknowledgement

Part of this work has been supported by the Fund for Scientific Research - Flanders (FWO-V) and by the National Research Council of Italy (CNR).

Appendix:

A The SOM Algorithm

The neural network model used for the simulations is of the Self-Organizing Map (SOM) type. The SOM algorithm has been developed in 1981-82 by Kohonen (1995) in the effort of modeling a learning principle responsible for the formation of maps in the brain cortex, such as visual and somatotopic maps. These maps develop progressively and autonomously from the action of external stimuli and a process of self-organization. A well defined topographical layout is acquired after learning.

The SOM defines an adaptive network of reference vectors whose values are modified to ultimately reflect the internal structure of the input data space in an ordered way. In other words the SOM realizes a non-linear mapping from a multidimensional data space $\Re_n$ onto a lower dimensional array. In the present case the input signal space is composed of the 56 dimensional vectors of the context images and the output layer is a 100 by 100 neurons square array shaped in a torus surface (the opposite edges of the array are connected). Each neuron m_i in the array is associated with a n-dimensional weight vector $W_i = (w_{i1}, w_{i2}...w_{in})$. The main steps of the algorithm are the following:

- The weight vectors are initialized by choosing the initial weights values. A common choice is to supply random numbers.
- At each discrete interval of time t = 1,2, ... an input vector $x(t)$ is presented to the network.
- The input vector $x(t) = (x_1(t), x_2(t)...x_n(t))$ is compared with all the $W_i(t)$ in the array. The index $c(t)$ of the neuron which is associated with the $W_i(t)$ having the minimum distance with the input vector $x(t)$, according to a predefined metric, is selected. The metric most often used is the Euclidean distance metric. The index of the closest reference neuron is:
  $c(t) = argmin(i)\{||x - W_i||^2\}$
- During this phase of learning, the winning neuron and those within a neighborhood set N have the associated vectors updated by an amount that is a decreasing function of the distance of a neuron from the winning one. The weights of all the neurons outside the neighborhood set are kept constant. Accordingly the plasticity of the network is gradually modified. A general updating rule is of the form:

$$W_i(t) = W_i(t-1) - h(t) * (x - W_i(t-1)) \qquad if\ i \in N,$$
$$W_i(t) = W_i(t-1) \qquad if\ i \notin N.$$

The value of the parameter $h(t)$ depends both on the distance of the i-th neuron from the winner neuron and on the learning rate. The latter rules the amount of change that must be applied to the weights. In the previous simulations the following function was adopted:

$$h(t) = learning\_rate * adaptation\_parameter$$

Where the learning rate is a function of the number of input vectors so far processed $f(t)$ and decreases to a fixed threshold of 0.02 while processing a predefined number of input vectors M.

$$learning\_rate = 0.9 * (1 - f(t)/M) \qquad if > 0.02$$
$$learning\_rate = 0.02 \qquad if \leq 0.02$$

The adaptation parameter regulates the amount of change as a function of the geometrical distance d from the selected neuron when the maximum adaptation radius is D:

$$adaptation\_parameter = exp(-d/2 * D^2) \qquad d \leq D$$

The process ends either when the predefined maximum number of input vectors to be processed is exhausted or when the value of a threshold error parameter is reached.

B Sound Examples (CD-tracks 11-15)

- Sound Example #1: The chords of the C-major context (I IV V). (This sound example, as well as the following sound examples are built using Shepard-spectra.)
- Sound Example #2: Context in C-major, followed by a C-major triad probe. Fig.5a shows the network output after processing this chord sequence.
- Sound Example #3: Context in C-major, followed by a C♯-major triad probe. Fig.5b shows the network output after processing this chord sequence.
- Sound Example #4: Context in C-major, followed by a C-minor triad probe.
- Sound Example #5: Context in C-major, followed by a C♯-minor triad probe.

References

Arbib, M. (1995). Schema theory. In M. Arbib (Ed.), *The handbook of brain theory and neural networks.* Cambridge, MA: The MIT Press.

Bregman, A. (1990). *Auditory scene analysis: The perceptual organization of sound.* Cambridge, MA: The MIT Press.

Bruhn, H. (1988). *Harmonielehre als Grammatik der Musik.* Munich: Psychologie Verlags Union.

Carreras, F., & Leman, M. (1996). Distributed parallel architectures for the simulation of cognitive models in a realistic environment. In E. D'Hollander, G. Joubert, F. Peters, & D. Trystram (Eds.), *Parallel computing: State-of-the art perspective.* Amsterdam: Elsevier.

Herreman, B. (1996). *Stageverslag: Stage Experimentele Psychologie aan het Instituut voor Psychoacustica en Elektronische Muziek Universiteit Gent* (Tech. Rep.). Brussels: Vrije Universteit Brussel.

Janata, P. (1995). ERP measures assay the degree of expectancy violation of harmonic contexts in music. *Journal of Cognitive Neuroscience, 7,* 153–164.

Koffka, K. (1936). *Principles of Gestalt psychology.* London: Kegan Paul, Trench, Trubner & Co.

Köhler, W. (1924). *Die physischen Gestalten in Ruhe und im stationären Zustand.* Erlangen: Verlag der Philosophischen Akademie Erlangen.

Köhler, W. (1971). *Die Aufgabe der Gestaltpsychologie.* Berlin: Walter de Gruyter.

Kohonen, T. (1995). *Self-organizing maps.* Berlin, Heidelberg: Springer-Verlag.

Kraaijveld, M., Mao, J., & Jain, A. (1992). A non-linear projection method based on Kohonen's topology preserving maps. In *Proceedings of the 11th ICPR, International Conference on Pattern Recognition* (pp. 41–45). Los Alamitos, CA: IEEE Comput. Soc. Press.

Krumhansl, C. (1990). *Cognitive foundations of musical pitch.* New York, NY: Oxford University Press.

Langner, G. (1992). Periodicity coding in the auditory system. *Hearing Research, 60,* 115–142.

Langner, G. (1997). Temporal processing of pitch in the auditory system. *Journal of New Music Research, 26.*

Leman, M. (1995). *Music and schema theory: Cognitive foundations of systematic musicology.* Berlin, Heidelberg: Springer-Verlag.

Leman, M. (1997). The convergence paradigm, ecological modelling and context-dependent pitch perception. *Journal of New Music Research, 26.*

Leman, M., & Carreras, F. (1996). The self-organization of stable perceptual maps in a realistic musical environment. In G. Assayah (Ed.), *Pro-*

ceedings of the Journées d'Informatique Musicale 1996. Caen: Univ. de Caen – IRCAM.

Moisala, P. (1995). Cognitive study of music as culture: Basic premises for cognitive ethnomusicology. *Journal of New Music Research, 24*, 8–20.

Scheerer, E. (1994). Psychoneural isomorphism: Historical background and current relevance. *Philosophical Psychology, 7*, 183–210.

Schreiner, C. E., & Langner, G. (1988). Coding of temporal patterns in the central auditory nervous system. In G. Edelman, W. Gall, & W. Cowan (Eds.), *Auditory function: Neurobiological bases of hearing.* New York, NY: John Wiley and Sons.

Seifert, U. (1991). The schema concept: A critical review of its development and current use in cognitive science and research on music perception. In A. Cammuri & C. Canepa (Eds.), *Proceedings of the IXth Colloquium on Musical Informatics.* Genova: AIMI/DIST.

Shepard, R. (1964). Circularity in judgments of relative pitch. *The Journal of the Acoustical Society of America, 36*, 2346–2353.

Shepard, R., & Chipman, S. (1970). Second-order isomorphism of internal representations: Shapes of states. *Cognitive Psychology, 1*, 1–17.

Terhardt, E., Stoll, G., & Seewann, M. (1982). Algorithm for extraction of pitch and pitch salience from complex tonal signals. *The Journal of the Acoustical Society of America, 71*, 679–688.

Van Immerseel, L., & Martens, J. (1992). Pitch and voiced/unvoiced determination with an auditory model. *The Journal of the Acoustical Society of America, 91*, 3511-3526.

Self-organizing Neural Nets and the Perceptual Origin of the Circle of Fifths

Nicola Cufaro Petroni[1] and Matteo Tricarico[2]

[1] Dipartimento di Fisica dell'Universtitá di Bari, via Amendola 173, I-70126 Bari, Italy
[2] Via G. Pazienza 31/A, I-71016 S. Severo, Italy

Abstract. One of the more surprising results of the use of artificial neural networks in the analysis of acoustic perceptions is the regularity of dispositions, on a Kohonen self-organizing map, of chord images: in fact these images clearly show the bent to spontaneously dispose themselves along closed, circular patterns reminiscent of the circle of fifths. In this paper we firstly reproduce these results and then test them against the hypothesis that the regularity of patterns could be the effect of the preprocessing and of the circular transpositions of the original signal images, rather than of their informative content referred to the tonal relations among the chords.

1 Introduction

In recent years, we have witnessed a consistent expansion of research devoted to the use of artificial neural networks in the study of acoustic perceptions (Bharucha, 1987, 1989; D'Autilia & Guerra, 1991; Cufaro Petroni, Guarino, & Pasquariello, 1995; Cufaro Petroni, Degrassi, & Pasquariello, 1994). Particular interest has been aroused by the works of M.Leman, devoted to an understanding of the classification of auditory images by means of self-organizing Kohonen feature maps (Leman, 1989, 1990, 1991, 1995). In fact the ability of these neural nets in elaborating the signal information in characteristic patterns seems to give very stimulating cues about the tonal structure of (western) music. In Bharucha (1987), for example, a network which *a priori* represents in its structure the relations among tones, chords and keys is activated by a musical context, and the pattern of activation at the equilibrium is analyzed: the results then show that the activation of units representing chords and keys decreases with increasing harmonic distance. On the other hand, in Leman (1990, 1991, 1995), similar results are found with no initial hypothesis either on the structure of the net or on the relations among chords and keys. Here the computing units of the net do not *a priori* represent musical elements, but that notwithstanding an amazing regularity among the locations of the chord images fed to a Kohonen feature map has been found: in fact these images clearly show the bent to spontaneously dispose themselves on the map along characteristic circular paths which are strongly reminiscent of the well-known *circle of fifths*. This may be the result of the fact that the 12-numbers templates used as chord images (see Sect.2) already contain inscribed in their structure all the relevant information about the more fundamental relations of tonal music, namely: the similarity and dissimilarity relations among the major (or minor) triads.

Moreover the results of these simulations seem to say even more since one of the interesting implications of this approach is the fact that, albeit the chord images can be calculated by means of several different pitch models, the outcomes of the experiments are always perfectly coherent. In other words "the conceptual difference between these [pitch] models ... are great, but their use in the tone semantic model have led to consistent and often surprisingly similar results"(Leman, 1995, pp.44,71). No matter how you calculate your chord images to supply to the Kohonen feature map - provided that you use a reasonably musically meaningful pitch model - you always find something similar to a circle of fifths. Of course this is very interesting since one possible interpretation of this fact could be that the relevant information about the relations among triads is preserved in the preprocessing of the signals based on every one of these models: the peculiarities of every model seem in fact to be of no consequence for the outcomes of the simulation, although it is possible to show that the use of one of the accepted pitch models is, after all, necessary.

However, a deeper analysis of the way in which the numerical images of the chords to feed into the Kohonen feature map are produced will inevitably arouse the suspicion that the regularity of the map configurations could be a consequence of the cyclical rotations of an original template used to transpose it on the notes of the chromatic scale (Leman, 1995, p.73), rather than of the informative content of the signal itself. In this paper we will scrutinize this hypothesis in order to show that in fact these suspicions are groundless: in particular in the following sections we will first of all reproduce in a simplified form the experiment of the original papers and then we will compare their outcomes with the results of a few simulations specially designed to put in evidence the meaning of the previous ones.

2 Simulation

We will describe here a computer simulated experiment about the learning of the relations among chords by means of a particular neural network known as Kohonen feature map. This is not the place to describe in detail how this algorithm in fact works: the interested reader is referred to an already wide literature about this type of computing devices (Hecht-Nielsen, 1990; Hertz, Krogh, & Palmer, 1991; Murtagh, 1994; Ripley, 1996). We will just remember here that these feature maps are self-organizing neural networks able to associate a spatial structure to statistical data organized in clusters. The idea is that the centers of the clusters are spatially correlated, so that their representatives on the grid become nearer when the clusters are more similar.

A Kohonen feature map usually is a set of N processing elements (neurons), disposed in a (usually) square grid, which receive as inputs a sequence of n-dimensional vectors $\mathbf{x}$ (considered as values drawn from a given random vector $\mathbf{X}$). Each i-th processing element then calculates the Euclidean distance $|\mathbf{x} - \mathbf{w}_i|$ between the example $\mathbf{x}$ and a given initial weight vector $\mathbf{w}_i$. At this point a competition starts to see which neuron has the $\mathbf{w}_i$ closest to $\mathbf{x}$: if this is the j-th neuron a weight modification takes place in accordance with the so called Kohonen learning law:

$$\mathbf{w}'_k = \mathbf{w}_k + \alpha(\mathbf{w}_k - \mathbf{x})$$

for all the neurons k which are neighbors of j on the grid. Both the constant α and the definition of *neighbor* are allowed to change with time. This rule in some way draws the weight vector of the winning element closer to the input vector $\mathbf{x}$, and the topological aspect enters into the game since not just the best matching weight vector is updated, but also the weight vectors in the immediate vicinity. After several rounds of iterations are performed on a given number of $\mathbf{x}$ samples, the $\mathbf{w}_i$ reach an equilibrium value disposing themselves in the space of the weight vectors in a way reproducing the original distribution of the $\mathbf{x}$'s. In other words the weight vectors become densest where the $\mathbf{x}$'s are more common, and become least dense where the $\mathbf{x}$'s hardly appear. The overall result is that every class of (in general n-dimensional) examples will be associated with a particular (in general 2-dimensional) region on the grid of computing elements following the pattern of the proximity between the weights $\mathbf{w}_i$ and the examples $\mathbf{x}$. In our simulation (performed

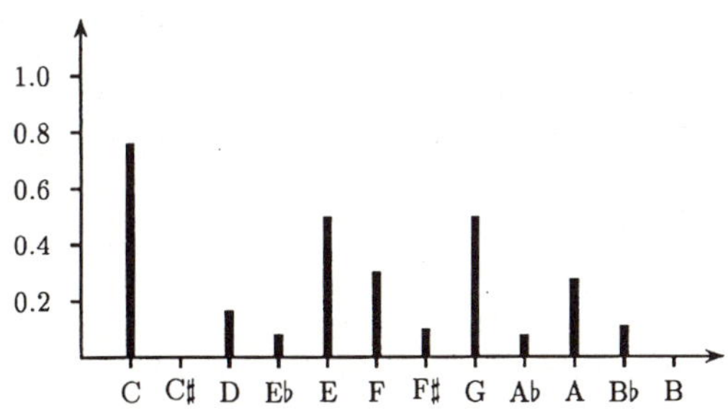

Fig. 1. Image (A) of the C major triad drawn from the Krumhansl measurements about similarities among keys (Krumhansl, 1990; Leman, 1995, p.14)

using a program in C language written by us, on a MS-DOS computer with CPU Intel 80386sx 25 MHz) the examples $\mathbf{x}$ are drawn from a 12-dimensional space ($\mathbf{R}^{12}$) of 12-number vectors which represent the images of tone centers obtained by means of different pitch models or, to check the results, by a suitable random generation. Following the discussion contained in Leman

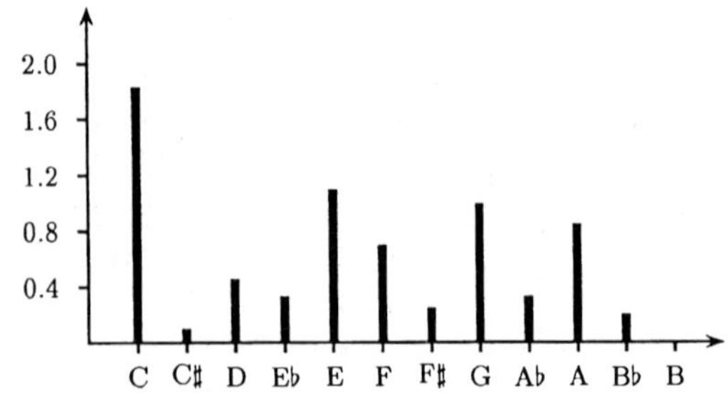

Fig. 2. Image (B) of the C major triad calculated by means of a simple pitch model as the subharmonic sum of Shepard tones (Parncutt, 1988; Leman, 1995, p.44)

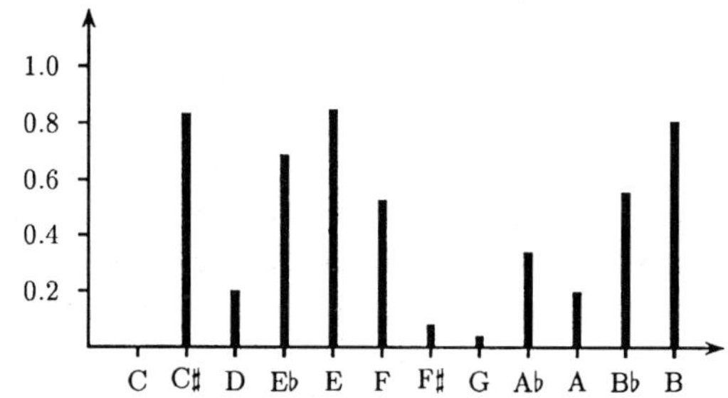

Fig. 3. Image (C) of the C major triad calculated by means of a random extraction of numbers.

(1991, 1995) we have first of all prepared two different sets of examples just to reproduce the results of the previous simulations: the first (A) has been produced taking into account the results of psychoacoustic measurements (Krumhansl, 1990; Leman, 1995, p.14) about similarities among keys; the second (B) comes from calculations based on a simple pitch model (Parncutt, 1988; Leman, 1995, p.44) where the resulting image is calculated as the subharmonic sum of the original representation of Shepard tones. The templates of 12 numbers to be used as inputs are then obtained by transposing the template obtained for one (major, minor, diminished and so on) triad on the 12 steps of a chromatic octave. Figures 1 and 2 show examples of the templates of the C major triad: other templates have similar structure.

The set of examples is then built as a suitable mixture (taking into account the occurrence of chords in western, tonal music) of tone center images. In the original papers the set of different examples amounted to 115 different chord images: in order to simplify the simulation we decided to reproduce the experiments with just 12 major and 12 minor triads plus 12 dominant seventh chords. This reduction in the number of proposed different examples is in fact justified by the remark that in any case the chosen chords can be

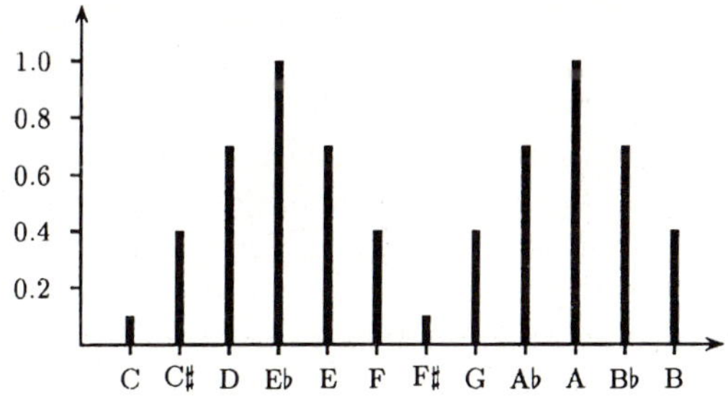

Fig. 4. Image (D) of the C major triad by means of an arbitrary, periodic choice of a template of numbers.

credited for about the 84% of the total weight of the chords (Bruhn, 1988; Leman, 1991). Indeed it must be pointed out that the set of examples is built by mixing up the chords in suitable proportions in order to take into account the weight of every particular chord in the usual musical production. Our simplification, which reduces the chords to the most relevant 36, allows one to correspondingly reduce the dimensionality of the grid of processing elements in the Kohonen feature map: instead of a 400 elements grid (a square of side 20) we could use a 100 elements grid (a square of side 10) because of the reduced number of objects to classify.

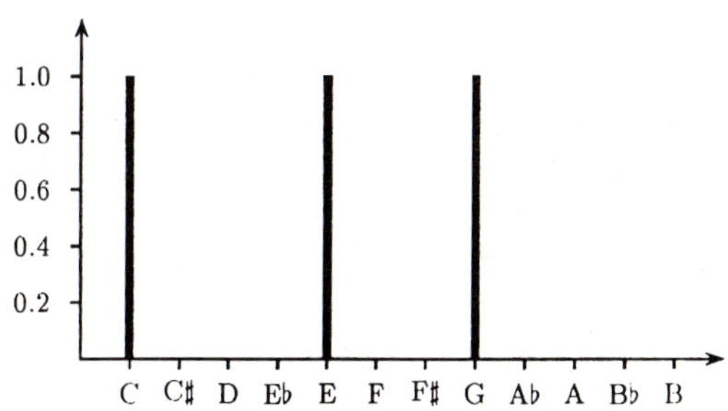

Fig. 5. Image (E) of the C major triad in the simple local representation which indicates just the pitches activated in the triad

As will be seen in the next section the results are similar to that of previous simulations (Leman, 1991, 1995); however the way of production of the sets of examples seems to be open to a criticism. We generated the tone center images by means of some model (either Krumhansl's data or a pitch model) just once for every set of the 12 shifts of the template along the steps of a chromatic octave. This means that in reality the 12 templates of, for instance, the major triads are always the same template just transposed on another degree of the chromatic octave (with, of course, a folding of the exceeding elements on the opposite side of the template). Since the interesting result of these simulations is the fact that a particular *circular order* emerges among the activation regions of the triads (for example the the major triads are distributed along a closed path strongly reminiscent of the circle of fifths), a legitimate question is whether this underlying order of the chords could just be the product of the *circularity of the transpositions* introduced by hand in our samples to represent every step of the chromatic scale. In other words it would be interesting to have a way to verify whether the order emerging out of the data is a property of the information encoded in the templates of numbers used as tone center images, or rather only a trivial consequence of our manipulation and preprocessing of the incoming signal. Moreover, since the results of the simulations seem to be largely independent from the particular model used to produce the input templates (Leman, 1995), a second, and related, question is whether the order of data can be achieved without any recourse to a model whatsoever, for example by feeding to the neural network a simple local representation of triads and chords.

In order to check these points, besides the examples of class (A) and (B) previously described (Fig.1-2), we have produced three other different sets of templates similar to the others in every respect except for the fact that now the first template in the series of 12 is not produced on the basis of a pitch model whatsoever. More precisely in the first series (C) we have produced an artificial template by means of 12 random numbers which are arbitrarily attributed as template for the C major triad; then the minor triad and the dominant seventh are generated from the major triad simply by reproducing the same differences among the corresponding templates in the Krumhansl data. Finally these three archetypal templates are transposed along the 12 steps of the chromatic octave. In the second series (D) we have calculated the templates in a way identical with (C), except for the fact that the first profile was not drawn at random, but was arbitrarily chosen to be highly regular and periodic, albeit dissimilar from the (A) and (B) cases. Finally the series (E) consists of images in simple local representation, with no pitch model or measurements involved, with the only indication of the pitches present in the triad or chord. Examples of these new type of templates are shown in Fig.3-5. We are now in the position to check the results of this simulation by comparing the activation patterns produced on the Kohonen feature maps by the sets of examples (A) and (B), generated either by an acceptable pitch model or by empirical measurements, with the corresponding activation patterns produced by the sets (C), (D) and (E).

3 Results and Discussion

We performed the simulation four times with different sets of examples (A), (B), (C), (D) and (E) on a square grid of processing elements (side 10 el-

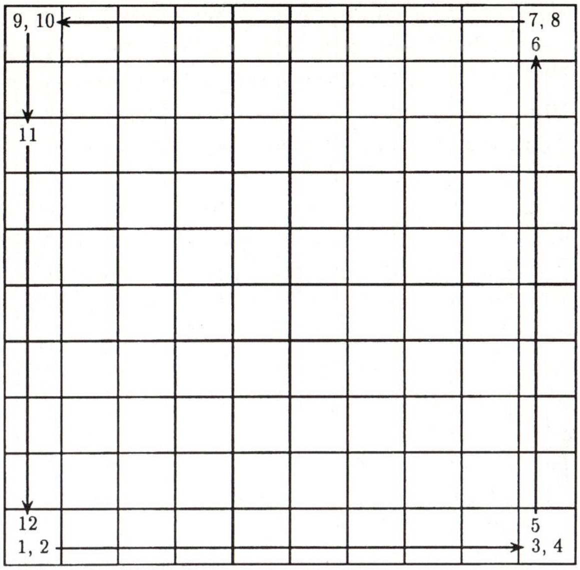

Fig. 6. Locations of the most activated neurons for major triads of the series (A) and indication of the sequence of triads in the circle of fifths

ements). Every set is composed of 120 replicas of the 36 chosen chords in a mixture which respects the composition used in the previous simulations (major triads about 50%, minor triads about 20% and dominant sevenths about 30%). We performed the training of the net varying a few parameters: number of iterations (from 100 to 200); learning rate α (from 0.2 to 0.3) and initial random values of the weight vectors, but we have always consistently found the same basic results, namely: *when the examples are generated either by means of a pitch model or by real measurements, the activation regions tend to dispose themselves around closed circular paths which follow the well known circle of fifths.* In other words the Kohonen feature map recognizes a structure of *topological* similarities among the examples which is strongly reminiscent of *musical* similarities. On the other hand *when the examples are either drawn at random or arbitrarily fixed, or no model is used at all the similarity structure among the chords is completely scrambled up.* As a consequence, a line joining, for example, the location of the major triads along the circle of fifths will appear as a closed but twisting curve on the self organizing map.

We will not show here the complete results of the four rounds of simulations, and we will rather limit ourselves to the picture of the activation regions of the 12 major triads for the five classes (A), (B), (C), (D) and (E) of examples examined, the disposition of minor triads and of dominant sevenths

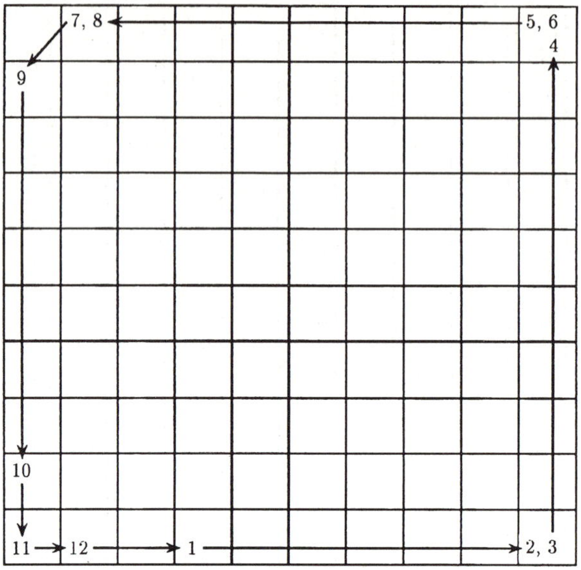

Fig. 7. Locations of the most activated neurons for major triads of the series (B) and indication of the sequence of triads in the circle of fifths

being quite similar. Remark that in the pictures displayed, we indicate just the location of the most activated (nearest to the particular class of examples) processing element (neuron) corresponding to every triad or chord. A more precise shadowing of the activation regions would inevitably hide the patterns. The association triad/most-activated-neuron is shown by numbers indicating the position of the chord in the circle of fifths according to the following scheme:

1	2	3	4	5	6	7	8	9	10	11	12
C	G	D	A	E	B	F♯	C♯	A♭	E♭	B♭	F

The lines drawn by us on the grid join in succession the location of the most activated neurons following the circle of fifths and hence give an idea of the order implied by the disposition on the map. The most remarkable results are shown in Fig.6-10. As can be seen from these diagrams the results are radically different in the (A) and (B) cases with respect to the (C), (D) and (E) cases. In fact when the chord images are calculated either from a pitch model or from real psychoacoustic data the major triads (but the same is true for the other types of chords) show the clear bent to dispose themselves around the closed, circular paths characteristic of the circle of fifths, indicating a clear relation between the topological proximity on the grid and the psychological similarity of the chords. On the other hand, when the examples are prepared either by randomly drawing the values of the first template, or by arbitrarily

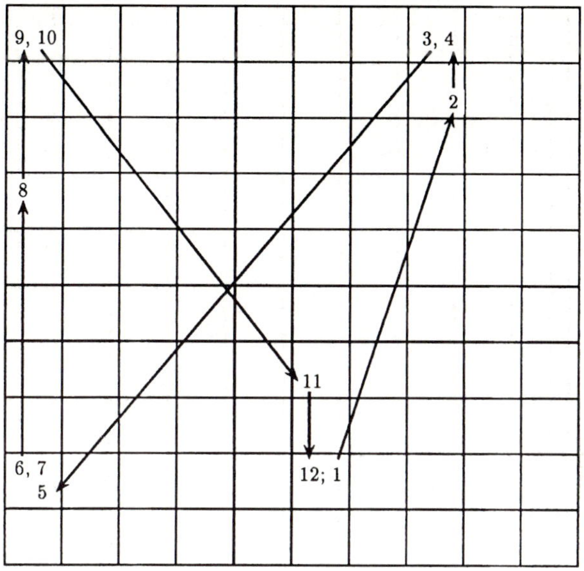

Fig. 8. Locations of the most activated neurons for major triads of the series (C) and indication of the sequence of triads in the circle of fifths

giving a regular profile, or just with no model at all, the line joining the 12 triads in the sequence of the circle of the fifths will show a twisted form which either criss-crosses its path or visits several times the same locations: this is a behavior which is not coherent with a straight relation between topological proximity and psychological similarity.

Remark also that the structure of the circle of fifths in these simulations is different from that of previous publications (Leman, 1991, 1995) which suggested a spiral structure, since here we have a plain circular structure. Moreover in our simulations (A) and (B) the chords show also the inclination to regroup themselves at the four corners of the square grid. Beside the oversimplification of our model, the reason for these differences can be found in the fact that in our feature map the four borders of the square grid are not folded together making the upper and lower borders, and the left and right borders coincide in a toroidal structure as in the previous papers. It is evident that this toroidal structure would give rise to a different topology on the grid so that the circle of fifths would develop along a spiraling line rather than along a simple circular path.

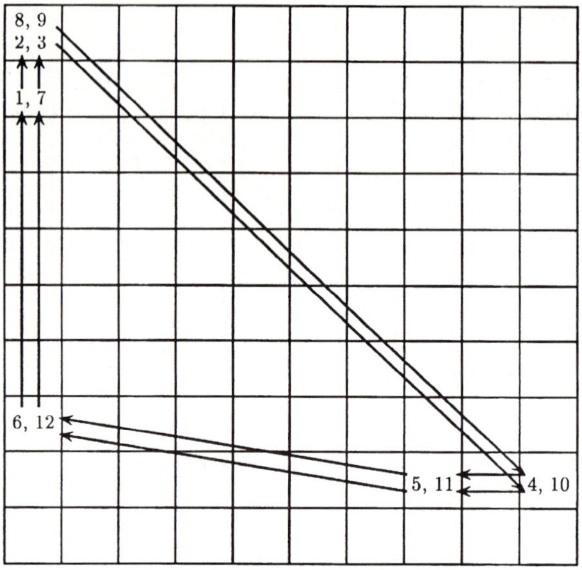

Fig. 9. Locations of the most activated neurons for major triads of the series (D) and indication of the sequence of triads in the circle of fifths

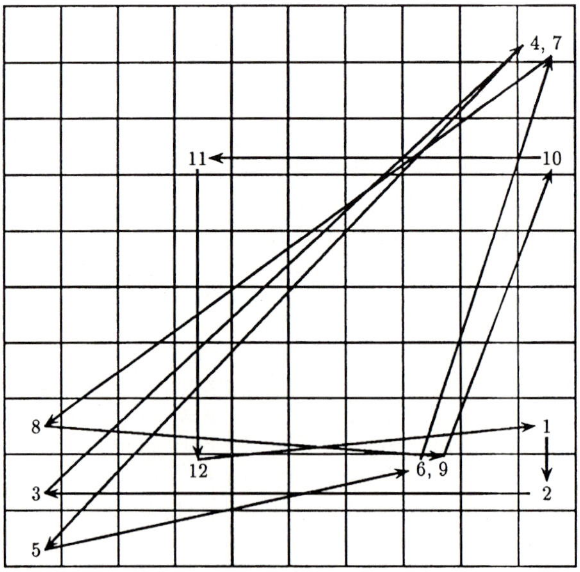

Fig. 10. Locations of the most activated neurons for major triads of the series (E) and indication of the sequence of triads in the circle of fifths

4 Conclusion

The results summarized in this paper seem not equivocally indicate that the informational content carried either by the Krumhansl templates (A), or by the templates produced by a simple pitch model (B) is in fact musically relevant since their regular disposition in the analogous of the circle of fifths on a Kohonen feature map cannot be achieved unless the archetypal template (which, by transpositions, generates all the templates of our examples) are the right ones. In fact we have tried to reproduce this regularity by generating anomalous examples which respect all the features of the usual examples, except for the fact that the first template is not generated by means of a pitch model or on the basis of empirical data, and we have shown that this is in fact impossible. Since a complete mathematical theory of the Kohonen feature maps is still not at hand (Hecht-Nielsen, 1990, p.141) it could be risky to draw too sharp conclusions on a matter as complicated as this one. However in the opinion of the authors the results of the previous simulations (Leman, 1991, 1995) and of this paper prove, beyond any reasonable doubt, the importance of the pitch models in this sort of investigations: indeed the relative insensitivity of the results to the details of these models (*robustness*), as long as they are musically relevant, does not imply that these models are irrelevant (as the series (D) of simulations clearly indicates), but rather that the information about the tonal relations contained in every one of them is roughly the same.

References

Bharucha, J. (1987). Music cognition and perceptual facilitation: A connectionist framework. *Music Perception*, *5*, 1–30.

Bharucha, J. (1989). Neural nets and perceptual learning of tonal expectancies. In *Proceedings of the Conference on Music Perception and Cognition.* Kyoto.

Bruhn, H. (1988). *Harmonielehre als Grammatik der Musik.* Munich: Psychologie Verlags Union.

Cufaro Petroni, N., Degrassi, F., & Pasquariello, G. (1994). Detection of pitch in random acoustic signals by neural networks. *Journal of New Music Research*, *23*, 369–399.

Cufaro Petroni, N., Guarino, M., & Pasquariello, G. (1995). Neural networks and pitch detection in random acoustic signals. In *Proceedings of the Conference on Engeneering Applications of Neural Networks, Otaniemi, Helsinki, 1995.* Helsinki.

D'Autilia, R., & Guerra, F. (1991). Qualitative aspects of signal processing through dynamical neural networks. In G. De Poli, A. Piccialli, & C. Roads (Eds.), *Representations of musical signals.* Cambridge, MA: The MIT Press.

Hecht-Nielsen, R. (1990). *Neurocomputing.* Reading, MA: Addison-Wesley.

Hertz, J., Krogh, A., & Palmer, R. (1991). *Introduction to the theory of neural computation.* Reading, MA: Addison-Wesley.

Krumhansl, C. (1990). *Cognitive foundations of musical pitch.* New York, NY: Oxford University Press.

Leman, M. (1989). Symbolic and subsymbolic information processing in models of musical communication and cognition. *Interface – Journal of New Music Research*, *18*, 141–160.

Leman, M. (1990). Emergent properties of tonality functions by self-organization. *Interface – Journal of New Music Research*, *19*, 85–106.

Leman, M. (1991). The ontogenesis of tonal semantics: Results of a computer study. In P. Todd & G. Loy (Eds.), *Music and connectionism.* Cambridge, MA: The MIT Press.

Leman, M. (1995). *Music and schema theory: Cognitive foundations of systematic musicology.* Berlin, Heidelberg: Springer-Verlag.

Murtagh, F. (1994). Neural networks and related massively parallel methods for statistics: A short overview. *International Statistical Review*, *62*, 275–288.

Parncutt, R. (1988). Revision of Terhardt's psychoacoustical model of the roots of a musical chord. *Music Perception*, *6*, 65–94.

Ripley, B. (1996). *Pattern recognition and neural networks.* Cambridge: Cambridge University Press.

A Model of the Perceptual Root(s) of a Chord Accounting for Voicing and Prevailing Tonality

Richard Parncutt

Department of Psychology, Keele University, Staffordshire ST5 5BG, U.K.

Abstract. A theory of harmony is presented that predicts the perceptual root of any simultaneity within the chromatic scale, embedded in the context of any chord progression. The theory takes into account the pitch classes (pcs) of the chord, its voicing, the prevailing tonality, and local voice-leading. Pcs: the perceptual root of an isolated, octave-generalized chord is determined by the root-support intervals P1, P5, M3, m7 and M2. Voicing: other things being equal, the bass note is more likely to function as the root than the other notes. Tonality: the harmonic profile of a chord in a tonal context is given by the sum of the chord's profile in isolation and the pc-stability profile of the prevailing tonality. Voice leading: parallel movement of parts increases the likelihood that the roots of two sonorities will be heard to move in parallel with the voices. Predictions of the theory for the perceptual roots of diatonic triads and 7ths and of common chromatic chords in major and minor keys are intuitively reasonable, but will need to be tested against the results of perceptual-cognitive experiments and systematic studies of scores and improvised performances.

1 Introduction

Existing theories of the root of a chord can be inconsistent and contradictory. Consider, for example, the sonorities D-F-A-C (D in the bass) and F-A-C-D (F in the bass) in the key of C major. Rameau's theory of stacked thirds would have it that both chords are versions of II^7, and so have the same root – the second degree of the scale, $\hat{2}$ (here, D). (When taking voice leading into account, Rameau held that the root of the chord is F if the chord is followed by a C chord, and D if it is followed by a G chord.) According to Riemann's theory of function, both chords may be regarded as subdominants, so both have the root F. In jazz, it is standard practice to notate D-F-A-C as Dm7 (not F/D) and F-A-C-D as F6 (not Dm7/F), implying that the root depends upon which note is voiced in the bass.

The present article develops a harmonic theory that attempts to account for the root of any chord in any tonal context (by *context*, I am thinking primarily of the chords that immediately precede a chord in a progression). The theory addresses primarily how a chord sounds, not how it is notated; to emphasize this difference, the predictions of the theory may be referred to as *perceptual roots*. Incorporating both music-theoretic and perceptual aspects, the theory demonstrates the considerable and still barely tapped potential of perceptual theory as a basis for the theory of tonal music. The juxtaposition of *artistic* (music theory) and *scientific* (music perception) approaches opens up a number of issues that are not yet satisfactorily resolved, but await further research.

Both scale-step and functional theory can feasibly accommodate most diatonic chords. But it is not a straightforward matter to apply these theories to chords such as the cadential $^6_4$ (or second-inversion tonic triad Ic, resolving to a root-position dominant Va) and the augmented 6ths (built by superposing major 3rd and augmented 6th intervals, plus an optional perfect 5th or augmented 4th, on a bass that is usually a minor 6th interval above the tonic). According to the theory of stacked thirds, the root of the cadential $^6_4$ is the tonic ($\hat{1}$); but in Schenker-inspired harmonic theory (e.g. Forte, 1974, it is the dominant ($\hat{5}$)). Within the functional paradigm, arguments may be conceived for classifying the cadential $^6_4$ as a species of sub- or pre-dominant (if it is thought of as leading to or preparing for V^5_3), as a kind of dominant (as a retrospective prolongation of V), or as a tonic (after all, it contains the same three pitch classes, or pcs, as I). Applying the stacked-thirds approach to the augmented 6th chords (in their correct enharmonic spellings) yields the scale-degrees $\sharp\hat{4}$ (for the *Italian* and *German* versions of the chord) and $\hat{2}$ (for the *French*) as root candidates. But these chords usually appear with $\hat{6}$ in the bass, suggesting that $\hat{6}$ is a root candidate; and the most commonly doubled note – at least in the *Italian version* of the augmented 6th chord – is $\hat{1}$, suggesting that $\hat{1}$ is another root candidate.

The aim of the present paper is to develop a theory of the perceptual root that deals more plausibly and consistently with these issues than previous theories. The new theory will be:

- octave generalized (primarily for reasons of convenience, and ease of music-theoretical application);
- implemented in the form of an algorithm whose input can be any chord or chords in the chromatic scale;
- entirely *bottom up*, so that the root-determination procedure is applied in exactly the same way to any chord (giving the theory a clear advantage over the theory of harmonic dualism (von Oettingen, 1866), and later refinements thereof); and
- systematically testable, and hence falsifiable.

The theory will strive to satisfy the following additional criteria:

- to reliably predict the root whenever it is clearly and unambiguously defined in existing music theory;
- to achieve a suitable compromise where existing music theories (stacked thirds, harmonic function) disagree; and
- to plausibly and parsimoniously account for effects of voicing (registral positions of tones in the chord) and tonal context on the perceptual root of a sonority.

2 Enharmonic Spellings and Intonation

Before proceeding with an account of the theory, I should briefly address two important aspects of tonal harmony that the theory does not consider in any way.

First, the theory ignores enharmonic spellings. This somewhat radical omission may be justified as follows. Enharmonic spellings are primarily determined by the context of preceding and following notes (for a concise account of the underlying rules, see Parncutt & Stuckey, 1992). But in the present theory the effect of context on the root is accounted for independently of enharmonic spellings. Consider, for example, the functional and perceptual difference between the dominant 7th and the German augmented 6th chords. Relative to the chromatic scale and taken out of context, these two chords are identical: both have intervals of 4, 7 and 10 semitones above the root (and/or octave equivalents). The augmented 6th chord sounds differently because it is embedded in a different tonal context (represented quantitatively in the present model by the appropriate key profile, see Krumhansl & Kessler, 1982), and because it resolves differently (again, a matter that can be addressed separately in the model). Clearly, there is not enough information present in enharmonic spellings alone to capture the richness of the difference between these two chords, or between any other chords which are identical except for the context in which they are embedded (e.g., diatonic II^7 in a minor key, and the Tristan chord). In sum, enharmonic spellings are primarily notational conventions. They are not directly relevant to the present model, which is primarily about perception.

The theory also ignores intonation. The results of several published studies on intonation suggest that, contrary to musical intuition, sharps are not consistently intoned sharper or flatter than enharmonically equivalent flats (Parncutt & Stuckey, 1992). Assuming that sharps relative to a key signature or scale lie at major 3rd intervals above other scale degrees or salient pitches, and that flats lie at major 3rd intervals below other scale degrees, sharps are *sharper* than enharmonically equivalent flats in *Pythagorean* tuning (where major thirds have frequency ratios of 81:64, or 408 cents), and *flatter* in *just* tuning (where major thirds have frequency ratios of 5:4, or 386 cents). Numerous experimental studies of intonation have shown that good performers tend to tune between these two extremes, tending toward Pythagorean tuning in some musical situations (e.g., fast melodic passages) and toward just tuning in others (e.g., slow harmonic progressions). Thus, intonation is not related in any simple way to enharmonic spelling, but it is instead determined by musical context (for experimental data on this point see Fyk, 1995). The present model consequently makes no reference to intonation; but if a model of intonation were to be built – possibly on the basis of this model, or along the lines of (Agmon, 1993) – available experimental data on intonation suggest that the model should attempt to account for effects of context, but ignore enharmonic spellings.

3 Definitions

Although the concept of root is of central importance to harmonic theory, it is surprisingly poorly defined (Thomson, 1993). At the most simplistic level, the root is the name most commonly given to a chord (e.g., we call C-E-G a *C chord*, not an *E chord* or a *G chord*). But this definition gives no insight into the root's nature and origins. According to scale-step theory (Rameau), the root is the lowest note in a stack of thirds, while functional theory (Riemann) suggests that the root of just about any chord in tonal music corresponds to $\hat{1}$, $\hat{4}$, or $\hat{5}$, depending on whether a chord functions as a tonic, sub- or pre-dominant, or dominant, respectively.

The *perceptual root* may be defined in a number of other ways. One is based on the well-known study of Krumhansl and Kessler (1982). Although the authors did not say so directly, this study clearly demonstrated the psychological reality of chord roots, an achievement that has escaped others with that aim (e.g. Pritschet, 1992). Krumhansl and Kessler presented musically trained listeners with chords constructed from Shepard (octave-complex) tones, both in isolation and in progressions. They then asked the listeners to rate how well a probe tone followed the chord. For isolated major and minor triads, the peak of the resultant profile corresponded to the stacked-thirds root. In Parncutt (1993), I determined the perceptual root of a wider range of chords, by a similar experimental method. Results again agreed broadly with music theory.

Further possible definitions of the root of a chord are based on the statistical analysis of musical scores, which nowadays is best carried out by computer (Huron, 1994). The root may be defined either as the pc most commonly voiced in the bass, or as the pc which is most often doubled in a chord. Similarly, the root may be defined as the pc which is most often played against a given chord during melodic improvisation although in jazz, performers may deliberately avoid playing the root (Järvinen, 1995). I favour these notation and performance based definitions, because they are the most straightforward to implement, apply and test.

As we have seen for the augmented 6th chord, music theories based on notation may yield different root predictions from theories based on perception. The present theory is of the perceptual kind, as are Krumhansl's experiments and Huron's analyses. Predictions of the present theory are intended primarily to be compared with experimental data. If predictions of the theory agree with experimental data but disagree with notation-based music theory, then we may reasonably suspect that notation-based theory is incorrect.

4 Theory

In Parncutt (1988), I developed a model of the perceptual root that navigated a middle way between Terhardt's simple (over-simplified?) model of the root

(Terhardt, 1982) and his much more general pitch algorithm (Terhardt, Stoll, & Seewann, 1982). My aim was to retain only enough detail to predict the perceptual roots of commonly used chords. The model incorporated five intervals called *root-supports* (a term originally suggested by R. Stuckey): perfect unison (P1), perfect fifth (P5), major third (M3), minor seventh (m7) and major second (M2), plus enharmonic and octave equivalents. These intervals, listed here in decreasing order of importance for root determination, correspond to the intervals between the first ten harmonics and the fundamental of a harmonic complex tone, when they are collapsed into a single octave. The algorithm successfully predicts the perceptual roots of commonly used chords presented in isolation (e.g., major, minor, diminished, augmented and suspended triads, and Mm7, mm7, MM7, hd7 and d7 chords). Note that voicing is ignored; predictions may be regarded as averaged over various different voicings.

In the present article, the theory of root-supports will be regarded as only one (i) of four separate factors influencing the root of a chord. The three additional perceptual effects are (ii) voicing (addressed qualitatively in Parncutt, 1988), (iii) prevailing tonality (involving key profiles Krumhansl & Kessler, 1982), and (iv) melodic streaming (based on experiments performed by Bregman and colleagues (Bregman, 1990)). Effects (i) and (ii) apply to a chord presented in isolation, while effects (iii) and (iv) depend upon the contrapuntal-harmonic progression in which a chord is embedded.

(ii) *Voicing.* In Parncutt (1988), I assumed that, other things being equal, the bass note is more likely to function as a chord's perceptual root than the other notes. Consider, for example, the minor-7th chord D-F-A-C, or II^7 in the key of C major. Its first inversion, IV6, can also function as a root-position chord (Rameau's "sixte ajoutée"). The theory of root-support intervals predicts that both D and F may act as perceptual roots; the actual perceptual root is determined by which of these two is in the bass. Another familiar example of the voicing principle is the second inversion of the tonic triad (cadential $^6_4$). The theory of root-support intervals predicts that the 4th above the bass (the tonic) is the perceptual root, but according to the voicing principle the bass is the root. The tension between these two root candidates may explain the chord's dissonance, the chord's relative rarity by comparison to $^5_3$ and $^6_3$ chords, and the special procedures by which $^6_4$ chords are typically prepared and quitted.

(iii) *Prevailing tonality.* A computer-implemented analysis of a representative sample of tonal scores (D.Huron, personal communication) has revealed that the distribution of doublings of chromatic scale-steps within a given major/minor tonality corresponds remarkably closely with the corresponding Krumhansl key profile. In other words, a note's suitability for doubling corresponds to the perceived goodness of fit between it and the passage in which it is embedded. Huron's finding is consistent with the following two hypotheses: (i) the perceptual root of a chord is generally the best note in a chord to double; and (ii) a note is more likely to be the root if it corresponds to a strong scale-degree such as $\hat{1}$ or $\hat{5}$, whereas it is unlikely to act as a perceptual root if it is tonally unstable (leading note or non-diatonic note). Thus, for example, the root of a Neapolitan 6th chord is normally regarded as $\hat{4}$ (consistent with its interpretation as a sub- or pre-dominant harmony) rather than $b\hat{2}$, even though according to (i) above the latter is the only pitch whose root function is *supported* by the other notes in the chord. Another factor pointing to $\hat{4}$ as the perceptual root of the Neapolitan is voicing – the root is in the bass.

To recapitulate briefly, the theory of stacked thirds is incapable of explaining the root of the Neapolitan 6th, or of the cadential $^6_4$. Instead, these examples demonstrate that any note of a major triad and, by extension, any note of any chord can function as the root, provided the chord appears in a suitable context. Only two conditions would appear to need be satisfied: first, put that note in the bass; and second, line it up with a stable pitch a major/minor key.

(iv) *Local voice leading.* To understand how voice leading can affect the perceptual root, it is instructive to re-examine relevant experiments on melodic streaming from a music-theoretic viewpoint. Bregman (1990) has enumerated several factors that cause the pure-tone components of time-varying sounds to fuse into single streams. One of these is coherent frequency variation: If two pure tones move in parallel motion, maintaining a constant frequency ratio, they are likely to be perceived as a single *event*, even if the ratio between them is not harmonic (although harmonicity enhances the effect). The effect is strongest when the frequency ratio is held strictly constant. Musically, this means maintaining a constant number of semitones (a well-known example is the parallel motion that occurs in the melody line of Ravel's *Bolero*). Fusion is also observed experimentally, although less strongly, when the frequency ratio is almost constant. Musically, this may occur in diatonic stepwise progressions of triads (either in root position, as in Debussy's *La cathédrale engloutie*, or in first inversion, as in Renaissance Fauxbordon).

Parallel movement has at least two separate perceptual effects. First, the voices involved tend to lose their perceptual independence; so if perceptual independence is regarded as desirable, parallel 5ths and 8ves should be avoided. Second, and most relevant to the present argument, the parallel movement tends to increase the likelihood that roots of the two sonorities will be heard to move in parallel with the voices. For example, the chord A-C-F tends to be heard as an A chord if it is followed by A-C-E, but as an F chord if followed by G♯-B-E, producing the sliding root progression F-E. In the case of parallel 5ths, the lower tones of the fifths tend to be heard as roots.

5 Model

A model based on the first principle (root-support intervals) has been presented and tested elsewhere (Parncutt, 1988, 1993). Here, that model is retained with minor amendments, and extended to encapsulate, as parsimoniously as possible, for the second (voicing) and third (key) principles described above. The fourth principle (local voice leading) is not yet implemented.

Input. The notes of a chord are input to the model as a vector *N(p)*. The integer variable p denotes pitch class (pc, or chroma), and varies between 0 (corresponding to the note C) and 11 (B). If $N(p) = 1$, pitch p is included in the chord; if $N(p) = 0$, pitch p is not included. For example, a C-major triad is given by $N(0) = N(4) = N(7) = 1$, and $N(p) = 0$ for all other p. Equivalently, we may write that $N = \{1, 0, 0, 0, 1, 0, 0, 1, 0, 0, 0, 0\}$.

Root-support intervals. The theory presented in Parncutt (1988) is implemented by means of a vector of root-support weights *w(i)*, where *i* is interval in semitones and varies from 0 to 11. Previously, I estimated the relative importance of the root-support intervals by assigning the following weights to them: P1=1, P5=1/2, M3=1/3, m7=1/4 and M2=1/5. That is, I set $w = \{1, 0, 1/5, 1/10, 1/3, 0, 0, 1/2, 0, 0, 1/4, 0\}$. These weights have since been adjusted in three ways. (i) My argument supporting the assignment of a non-zero weight to the m3 interval (Parncutt, 1988, p.75) later turned out to be flawed, so I reduced $w(3)$ to zero. (ii) While comparing model predictions with experimental data (Parncutt, 1993), I found that my original values for the less important intervals (m7, M2) were somewhat too high by comparison to the values for the more important intervals (P1, P5). A similar conclusion followed from comparisons of calculated profiles for major and minor triads with data of Krumhansl and Kessler (Parncutt, 1994). The relative size of the corresponding values of $w(i)$ were adjusted accordingly. (iii) It is easier to remember and work with the weights if they are whole numbers, so I multiplied them by 10 and rounded. After these three adjustments, $w = \{10, 0, 1, 0, 3, 0, 0, 5, 0, 0, 2, 0\}$. I should stress that the values in this vector are still only estimates – it has so far proven impossible to measure them directly – but their inherent imprecision is not a significant problem for the theory, because small variations in the values have relatively little effect on the model's output.

The two vectors are now combined to produce a vector of pc-weights $W(p), p = 0, 11$. The first element of W is produced by vector multiplication: $W(0) = Nw$. In the case of a C-major triad, $W(0) =$ 1x10 + 0x0 + 0x1 + 0x0 + 1x3 + ... 0x0 = 10 + 3 + 5 = 18. Other elements of W are produced by performing cyclic permutations of w before multiplying. So $W(1) = Nw'$ where $w' = \{0, 10, 0, 1, 0, 3, 0, 0, 5, 0, 0, 2\}$; $W(2) = Nw''$ where $w'' = \{2, 0, 10, 0, 1, 0, 3, 0, 0, 5, 0, 0\}$; and so on. Another way of performing this procedure is to create a 12x12 *circulant matrix* (Ferrar, 1941) as follows:

$$w_c = \begin{matrix} 10 & 0 & 2 & 0 & 0 & 5 & 0 & 0 & 3 & 0 & 1 & 0 \\ 0 & 10 & 0 & 2 & 0 & 0 & 5 & 0 & 0 & 3 & 0 & 1 \\ 1 & 0 & 10 & 0 & 2 & 0 & 0 & 5 & 0 & 0 & 3 & 0 \\ 0 & 1 & 0 & 10 & 0 & 2 & 0 & 0 & 5 & 0 & 0 & 3 \\ 3 & 0 & 1 & 0 & 10 & 0 & 2 & 0 & 0 & 5 & 0 & 0 \\ 0 & 3 & 0 & 1 & 0 & 10 & 0 & 2 & 0 & 0 & 5 & 0 \\ 0 & 0 & 3 & 0 & 1 & 0 & 10 & 0 & 2 & 0 & 0 & 5 \\ 5 & 0 & 0 & 3 & 0 & 1 & 0 & 10 & 0 & 2 & 0 & 0 \\ 0 & 5 & 0 & 0 & 3 & 0 & 1 & 0 & 10 & 0 & 2 & 0 \\ 0 & 0 & 5 & 0 & 0 & 3 & 0 & 1 & 0 & 10 & 0 & 2 \\ 2 & 0 & 0 & 5 & 0 & 0 & 3 & 0 & 1 & 0 & 10 & 0 \\ 0 & 2 & 0 & 0 & 5 & 0 & 0 & 3 & 0 & 1 & 0 & 10 \end{matrix}$$

and to set $W = Nw_c$.

The vector W is a series of pc-weights whose values are only meaningful relative to each other. For a C-major triad, for example, $W = \{18, 0, 3, 3, 10, 6, 2, 10, 3, 7, 1, 0\}$; this means that, when a C-major triad is sounded, C (W=18) is typically more salient than E and G (W=10), and F (W=6) is more salient than F♯ (W=2).

The absolute values of the weights become meaningful in the present model by normalization, achieved by multiplying each pc-weight by a constant factor such that the mean of all 12 weights is 10 (an arbitrary value chosen for convenience). Some examples of normalized pc-salience profiles are presented below. Each element is rounded to the nearest whole number. The top line of the table shows the interval in semitones above the conventional (stacked-thirds) root. For clarity, salience values corresponding to actual notes are printed in bold.

semitones	0	1	2	3	4	5	6	7	8	9	10	11
maj. triad	**34**	0	6	6	**19**	11	4	**19**	6	13	2	0
min. triad	**29**	2	4	**25**	0	15	0	**19**	15	4	2	6
dim. triad	**19**	2	10	**19**	2	13	**19**	0	19	0	2	15

Voicing. The effect of voicing on the perceptual root is accounted for in the model by adding a constant to the normalized weight at the pc of the bass note. I have arbitrarily chosen the value 20, as it consistently gives results that agree with musical theory and practice and with my musical intuition. As an example, consider the major triad. If the triad is in root position, the weight assigned to the stacked-thirds root is boosted by from 34 to 54 in the above table. If it is in first inversion, the weight assigned to its 3rd is increased from 19 to 39.

It would of course be preferable to establish the value of the added constant more systematically, but the necessary experimental work has not yet been done. Profiles produced by the model would need to be compared with experimentally determined profiles of chords in the context of a chord progression in which each tone was harmonic-complex – not octave-complex, as in

(Krumhansl & Kessler, 1982) – so that chords could be presented in different inversions. Probe tones, on the other hand, would need to be octave-complex, because the model is octave-generalized.

Prevailing tonality. To account, in a simple but appropriate way, for the effect of the prevailing tonality on the perceptual root, it is assumed that the chord is embedded in an unambiguous tonal context, a major or minor key. A suitably normalized version of the pc-stability profile of that key (after Krumhansl & Kessler, 1982) is simply added to the chord's profile in isolation.

The normalization of the key profiles is carried out in two stages. First, a constant is subtracted from all 12 values such that the minimum value of the profile becomes zero. Second, all 12 values are multiplied by a constant value such that the mean of all profile elements is (again) 10. The resultant normalized stability profiles are shown below (again, with elements rounded to the nearest integer, and values corresponding to actual notes in bold):

semitones	0	1	2	3	4	5	6	7	8	9	10	11
major key	$\hat{1}$		$\hat{2}$		$\hat{3}$	$\hat{4}$		$\hat{5}$		$\hat{6}$		$\hat{7}$
profile	**33**	0	**10**	1	**17**	**15**	2	**24**	1	**11**	0	**5**
minor key	$\hat{1}$		$\hat{2}$	$\hat{3}$		$\hat{4}$		$\hat{5}$	$\hat{6}$			$\hat{7}$
profile	**28**	3	**9**	**21**	3	**9**	2	**17**	**12**	3	8	**6**

To account for the effect of prevailing tonality on the root, the pc-salience profile of a chord is rotated around the pc-cycle until its first element corresponds to the prevailing tonic, and then added to the stability profile of the prevailing tonality. The resultant profile is the predicted goodness-of-fit tone profile of the chord in context. The peak of the resultant profile is the predicted root of the chord in context.

The procedure is inspired by experimental results of Järvinen (1995). He obtained 12-element profiles for the frequency-of-occurrence of pcs that were played during improvised jazz solos over a blues progression. Profiles for individual chords were influenced by a combination of the chord itself (the local hierarchy) and the prevailing tonality (the global hierarchy).

If the tonality of a chord progression is ambiguous, modal, or in a process of change (during a modulation), the global hierarchy will not correspond simply to one of the Krumhansl profiles. In this case, the model may be adapted by replacing the Krumhansl profiles by *aggregate pc-salience profiles*, calculated by summing across time the profiles of several preceding chords and accounting appropriately for the decay of sensory memory (Huron & Parncutt, 1993).

Local voice leading. This effect has not yet been formulated quantitatively in the model and will not be considered further in the present article.

Interpretation of output. Elements in the profiles output by the above procedure typically vary in the range 0 to 80. Given their inherent imprecision, different elements should not always be regarded as significantly different. In the following applications, I will assume that values are different only if they differ by 5 or more. Values differing by 4 or less will be regarded as the same. So the root of a chord will be regarded as ambiguous if the profile has two or more peaks whose heights differ by less than 5 units. The value 5 is arbitrary and may be estimated more precisely on the basis of future experimental data.

6 Predictions

Let us first of all check that the model makes reasonable predictions for diatonic triads. The table below shows the profile generated by the model for each of the seven root-position diatonic triads in a major key. The top line of the table (*sem*) shows the number of semitones above the tonic. The second line shows major scale degree numbers. Calculated salience values for the notes of each chord are printed bold, bass notes are italicized, and predicted roots are underlined:

sem	0	1	2	3	4	5	6	7	8	9	10	11
	$\hat{1}$		$\hat{2}$		$\hat{3}$	$\hat{4}$		$\hat{5}$		$\hat{6}$		$\hat{7}$
Ia	***87***	0	16	7	**36**	26	6	**43**	7	25	2	5
IIa	35	6	***59***	3	21	**40**	2	39	1	**30**	16	9
IIIa	48	4	12	7	***66***	17	6	**48**	1	27	0	**24**
IVa	**52**	6	23	3	17	***69***	2	29	7	**30**	12	9
Va	44	4	**29**	7	30	17	2	***78***	1	17	6	**24**
VIa	**58**	0	25	1	**36**	30	6	26	7	***60***	2	9
VIIa	35	10	**29**	3	30	**34**	2	43	1	13	16	***44***

Note: The chord symbols in the left-hand column are commonly used in Britain, but with the gradual acceptance of the Schenkerian approach they are becoming less popular. The symbols are based purely on the principle of stacked thirds: the Roman numeral corresponds to the scale-step where the stacking begins (here, only upper case is used (I, II, III, etc.) regardless of the quality of the chord), and the letter shows which note is in the bass (a = root position, b = first inversion, etc.). For present purposes, the symbols are intended to specify only the notes of which the triads are composed (e.g., VIIa consists of the notes $\hat{7}$, $\hat{2}$ and $\hat{5}$, with $\hat{7}$ in the bass), and the symbols are not intended to imply roots or functions (e.g., the root of VIIa is not necessarily $\hat{7}$).

Summarizing the values in the above table, the model predicts that the perceptual root corresponds to the bass for triads Ia, IIa, IIIa, IVa and Va. The perceptual root of VIa is predicted to be ambiguous (it is either the stacked-thirds root $\hat{6}$ or the tonic $\hat{1}$), consistent with Riemann's idea that VIa is a kind of tonic (Tp). Also ambiguous is the root of VIIa (it is either the bass note $\hat{7}$ or the missing stacked-thirds root $\hat{5}$), supporting the notion that VIIa is a species of dominant.

Perhaps the easiest way to interpret these predictions, and further predictions to be presented below, is to define the predicted root as the most appropriate note to double in the chord, in the absence of considerations of voice leading. The table may thus be interpreted as a set of recommended doublings. Doubling the predicted root tends to render the chord more stable or consonant, whereas doubling some other note might make it less stable; so doubling the predicted root is less likely to upset a chord's harmonic character. In this interpretation, the table suggests that the best note to double in a root-position diatonic triad in a major key is its stacked-thirds root, except if the triad is built on the submediant or leading note of the major scale. This is in broad agreement with conventional harmonic theory and the tonal repertoire. Of course, the leading note cannot be doubled (regardless of what the model predicts) if it is constrained to rise to the tonic and parallel octaves are forbidden. The model also suggests in the case of the diminished triad that an implied root on the dominant may be added without changing the chord's harmonic character.

On the basis of dualist harmonic theories (von Oettingen, 1866, and later adherents) – not to mention the many other historic theories based on the harmonic series that, for one reason or another, have regarded the major triad as somehow more natural than the minor – one might expect the salience profiles for diatonic chords in minor tonalities to look quite different from those in major tonalities. But in actual music, the rules according to which scale degrees function in major and minor are surprisingly similar. For example, major and minor modes may be mixed on the same tonic without upsetting the prevailing tonal hierarchy (for examples see Forte, 1974, p.464-465), and the patterns of relative prevalence of specific chord doublings and progressions (defined relative to the diatonic rather than the chromatic scale) are much the same in major and minor tonalities (McHose, 1947).

The predictions of the present model are more consistent with expectations based on real music than are the predictions of more simplistic theories based the harmonic series. Root-position diatonic triads in a minor key turn out to be rather similar to analogous predictions for the major key. In fact, the perceptual roots of all root-position diatonic triads in a minor key are predicted to lie in the bass:

sem	0	1	2	3	4	5	6	7	8	9	10	11
	$\hat{1}$		$\hat{2}$	$\hat{3}$		$\hat{4}$		$\hat{5}$	$\hat{6}$			$\hat{7}$
Ia	***76***	5	13	**46**	3	24	2	**36**	27	7	9	12
IIa	30	18	***48***	23	12	**28**	4	30	**31**	3	27	6
IIIa*	41	5	9	***76***	3	15	8	**36**	23	7	**27**	12
IVa	**47**	18	13	23	8	***57***	4	21	**37**	3	23	6
Va	39	7	**28**	27	16	11	2	***71***	12	9	13	**25**
VIa	**47**	14	13	**40**	8	22	4	17	***66***	3	13	12
VIIa	30	13	**28**	23	16	**28**	2	36	12	5	23	***45***

Here and in the following the asterix (*) against chord III means that its 5th is perfect – corresponding to the natural 7th degree of the minor scale – not augmented. In all other chords, $\hat{7}$ is sharpened (from 10 to 11 semitones above the tonic).

We turn now to inverted triads. Here are predictions for first-inversion diatonic triads in a major key:

sem	0	1	2	3	4	5	6	7	8	9	10	11
	$\hat{1}$		$\hat{2}$		$\hat{3}$	$\hat{4}$		$\hat{5}$		$\hat{6}$		$\hat{7}$
Ib	**67**	0	16	7	***56***	26	6	**43**	7	25	2	5
IIb	35	6	**39**	3	21	***60***	2	39	1	**30**	16	9
IIIb	48	4	12	7	**46**	17	6	***68***	1	27	0	**24**
IVb	**52**	6	23	3	17	**49**	2	29	7	***50***	12	9
Vb	44	4	**29**	7	30	17	2	**58**	1	17	6	***44***
VIb	***78***	0	25	1	**36**	30	6	26	7	**40**	2	9
VIIb	35	10	***49***	3	30	**34**	2	43	1	13	16	**24**

The perceptual roots of Ib and Vb correspond to their stacked-thirds roots. All three notes of IVb are potential roots. The perceptual roots of the first-inversion minor triads IIb, IIIb and VIb correspond to their bass notes. The root of VIIb is also in the bass.

The corresponding predictions for the minor key are as follows:

sem	0	1	2	3	4	5	6	7	8	9	10	11
	$\hat{1}$		$\hat{2}$	$\hat{3}$		$\hat{4}$		$\hat{5}$	$\hat{6}$			$\hat{7}$
Ib	**56**	5	13	***66***	3	24	2	**36**	27	7	9	12
IIb	30	18	**28**	23	12	***48***	4	30	**31**	3	27	6
IIIb*	41	5	9	**56**	3	15	8	***56***	23	7	**27**	12
IVb	**47**	18	13	23	8	**37**	4	21	***57***	3	23	6
Vb	39	7	**28**	27	16	11	2	**51**	12	9	13	*45*
VIb	***67***	14	13	**40**	8	22	4	17	**46**	3	13	12
VIIb	30	13	***48***	23	16	**28**	2	36	12	5	23	**25**

Here, the perceptual roots of Ib, IIb, IVb, VIb and VIIb are all predicted to lie in the bass. To take one of many examples of how the bass of these chords can function as roots in music: the chord progression VIb-I in a minor key occurs repeatedly in Debussy's *La Mer*; the VIb chord functions as I (that is, with root $\hat{1}$) with a suspended 6th that resolves to the 5th of the following I chord. The root of IIIb* is ambiguous: it is either the stacked-thirds root $\hat{3}$ or the bass $\hat{5}$. Only the root of Vb corresponds to stacked-thirds theory.

According to the model, then, voicing the third instead of the root in the bass affects the root of all diatonic triads except I and V in major keys and V in minor. In this sense, the model is more consistent with a figured-bass approach to harmony (such as Schenker's) than with a stacked-thirds approach (after Rameau) – but the balance could easily be tipped back in the direction of Rameau by reducing the size of the added constant for the bass note from 20 to, say, 10 or 15.

The predicted profiles for second-inversion triads in major keys are as follows:

sem	0	1	2	3	4	5	6	7	8	9	10	11
	$\hat{1}$		$\hat{2}$		$\hat{3}$	$\hat{4}$		$\hat{5}$		$\hat{6}$		$\hat{7}$
Ic	**67**	0	16	7	**36**	26	6	***63***	7	25	2	5
IIc	35	6	**39**	3	21	**40**	2	39	1	***50***	16	9
IIIc	48	4	12	7	**46**	17	6	**48**	1	27	0	***44***
IVc	***72***	6	23	3	17	**49**	2	29	7	**30**	12	9
Vc	44	4	***49***	7	30	17	2	**58**	1	17	6	**24**
VIc	**58**	0	25	1	***56***	30	6	26	7	**40**	2	9
VIIc	35	10	**29**	3	30	***54***	2	43	1	13	16	**24**

The predicted profiles for second-inversion triads in minor keys are:

sem	0	1	2	3	4	5	6	7	8	9	10	11
	$\hat{1}$		$\hat{2}$	$\hat{3}$		$\hat{4}$		$\hat{5}$	$\hat{6}$			$\hat{7}$
Ic	**56**	5	13	**46**	3	24	2	***56***	27	7	9	12
IIc	30	18	**28**	23	12	**28**	4	30	***51***	3	27	6
IIIc*	41	5	9	**56**	3	15	8	**36**	23	7	***47***	12
IVc	***67***	18	13	23	8	**37**	4	21	**37**	3	23	6
Vc	39	7	***48***	27	16	11	2	**51**	12	9	13	**25**
VIc	**47**	14	13	***60***	8	22	4	17	**46**	3	13	12
VIIc	30	13	**28**	23	16	***48***	2	36	12	5	23	**25**

In both major and minor modes, the perceptual root of Ic (the cadential $^6_4$) is predicted to be ambiguous – it is either the stacked-thirds root $\hat{1}$ or the bass $\hat{5}$. In the major key, the root of Vc is $\hat{5}$, as expected from stacked-thirds theory; the root corresponds to the bass for IIc, IVc and VIIc; all three notes of IIIc, as well as the tonic, are root candidates; and the root of VIc is the tonic $\hat{1}$ or the bass $\hat{3}$. In the minor key, the root of IIIc* is $\hat{3}$ as expected from stacked-thirds theory; of Vc, it is either $\hat{5}$, the stacked-thirds root, or $\hat{2}$, the bass; and the roots of IIc, IVc, VIc and VIIc all lie in the bass.

In general, the perceptual roots of second-inversion diatonic triads are predicted to be more ambiguous than the roots of first-inversion triads. This observation alone cannot, however, account for the rarity of second inversions in common-practice tonal music. A fuller explanation of that phenomenon is beyond the present scope.

The root of the dominant-7th chord in a major key is predicted to be the stacked-thirds root $\hat{5}$ in all inversions except V^7d, for which the root is ambiguous (either $\hat{5}$ or the bass, $\hat{4}$):

sem	0	1	2	3	4	5	6	7	8	9	10	11
	$\hat{1}$		$\hat{2}$		$\hat{3}$	$\hat{4}$		$\hat{5}$		$\hat{6}$		$\hat{7}$
V^7a	41	7	**24**	7	27	**31**	2	***72***	1	16	12	**19**
V^7b	41	7	**24**	7	27	**31**	2	**52**	1	16	12	***39***
V^7c	41	7	***44***	7	27	**31**	2	**52**	1	16	12	**19**
V^7d	41	7	**24**	7	27	***51***	2	**52**	1	16	12	**19**

Much the same applies for the minor key. The root is predicted to be $\hat{5}$ in all cases, but with increasing competition from the bass note in the higher inversions:

sem	0	1	2	3	4	5	6	7	8	9	10	11
	$\hat{1}$		$\hat{2}$	$\hat{3}$		$\hat{4}$		$\hat{5}$	$\hat{6}$			$\hat{7}$
V^7a	36	10	**23**	27	13	**25**	2	***66***	12	7	19	**21**
V^7b	36	10	**23**	27	13	**25**	2	**46**	12	7	19	***41***
V^7c	36	10	***43***	27	13	**25**	2	**46**	12	7	19	**21**
V^7d	36	10	**23**	27	13	***45***	2	**46**	12	7	19	**21**

These results are consistent with the free use of all inversions of dominant 7ths in common-practice tonal music, and the maintenance of the chord's *dominant function* in all inversions.

Predictions look quite different for the diatonic supertonic 7th chord in a major key (II^7, a minor-7th chord). The chord is tonally weaker and more ambiguous than the dominant seventh, and the root is predicted to change whenever the chord is inverted. The predicted root corresponds to the bass in all inversions except for II^7c, where the root may be either the bass $\hat{6}$ or the tonic $\hat{1}$. The principle of stacked thirds, according to which the root should always be $\hat{2}$, is completely overpowered by the competing principles of voicing and tonality:

sem	0	1	2	3	4	5	6	7	8	9	10	11
	$\hat{1}$		$\hat{2}$		$\hat{3}$	$\hat{4}$		$\hat{5}$		$\hat{6}$		$\hat{7}$
II^7a	**49**	4	***54***	2	20	**41**	2	35	6	**26**	13	8
II^7b	**49**	4	**34**	2	20	***61***	2	35	6	**26**	13	8
II^7c	**49**	4	**34**	2	20	**41**	2	35	6	***46***	13	8
II^7d	***69***	4	**34**	2	20	**41**	2	35	6	**26**	13	8

A similar picture emerges in the minor key, where the root of the II^7 (now a half-diminished 7th chord) is predicted to correspond to the bass in all inversions, with some competition from the tonic in the root-position chord II^7a:

sem	0	1	2	3	4	5	6	7	8	9	10	11
	$\hat{1}$		$\hat{2}$	$\hat{3}$		$\hat{4}$		$\hat{5}$	$\hat{6}$			$\hat{7}$
II^7a	**43**	14	***46***	23	10	**30**	4	27	**30**	3	23	6
II^7b	**43**	14	**26**	23	10	***50***	4	27	**30**	3	23	6
II^7c	**43**	14	**26**	23	10	**30**	4	27	***50***	3	23	6
II^7d	***63***	14	**26**	23	10	**30**	4	27	**30**	3	23	6

Let us move on now to some of the more common chromatic chords. The perceptual root of the Neapolitan 6th is predicted to lie in the bass in both major and minor key contexts:

sem	0	1	2	3	4	5	6	7	8	9	10	11
major key	33	**34**	10	7	23	***54***	14	27	**20**	17	14	7
minor key	28	**37**	9	27	8	***48***	14	21	**31**	9	21	8

In a major-key context, the root of the augmented 6th chord is predicted to be ambiguous. It is either the stacked-thirds root $\hat{6}$, or the tonic $\hat{1}$:

sem	0	1	2	3	4	5	6	7	8	9	10	11
	$\hat{1}$		$\hat{2}$		$\hat{3}$	$\hat{4}$		$\hat{5}$		$\hat{6}$		$\hat{7}$
Italian	**52**	10	20	1	25	24	**23**	24	***50***	11	6	15
German	**47**	9	17	**15**	23	25	**18**	24	***50***	11	5	17
French	**49**	7	**31**	1	26	22	**18**	31	***43***	11	9	12

In a minor key, the root of the augmented 6th corresponds to the bass, $\hat{6}$:

sem	0	1	2	3	4	5	6	7	8	9	10	11
	$\hat{1}$		$\hat{2}$	$\hat{3}$		$\hat{4}$		$\hat{5}$	$\hat{6}$			$\hat{7}$
Italian	**47**	13	18	21	10	18	**23**	17	***60***	3	13	16
German	**42**	12	16	**36**	8	19	**18**	17	***60***	3	12	18
French	**43**	10	**30**	21	11	16	**18**	24	***53***	3	16	13

In a major key, the root of the diminished-7th chord on $\hat{7}$ lies in the bass, with competition from both $\hat{5}$ (*missing* stacked-thirds root) and $\hat{1}$ in the case of VII^7a and VII^7d:

sem	0	1	2	3	4	5	6	7	8	9	10	11
	$\hat{1}$		$\hat{2}$		$\hat{3}$	$\hat{4}$		$\hat{5}$		$\hat{6}$		$\hat{7}$
VII^7a	34	14	**24**	2	31	**29**	4	38	**16**	13	15	***39***
VII^7b	34	14	***44***	2	31	**29**	4	38	**16**	13	15	**19**
VII^7c	34	14	**24**	2	31	***49***	4	38	**16**	13	15	**19**
VII^7d	34	14	**24**	2	31	**29**	4	38	***36***	13	15	**19**

In the minor key, the root of the diminished 7th on $\hat{7}$ is less ambiguous, with the bass note predominating in all cases:

sem	0	1	2	3	4	5	6	7	8	9	10	11
	$\hat{1}$		$\hat{2}$	$\hat{3}$		$\hat{4}$		$\hat{5}$	$\hat{6}$			$\hat{7}$
VII^7a	29	17	**23**	23	17	**23**	4	31	**26**	5	22	***41***
VII^7b	29	17	***43***	23	17	**23**	4	31	**26**	5	22	**21**
VII^7c	29	17	**23**	23	17	***43***	4	31	**26**	5	22	**21**
VII^7d	29	17	**23**	23	17	**23**	4	31	***46***	5	22	**21**

Finally, consider the (diatonic) half-diminished 7th chord in root position on $\hat{7}$ in a major key. The root of VII^7a is predicted to be either the bass $\hat{7}$ or the *missing* stacked-thirds root $\hat{5}$:

sem	0	1	2	3	4	5	6	7	8	9	10	11
	$\hat{1}$		$\hat{2}$		$\hat{3}$	$\hat{4}$		$\hat{5}$		$\hat{6}$		$\hat{7}$
VII^7a	34	7	**31**	2	27	**33**	2	39	1	**27**	12	***42***

The half-diminished 7th chord in root position occurs diatonically in a minor key as II^7a. The predicted root is either the stacked-thirds root $\hat{2}$ or the tonic $\hat{1}$:

sem	0	1	2	3	4	5	6	7	8	9	10	11
	$\hat{1}$		$\hat{2}$	$\hat{3}$		$\hat{4}$		$\hat{5}$	$\hat{6}$			$\hat{7}$
II^7a	**43**	14	***46***	23	10	**30**	4	27	**30**	3	23	6
TC	29	13	**27**	21	18	9	**18**	28	***49***	5	15	**28**

The perceptual root of the Tristan chord (TC in the above table, enharmonically equivalent to a half-diminished 7th in root position) is predicted to correspond to its bass $\hat{6}$. On the one hand, this result should be taken with a grain of salt, because the model assumes an established minor key context, and the context in which the Tristan chord appears is arguably too tonally uncertain to warrant a Krumhansl minor-key profile. On the other hand, the argument that the bass is the root is strengthened by the observation that the bass and alto voices move in parallel (augmented 6th to minor 7th) as the chord resolves. According to principle (iv) above, this parallel motion reinforces the bass note as a perceptual root.

7 Conclusion

The model systematically makes predictions that are broadly in accord with music theory, and that promise to allow conventional diatonic theory to be extended to accommodate any chord in the chromatic scale, in any chromatic context. The theory may thus be able to bridge the gap tonal and atonal theory, forging a smooth transition where previously there was impassable abyss. It would appear to have considerable potential for shedding light on the harmonic procedures of tonal modernists such as Debussy and Stravinsky, and on the harmony of jazz – particularly that of the bebop period.

The model suggests a new, systematic approach to harmonic function in the Riemann tradition. Riemann pointed out that the function of a chord is often not clear-cut: A chord may have a mixture of tonic, dominant and subdominant functions, in varying proportions. The present model offers a new means of estimating these proportions. The degree to which a given chord represents one of the three primary functions may be predicted by correlating the pc-salience profile of the chord with the pc-salience profiles of suitable tonic, dominant, and subdominant archetypes. For practical purposes, these archetypes may be represented acoustically as triads (I, V, IV) or tetrads (IV^6, V^7) of octave-complex tones. In a similar vein, the theory may shed light on the theory of chord substitutions, especially in jazz. It may be that a chord is a good substitution for another chord if the correlation coefficient between the pc-salience profiles of the two chords exceeds a certain value. Quantitative exploration of these ideas are beyond the present scope.

The main problem with the model as it stands is the paucity of systematic tests. Further improvements are unlikely without detailed comparisons between model predictions and the results of perceptual and cognitive experiments, and of computer-based analyses of musical scores.

The model currently neglects the soprano voice. McHose (1947, p.72) observed that, in major and minor triads in the Bach Chorales, the soprano is doubled more often than the bass, and the bass more often than the middle parts. Given that doubling is one basis for a definition of the root, this observation strongly suggests that the soprano, like the bass, is a root candidate by virtue of its voicing alone, consistent with the idea that the soprano and bass voices are the most perceptually salient voices in a 4-part texture (presumably due to relative lack of masking by other voices). The present model's addition of a constant to account for the salience of the bass should, therefore, be accompanied by similar acknowledgment of the added salience of the soprano. The reason for neglecting the soprano in the present model is that it soprano voicing is also neglected in mainstream music theory, with which the predictions of the model are intended to be compared. It is anticipated that the calculated salience of the soprano voice will need to be boosted when comparing the predictions of the model with future experimental data.

A final drawback of the present model is its neglect of local voice leading – clearly an important influence on the root. It is not yet clear how local voice leading may be appropriately formulated.

References

Agmon, E. (1993). Towards a theory of diatonic intonation. *Interface – Journal of New Music Research, 22*, 151–163.

Bregman, A. (1990). *Auditory scene analysis: The perceptual organization of sound.* Cambridge, MA: The MIT Press.

Ferrar, W. (1941). *Algebra: A text-book of determinants, matrices, and algebraic forms.* Oxford: Oxford University Press.

Forte, A. (1974). *Tonal harmony in concept and practice* (2nd ed.). New York, NY: Holt, Rinehart and Winston.

Fyk, J. (1995). *Melodic intonation, psychoacoustics, and the violin.* Zielona Gora: Organon Press.

Huron, D. (1994). *The Humdrum toolkit reference manual.* Menlo Park, CA: Centre for Computer Assisted Research in the Humanities.

Huron, D., & Parncutt, R. (1993). An improved model of tonality perception incorporating pitch salience and echoic memory. *Psychomusicology, 12*, 152–169.

Järvinen. (1995). Tonal hierarchies in jazz improvisation. *Music Perception, 12*, 415–437.

Krumhansl, C., & Kessler, E. (1982). Tracing the dynamic changes in perceived tonal organization in a spatial representation of musical keys. *Psychological Review, 89*, 334–368.

McHose, A. (1947). *The contrapuntal harmonic technique of the 18th century.* Englewood Cliffs, NJ: Prentice-Hall.

Parncutt, R. (1988). Revision of Terhardt's psychoacoustical model of the roots of a musical chord. *Music Perception, 6*, 65–94.

Parncutt, R. (1993). Pitch properties of chords of octave-spaced tones. *Contemporary Music Review, 9*, 35–50.

Parncutt, R. (1994). Template-matching models of musical pitch and rhythm perception. *Journal of New Music Research, 23*, 145-168.

Parncutt, R., & Stuckey, R. (1992). Towards a standard alternative notation and terminology based on the chromatic scale. *Musikometrika, 4*, 117–143.

Pritschet, H. (1992). *Wahrnehmung des Grundtons von Dreiklängen: Eine experimentelle Untersuchung.* Regensburg: S. Roderer.

Terhardt, E. (1982). Die psychoakustischen Grundlagen der musikalischen Akkordgrundtöne und deren algorithmischen Bestimmung. In C. Dahlhaus & M. Krause (Eds.), *Tiefenstruktur der Musik.* Berlin: Technical University of Berlin.

Terhardt, E., Stoll, G., & Seewann, M. (1982). Algorithm for extraction of pitch and pitch salience from complex tonal signals. *The Journal of the Acoustical Society of America, 71*, 679–688.

Thomson, W. (1993). The harmonic root: A fragile marriage of concept and percept. *Music Perception, 10*, 385–416.
von Oettingen, A. (1866). *Harmoniesystem in dualer Entwicklung: Studien zur Theorie der Musik.* Dorpat, Leipzig: W. Gläser.

'Good', 'Fair', and 'Bad' Chord Progressions: A Regression-Analysis of Some Psychological Chord Progression Data Obtained in an Experiment by J. Bharucha and C. Krumhansl

Daniel Werts

Suite H35, 332 Bleecker Street New York, NY 10014 USA

Abstract. This paper attempts to justify the psychological ratings determined in an experiment reported by Bharucha and Krumhansl (1983) and Krumhansl (1990a, pp.192-195) for the 42 possible triad-progressions of the major key. The paper shows how a set of six musical factors can be used in a regression formula to generate an extremely close fit to this experimental data. Of these six factors, the most important is the psychological weight of the individual chords; this dimension alone accounts for .541 of the explained variance in the data. The remaining five factors pertain to chord-order, root progression by an upward or downward fifth, the possibility of a progression being misread in the relative minor, the treatment of the leading tone, and a condition concerning the juxtaposition of a subdominant chord (IV or II) with a dominant chord (V or VII). The paper ends with a derivation of the chord weights generated by the regression; this derivation is based on a series of hypothesized tone weights, and assumes that the psychological weight of a chord is dependent on the summed psychological weights of that chord's component tones. It is shown that the proposed tone weights exhibit a high correlation with two sets of empirically-determined tone weights, one reported by Krumhansl (1990a) and the other by Cuddy and Thompson (1992).

1 Introduction

My title alludes to the kinds of qualitative statements about chord progressions that writers of harmony books used to make so matter-of-factly only a couple of generations ago. I am referring to authors such as P. Goetschius (1895, 1931, 1970) and E. Prout (1900), who routinely characterized progressions with terms like 'good', 'fair', and 'poor'. Such descriptions, if no longer stylish, are of considerable interest to music psychology because they represent an attempt on the part of their authors to articulate often strongly-felt, and sometimes widely-shared intuitions.

Today it is possible to test such intuitions experimentally, and thereby to furnish a much sturdier grip on them than one had in the past. In an experiment reported by Bharucha and Krumhansl (1983), psychological ratings based on the subjective responses of sixteen subjects were determined for the 42 possible ordered pairs of diatonic triads in the major key; these values, provided by Krumhansl (1990a, Table 8.2), are seen in the second column

Table 1. Regression Data, Source for Y-values: Krumhansl (1990a, Table 8.2)

	Y	X1	X2	X3	X4	X5	X6	Y_{est}	$Y-Y_{est}$
G-C	6.19	418.0	1	0	1	0	0	6.40	-0.21
Dm-G	6.10	378.0	1	1	1	0	0	6.15	-0.05
F-G	6.00	383.7	1	1	0	0	0	6.00	0.00
C-G	5.94	418.0	0	0	1	0	0	5.97	-0.03
F-C	5.94	389.7	1	0	1	0	0	6.04	-0.10
C-F	5.91	389.7	0	0	1	0	0	5.61	0.30
Bo-C	5.85	361.4	1	0	0	0	0	5.45	0.40
Dm-C	5.69	384.0	1	0	0	0	0	5.74	-0.05
Am-G	5.56	371.1	1	0	0	0	0	5.57	-0.01
G-F	5.51	383.7	0	0	0	0	0	5.30	0.21
Am-Dm	5.44	337.1	1	0	1	0	1	5.09	0.35
Dm-Bo	5.41	321.4	1	1	0	0	1	4.93	0.48
Em-C	5.38	349.5	1	0	0	0	0	5.29	0.09
C-Am	5.26	377.1	0	0	0	0	0	5.22	0.04
G-Am	5.19	371.1	0	0	0	0	0	5.14	0.05
Bo-G	5.16	355.4	1	0	0	1	0	5.13	0.03
C-Dm	5.10	384.0	0	0	0	0	1	5.04	0.06
Am-F	5.07	342.8	1	0	0	0	1	4.94	0.13
Am-C	5.04	377.1	1	0	0	0	1	5.38	-0.34
Em-G	5.03	343.5	1	0	0	1	0	4.97	0.06
F-Dm	5.00	349.7	1	0	0	0	1	5.03	-0.03
Dm-Am	4.97	337.1	0	0	1	0	1	4.66	0.31
G-Bo	4.85	355.4	0	0	0	1	0	4.69	0.16
F-Bo	4.79	327.1	1	1	1	0	1	5.23	-0.44
G-Dm	4.79	378.0	0	0	1	1	1	4.94	-0.15
C-Em	4.78	349.5	0	0	0	0	0	4.86	-0.08
Dm-F	4.76	349.7	0	0	0	0	0	4.86	-0.10
Am-Em	4.72	302.6	1	0	1	0	0	4.92	-0.20
Em-F	4.63	315.2	1	0	0	1	0	4.61	0.02
Em-Am	4.60	302.6	0	0	1	0	0	4.48	0.12
C-Bo	4.57	361.4	0	0	0	0	1	4.75	-0.18
Bo-F	4.53	327.1	0	0	1	1	0	4.55	-0.02
Am-Bo	4.50	314.5	1	0	0	0	1	4.58	-0.08
Em-Dm	4.47	309.5	1	0	0	1	0	4.53	-0.06
Em-Bo	4.47	286.9	1	0	1	1	0	4.47	0.00
G-Em	4.47	343.5	0	0	0	1	0	4.54	-0.07
F-Am	4.35	342.8	0	0	0	0	1	4.51	-0.16
F-Em	4.22	315.2	0	0	0	0	0	4.42	-0.20
Bo-Am	4.19	314.5	0	0	0	0	1	4.14	0.05
Bo-Dm	4.16	321.4	0	0	0	1	0	4.25	-0.09
Bo-Em	4.16	286.9	0	0	1	1	0	4.03	0.13
Dm-Em	4.00	309.5	0	0	0	0	0	4.34	-0.34

(marked *Y*) of Table 1; to the left are the 42 progressions, given here in C major. We may view these values as an updated, far more precise version of the old 'good', 'fair', and 'poor' categories.

My goal in this paper is to provide a musical rationale for these numbers. At the heart of this rationale is a set of six musical properties, or factors, which, when quantified and inserted into the regression formula below, lead to an extremely close approximation of the psychological data. The symbol Y_{est} stands for each chord progression's approximated, or estimated value; the values for *X1* through *X6*, representing the six musical properties, are given in the middle columns of Table 1.

$$Y_{est} = 0.320 + .013(X1) + .433(X2) + .262(X3) + .227(X4) - .244(X5) - .266(X6)$$

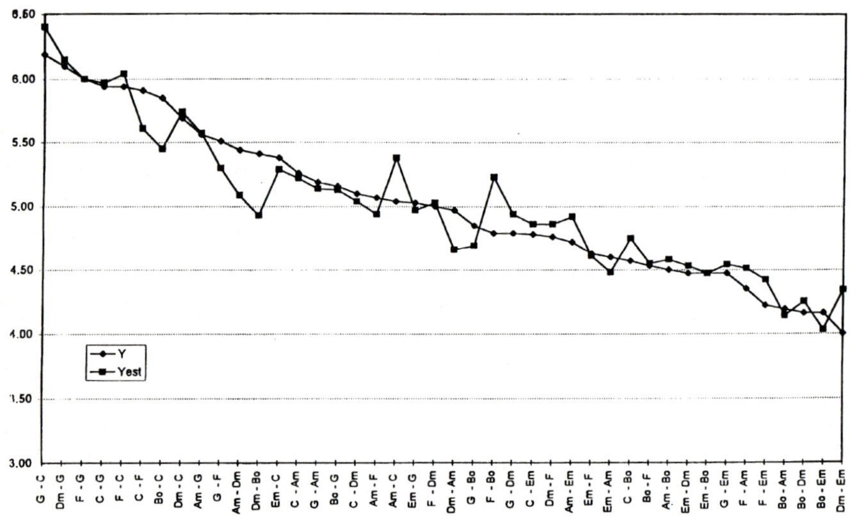

Fig. 1. Picture of the regression formula's results

Figure 1 compares the psychological values (Y) with those generated by the equation (Y_{est}); it gives a clear visual picture of the regression formula's results; the correlation, as can be sensed, is high, .95 (R^2=.90). Table 2 provides some additional technical information of a statistical nature; from the T-values shown here, it will seen that all of the factors involved in the regression are highly significant. In every case except one, $p < .01$; for the exception, $p < .05$.

Table 2. Statistics

Y	Krumhansl Chord Progression Ratings	β	s.e.β	T-value
	Constant	0.350	0.390	0.896
X1	Sum of Ideal Chord Weights	0.013	0.001	12.067**
X2	Order Condition	0.433	0.068	6.351**
X3	S-D Progression Condition	0.262	0.120	2.187*
X4	Upward or Downward Fifth Condition	0.227	0.069	3.284**
X5	Problematic Leading Tone Condition	-0.244	0.084	-2.911**
X6	Parallel Minor Condition	-0.266	0.074	-3.582**

* Significant at the .05 level
** Significant at the .01 level
$R^2 = .900$

2 Description of the Experiment

The psychological values (*Y*) were obtained by averaging two sets of experimental data. One set was determined for the 42 possible triad-pairs in C major, heard after a IV-V-I cadence in C major; the other set was determined for the 42 possible triad-pairs in F♯ major, heard after a IV-V-I cadence in F♯ major. After listening to the cadence, the subjects stated how well the second chord in each pair seemed to follow the first. Generally speaking, higher values were given to chord pairs commonly considered characteristic of the key, and lower ones to those considered uncharacteristic. For instance, II-V and IV-V received high ratings (respectively, 6.10 and 5.94), while IV-VI and II-III received low ones (4.35 and 4.00). A rating scale of one to seven was used, with seven as the highest possible score. The examples were performed with octave-complex tones so that the extreme top and bottom tones in any chord were not ascertainable; that is, neither chordal inversion, nor melodic position, could be determined. As part of the same experiment, subjects also had to rate all possible chord pairs involving one chord from C major and one chord from F♯ major. Although values for these bitonal pairs were not available, Krumhansl (1990a, p.192) remarks that these numbers were substantially lower than those for the diatonic pairs; she theorizes that the presence of the bitonal pairs may help explain why the diatonic ratings are, generally rather high (4.00 to 6.19), relative to the rating scale (1.00 to 7.00).

3 Rejected Factors

The six factors finally chosen for the regression were found after a long period of trial and error in which many factors, all of which seemed musically relevant, were tested. Certain factors were rejected because they were statistically insignificant. Falling in this category was a hypothesized tritone condition, which stipulated that progressions introducing the 4th degree of the scale before the 7th degree should tend to receive higher values, while progressions introducing the 7th degree before the 4th degree should tend to receive lower ones. Similar views concerning the tritone's key-defining properties have been defended by Ratner (1962) and by Butler & Brown (1984), but called into question by Krumhansl (1990b, p.321). When the data was examined for a statistical trend supporting this hypothesis, none could be detected. The presumed positive effect of the 4th degree preceding the 7th degree was spoiled by Dm-Em, which receives the overall lowest value in the data, 4.00; the presumed negative effect of the 7th degree preceding the 4th was spoiled by Em-F and Em-Bo, which received higher values (4.63 and 4.47) than F-Em and Bo-Em (4.22 and 4.16). The factor was thus eliminated.

Other rejected factors proved significant when tested alone, but detracted from the overall correlation when tested in combination with other factors. A factor defined by root-movement by an upward third fell into this category. A highly significant negative factor when considered by itself, it worsened the correlation with the experimental data when other factors were present. Many such factors were encountered.

Of all those rejected, the most problematic factor was one meant to account for the influence of the subjects' listening habits on the ratings. The need for such a factor was grounded in the assumption that listeners would tend to give higher values to progressions they had often heard, and lower ones to progressions they had less often heard; this assumption was consistent with observations by Krumhansl (1990b, p.315) about the relevance of statistical distributions of musical elements to psychological data. Unfortunately, precise information about the subjects' listening repertoire was not available. It was argued, nonetheless, that most of the subjects (sixteen members of the Cornell summer community) would be familiar with well-known classical pieces composed in the period 1800-1750, and so this repertoire could be taken as a good starting point. That Krumhansl herself had shown significant correspondences between psychological data and statistical properties of such pieces (Krumhansl, 1990a, pp.67-69, 135, 180-182) buttressed this line of reasoning. Surprisingly, the results of the subsequent inquiry were disappointing. For example, statistical data from Norman's (1945, Table III) frequency-of-usage analysis of major-key chord progressions in the first 8 *Sonatas* of Beethoven produced a correlation of only .480. This compared poorly with the correlations, ranging from .67 to .887, which Krumhansl reported for other dimensions such as the distributions of tones (.887), intervals(.67), and chords (.75) (Krumhansl, 1990a). A particularly troublesome point was that the data in all the pieces examined revealed an extremely skewed distribution not reflected in the psychological data. In the Beethoven analysis, for example, four progressions (V-I, I-V, I-IV, IV-I) occur 28.15 per cent of the time, while 31 progressions occur less than .41 per cent of the time. Nothing even vaguely resembling this kind of division can be observed in the psychological values; indeed, as seen from Fig.1, they produce an even, gently descending curve without any significant breaks. Further investigation of pieces from the classical repertoire kept producing similar results, and it was theorized that the sampling base was probably too narrow. Several informal investigations then showed that correlations improved if data from representative examples of folk-music, rock, jazz and early music were averaged in with data for the classical pieces. Although promising, this approach was not pursued systematically, because the difficulties involved in objectively selecting and analyzing a sufficiently large number of tonal contexts proved unwieldy for the current project. However, a systematic study was undertaken of chord-progression statistics in 82 major-mode hymns from *The Methodist Hymnal* (1939). A condition specifying that each hymn be entirely diatonic–that is, contain no accidentals–simplified the analysis; another condition, requiring that all hymns meeting this first condition be considered, assured objectivity in the selection process. When a comparison was made with the psychological data, the resulting correlation was .61, which is higher the the .48 seen before, but still unconvincing. (What slightly elevated the correlation was the presence of a number of hymns drawn from folk-music and pre-tonal sources.) On the basis of this and the other statistical studies, it was speculated that, even if information about the subjects' listening habits had been available, a factor representing this dimension would not have been significant enough to merit inclusion in the regression.

4 The Six Accepted Factors

4.1 Factor X1: Sum of Ideal Chord Weights.

Of all six factors, *X1* is by far the most powerful, accounting as it does for 54.1 per cent of the explained variance in the data. For each chord pair, the *X1* value is found by the equation, $X1 = W1 + W2$; *W1* is the assumed psychological weight for the first chord in the progression, and *W2* is the assumed psychological weight for the second chord. The weights are "ideal" in the sense that they were generated from the data itself, by treating each of the seven chords as an dummy variable and then regressing them against the experimental data. Here are the weights found for the I-VII triads in the major key: 212, 172, 137.5, 177.7, 206, 165, 149.4. Near the end of the paper, I shall show how it is possible to derive these weights, or at least a close approximation of them, from another source.

One noteworthy aspect of these weights is their high correlation, .931, with a set of chord-frequency values for the I-VII triads of the major scale, which was provided by Krumhansl (1990a, Table 7.9): 41.79, 8.35, 1.35, 8.45, 31.68, 5.27, 2.76. These values represent each chord's percentage of occurrence in a sampling of nearly 66,000 chords and were extracted from an elaborate chord-frequency study by Budge (1943, Tables IX and X). An even higher rank correlation–1.00–can be observed between the assumed weights and those just cited. Here is the ranking for both series: I V IV II VI VII III. It is of interest that Durand (1890?, p.20) ranks the chords almost identically, reversing only the order of the last two chords: I V IV II VI III VII. He comments that I, V, and IV are "les plus usités":, II and VI are "très usités" and III and VII "moins usités". This same ranking was used by Melcher and Warch (1966, Table of Contents), to introduce the chords in the scale; it can also be observed, without the VII, in Wood (1936, Table of Contents) and in many other sources. From all this we see that the ideal chord weights correspond well with both statistical properties of the repertoire and with accepted music-theoretical notions about chord rank.

4.2 Factor X2: Chord-Order Condition.

This factor accounts for 13.9 per cent of the explained variance in the data. For the 21 progressions meeting the condition, $X2$=l; for all others, $X2$=0. To meet the condition, a chord pair must be extractable by reading from left to right in the following array: Am Em F Dm Bo G C. It will be seen that almost all cadential progressions are derivable from this array by starting on the tonic, skipping back to a chord on the left, and then progressing gradually or by skip to the tonic. Some typical examples derivable in this way are: C-F-G-C; C-Dm-G-C; C-F-Bo-C; C-Dm-Bo-C; C-Am-Dm-G-C; C-F-Dm-G-C; C-Am-F-Dm-G-C. An important characteristic of the model is that the seventh in those chords that commonly receive a 7th–G, Bo, Dm and F–can be prepared by at least one chord to the left, and resolved in at least one chord to the right. It is of historical interest that a practically identical model was proposed by Marpurg (1749, p.265); in Marpurg's model, however, Em is located between G and C, so that a series of uninterrupted descending third-progressions results. The anomalous placement of Em in our model is justified quite clearly from the data: listeners preferred progressions such as Em-G, Em-Bo, Em-Dm, and Am-Em to their retrogrades.

4.3 Factor X3: Subdominant-Dominant Condition.

The factor has a fairly strong positive effect, explaining 8.4 per cent of the explained variance in the data. For chord-pairs meeting the condition, $X3$=1; for all others, $X3$=0. To meet the condition, a chord-pair has to contain one chord from the subdominant group of chords (F, Dm) and one from the dominant group (G, Em). One possible explanation of this factor's importance is that, in many typical progressions derived in the manner just described (and indeed in all progressions excluding the presence of both Dm and Em), it is at the transition from the subdominant group to the dominant group that the introduction of all seven tones of the scale is achieved and the key thus unambiguously defined. In the series C-F-G-C, this key-defining moment occurs as F moves to G. In C-Am-F-Dm-G-C, it occurs as Dm moves to G. Since such progressions tend to be associated with key-definition, they generally receive higher than normal ratings.

4.4 Factor X4: Root-Movement by Upward or Downward Fifth.

This is a positive factor, and accounts for 7.3 per cent of the variation in the data. For all 14 progressions meeting this condition, $X4=1$; for the remaining 28 progressions, $X4=0$. To fulfill the condition, the root of the second chord must be readable as a fifth above or below that of the first chord. Kitson (1924, p. 51) and many other theorists have stressed the importance of fifth-related root-progressions, and so it comes as no surprise to find that such progressions exert a positive influence on listeners' perceptions. What is surprising, from a music-theoretic point of view, is that the best correlation with the experimental data was obtained when no distinction was made between upward and downward fifths; Rameau, for example, regards downward fifths as the more important of the two (Shirlaw, 1955, p.128).

4.5 Factor X5: Problematic Resolution of the Leading-Tone.

This is one of two negative factors in the regression; it accounts for 7.8 per cent of the explained variance in the data. For all 11 progressions meeting this condition, $X5=l$; for the remaining 22 progressions, $X5=0$. To fulfill the condition, the first chord must contain a leading tone that either cannot resolve to the tonic note because the second chord does not contain it (as in the progression Em-G), or conventionally does not (as in the progression Em-F, where the leading tone normally resolves downward to the sixth degree). The rule that the leading tone must resolve to the tonic note is such a powerful one that its disregard can easily awake the impression that a modulation has taken place. Chord-progressions where the rule is not observed thus tend to be perceived as being foreign to the key and therefore receive lower ratings. As an example illustrating the rule's power, I cite the opening of the chorale-melody *Vom Himmel Hoch*: c-b-a-b-g-a-b-c. It is almost impossible to harmonize this melody convincingly in C major, because within a span of only five notes that key's leading tone is twice incorrectly resolved. A purely C major solution, as found in *The Lutheran Hymnal* (1941, Hymn No. 85) will tend to sound modal; here are the chords used in this source: C-G-Am-G-C-Am-G-C. If part of the melody is heard in the dominant key, an appropriate tonal harmonization can be found, such as the following set of chords or some close variant: C-G-D-G-C-F-G-C; this is the solution chosen by the Moravian Church in its *Hymnal and Liturgies* (1969, Hymn No. 62); two particularly ingenious tonal solutions, including this one, can be found in the Bach's *371 Harmonized Chorales* (Riemenschneider, 1941).

4.6 Factor X6: Parallel Minor Condition.

For the 13 progressions fulfilling the condition, $X6=1$; for all others, $X6=0$. This a negative factor, one that accounts for 8.5 per cent of the explained variance in the data. To fulfill this condition, a progression must be part of, or lead smoothly to, a cadence in the parallel minor key, or otherwise create an impression of that key. It is easy to theorize why progressions easily interpretable in the parallel minor tend to receive lower than normal ratings in the experiment. Since the natural form of the scale in that key is identical in content to the major scale, progressions characteristic of the minor can easily imply a passing modulation, even if no accidentals are introduced; this is a possibility not existing, or existing to a far lesser extent, for other keys such as the dominant or subdominant. Progressions characteristic of the parallel minor thus uniquely threaten the stability of the major tonic. Earlier theorists, it should be pointed out, were quite sensitive to such matters, and on this point the reader is referred to Fétis (1844, p.20, paragraph 51), and to Geyer (1862, p.95), who show respectively that in a C major context, even progressions like C-Dm and C-Bo can carry with them a strong A minor implication. Fétis (p. 20) goes so far as to hear "un vague changement de tonalité", towards A minor (or D minor) in the C-Dm segment of an example clearly based on C-Dm-G-C; this is regarded by many theorists as an important cadential progression in C major! In an actual cadential situation, action is usually taken, such as adding 7ths to both the Dm and G chords, as well as inverting the Dm chord to its first inversion, so as to create a situation unambiguously characteristic of C major.

It is difficult, given the complexities of key-perception, to specify purely objective criteria for determining which progressions should be considered as evocative of the minor key and which should not. Ideally, membership on such a list would be quantified, according to empirical or theoretical criteria, along a multivalent scale recognizing varying intensities of membership, and not determined by a simple yes-no scheme as done here. In the absence of such quantified data, the suggested list of 13 progressions must suffice. Speaking in its favor is its high degree of consistency with accepted notions about progressions in the minor key. For example, 11 of the 13 progressions on the list are readable in A minor as that key is formulated by Rimsky-Korsakov (1943). However, the progression F-C–interpretable by Rimsky-Korsakov as VI-III in A minor–was excluded because it was felt that, when heard in a C major context, this would not normally evoke an impression of the minor key. In addition, two progressions not mentioned by Rimsky-Korsakov, C-Bo and F-Am, were added to the list because they fulfill the requirement of leading smoothly to a cadence; in both cases the minor key's dominant chord, E, could follow unproblematically. Further refinements made by adding or subtracting a progression here or there can be ignored; even if musically justifiable, they would probably have no more than a minimal effect on the regression.

5 Derivation of the Ideal Chord Weights

In closing, I shall show how the ideal chord weights can be derived, in the sense of being almost perfectly approximated, from set of other, seemingly unrelated, values. The derivation will confirm something we already know: that the melodic and harmonic dimensions of music are intricately and inseparately connected with one another. We begin the derivation by assuming a set of hypothetical tone weights, defined for degrees 1-7 of the major scale; these are as follows: 60 38.5 39 37 53 33 31. The rank ordering of the tones here, in C major–c g e d f a b–corresponds to one I proposed in an earlier work (Werts, 1983); it has at least one historical precedent: Kirnberger (1721-1783) tuned the notes of the major scale in this order (Kirnberger, 1982, p.16). As for the specific numbers I have chosen, I make no great claims for them except that they seem intuitively about right, and also lead to interesting results.

It is certainly worth noticing, however, that the weights correlate well with several sets derived empirically. The first, provided by R. Kluge of Humboldt University (personal communication, January, 1997), is a set of durational tone weights for the 82 Methodist Hymns mentioned earlier; these numbers represent the percentage of time allotted in the Hymns' 15053 note-events to the tones 1-7 of the major scale; 32.4 7.1 15.9 8.4 25.3 5.2 5.7. The correlation with the hypothetical weights is .970. Another set of weights is provided by Cuddy and Thompson (1992, Table 2): 6.15 4.20 4.15 4.15 5.40 3.70 3.05. These were obtained in a probe-tone experiment in which subjects were asked to rate tones heard against the tonic and dominant harmonies of a previously established major key; the numbers given represent the average values for the two contexts. The correlation with the hypothetical weights is .986. An even better correlation–indeed, a near perfect one, .991–is found with the next set, provided by Krumhansl (1990a, Table 5.1, major key context, column averages): 4.95 3.84 3.84 3.64 4.56 3.41 3.14. These were found when listeners were asked to judge how well the 132 possible ordered tone pairs fit into a previously heard major-key context; the numbers give the average weight, found for each tone of the major scale, of the 11 possible tone pairs containing that tone as their second element. (The value 4.95, for example, represents the average value of the weights for the following pairs: c♯-c, d-c, d♯-c, e-c, f-c, f♯-c, g-c, g♯-c, a-c, a♯-c, b-c. The methodology here implies an indirect, but still valid, measurement of tone weight.)

Returning to our demonstration, we shall now use the hypothetical tone weights to derive a set of chord weights; these are similar to the ones we are attempting to approximate, but will require some adjusting to bring them acceptably near the desired results. In proceeding, we shall assume that a chord's weight is largely dependent on the weights of its component tones. Thus the C chord's heavy weight in C major will be seen as reflecting the heavy weights of the tones c, e and g; the Bo chord's relatively low weight will be seen as reflecting the low weights of the tones b, d and f. Accepting this line of reasoning, we now add together, for each chord, the weights of its individual tones; since the root is usually doubled, we shall count its value twice, bringing the total number of values summed for each chord to four. The results obtained by this procedure are shown below in the first row; the second row gives the ideal chord weights. The correlation between the two series is .754, necessitating one further step in the derivation:

C (212)	Dm (147)	Em (162)	F (167)	G (175.5)	Am (165)	Bo (137.5)
(212)	(172)	(137.5)	(177.7)	(206.0)	(165.1)	(149.4)

Inspection reveals that two values in the first row, those for Dm and G, need considerable boosting; two others, for F and Bo, need some boosting, but not as much; those for C and Am can remain as they are; that for Em must be reduced. We now pose the question as to which musical characteristics would lead listeners to make such adjustments. One plausible answer involves noticing the rather striking parallels between the needed adjustments and the widely differing treatments, from chord to chord, of the interval of a dissonant 7th (for example: f in G7 or c in Dm7). It can easily be argued that variables such as this tone's frequency of usage and the degree of voice-leading strictness to which the tone is subjected differ considerably from chord to chord, and thus add to or detract from each chord's psychological weight. Surely part of what makes G in C major different from, let us say Am, is that the 7th in G7 occurs far more frequently, and is treated far more freely regarding its preparation, resolution, and registral placement than the 7th in Am7. While G7 is an essential component of the key, Am7 is not, and indeed, under certain circumstances can even be perceived as key-threatening; Kirnberger, for example, interprets Am7 in C major as implying a modulation to the dominant key (Shirlaw, 1955, p.320). The following qualitative statements, based on extensive observations by Goetschius (1895, 1931, 1970), Kitson (1924), McHose (1947), roughly summarize the differing treatments of the various 7ths in C major; it is of course understood that the treatment of the sevenths in each category is only similar, not identical: 7ths are added often and freely to G and Dm; less often and less freely to F and Bo; rarely and rather strictly to C and Am; very rarely, and very strictly to Em. (Some examples of *free* treatment of a seventh include its entering without preparation, resolving upwardly, resolving into a dissonance, resolving after a delay created by an inserted chord, resolving in another register, and not resolving at all; durational emphasis is another aspect.) Here is one possible quantification scheme for the above descriptions: *often and freely added*=25; *less often and less freely added*=12.5; *rarely and rather strictly added*=0; *very rarely and very strictly added* = -12.5. When these numbers are combined with the chord weights just obtained, an extremely close approximation of the ideal chord weights results; the correlation between these this new approximation and the ideal chord weights is extremely high, .988. The new results are shown in the first row; underneath are written the ideal weights:

C (212)	Dm (172)	Em (149.5)	F (179.5)	G (200.5)	Am (165)	Bo (150.0)
(212)	(172)	(137.5)	(177.7)	(206)	(165.1)	(149.4)

To summarize what we've done in this section: we started with a set of hypothetical tone weights that are highly correlated with several sets of empirical data; these tone weights were then used to generate some provisional chord weights; finally, these provisional chord weights were adjusted by adding to them values derived from qualitative statements about the usage of the dissonant 7th. A nearly perfect approximation of the ideal chord weights results.

6 Concluding Comment

It seems more and more conceivable that someday a comprehensive theory of harmony will be formulated that would help explain the empirical results of experiments such as the ones referred to in this study. Such a theory would quantify all basic concepts, thereby avoiding much of the vagueness and confusion of traditional theory. At the same time, it would be consistent with rules and observations formulated by exact analyses of actual pieces. I hope the results presented here will be seen as contributing towards the construction of such a theory.

Acknowledgement

For help with computational work, I'm most grateful to a friend and colleague, Stanley Noval, of Marketing, Strategy & Planning of New York City.

References

Bharucha, J., & Krumhansl, C. (1983). The representation of harmonic structure in music: Hierarchies of stability as a function of context. *Cognition, 13*, 63–102.

Budge, H. (1943). *A study of chord frequency.* New York, NY: Bureau of Publications, Teachers College, Columbia University.

Butler, D., & Brown, H. (1984). Tonal structure versus function: Studies of the recognition of harmonic motion. *Music Perception, 2*, 6–24.

Cuddy, L., & Thompson, W. (1992). Asymmetry of perceived key-movement in chorale sequences: Converging evidence from a probe tone analysis. *Psychological Research, 54*, 51–59.

Durand, E. (1890?). *Abrégé du cours d'harmonie theorique et pratique.* Paris: Alphonse Leduc.

Fétis, F. (1844). *Traité complet de la theorie et de la pratique de l'harmonie.* Brussels: Au Conservatoire.

Geyer, F. (1862). *Musikalische Compositionslehre.* Berlin: Verlag von A. Vogel & Comp.

Goetschius, P. (1895). *Materials used in musical composition.* New York, NY: G. Schirmer.

Goetschius, P. (1931). *The theory and practice of tone-relations.* New York, NY: G. Schirmer.

Goetschius, P. (1970). *The structure of music.* Westport, CT: Greenwood Press. (Reprint of 1934 ed.)

Hymnal and liturgies of the Moravian Church. (1969). Elk Grove, IL: W. Carqueville.

Kirnberger, J. (1982). *The art of strict composition* (D. Beach & J. Thym, Trans.). New Haven, CT: Yale University Press.

Kitson, C. (1924). *The evolution of harmony* (2nd ed.). Oxford: Clarendon Press.

Krumhansl, C. (1990a). *Cognitive foundations of musical pitch.* New York, NY: Oxford University Press.

Krumhansl, C. (1990b). Tonal hierarchies and rare intervals in music cognition. *Music Perception, 7*, 309–324.

The Lutheran Hymnal. (1941). St. Louis and London: Concordia Publishing House.

Marpurg, F. (1749). Fortsetzung des Versuchs über die Zeugung der Intervallen, der Harmonie und der daher fliessenden Melodie. *Der kritische Musicus an der Spree, 33*, 263–268.

McHose, A. (1947). *The contrapuntal-harmonic technique of the 18th century.* New York, NY: Appleton-Century-Crofts.

Melcher, R., & Warch, W. (1966). *Music for keyboard harmony.* Englewood Cliffs, NJ: Prentice-Hall Inc.

The Methodist Hymnal. (1939). New York, Nashville: The Methodist Publishing House.

Norman, P. (1945). *A quantitative study of harmonic similarities in certain specified works of Bach, Beethoven and Wagner.* New York, NY: Carl Fischer.

Prout, E. (1900). *Counterpoint: Strict and free* (7th ed.). New York, NY: G. Schirmer.

Ratner, L. (1962). *Harmony, structure and style.* New York, NY: McGraw Hill.

Riemenschneider, A. (Ed.). (1941). *J.S. Bach, 371 harmonized chorales.* New York, NY: G. Schirmer.

Rimsky-Korsakov, N. (1943). *Practical manual of harmony* (7th English ed.; J. Achron, Trans.). New York, NY: Carl Fischer.

Shirlaw, M. (1955). *The theory of harmony.* Dekalb, IL: Dr. Birchard Coar.

Werts, D. (1983). *A theory of scale references.* Unpublished doctoral dissertation, Princeton University.

Wood, C. (1936). *Ear tests in harmony.* New York, NY: American Book Company.

Problems of Shape and Background in Sounds with Inharmonic Spectra

Jenö Keuler

Institute of Musicology, Syst. Musicology, H-1250 Budapest Pf.28,
Táncsics Mihály u. 7, Hungary

Abstract. In this paper, we show that it is possible to synthesize inharmonic sounds which exhibit certain tonal properties. Their similarity, consonance, and dissonance depend on the spectral configuration while their use in a musical context depend on the scaling used as well. The sounds are perceived as Gestalts in that their pitch and timbre belong to the whole sound and not to the partials that constitute the sound. The sounds can be configured to form melodies, voices, chords, accompaniment, and they may form tonal fields that influence the behavior of the sounds. The tonal field manifests itself as a holistic form having characteristics depending on the structure of the sound system. It is argued that the exploration of inharmonic sounds from a musical point of view must be completed with knowledge about psychophysiological processes.

1 Introduction

In this paper, I intend to provide some examples of the musical behavior of synthesized sounds having particular properties that are related to C. von Ehrenfels' notion of "Gestaltqualitäten". In the traditional European music theory, it is taken for granted that musical sounds, when played an octave apart, exhibit the same pitch quality. This assumption has almost received the status of an axiom, and it seems to be so evident that sounds at different octaves have got the same pitch name. Octave equivalence is evident on psychoacoustical grounds: musical sounds at a distance of one octave share a great number of common frequencies and that is why the qualities of their sensation is so similar. It is possible, however, to synthesize sounds which have common frequencies using intervals other than the octave. In such case, one perceives another interval as the most consonant relationship. Interestingly, the sounds have the same pitch quality at this interval, which then takes over the role of an octave. The sound examples described in this paper are available on the CD which is included within this book. They are summarized in the Appendix.

2 Sound Synthesis Method

All examples have been realized with a Yamaha DX7-II synthesizer using additive synthesis in *dual mode* (example #14), and a combination of additive synthesis with frequency modulation in *single mode* for the examples #1-3 and #5-7. In these examples, a voice-algorithm has been used where the frequency of the 3rd operator is modulated by that of the 4th one. The 5th operator's frequency is also a carrier which is modulated by the 6th operator. For example, in sound example #5, the spectrum has been built up step by step using Op1 = 466, Op2 = 880, Op3 = 1661, Op5 = 5920, Op5+3 = 5920 & 1661, Op5+(3+4) = 5920 & (derived frequencies), Op(5+6)+(3+4) = (derived frequencies), Op(5+6)+(3+4)+2 = (derived frequencies) & 880, Op(5+6)+(3+4)+2+1 = (derived frequencies) & 880 & 466. The frequency setting of the operators is adjusted to the values of the pillar frequencies (see below), likewise as in the case of additive synthesis. In sound example #4, a more complicated sound is presented, composed of two *voices* in *dual mode* and one of these has the same type of spectrum as the ones of the examples #1-3 and #5-7.

The direct frequencies in additive synthesis can be tuned very precisely. On the other hand, the intensity values of the frequencies cannot be set very precisely. The 99 signal levels of the operator do not represent physical quantities, and the levels that are set become moreover altered by keyboard scaling. They also change in time according to the envelope of each operator. The frequencies of a spectrum based on frequency modulation depend not only on the frequency values of the carrier and the modulator signals, but also on their intensity values. For this reason, it is a hopeless matter to calculate the exact values of the spectrum created in this way on the DX7 synthesizer. Therefore, in the examples that follow, only the direct frequencies can be properly given. In addition, some frequencies are given that could be deduced from a spectrogram analysis.

All the data given for the sound examples #9-14 are exact (inclusive also the pillar alien frequencies of the example #14). Additive synthesis was used here. On the other hand, the frequency values derived from the frequency modulation are less precise.

3 Sound Examples

3.1 Sound Example #1

The first example demonstrates a sound which is consonant with the same sound at interval distance of a major seventh, but *not* with its octave transposition. The spectra of the sounds can be read as follows:

Spectrum:			(23679)
		12544	12544
	6645	6645	6645
		5920	
	3520	3520	3520
	3136		*3136*
	1865	1865	1865
		1661	
	988	988	988
	880		
	523	523	
	277		
Keyboard notation:	C♯3	C4	B4

As for all the other examples, the table should be read as a sort of spectrographical chart in which the numbers in a column represent the frequencies of the spectral tone components. The sounds are labeled using a common keyboard notation. The precision of the sounding result, however, is restricted to the possibilities of the synthesizer used. As such, the numbers within parenthesis cannot be executed. The numbers in italics, on the other hand, indicate approximate values of frequency modulations. They are not calculated but measured values.

3.2 Sound Examples #2 and #3

The second and third sound example demonstrate the effects of consonance and dissonance. The sound of the previous example is first confronted with its consonant major seventh and dissonant octave. In the next example, it is confronted with its consonant major seventh and dissonant minor seventh. The spectra of sound example #2 are:

Spectrum:		(23679)		(25088)
	12544	12544	12544	(13290)
	6645	6645	6645	7040
	5920		*5920*	
	3520	3520	3520	3730
		3136		*3323*
	1865	1865	1865	1978
	1661		*1661*	
	988	988	988	1046
	523		523	
Keyboard notation:	C4	B4	C4	C5
	Consonant		Dissonant	

The spectra of sound example #3 are:

Spectrum:				
		(23679)		(22350)
	12544	12544	12544	11840
	6645	6645	6645	6272
	5920		*5920*	
	3520	3520	3520	3322
		3136		*2960*
	1865	1865	1865	1760
	1661		*1661*	
	988	988	988	932
	523		523	
Keyboard notation:	C4	B4	C4	A♯4
	Consonant		Dissonant	

The interval of the major seventh is called *major seventh type octave.* The minor seventh gets the name *10/11 major seventh type octave.* In some respects, the sounds based on these spectra can behave themselves in a way similar to the traditional musical octave-based sounds. Melodies can be composed, they can be used for accompaniment and to create tonal relations which remind the traditional functional relationships between chords.

3.3 Sound Example #4

The fourth sound example presents a short composition entitled *Major Seventh Type Octave* (Fig. 1). The composition uses the consonant major sevenths, but the notation marks the keys of a DX7 synthesizer. The spectra of the sounds contained in the final chord are enriched by frequency modulation, but in the spectra below, only the direct or so-called *pillar* frequencies are shown. The values derived from modulation are omitted. The next table shows the spectra of the sounds in the final chord of the composition. This chord consists of sounds notated as: C♯1–G1–C2–A♯3–E4–A4.

Spectrum:						
						(21096)
					(15798)	
				11175		11175
					8369	
				5920		5920
					4433	
			3136	3136		3136
		2349			2349	
	1661		1661	1661		1661
		1244			1244	
	880		880	880		880
		659			659	
	466		466	466		466
		349			349	
	247		247	247		
		185				
	131		131			
		98				
	69.3		69.3			
		52				
	36.5					
Keyboard notation:	C♯1	G1	C2	A♯3	E4	A4

3.4 Sound Examples #5, #6, and #7

Obviously, in this music, the spectrum of the sounds is not composed of harmonic overtones but the frequency components stand in the relation of a tempered major seventh $2^{11/12}$. They constitute the pillars of the spectrum but there are often some pillar-alien frequencies used as well. In the fifth sound example, one can observe such an enriched sound spectrum being composed of its components step by step:

Spectrum:							11175	11175	11175
				5920	5920	5920	5920	5920	5920
							5480	*5480*	*5480*
			1661		1661	1661	1661	1661	1661
						1537	*1537*	*1537*	*1537*
		880						880	880
	466								466
Keyboard notation: A♯3									

The numbers in italics are pillar alien frequencies derived from frequency modulation and read out from a spectrogram.

When listening to these kinds of sounds, the first impression one gets is that of a single sound having a quasi-definite pitch. Yet it is not certain which part tone of the spectrum takes the role of the salient tone. Sound example #6 gives a succession of paradoxical sounds. The sounds behave as wholes. Their pitch and timbre are characteristic Gestalt properties, and is not readily related to the partials. Sound example #7 gives a melody built up of paradoxical sounds. First, the structure of the spectra is similar to one described in sound example #5, then the same melody is played with spectra that have no pillar alien frequencies.

3.5 Sound Example #8

In sound example #8, the paradoxical sounds are superimposed to form chords. The timbre of the chord perceived, however, depends on the chord configuration. The sounds remain more or less recognizable as Gestalt wholes even when they are sounding together in chords. The chords used are:

F♯3	A3	G3	A3	F♯3
D♯3	D♯3	E3	D♯3	D♯3
C3	A♯2	B2	A♯2	C3

Although there are no harmonic relations in the spectra of these sounds, the factors of the *harmonic* and *melodic* affinity are effective in one form or another. This is the basis by which sounds can have a vertical and horizontal coherence. It is an important factor in the relationship between their *shape* and *background*.

3.6 Sound Example #9

The next sound example gives an example of sounds that do not have identical pitch quality with their major seventh, but with their minor ninth instead.

These sounds are dissonant with respect to the traditional octave, as well as with the major ninth. The spectra built up of only pillar frequencies are the following:

Spectrum:				
				(63191)
			(29822)	(29822)
		(14074)	(14074)	(14074)
	6642	6642	6642	6642
	3135	3135	3135	3135
	1479	1479	1479	1479
	698	698	698	698
	329.5	329.5		
	155.5			
Keyboard notation:	D♯2	E3	F4	F♯5

3.7 Sound Example #10

Sound example #10 illustrates the consonance of the minor ninth, and dissonance of the octave and the major ninth:

Spectrum:						
		(28148)		(29822)		(31595)
	(14074)	(13284)	(14074)	(14074)	(14074)	(14910)
	6642	6270	6642	6642	6642	7337
	3135	2958	3135	3135	3135	3321
	1479	1396	1479	1479	1479	1567
	698	659	698	698	698	740
	329.5		329.5		329.5	
Keyboard notation:	E3	E3	E3	F4	E3	F♯4
	Dissonant		Consonant		Dissonant	

In the above cases, the minor ninth behaves as an octave. The name of the original octave becomes: *12/13 minor ninth type octave*, and that of the major ninth: *14/13 minor ninth type octave.*

3.8 Sound Example #11

The spectrum of the sound in the next example is also composed purely of pillar frequencies calculated using the relative numbers of a tempered minor ninth ($2^{13/12}$=2.1189262) and all the frequencies have the same intensity level. The partial tones are:

G♯1, A2, A♯3, B4, C6, C♯7 (103.8, 220, 466, 988 ,2093, 4434)

In adopting a global listening attitude, rather than an analytical one, the sound is perceived as a whole with a quasi-definite pitch although the pitch cannot be related to a concrete frequency in the spectrum.

3.9 Sound Examples #12 and #13

In sound example #12 a succession of octaves is played upwards, while a chromatic movement is heard downwards.

Spectrum:			(53136)
		(26568)	(25076)
	(13284)	12538	11836
	6269	5918	5584
	2959	2792	2636
	1396	1318	1244
	659	622	
	311		
Keyboard notation:	D♯2	D♯3	D♯4

The lengthy decay of the sounds works like a *background* of the chromatic movement. If followed attentively, it is possible to notice the traditional octaves that form a motionless background that alters only in brilliance and against which the chromatic movement is perceived.

In sound example #13, a succession of octaves is played downwards while the chromatic movements are perceived upwards. In the *background*, it is possible to perceive the traditional octaves as well.

Spectrum:	(31608)			(39824)		
	(14916)	(15804)		(18796)	(19912)	
	7040	7454	7902	8868	9398	9956——
	3324	3520	3729	4188	4434	4699——4699
	1568	1662	1760	1976	2094	2217——2217
	740	784	831	932	988	1047——1047
		370	392		466	494——494
			185			223——223
						110
						(together)
Keyboard notation:	F♯4	F♯3	F♯2	A♯4	A♯3	A♯2——A1

The chromatic movement is perceived as a sequence of sounds having definite pitches. The ear seems to take other partials as the dominant partials at every step. In sound example #12, it was the third partial that dominated the first sound, then the second partial that dominated the second sound, and finally the first partial that dominated the third sound.

3.10 Sound Example #14: Pseudo Twelve Tone Tunings

Sound example #14 is a short three part composition, having as title *Pseudo Twelve Tone Tunings*. In the first and second part of the composition the traditional octave interval is divided into 11 equal distances. The 12/11 octave interval behaves as a pseudo-octave. The spectra, obtained by additive synthesis, involve also pillar alien frequencies, and the relations among the components imitate the structure of the overtone row. All the data given, inclusive pillar alien frequencies printed in italics, are exact values. The keyboard notation and the fundamental frequencies of the sounds at the beginning of the composition are:

D♯5 = 1368 Hz (F5 - 36 cent [1])

D♯4 = 642 Hz (E4 - 46 cent)

[1] The cent values are calculated on the basis of segment values used by the DX7.

D♯3 = 301 Hz (D3 + 45 cent)
D♯2 = 141.5 Hz (C♯2 + 35 cent)

The spectra obtained by additive synthesis (pillar alien frequencies are in italics) are:

Spectrum:	6205	(13217)	(28153)	(59969)
	2913	6205	(13217)	(28153)
	—	—	—	—
	1759	*3748*	*7983*	*(17004)*
	1551	*3304*	*7038*	*(14991)*
	1368	2913	6205	(13217)
	1206	*2568*	*5470*	*11650*
	998	*2125*	*4527*	*9644*
	826	*1759*	*3748*	*7983*
	642	1368	2913	6205
	468.5	*998*	*2125*	*4527*
	301.4	642	1368	2913
	141.5	301.4	642	1368
Keyboard notation:	D♯2	D♯3	D♯4	D♯5

In the second part of the composition the traditional octave interval is divided into 13 equal distances. The 12/13 octave interval behaves as a pseudo-octave. Some keys and fundamental frequencies belonging to this division are:

D♯5 = 1149 Hz (D5 - 39 cent)
D♯4 = 606 Hz (D♯4 - 46 cent)
D♯3 = 319.5 Hz (D♯3 + 46 cent)
D♯2 = 141.5 Hz (E2 + 38 cent)

The spectra are:

Spectrum:	4130	7831	(14850)	(28157)
	2178	4130	7831	(14850)
	—	—	—	—
	1422	*2696*	*5113*	*9695*
	1278	*2423*	*4581*	*8714*
	1149	2178	4130	7831
	1032	*1958*	*3713*	*7040*
	880	*1668*	*3164*	*6000*
	750	*1422*	*2696*	*5113*
	606	1149	2178	4130
	464	*880*	*1889*	*3164*
	319.5	606	1149	2178
	168.5	319.5	606	1149
Keyboard notation:	D♯2	D♯3	D♯4	D♯5

In this composition, the sounds of a pseudo-octave belong to the keys of a true octave on the keyboard.

4 Discussion

In all the examples, it was convincingly demonstrated that our hearing system interprets the information mediated by the acoustic stimuli according to a multiple code system. The ear indeed can deal with unknown and unusual

OKTAVENPARADOXIEN

Septimartige Oktaven

Die temperierten grossen Septimen dienen als Oktaven.
Die Noten bezeichen die Tasten.

Jenö KEULER
1991

Fig. 1. *Major Seventh Type Octave* by J. Keuler

acoustical information. It finds features among acoustic quantities, finds degrees of relationships between them, and according to this decoding process we perceive lines, masses, figures, backgrounds, integrated, and differentiated sound phenomena.

In this context, it is important to mention that the study of Gestalt properties of sounds is much related to the relationships between the sounds. When listening to a particular sound, this sound is kept in memory for a while, and the next sound we hear is qualified with respect to the traces of the first sound. As such, it is possible to establish a tonality which may define a durable field of musical movement. The tonal field is invisible as background, but its presence is experienced as a field. Within a sequence of sounds, a particular sound may take on properties which are not present when the sound is considered as an isolated sound object. Depending on their places in the tonal field, certain sounds seem to be more stable than other sounds. As such, tensions can be created and sounds may strive towards resolution of these tensions. Observing musical processes as sounding pictures, it is evident that the forces of the tonal field does not manifest itself as a pictorial form or background. Nevertheless, it should be considered as a holistic form, having characteristics that depend on the *structure* of the sound system that is considered. The tonal field may influence the behavior of the sounds and as such we can say that it has an effect on the Gestalt qualities of the musical configuration.

My composition *Major Seventh Type Octaves* (Fig. 1) illustrates the fact that tonal fields can be created using artificial sounds that are based on an inharmonic structure. Obviously, we may assume that the auditory decoding processes are the same as with traditional musical sounds (using sounds with an harmonic overtone structure). The auditory system analyses the effects of the stimuli and creates an internal model of the sound system. The new sounds get their meaning accordingly.

The next composition *Pseudo Twelve Tone Tunings* tests our sense of hearing. It is a short three part composition in which neither the spectra nor the tuning of the chromatic scale is traditional. At first, the tuning of the traditional octave is divided into 11 equal distances, later on, in the middle section, it is divided into 13 distances. The pillar frequencies of the spectra are settled at first according to the relative number of the 12/11 octave (approximately a minor ninth), later on, to that of the 12/13 octave (approximately a major seventh). The distances of the pseudo-octaves are divided into 12 equal sections. Since our auditory system tends to orient itself according to the pseudo-octaves, one perceives the pieces as if it was written in the traditional 12 tone tuning. In all cases in which steps or passages of octaves are perceived, they are in fact sounding pseudo-octaves (roughly minor ninth or major seventh). At the same time, one gets a feeling of a tonal system, of the freshness and directions of the sounds, the melodies and other musical formations developing in time. Musical motives get imprinted into mind as well.

5 Conclusion

The psychology of Gestalt perception directed our attention to the fact that a whole cannot be understood from the side of its parts alone. Rather, the parts have to be understood from the side of the wholes. In this paper, we have explored this basic principle of Gestalt psychology using artificial sounds based on inharmonic scaling and tuning. Our focus on these aspects is limited, but highly related to the domains of music psychology and cognitive musicology. We have mainly shown here that the study of musical cognition can profit from the interaction with musical praxis.

Appendix

Sound examples (CD-tracks 16-29)

- Sound Example #1: A sound which is consonant with the same sound at interval distance of a major seventh, but *not* with its octave transposition.
- Sound Example #2: The effects of consonance and dissonance. The sound of the example #1 is confronted with its consonant major seventh and dissonant octave.
- Sound Example #3: The sound of the example #1 is first confronted with its consonant major seventh and dissonant minor seventh.
- Sound Example #4: A short composition entitled *Major Seventh Type Octave*.
- Sound Example #5: Enriched sound spectrum.
- Sound Example #6: A succession of paradoxical sounds.
- Sound Example #7: A melody built up of paradoxical sounds.
- Sound Example #8: The paradoxical sounds are superimposed to form chords.
- Sound Example #9: Sounds that do not have identical pitch quality with their major seventh, but with their minor ninth instead.
- Sound Example #10: The consonance of the minor ninth, and dissonance of the octave and the major ninth.
- Sound Example #11: Inharmonic complex sound based on the tempered minor ninth.
- Sound Example #12: A succession of octaves is played upwards, while a chromatic movement is heard downwards.
- Sound Example #13: A succession of octaves is played downwards while the chromatic movements are perceived upwards.
- Sound Example #14: A short three part composition, having as title *Pseudo Twelve Tone Tunings*.

A Method of Analysing Harmony, Based on Interval Patterns or "Gestalten"

Roland Eberlein

Musikwissenschaftliches Institut, Universität Köln, Albertus-Magnus-Platz, D-50923 Köln (Cologne), Germany

Abstract. Although in the last decades the traditional theory of harmony in tonal music has often been sharply criticized, it is still in use and a better alternative has not yet been elaborated. Since it is used today mainly for the purpose of describing and teaching tonal harmony, a modern alternative should offer, above all, a better method of describing any succession of harmonies. Starting from a recently evolved system of encoding harmony objectively as a succession of harmonic interval combinations and melodic bass steps, a similar shorthand writing of harmonies with additional mnemonic qualities is suggested. This shorthand may be used in describing and teaching tonal harmony. Moreover, its use within the scope of a modern Gestalt theory of musical perception concentrating on the learning and recognizing of patterns of harmonic and melodic intervals is outlined.

1 Introduction

Today, the traditional teaching of tonal harmony – based either on the scale step theory of G.Weber or on H.Riemann's theory of harmonic functions – is often regarded with deep scepticism. A great number of articles published in the last decades (Borris, 1950, 1967; Dahlhaus, 1969; Ganter, 1983; Jeppesen, 1949; Hoyt, 1990; Möllers, 1976; de la Motte, Birnstein, & Kühn, 1972; Rischel, 1988/1989; Smith, 1974; Wünsche, 1965) subject it to an often scathing criticism. Nevertheless, this criticism did not alter the practice of teaching tonal harmony in the least: in Germany, tonal harmony is analyzed and taught today hardly any differently than 65 years ago when Maler published his *Beitrag zur durmolltonalen Harmonielehre* for the first time (Maler, 1931). One of the main reasons for this persistence might have been the lack of better alternatives: all the criticism of the traditional teaching of harmony did not lead to the elaboration of a fundamentally different method of harmony description avoiding the criticized flaws. This is a serious failure of systematic musicology and the present article presents some thoughts on how it might be corrected.

2 Criticism of the Traditional Theories of Harmony

The first step on the path to a better alternative must be a collection and evaluation of the criticism uttered till now. At this place only the most important points of earlier criticism can be mentioned shortly without discussing them:

- The traditional theories of harmony have been accused of being the product of unsubstantiated speculations all the way from the first axioms up to the last conclusions. In a lecture at the congress of the International Society of Musicology in Bâle 1949, Jeppesen (1949) briefly described the ideas fundamental to the traditional theories of harmony and continued then with the words (p.30):

  "A haughty, solid edifice of thoughts ... on the basis of unrestraint impulses of fantasy. If one asks what its empirical basis is, it turns out that this basis is altogether meaningless or at any rate ambiguous. Neither Riemann nor his predecessors were able to prove their theorems, indeed, they did not even seriously try to prove them. The whole system has been elaborated on the basis of mere speculations."[1]

- Some axioms fundamental to the traditional theories of harmony are in contradiction with the practice of composition. In the lecture just mentioned, Jeppesen (1949)called attention to the fact that e.g. the idea of free invertibility of chords is in crass contradiction with the considerably differing use of chord inversions (p.33):

  "For example, the use of harmony in the 17th century shows us especially clearly that triads in root position, sixth chords and six-fourth chords are essentially differing objects which demand an extremely differentiated treatment. Whereas triads in root position are used with much greater harmonic freedom, sixth chords are always horizontally motivated resulting from voice leading, not to mention the six-four chord."[2]

  More recently, Hoyt (1990), for one, has advocated doing away with the inversion principle in harmony altogether.

- Some axioms fundamental to the theories of harmony are in contradiction with musical perception. Already in 1974 F.J.Smith wrote:

  "The sad fact is, that in the light of both historical studies and research into musical perception the framework of "traditional harmony" and the attitudes derived therefrom simply do not stand up. One wonders why such a system is allowed to persist (Smith, 1974)."

---

[1] "Ein stolzes, festgefügtes Gedankengebäude ...auf Grundlage von freien Eingebungen der Phantasie. Fragt man sich nämlich, was an Empirik daran haftet, zeigt es sich, daß dies ganz nichtig oder jedenfalls zweideutig ist. Weder Riemann noch seine Vorgänger haben es vermocht, ja nicht einmal ernstlich versucht, ihre Lehrsätze zu beweisen. Das ganze System ist rein spekulativ entwickelt."

[2] "So werden wir besonders deutlich von der Harmonik des 17. Jahrhunderts (...) darüber belehrt, daß Grunddreiklang, Sextakkord und Quartsextakkord wesensverschiedene Tatsachen sind, die eine äußerst differenzierte Behandlung verlangen. Während die Grunddreiklänge mit viel größerer harmonischer Freiheit gesetzt werden, sind die Sextakkorde immer horizontal motiviert als Ergebnisse der Stimmführung, vom Quartsextakkord gar nicht zu reden"

For instance, the series of *undertones* proposed by Riemann (1873) was recognized early on as being non existing. Some years ago the experimental results of Cook (1987) caused a sensation since they showed that the return to the original key at the end of a composition does not influence the esthetic judgement of musically trained subjects, at least if the composition lasts more than half a minute. The concept of tonality perception accepted as a fact in textbooks on harmony is thus highly questionable. Riemann's thesis that the cadence effect is based on the tonal relationship between the root tones of chords has been experimentally refuted (Eberlein & Fricke, 1992, pp.123-124,139). These incorrect axioms, and others more, hamper not only the current scientific research – Cook (1987) justly deplores the tendency in music psychology to misunderstand concepts of traditional music theory as descriptions of musical perception – but also the propagation of new experimental results on musical phenomena to the broad public, since the latter of course assumes that the assertions heard during music instruction are assured findings and therefore it tends to refuse opposed maintainings even if these are the results of scientific investigations.

- Traditional textbooks on harmony do not refer to a concrete historic style of music, but teach an ahistoric method of composition with rules abstracted from many different styles. Therefore, they do not help to acquire an intimate knowledge of historic styles; in fact they impede the acquisition of such a knowledge. Actually, music written according to these rules considerably differs from any historic style. This reproach was made already by Borris (1950) and restated by D. de la Motte, R. Birnstein and C. Kühn (1972). In his textbook on harmony de la Motte tried to take it into account (de la Motte, 1976). Therefore, this textbook no longer teaches a single abstract system of harmony detached from any concrete style; rather, it explains the compositional practices of nine periods or composers. No rules are given, instead, numerous examples drawn from the period or composer under discussion illustrate the practice. Nevertheless, in describing practices the book uses traditional nomenclature and harmony symbols, thus implying inappropriate concepts of a later time.
- The traditional methods of harmony analysis are unable to describe music of the 20th century adequately, since their fundamental axioms (e.g. all chords are built up of thirds, all chords are representatives of either certain scale steps or three different musical functions, cadences are the foundation of all harmonic progressions etc.) mostly do not hold true in this music. This objection has been raised by Wünsche (1965). I would like to add some further points to the objections of other authors mentioned so far:
- The traditional methods of harmony analysis are unable to describe medieval music, since all of their terms are related to major and minor triads and their ordering according to tonal cadences, whereas medieval music is based on fifth-octave chords and the opposition between perfect and

imperfect consonances. Triadic harmony and tonal cadences are totally extraneous to it. Nevertheless, a modern tool apt at describing medieval harmony is urgently required since the original medieval music theory concentrates on two-part counterpoint; thus it is hardly able to deal satisfactorily with successions of multi-voiced harmonies.

– The traditional teaching of harmony is by no means as simple as it should be from a didactical point of view. This especially concerns the variant predominantly used in Germany, which is based on Riemann's theory of musical functions. In this variant, already the symbol and name of the chord often causes considerable brain acrobatics since first one has to search for a - frequently missing – root and then to translate the roots name into terms based on its – often complicated – relation to the tonic. Therefore, theoretical interpretations of harmonic successions are not seldom difficult to understand. Moreover, some interpretations of chord progressions are in contradiction with intuition (e.g. it is difficult to believe that the root of the chord C/E/G/A♯, resolved in a B-major triad, is not C, but F♯). Consequently, harmony is not as easy to teach as would be desirable.
– It is nearly impossible to produce faultless traditional harmonic analyses by using a computer program, since in many cases there is a multitude of possible interpretations of one and the same succession of chords. The multitude of possible interpretations of a single chord structure has been strikingly shown by Piston (1957).

3 Tasks of an Alternative Method of Harmony Analysis

The search for an alternative to the traditional description of harmony must be preceded by a definition of what an alternative method of harmony analysis has to achieve. On principle, I see at least three different goals which might be strived for in producing an analysis of harmonic successions:

– one might try to show how the composer produced his composition and how he might have interpreted it in terms of the compositional theory of his time;
– one might try to show how a listener perceives that music;
– one might try to describe harmonic successions in qualified terms in order to facilitate the communication about music and the teaching of tonal syntax to pupils.

Analyses of the first type (describing the composer's way of working and thinking) require in every case the music theory of the composer's time. In today's practice, this type of analyses is rather exceptional. The second goal (description of musical perception) is of a rather Utopian nature and, therefore, analyses of this type are seldom in today's musical and scientific practice. Originally, the harmony theories of J-P. Rameau and H.Riemann pursued this goal. Today, such an analysis would have to dispense with a great number of untenable assumptions made in the traditional theories of harmony. Moreover, we know today that musical perception depends on musical experience and, therefore, that it varies from individual to individual, from culture to culture, from epoch to epoch. A theory aimed at that goal would have to be able to deal with this variability of musical perception. Today, harmony analyses are generally analyses of the third type (producing a description of harmony in order to facilitate communication and instruction). An alternative to the traditional theory of harmony would therefore have to be orientated towards the third goal, at least in the first place. From this goal a number of requirements can be derived which should be met by a new method for analyses of the third type:

- It should describe harmonies as objectively as possible in order to be acceptable as a communication tool to persons with different theoretical convictions.
- It should be well suited to didactical aims, that means, it should be easy to teach, simple to use, well adapted to the needs and the way of thinking of musicians, and it should provide a mnemonic shorthand writing of harmonies.
- It should be communicable by spoken speech, that means, it must not be solely graphical and it should allow verbal descriptions of even complex simultaneous and sequential musical structures.
- It should be suited to music of the most different ages.
- It should make possible the production of harmony descriptions by computer – a demand certainly of great importance for the future.

4 Earlier Attempts to Describe Harmony in a New Way

An overview of earlier attempts to reform music notation may be found in a book of G. Read (1987). There already exist new methods of describing harmony in music that meet some of the above mentioned requirements. For example, R. Parncutt (Parncutt, 1990; Parncutt & Stuckey, 1992, 1995) has recently suggested *chromatic chord symbols* analogous to the conventional chord symbols of the scale-step theory, but based on the chromatic scale instead of the diatonic scale. In these chromatic chord symbols, the root of any chord is indicated by its distance in semitones above the tonic, and right-hand superscripts specify the chord notes distances in semitones above the root, ignoring the octave position of the notes. For instance, the major triad of the dominant in root position is signalled by the chord symbol 7^{47}. For chords in inversion, the distance of the bass note above the root is written as a left-hand subscript, e.g. ${}_47^{47}$ indicates a first-inversion dominant major triad. This method is somewhat more objective than traditional harmony descriptions, but its use has to be trained for some time: as Parncutt remarked, the symbols look like hieroglyphics at first, since the value of the figures does not show an obvious relation to conventional interval names. A second problem arises from the indication of roots as distances above the tonic: in many cases the question as to which note constitutes the tonic may lead to individually varying answers, and for many dissonant chords there may be divergent opinions concerning the root of the chord. Parncutt's method of describing harmonies is not as objective as could be desired.

A largely objective system of describing harmonic successions has been used by R. Eberlein in his study on the emergence of tonal syntax (Eberlein, 1994, pp.123-124,139) in order to produce statistics of harmonic successions in tonal compositions. In his system the figures of chord codes specify distances in semitones above the bass, ignoring the octave position of the notes by assuming the smallest equivalent interval. For example, a major triad in root position is signalled by the figures 47, a first-inversion major triad by 38. Between the codes of two successive chords, a figure is inserted denoting the melodic bass interval in semitones; a heading plus or minus sign indicates the direction of the interval. Again the octave position of the notes is ignored and the smallest equivalent interval is assumed. For instance, a cadence consisting of a G-major triad and a C-minor triad gets the figures 47 +5 37. A great advantage of this code system is the absence of any assumptions concerning the scale or tonal center fundamental to the music. Moreover, it does not contain any assumptions about the root note of the chords. The system is simply descriptive. It is therefore well suited for the computer-aided generation of statistics of harmonic successions. Nevertheless, it is certainly not suited for being used as a substitute for conventional descriptions of harmony in teaching tonal harmony to pupils: since the codes do not have any mnemonic quality, their use requires much training. Besides, the codes are difficult to translate into spoken speech. The figures 47 +5 37, for example, misleadingly suggest the foolish and absolutely unintelligible pronunciation "fortyseven, plus five, thirtyseven".

5 Suggestion of an Improved System of Harmony Description

The usability of the system just described might be improved by altering the nature of the codes: instead of figures denoting distances in semitones one might write down two-letter abbreviations of the conventional interval names using capital letters for major intervals and small letters for minor intervals. The abbreviations should be understandable in different european languages, and to achieve this goal it might be best to choose the letters out of the corresponding latin name: As in the system used by Eberlein (1994), intervals greater than a major seventh are given the code of the corresponding interval

Table 1. Interval abbreviations used in the proposed coding system

Interval in semitones	Abbre-viation	Interval names Latin	German	English	Italian	French
0	UN	unisonus	Unisono	unison	unisson	unisono
1	sc	secunda	Sekunde	second	seconde	seconda
2	SC					
3	te	tertia	Terz	third	tierce	terza
4	TE					
5	QA	quarta	Quarte	fourth	quarte	quarta
6	Qq		(augmented fourth/diminished fifth)			
7	QI	quinta	Quinte	fifth	quinte	quinta
8	sx	sexta	Sexte	sixth	sixte	sesta
9	SX					
10	se	septima	Septime	seventh	septiéme	settima
11	SE					

within an octave. For the sake of simplicity, the doubling of notes in unison or the octave might be ignored. In this case a cadence in a minor key – e.g. a G-major triad followed by a C-minor triad – gets the code TEQI +QA teQI. Depending on the purpose of the coding it might sometimes be wise to code unisons as well as octaves and to order the codes according to the actual pitch of the notes so that the coding retains informations about the voicing of chords and the leading of voices. To be sure, such a coding rule results in somewhat complicated codes: e.g. the cadence just mentioned might be given the code QIUNTEQI +QA teQIUNUN. The suggested description of harmonic successions – especially the first coding strategy – shows a number of advantages when compared to already existing systems of harmony description:

1. The new description system makes possible a rather objective description of harmony: it does not make use of questionable hypotheses about, for instance, the tonal center of a piece or the root note of a chord and it does not depend on any special theory of harmonies, since its sole theoretical assumption is the harmonic equivalence of notes in octave distances. Therefore, the system should be acceptable to persons with all kinds of theoretical convictions.
2. The new description system is well suited for didactical purposes: it is simple to learn and to use, since it solely consists in mnemonic abbreviations of already known interval terms which are fundamental to music, and since its abbreviations do not elicit misleading mathematical associations as do the figures used by Parncutt (1990) and Eberlein (1994). Moreover, the new description system drastically simplifies the teaching of tonal harmony since the brain acrobatics of conventional harmony descriptions are done away with.

3. The codes can be translated into spoken speech without any difficulties (e.g. TEQI +QA teQI = a major third-fifth chord, proceeding by an ascending fourth in the bass to a minor third-fifth chord). Thus, the description system fulfills an important requirement for being used as a communication tool.
4. Owing to its objectivity, the description system can be applied to music of any style and epoch. It can describe, for example, non-triadic medieval harmony just as well as tonal harmony.
5. Harmony descriptions can be produced by using rather simple computer programs, since there is a simple relation between differences in MIDI pitch numbers and the symbols suggested above and since the description of a given chord does not depend on any theoretical interpretations of the harmonic context.

Thus, the new description system meets all the above stated requirements of an analysis method aiming at a facilitation of communication and music instruction. For instance, one might select a repertory of harmony progressions often used in tonal music and teach these progressions (cadences, typical resolution of dissonances and so on) to pupils, describing them with the above proposed system. The choice should be based on the results of frequency countings like that published by Eberlein (1994). With the help of Table 1, the harmony descriptions in his tables can easily be translated into the description system proposed here. Today, such countings can be executed with the aid of a computer system. For questions on how to obtain statistics of music by means of a computer system, see Huron (1995).

6 Some Thoughts Concerning a Harmony Analysis Aiming at a Description of Music Perception

At this point, one might argue that conventional analyses of harmony do not only serve the purpose of describing harmonic successions, but also that of interpreting unusual harmonic successions like the famous *Tristan chord*, i.e. they try to show how the listener relates one chord to another. In comparison with these, the suggested method of analysis restricts itself to a mere description of the harmonic succession. One might therefore argue that it cannot produce any new findings, thus being uninteresting for musicologists (a similar reproach has been made by Billeter, 1971, with regard to the conventional scale step theory of harmony). But although it is not its primary aim, the suggested method of harmony description can nevertheless be used for producing interpretations of harmonic successions aiming at a description of the cognition process, if it is supplemented by a theory of music perception. A fitting theoretical framework might be the theory proposed by Eberlein (1994), and Eberlein and Fricke (1992), dealing with the recognition of harmonic-melodic interval patterns by listeners. Experiments concerning the recognition of cadences and other often repeated harmonic successions in tonal music led Eberlein to the thesis that experienced listeners of tonal music have learnt a great number of harmonic successions as patterns of harmonic and melodic intervals and that they try to recognize these learnt patterns or "Gestalten" in newly sounding music by comparing the intervallic features of the passage they are hearing to the intervallic features of hitherto learnt patterns (Eberlein, 1994, pp.197-204, Eberlein & Fricke, 1992, pp.36-43,379-393). In a model of this recognition process, the suggested description of harmonic successions might be used in order to describe intervallic patterns. The following schema shows the rough structure of such a model (Fig.1):

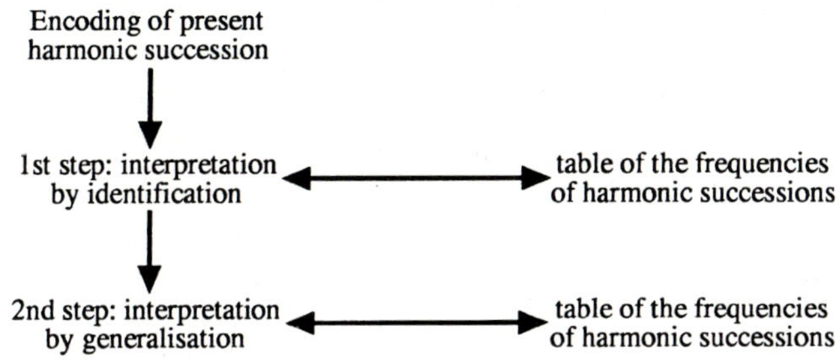

Fig. 1. Schema of the proposed process of interpreting harmonic successions

The model implies a two-step process of interpreting harmonic successions: after encoding a harmonic succession, the resulting code may be compared with a table showing the frequencies of harmonic successions in tonal music. With this frequency information in mind, it is possible to discriminate between harmonic successions which can be recognized as an often heard pattern, and other harmonic successions which will sound strange even to an experienced listener. In this first step, often heard harmonic successions are interpreted by identification with a learnt pattern. For example, the beginning of the Bach choral *Du Friedefürst, Herr Jesu Christ* BWV 67/7 is characterized by the following harmonic code: TEQI +QA TEQI -sc scQAsx -SC teSX -SC TEQI -sc tesx. The first harmonic succession TEQI +QA TEQI is the harmonic succession most often occurring in tonal music; according to the frequency countings published by Eberlein (1994, p.422), its relative frequency is about 3%. An experiment has shown that subjects rate this harmonic succession as sounding very familiar ("vertraut klingend"), more familiar than any other harmonic successions presented to them (Eberlein, 1994, p.387). So it may be assumed that this interval pattern has been learnt by experienced listeners and that they will identify the beginning of the Bach choral just mentioned with this learnt pattern or *Gestalt*. The second step consists in a search for learnt patterns which are similar to the present harmonic succession, i.e. which have some intervallic features in common with it. In this step, unfamiliar harmonic successions can be interpreted as a variant of an already known pattern. For example, the chord progression TEQI -sc TEQI (e.g. F major - E major) which occurs very seldom in tonal music (it was not present within the sample of tonal pieces analyzed by Eberlein) was rated amazingly high in the experiment just mentioned concerning the familiarity of harmonic progressions (mean 6,13 on a nine-step scale of familiarity Eberlein, 1994, p.387). This might be due to the similarity of its intervallic pattern to the often used Phrygian cadence TESX -sc TEQI: it might be that experienced listeners interpret the unfamiliar succession TEQI -sc TEQI as a variant of the learnt pattern TESX -sc TEQI. Similarly, experiments concerning the recognition of cadences showed that any strange harmonic succession may be accepted as a cadence if it contains harmonic and melodic intervals characteristic of conventional cadences, especially of conventional two-part cadences (Eberlein & Fricke, 1992, pp.190-192,197-201; Eberlein, 1994, pp.40-43,379-393). It is of course necessary in this second interpretational step to take into account all melodic and harmonic intervals between the different voices, not only the intervals related to the bass that are named in the code. This is possible if the more extensive encoding strategy is adopted which allows for the doubling of notes and which orders the codes according to the pitch of the notes. It should perhaps be mentioned that in this interpretational step intervallic similarities between different inversions of a chord progression may also be detected. In contrast with traditional methods of harmony analysis, the procedure described here is able to take into account the effect of differences in musical experience and cultural background by computing the frequency table on the basis of appropriately chosen musical repertories.

7 Conclusion

In the present article it is not possible to give more than a rough outline of the possible use of the suggested harmony description within the framework of a modern theory of music perception and within music instruction. But I hope this brevity encourages the reader to elaborate further ideas for a pragmatic description of harmony avoiding the flaws of the traditional description. What we need today is a general discussion about an up-to-date method of harmony description and harmony teaching, since only in the course of a broad and public discussion will it be possible to break the fossilized habits of thinking which caused the living on of a long ago obsolete theory of harmony.

Acknowledgement

Thanks to Stanley Hanks for improving the English of this article.

References

Billeter, B. (1971). *Die Harmonik bei Frank Martin. (= Publikationen der Schweizerischen Musikforschenden Gesellschaft Serie II, 23).* Bern: Paul Haupt.

Borris, S. (1950). Die Widersprüche zwischen älterer Harmonielehre und neueren Tonsatzlehren. In H. Albrecht, H. Osthoff, & W. Wiora (Eds.), *Kongreßbericht Gesellschaft für Musikforschung, Lüneburg, 1950* (pp. 211–213). Kassel: Bärenreiter.

Borris, S. (1967). Probleme der traditionellen Harmonielehre. In *Probleme des Musiktheoretischen Unterrichts. (= Veröffentlichungen des Instituts für Neue Musik und Musikerziehung Darmstadt 7).* Berlin: Merseburger.

Cook, N. (1987). The perception of large-scale tonal closure. *Music Perception, 5*, 197–206.

Dahlhaus, C. (1969). Zur Kritik der Harmonielehre. *Musik und Bildung, 1*, 375–382.

de la Motte, D. (1976). *Harmonielehre.* Kassel: Bärenreiter.

de la Motte, D., Birnstein, R., & Kühn, C. (1972). Plädoyer für eine Reform der Harmonielehre. *Musica, 26*, 233–238.

Eberlein, R. (1994). *Die Entstehung der tonalen Klangsyntax.* Frankfurt/M.: P. Lang.

Eberlein, R., & Fricke, J. (1992). *Kadenzwahrnehmung und Kadenzgeschichte: ein Beitrag zu einer Grammatik der Musik.* Frankfurt/M.: P. Lang.

Ganter, C. (1983). *Harmonielehre – ein Irrtum? Literaturbeispiele zur Dur-Moll-tonalen Harmonik.* Basel.

Hoyt, R. (1990). Harmonic function and the motion of the bassline. *Journal of Music Theory Pedagogy, 4*, 147–190.

Huron, D. (1995). *The Humdrum toolkit reference manual.* Menlo Park, CA: Center for Computer Assisted Research in the Humanities.

Jeppesen, K. (1949). Zur Kritik der klassischen Harmonielehre. In *Kongreßbericht Internationale Gesellschaft für Musikwissenschaft, Vierter Kongreß, Basel 1949* (pp. 23–34). Basel: Bärenreiter.

Maler, W. (1931). *Beitrag zur durmolltonalen Harmonielehre.* Leipzig: L. Leuckart.

Möllers, C. (1976). Vom Unsinn der Modulationslehre. *Die Musikforschung, 29*, 257–273.

Parncutt, R. (1990). Chromatic chord symbols. *Computer Music Journal, 14*, 13–14.

Parncutt, R., & Stuckey, R. (1992). Towards a standard alternative notation and terminology based on the chromatic scale. *Musikometrika, 4*, 117–143.

Parncutt, R., & Stuckey, R. (1995). Research in progress: Chromatic solfa. *British Journal of Music Education, 12*, 72–75.

Piston, W. (1957). *Thoughts on the chordal concept.* Cambridge, MA: Harvard University Press.

Read, G. (1987). *Source book of proposed music notation reforms.* New York, NY: Greenwood Press.

Riemann, H. (1873). *Musikalische Logik.* Leipzig: Kahnt.

Rischel, G. (1988/1989). Tonal analyse. *Musik & Forskning, 14*, 110–133.

Smith, F. (1974). Traditional harmony? A radical question. *The Music Review, 35*, 63–75.

Wünsche, W. (1965). Nur Terzenstruktur? *Musik und Gesellschaft, 15*, 160–161.

Neural Network Models for the Study of Post-Tonal Music

Eric Isaacson

School of Music, Indiana University, Bloomington, Indiana 47405, USA

Abstract. Neural networks are used to study two issues pertaining to atonal music. In the first part of the paper, feed-forward neural networks, using a variant of the backpropagation learning algorithm, try to learn a variety of abstract theoretical constructs from pitch-class set theory. First, learning the properties of individual sets is studied. Then a network's ability to learn various relationships between sets is examined. Based on the behavior of the network during learning, conclusions are drawn with regard to perceptual issues relating to pcset theory. In the second part of the paper, an interactive activation and competition (IAC) network is used to parse a musical passage into analytical objects. The paper concludes with suggestions for further research.

1 Introduction

The atonal music of the Second Viennese School continues to challenge both listeners and scholars. While this can be attributed to a number of factors, the principal reason is that, without the hierarchy inherent in the diatonic scales and the major/minor tonal system, relations between pitches in Schoenberg's post-tonal music are not defined systemically. In the scholarly community there has been considerable theoretical work on the organization of this music, particularly regarding pitch structure. Beginning with Babbitt's studies in the 1950s (which actually concerned the properties of hexachords in twelve-tone music), music theorists developed an ambitious appropriation and adaptation of mathematical set theory to study relations among unordered collections of pitch-classes (or pcsets) in the music. Although a number of theorists have contributed to this development, the formalizations of Forte are most widely discussed (Forte, 1973). Forte's catalog of pitch-class set classes and related theoretical apparatus have been used extensively in the more than 30 years since they were first introduced.

Forte's proposals have been hotly debated (Forte, 1985; Haimo, 1996; Perle, 1990; Taruskin, 1979). A consistent criticism of pcset theory is that its constructions lack perceptual validity. Unfortunately, there has been relatively little experimental research which tests predictions inferred from pitch-class set theory (Bruner, 1984; Gibson, 1986; Krumhansl, Sandell, & Sargeant, 1987; Stammers, 1994; Imberty, 1993). The conclusions that can be drawn from these studies which have been done are limited, however. Claims about the perceptual relevance of pcset theory are hard to verify experimentally because the context in which stimuli are presented can sharply affect perception. Registral and temporal ordering play a powerful role in the way we hear the stimuli and, if we are to ascertain how relations between pcsets contribute to our perceptual understanding of the music, such factors need to be controlled for. Another problem with some studies is that they test aspects of the theory that are outdated. And most of the studies work with a very restricted subset of the pcset universe (e.g., just trichords). The resulting predictions may, therefore, not generalize very well to the full space of pcsets. Significant questions relating to the perceptual validity of pcset theory thus remain unanswered.

This paper will look at ways in which neural networks might be used in the study these two issues. Artificial neural networks are well suited to tasks of this sort. They have already proven useful in the study and modeling of music perception, theory, and analysis. Leman (1992) provides an excellent summary of a number of musical applications of neural nets (as well as an accessible introduction to neural network theory). Todd and Loy (1991) describe additional applications of neural networks to musical problems. No existing applications deal with the atonal repertoire, however.

The first part of the paper takes up the issue of properties of, and relations between, pcsets. It examines the behavior of simple backpropagation neural networks (Rumelhart, Hinton, & Williams, 1986) as they try to *learn* abstract constructs from pcset theory. The second part of the paper explores perceptual parsing and Gestalt formation in post-tonal music. An interactive activation and competition (IAC) network is used to parse a musical surface into discrete analytical objects. This topic is important since it makes no sense to study relations between musical structures that are not valid in some way.

Before proceeding, I summarize the basic elements of pitch-class set theory. There are twelve pitch classes, corresponding to the twelve enharmonic – and octave equivalent pitches of the chromatic scale. By convention, the pitch classes are referenced by number, where C=0, C-sharp/D-flat=1, and so on. A pitch-class set is an unordered collection of pitch classes. A pc set class comprises those pcsets which are equivalent under transposition, inversion, or both. There are 224 set classes of size 0 to 12. Some theorists acknowledge equivalence only under transposition (Castrén, 1994), in which case there are 352 set classes. The seven interval classes (0-6) are the unordered intervals between pitch classes and can be thought of as the shortest distance between two pitch classes around a mod 12 number wheel. The interval-class content of a pcset is notated in the form of an interval-class vector, a six-place array showing the number of instances of each interval class, 1 through 6. (The number of interval-classes 0, trivially, reflects the size of the set and is normally ignored.) For example, the interval-class vector for the set C, C-sharp, D-sharp, E, G-sharp is [2 1 2 3 2 0], meaning there are two ic1s (C to C-sharp and D-sharp to E), one ic2 (C-sharp to D-sharp), and so on.

2 Learning the Properties of a Pitch-Class Set

In each of the following simulations, a backpropagation network with one hidden layer has been trained to learn some characteristic of an unordered pcset, or some relationship between two such sets. One advantage to using a neural network for this type of study is that a network can be designed to pay attention only to those aspects of the stimulus that are desired. A neural network can deal with pcsets as entities without the distractions created in human perception studies by the musical context. Understanding the effects of context on pcset perception is possible only after a context-free understanding of these sets exists. Watching how the network models learn will in some cases suggest hypotheses about human perception that could then be tested experimentally. Some of the simulations will provide a foundation for future neural network-based studies that will take musical context and human cognition more fully into account. And some of the results will simply reveal interesting properties of the pitch-class universe.

Little effort has been made to discover the optimal size of the hidden layer. In some cases the network has learned the problem incompletely, while in some others it has not learned very well at all. But studying the network's learning process and analyzing the types of patterns it has trouble learning is more instructive than is simply showing that a particular problem can be learned. All simulations described in this section of the paper use the Rprop (for *R*esilient back*prop*agation) algorithm (Riedmiller & Braun, 1993), an adaptation of the backpropagation algorithm that is generally faster than the standard backpropagation algorithm. All network units use a logistic activation function that constrains their values to the range 0.0 to 1.0. (The mathematical details of the Rprop algorithm are omitted for reasons of space. All network simulations described in this section of the paper were run using the Stuttgart Neural Network Simulator, version 4.1, running under Linux.)

2.1 Recognizing Interval-Classes

Music theorists have devoted substantial attention on the intervallic qualities of pcsets (Forte, 1964; Lewin, 1959, 1960). The interval class formed between two pitch classes can be learned using a three-layer network like that described above. In two separate runs, a network with twelve input units, twelve hidden units, and six output units (that is, a 12-12-6 network) learned to recognize the interval classes formed by all 66 pitch-class pairs (twelve each of interval-classes 1-5 and the six tritones which comprise interval-class 6). Output was considered correct if the activation of the unit representing the answer was ≥ 0.8 and the activation of the other units was ≤ 0.2. In the first run, 64 of the patterns were learned within 87 training epochs, while the two remaining patterns resisted classification until epoch 816. In the second run, all patterns were learned by epoch 157.

Because the connection weights in the network start at random values, the learning path a network follows will be different each time. On the two runs here, the network took two very different paths. On the first run, the network learned the patterns representing even interval classes noticeably faster than those representing odd ics. On the second run, the opposite was the case: over half of the odd-interval patterns were learned before even one even-interval pattern was learned (see Table 1). This suggests that there is an inherent linkage among the even and odd interval classes. It suggests further that claims regarding interval content (such as in measures of intervallic similarity; see section 3.3) should take into account this apparent *resonance* within these groups of ics. (For example, ics 3 and 5 might be less dissimilar than ics 4 and 5.)

Table 1. Numbers of each interval-class learned at selected points during the two training runs

Epoch	ic1	ic2	ic3	ic4	ic5	ic6
40	0	4	0	1	0	0
55	2	6	2	3	2	0
62	3	8	3	6	5	2
70	4	12	6	12	6	2
80	6	12	9	12	8	6
100	9	12	11	12	10	6
829	12	12	12	12	12	6

Epoch	ic1	ic2	ic3	ic4	ic5	ic6
35	1	0	3	0	0	0
44	4	0	8	0	8	0
54	5	0	8	2	8	0
63	9	2	9	4	7	0
70	10	4	11	4	8	1
80	12	7	11	8	11	3
122	12	9	12	12	11	6
157	12	12	12	12	12	6

2.2 Total Interval Content

Determining the ic between two pcs generalizes to the task of identifying the total ic content of larger collections. This can be tested with a network of the same architecture, though the hidden layer needs to be larger to accommodate the increased complexity of the learning problem. A network containing 90 units in the hidden layer was set upon the task of learning the interval-class vectors for 4095 collections (2^{12}-1, for the single twelve-note set class).

The twelve input units again represented the twelve pitch classes, while the six output units represented the six interval classes. The target values were (number of instances)/10. Since the learning function used only permits values between 0 and 1, the 12-pc set class was not used, since it would require target outputs of 1.2 for ics 1-5. A pattern was considered learned if the values of all output units were within 0.033 of their targets.

After 8000 epochs, the network had learned 80% of the patterns. It took another 15,000 epochs to reach 85%, and the total never reached 90%. Analysis of the patterns the network had trouble with showed that it had the most difficulty learning symmetrical sets – those which map onto themselves under transposition, inversion, or both. The network learned about 79% of the 1046 symmetrical collections, while it learned 93% of the 3048 non-symmetrical collections. One explanation may be that symmetrical sets tend to have interval-class vectors that are strongly skewed (for example, the vector for the diminished triad, two instances of which were among the ten patterns with the greatest errors, is [004002]). Since the entries in these sets' vectors are all far from the norm, the network had trouble reaching those extreme values. The effect was especially pronounced for larger symmetrical sets, such as the octatonic collection. The network had its best success with those sets that have, in Forte's terms, Z-partners, sets that have the same interval-class content, but are not related by transposition and/or inversion. The network was able to produce the correct vector for all instances of all 38 such set-classes. This is somewhat surprising, since, although the Z-related sets contain the same interval classes, those interval classes are combined in considerably different ways in the two sets. A possible explanation is that the intervals for all set classes that belong to Z-pairs tend to be fairly evenly distributed among the six interval classes, and that these smoother vectors are easier to learn. This may also be explained by the fact that there are fewer instances of the symmetrical set types to learn from. For example, there are just three {0,3,6,9} types, compared to twenty-four {0,1,4,6} types. The learning profile may have been different had each set class been represented in the training patterns the same number of times. In terms of human perception, these results seem to correlate with intuitions about symmetrical and non-symmetrical collections in the tonal system. Symmetrical structures are associated with ambiguity and stasis (the diminished seventh and octatonic collections, for example), while non-symmetrical structures are associated with harmonic function and tonal hierarchy (the dominant seventh chord and the diatonic scale, for example). What the implications are for perception of atonal music, however, is unclear.

2.3 Generalization

One of the things neural networks are especially good at is generalizing to new situations based on incomplete experiences. A 12-90-6 network was trained to learn interval-class vectors as in the previous experiment. This time the network was trained on only 2696 different pcsets, about two-thirds of the total possible. The network was able to learn the reduced training set to nearly the same degree of accuracy as when it had all the sets to learn from (79% of the patterns after 8000 epochs, peaking at about 88%). As training progressed, the ability of the network to generalize was tested by using the same network to analyze 1393 previously unseen sets. The network was very successful in working with these unfamiliar collections; accuracy on this validation set lagged less than 2% behind performance with the training set.

For both the training and validation sets, the effects of symmetry and membership in a Z-pair were the same as in the previous study when all sets were presented. For this network, the influence of set size was also examined. The network's performance varied markedly depending on the size of the set. In general, the network learned the ic vectors of larger sets better than those of smaller sets. With the training patterns it correctly determined vectors for all sets of sizes 7, 8, and 11; all but one nine-note collection; over 90% of the six- and ten-note sets; just 86% of the three-note sets; and 81% of the five-note sets. The tetrachords were the most surprising, however. The network correctly learned only 54% of these sets' interval-class vectors. Similar results obtained for the validation set. More extensive data analysis and additional trials will be required before an explanation can be offered for this.

2.4 Determining the Prime Form of a Collection

In Forte's theory of pitch-class sets, pcsets are considered equivalent under the operations transposition and inversion. One way in which the resulting equivalence classes are referred to is by their prime form, the member of the set class that is the most compact and begins on 0. It can be thought of as a prototype for the other members of the set class. The prime form {0,1,3} thus represents all three-note collections formed by a semi-tone followed by a whole tone in the same direction.

A 12-72-12 network was constructed to see if it could learn the prime forms of all 4096 pcsets. After 10,000 epochs only 30% of the patterns were correctly learned and training was stopped. It is noteworthy, however, that symmetrical sets were learned at a higher rate than non-symmetrical sets (39% to 26%). It was apparent that this problem required either a larger network or a smaller pattern set. The same network was then presented with only the 924 six- note collections, reaching 90% accuracy. Unfortunately, there was no apparent pattern to the incorrectly classified sets. Symmetrical and non-symmetrical sets were recognized with similar frequency; membership in a Z-related pair likewise seemed to play no role in the results. This indicates again that a larger hidden layer might prove helpful for this particular task. It might also reflect the difficulties associated with recognizing both transposed and inverted forms of a set as equivalent. To test for this possibility one would simply need to test for the ability to map sets onto their Tn-based prime forms (rather than the Tn/TnI forms used here).

3 Relations Between Pitch-Class Set

In addition to looking at properties of individual sets, theorists and analysts are often interested in the relationship between sets. There are many such relationships and these are, in fact, the source of many of the questions regarding the perceptual validity of pcset theory. Three relations were selected for study here: recognizing whether sets are Tn or TnI-related, whether a subset/superset relation holds between two sets, and learning a similarity relation between pairs of sets.

3.1 Tn and TnI Equivalence

The equivalence of pcsets under transposition (Tn) and inversion (optionally followed by transposition, hence TnI), is a central part of Forte's theory, and, in the case of the TnI-equivalence, one of its most controversial. A set of 9000 training patterns involving non duplicating pairs of sets of sizes 5-7 were chosen at random. One-third of the patterns involved sets from different set classes (and could be of different size), one-third were Tn-related members of the same set class, and one-third were TnI-related members of the same set class. No set pairs were used that were related by both Tn and TnI. In a 24-60-1 network, the inputs represented the two pcsets, while the target for the single output unit was 1.0 if the sets belonged to the same set class, and 0.0 if they were not.

Learning the training set turned out to be an easy task. The network was able to learn 99% of all patterns. Among the three categories of set pairs used, it learned to recognize the transpositionally related sets first and best, reaching 100% accuracy. It is noteworthy that its accuracy with both non-equivalent and TnI-equivalent sets was not as good. Though the network still reached about 96% with these, the rate at which it learned these patterns was slower than for the transpositionally related set pairs.

This network was able to generalize fairly well, correctly identifying 87% of the patterns in a validation set containing 1000 different patterns. As with the training set, the network was most successful in classifying patterns in which the sets were related by transposition (92%). It learned the correct relationship in about 75% of the unrelated sets and TnI-related the sets. These results seem to support the notion that inversional equivalence, criticized by many, is a more difficult perceptual concept than transpositional equivalence.

3.2 Subset Relations

Another way in which sets can be related is by inclusion. Inclusion comes in two forms: literal inclusion, in which the one set contains all the pitch classes of the second, and abstract inclusion, in which one set contains the pitches of the second set, but transformed by transposition and/or inversion. Abstract inclusion plays a critical role in many of Forte's writings. From a perceptual point of view, however, the concept is challenging–particularly when that inclusion involves both transposition and inversion. To test the ability of a network to learn to recognize abstract inclusion, a training set was generated consisting of 10,000 patterns with 24 input units representing two pcsets. The first of these contained between five and eight pitches and the second contained from two to n-1. In two-thirds of the patterns the second set was abstractly included in the first, while in one-third it was not. The single output unit was on if the inclusion relation held.

The network was able to learn to classify correctly 87% of the training patterns. It was substantially better at learning which sets were not in a subset relation (96%) than those that were (82%). The network's ability to generalize was tested with a set of 2000 previously unseen patterns. The network could correctly identify 65% of the patterns, a respectable degree of generalization. Striking, however, was that it was more accurate in classifying patterns that involved abstract inclusion (71%) than those that did not (52%). This is the opposite of its performance on the training data. An analysis of the data for the two pattern sets shows a negative correlation between the degree of inclusion, that is, the number of transforms of one set contained in some form of the other, and the mean squared error on the output unit. This is to be expected, since that is what the network is learning to discern, but the correlation is noticeably stronger in the validation set than in the training set, indicating that the network has somehow learned to use information about degree of inclusion to its advantage in recognizing abstractly included subsets.

Unlike the network which learned about transpositionally and inversionally related sets, this network was equally successful in discerning abstract inclusion that involved inversion with that involving just transposition and does not therefore support the intuition that abstract inclusion involving inversion should be harder to recognize than that involving only transposition. This is not altogether surprising, however, since the network knows nothing about up or down. It was, in fact, presented with patterns which exemplified both inverted and non-inverted forms, so it learned to deal with that reality. This may suggest that, if inversionally derived inclusion is in fact harder to discern, some factor external to idealized pitch classes is involved. For example, if the network knew something about register and frequency ratios–factors that influence our own preferences for contour-preserving operations–it might have yielded different results.

3.3 Similarity Relations

The last means of relating two sets to be discussed is a measure of their similarity. This is typically done through some mathematical function that compares the interval or total subset content of two sets. A number of such measures have been proposed (Castrén, 1994; Isaacson, 1990). Castrén's (1994) RECREL was chosen for testing as it strikes me as being among the most promising in terms of correlation with perceptual similarity. RECREL measures the total subset similarity of two sets by combining a measure of the similarity of their subset content with a measure the similarity of their non-shared subsets. RECREL values range from 100, indicating minimum similarity, to 0, indicating maximum similarity. The values were modified by the formula (1 - (RECREL / 100)) so the values would run from minimally similar = 0.0 to maximally similar = 1.0. Sets are minimally similar if they have no common subsets and maximally similar if their total subset content is proportionally identical. (RECREL does not consider inversionally related sets to be equivalent.)

The training set consisted of 40,000 pairs of sets of size 5-7, which is a fairly small sample, considering there are 924 possible six-note collections, and 792 each five- and seven-note sets, for a total of 2508^2or 6.3 million possible pairs. The sets were generated at random, with duplicates not allowed. A validation set of 2000 pairs was likewise generated. A network with 200 hidden units was trained for 4000 epochs, by which point the network could produce a value within 0.03 of the target for just 47% of the patterns. Training was stopped because learning had virtually stopped and generalization was not improving. Although the network was not able to learn the relation very well, its performance on the validation set was nearly the same, just below 44%, showing that what was learned generalized fairly well. If the criteria for correct identification is expanded to +/- 0.1, accuracy increases to over 92% for both pattern sets, however. Analysis of both the training and validation data (using the +/- 0.03 criteria) showed relatively strong correlations (as much as 0.28) between the error at the output unit and the RECREL values themselves, as well as with the degree of transpositional inclusion, meaning that set pairs that were more similar tended to have greater error at the output. Nevertheless, the network's relatively poor performance indicates need for further study of similarity measures. There has been some experimental work aimed at validating similarity measures (Bruner, 1984; Gibson, 1986; Stammers, 1994). Though these studies have shown similarity measures to be poor predictors of perceived similarity, there are problems with these studies. For example, they generally involve only the study of trichords, and they are all based on similarity measures that have been criticized in the theoretical literature. The effect of context is a significant issue in the study of similarity, and how this affects our judgement of similarity has only recently been considered in theoretical circles (Isaacson, 1996; Morris, 1995).

4 Musical Segmentation

The segmentation of a musical surface into analytic objects is an essential part of the analysis of music. In tonal music, the preference for tertian sonorities, the relatively clear distinction between melody and harmony, and the hierarchy among the diatonic and chromatic degrees of the scale make segmentation fairly easy. In post-tonal music, however, there are no systemically preferred sonorities. So parsing the musical surface into discrete analytical objects must be done according to new methods. Although the topic has been examined both formally and informally (Forte, 1973; Hasty, 1981; Roeder, 1988; Schaffer, 1991), there is no general agreement on the best way to proceed. Hasty's work in this area (Hasty, 1981) is the most promising. It makes significant use of Gestalt principles (though he does not cast his discussion in such terms). His approach informs the simulation described below.

One type of network that is appropriate to the segmentation problem is the interactive activation and competition model (McClelland & Rumelhart, 1988). The IAC model uses a localized representation of the data–each unit has a specific meaning. In the network to be discussed these include such meanings as *m. 4 beat 1*, *pizzicato*, and *slur group 4* (Fig.1), and the units are grouped by the feature they represent. The network described here explores segmentation possibilities in Webern's *Six Bagatelles*, Op.9, No.3 (see Fig.2). The central unit group represents the individual pitch events in the piece. For this simulation, ten feature groups have been identified. Excitatory connections between these feature groups and the event units, and inhibitory connections within each group, are set as shown in Table 2.

Table 2. Excitatory connections between feature groups and event units, and inhibitory connections within each group

Node Type	Excitatory Weight	Inhibitory Weight
Event		-1.0
Time	AP 1.0 / SP 0.6	-1.0 to 0
Pitch	1.0	-0.5 to 0
Part	0.5	-0.5
Slur Group	1.0	-0.5
Dynamic Group	1.0	-0.5
Dynamic	0.5	Var.
Attack	0.8	-1.0
Articulation	0.8	-1.0
Special Tech.	0.8	-1.0

Inhibitory weights within the Time, Pitch, and Dynamic groups are variable. In the Time group, for example, inhibition between units increases as the distance between the time points increase. The Pitch and Dynamic groups are treated similarly. So a pitch node inhibits other pitch nodes close to it in register less than it does pitches found at larger intervals.

Figure 2 shows the musical segments (drawn in boxes) determined by the network when the circled event units were clamped on. In each case the segmentation is musically defensible. Interestingly, when the C♯ in m. 4, vln. 2, was clamped on, the network attended for quite a while, though

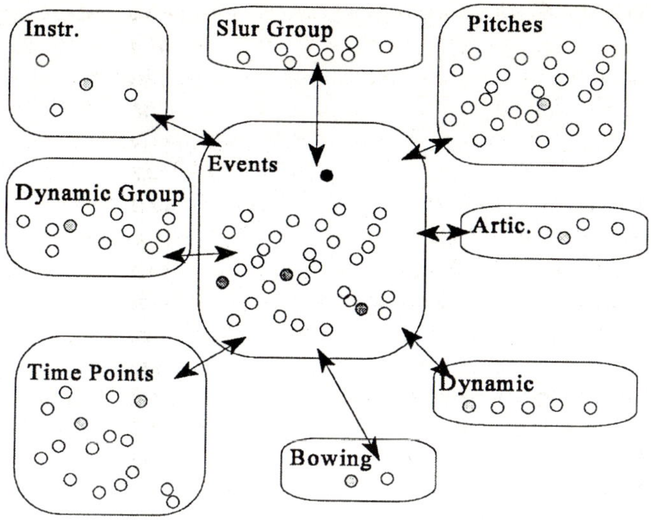

Fig. 1. Interactive Activation and Competition Netowrk for Webern, *Op.9, No.3*

fairly weakly, to the sustained notes of the cello and viola of mm. 2 and 3. Both events eventually succumbed to inhibitory pressures before the network stabilized.

With this network architecture, it is a simple matter to experiment with various weightings of the feature sets. It provides an easy means of testing various hypotheses about the relevance of various musical features. Networks of this type have a number of limitations, however. First, although this network appears to settle into a stable state on its own,this was possible only after the weights between and within node groups were adjusted until plausible results emerged. Second, the selected feature sets are not the only musically significant features. Further, there is no mechanism for letting prior segmentations inform subsequent ones. So, while similarity among discrete events influences the grouping of these events into musical segments, Gestalt principles suggest an important role for higher-level mechanisms than this network architecture allows (see Goldstone, 1996; Hasty, 1981). The "Grundgestalt" principle and Schoenberg's interest in developing variation (Haimo, 1996) are suggestive in this regard. A more effective approach would be to generate a much larger feature set and to allow a self-organizing network to discover patterns on its own (Gjerdingen, 1991). Such a network might be able to find a hierarchical structuring in atonal music that has so far proven elusive (see Lerdahl, 1989; Dibben, 1994).

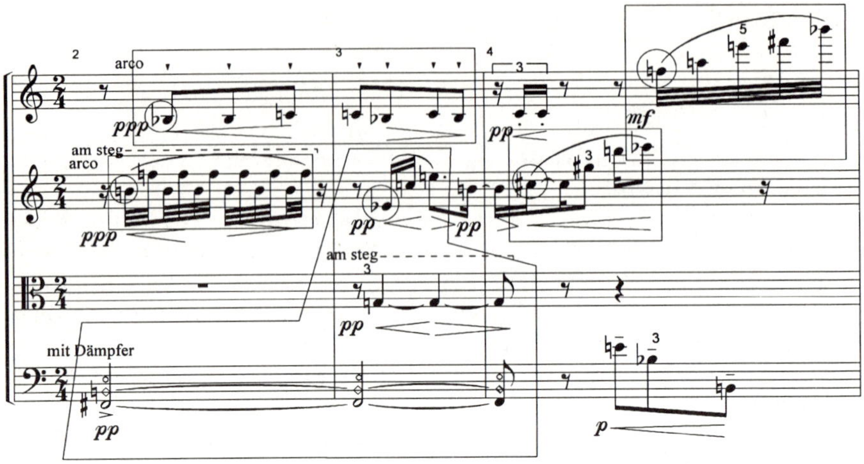

Fig. 2. Webern *Six Bagatellen*, Op.9, No.3, mm. 2-4. Copyright 1924 by Universal Edition AG., Wien. Copyright renewed. All rights reserved. Used by permission of the European American Music Distributers Corporation, sole U. S. and Canadian Agent for Universal Edition

5 Suggestions for Future Study

The networks described in this paper represent first steps into the use of neural networks to study atonal music. The results suggest a number of directions for future work, particularly as related to human perception. In perception studies dealing with pcsets, the myriad dimensions involved in a musical context can easily confound efforts at obtaining reliable data. With neural network models, contextual effects could be brought on-line in a controlled fashion, which might lead to a better understanding of the role various musical parameters play in how we hear post-tonal music. Finally, since we listen to both tonal and atonal music with the same pair of ears, it would be interesting to see how a self-organizing network might deal with atonal music with and without having learned something about tonal music.

References

Bruner, C. (1984). The perception of contemporary pitch structures. *Music Perception, 2*, 25–39.

Castrén, M. (1994). *RECREL: A similarity measure for set-classes, Studia Musica No.4.* Unpublished doctoral dissertation, Sibelius Academy, Helsinki.

Dibben, N. (1994). The cognitive reality of hierarchical structure in tonal and atonal music. *Music Perception, 12*, 1–25.

Forte, A. (1964). A theory of set-complexes for music. *Journal of Music Theory, 8*, 136–83.

Forte, A. (1973). *The structure of atonal music.* New Haven, CT: Yale University Press.

Forte, A. (1985). Pitch-class set analysis today. *Music Analysis, 4*, 29–58.

Gibson, D. (1986). The aural perception of non-traditional chords in selected theoretical relationships: A computer generated experiment. *Journal of Research in Music Education, 34*, 5–23.

Gjerdingen, R. (1991). Using connectionist models to explore complex musical patterns. In P. Todd & D. Loy (Eds.), *Music and connectionism.* Cambridge, MA: The MIT Press.

Goldstone, R. (1996). *Hanging together: A connectionist model of similarity* (Vol. 185; Tech. Rep.). Bloomington: Indiana University.

Haimo, E. (1996). Atonality, analysis, and the intentional fallacy. *Music Theory Spectrum, 18*, 143–166.

Hasty, C. (1981). Segmentation and process in post-tonal music. *Music Theory Spectrum, 3*, 54–73.

Imberty, M. (1993). How do we perceive atonal music? Suggestions for a theoretical approach. *Contemporary Music Review, 9*, 325–337.

Isaacson, E. (1990). Similarity of interval-class content between pitch-class sets: The IcVSIM relation. *Journal of Music Theory, 34*, 1–28.

Isaacson, E. (1996). Issues in the study of similarity in atonal music. *Music Theory Online, 2.*

Krumhansl, C., Sandell, G., & Sargeant, D. (1987). Tone hierarchies and mirror forms in serial music. *Music Perception, 5*, 31–78.

Leman, M. (1992). Artificial neural networks in music research. In A. Marsden & A. Pople (Eds.), *Computer representations and models in music.* London: Academic Press.

Lerdahl, F. (1989). Atonal prolongational structure. *Contemporary Music Review, 4*, 65–87.

Lewin, D. (1959). Re: Intervallic relations between two collections of notes. *Journal of Music Theory, 3*, 298–301.

Lewin, D. (1960). Re: The intervallic content of a collection of notes and its complement: An application to Schoenberg's hexachordal pieces. *Journal of Music Theory, 4*, 98–101.

McClelland, J., & Rumelhart, D. (1988). *Explorations in parallel distributed processing: A handbook of models, programs, and exercises.* Cambridge, MA: The MIT Press.

Morris, R. (1995). Equivalence and similarity in pitch and their interaction with pcset theory. *Journal of Music Theory, 39*, 207–243.

Perle, G. (1990). Pitch-class set analysis: An evaluation. *Journal of Musicology, 8*, 151–72.

Riedmiller, M., & Braun, H. (1993). A direct adaptive method for faster back propagation learning: The RPROP algorithm. In *Proceedings of the IEEE International Conference on Neural Networks 1993.* San Francisco, CA.

Roeder, J. (1988). A declarative model of atonal analysis. *Music Perception, 9*, 21–34.

Rumelhart, D., Hinton, G., & Williams, R. (1986). Learning internal representations by error propagation. In D. Rumelhart & J. McClelland (Eds.), *Parallel distributed processing: Explorations in the microstructure of cognition* (Vol. 1). Cambridge, MA: The MIT Press.

Schaffer, S. (1991). *Segmentation issues in the analysis of selected non-serial atonal works of Schoenberg, Berg, and Webern, 1908–1923.* Unpublished doctoral dissertation, Indiana University.

Stammers, D. (1994). *Set theory in the perception of atonal pitch relations.* Unpublished doctoral dissertation, Darwin College, Cambridge, UK.

Taruskin, R. (1979). Review of the structure of atonal music. *Current Musicology, 28*, 114–129.

Todd, P., & Loy, D. (1991). *Music and connectionism.* Cambridge, MA: The MIT Press.

III

From Rhythm to Expectation

Tempo Relations: Is There a Psychological Basis for Proportional Tempo Theory?

Marek Franěk[1] and Jiří Mates[2]

[1] Inst. of Musicology, AS CR, Puškinovo nám. 9, 160 00 Prague 6, Czech Republic
[2] Inst. of Physiology, AS CR, Vídeňská 1083, 142 00 Prague 4, Czech Republic

Abstract. This paper is devoted to experimental research in the field of tempo performance. Epstein (1979, 1995) proposed that in a piece of music all tempo relations can be expressed by low order whole number ratios 1:1, 1:2, 2:3, and 3:4 (or the inverse). This theory relates to the Gestalt concept, in that it deals with part/whole relations and coherence of multipart pieces of music. We designed our study to look for possible underlying psychological mechanisms. The experiment addressed the question, which temporal ratios between neighboring tempi occur when subjects are asked to make arbitrary tempo modulation. Subjects were instructed to make tempo changes defined by vague terms like *small*, *large*, and *arbitrary* modulation. The finger tapping method was employed. Theoretically, a random continuum of ratios can appear. However, the results showed that tempo ratios were characterized by narrow distributions of the values. Subjects rarely responded with whole number ratios. Averaged values show that in slow down conditions (except *arbitrary*), the responses were close to whole number ratios; while in speed up conditions, deviations from whole number ratios were greater. We assume that results for speed up conditions were influenced by systematic temporal error. We propose that the relations between neighboring tempi may better be described in terms of categories of tempo difference.

1 Introduction

Much attention has been paid to timing in music performance. A large amount of research has focused on the temporal analysis of musical performance. Some authors, such as Povel (1977), studied rhythm and tempo fluctuations. Other systematic research was directed to the study of fluent modulations of a tempo, such as musical ritardando and accelerando (Feldman, Epstein, & Richard, 1992; Sundberg & Verillo, 1980). However, little attention has been paid to proper *tempo modulation*. How precise can the performers be in an exactly defined alteration of a tempo? What are the relations between tempi of two adjacent movements? Can the performer then change the tempo in an arbitrary way or not? Are there any preferred tempo ratios?

An important contribution to our knowledge about tempo relations in Western art music and non-Western music is the study by the musicologist and conductor D. Epstein (1985). He proposed that in a piece of music, tempo relations can be expressed by some low order whole number ratios, such as

1:1, 1:2, 2:3, and 3:4 (and the inverse). The concept of *proportional tempo* is based on the theory of continuous pulse (Epstein, 1979). The theory suggests that in works which contain different movements (or in single movement works that contain different tempi), all tempi are intrinsically related. These relationships can be concisely expressed by whole number low order ratios. Epstein's support of the theory lies in an analysis of performances of Mozart's, Haydn's, and Brahms' symphonies. To confirm the general validity of the theory, he also analysed various examples of music from non-Western cultures. He studied tempo ratios in musical excerpts from seven cultures: one Pacific, two Asian, two African and one South American. Surprisingly, in all excerpts, only low order whole number ratios occurred between any two contiguous movements.

Epstein's proportional tempo theory is relevant to the Gestalt concept in that it deals with the classical Gestalt problem of the part/whole relation. It assumes that a multipart piece of music can be perceived as a whole instead of a set of isolated excerpts. As such, not only local relationships between the durations of successive notes, but also global relationships on higher levels such as tempo proportions between particular parts, can be responsible for perceptual coherence of a whole piece. Epstein has suggested that an aesthetically and communicatively sufficient musical performance requires low order whole number tempo ratios, while tempo changes in other ratios are suggested to be unnatural. This idea is also connected to the perceptual coherence of a piece.

The investigation of temporal coherence in music raises several complex problems at multiple levels. Detailed comments discussing various methodological problems of the research in the area of tempo proportionality has been given by Repp (1996). If we consider music only in a written form, then we believe that his theory can be simply confirmed. However, the sounding form of music introduces much complexity. The important point is that the temporal structure of music can be considered in terms of global and local tempi. Global tempo is related to a tempo marking at the beginning of the piece, and it is computed by averaging over a number of beats. However, the timing microstructure of a performed movement consists of continuous modulations of local tempi, resulting in unequal intervals between successive tone onsets, even if corresponding notes have the same value in the score. From this point of view, the mean value of the speed of tempo – which was used in Epstein's measurements – could be very misleading. A suspicious example of the difference between local and global tempi is ritardando before the end of a movement. Local tempo at the end of such a piece then differs substantially from the global tempo of the piece. Shall we consider tempo proportions between successive movements as ratios of global tempi or shall we take into account also local changes of tempo such as ritardando?

Another important point is related to the estimation of the region of uncertainty around observed tempo ratios. In sounding music, exact temporal ratios (1:2, 2:3, 3:4) hardly occur, but ratios with fractional approximations in this range are tolerated. How small are the perceptual differences that subjects accept as random errors or *noise*? Perceptual differences are based mostly on perceptual thresholds. For instance, if several slightly different tempi in some speed range do not overpass a certain threshold value, then they can be perceived as equal. Discrimination thresholds for tempo difference perception are generally between 5 and 8% of stimulus duration, i.e. beat interval (Ellis, 1991). But what about smaller differences? Epstein took 5% deviation as confidence limit. Repp (1996) pointed out that in this case there is a relatively high probability of finding evidence for tempo proportionality. The probability that any randomly chosen tempo ratio will support the proportionality theory is 0.43!

Accordingly, Epstein's hypothesis needs further empirical data, especially for performance analysis of a large sample of Western as well as non-Western music. An experimental study on the psychology of time might also prove productive. Time psychologists point to various timing constraints in producing rhythmical patterns (Fraisse, 1978; Povel, 1981). Probably, certain constraints in the area of tempo modulations, given by operational principles of an internal clock, could exist. These could represent the underlying psychological mechanisms for proportional tempo theory.

In order to gain insight into these phenomena, we designed a simple experiment which addressed the question which temporal ratios between neighboring tempi occur when subjects are asked to make various levels of arbitrary tempo modulation. Theoretically, a random continuum of tempo ratios could be observed. However, on the basis of Epstein's findings, we hypothesize that subjects will prefer only certain intertempi ratios and that these ratios will be close to low order whole numbers. To encourage subjects to make truly arbitrary tempo modulations, we asked them to make tempo change by means of vague terms like *small*, *large*, and *arbitrary*. As control, double tempo change was also included. Since we intended to study basic constraints, we made the experimental condition as simple as possible in order to avoid the influence of musical structure. Therefore, simple pulse sequences were used for stimuli, and subjects were asked to tap the pulse sequences on the response key.

2 Method

2.1 Subjects

Thirteen subjects participated in the experiment. They were musicology and music psychology students from 18 to 25 years old.

2.2 Stimuli and Procedure

Subjects were asked to modulate the tempo of their finger tapping onto a response key. The motor task consisted of two periods: first the establishment of the base tempo, then the introduction of tempo modulation (see Fig.1).

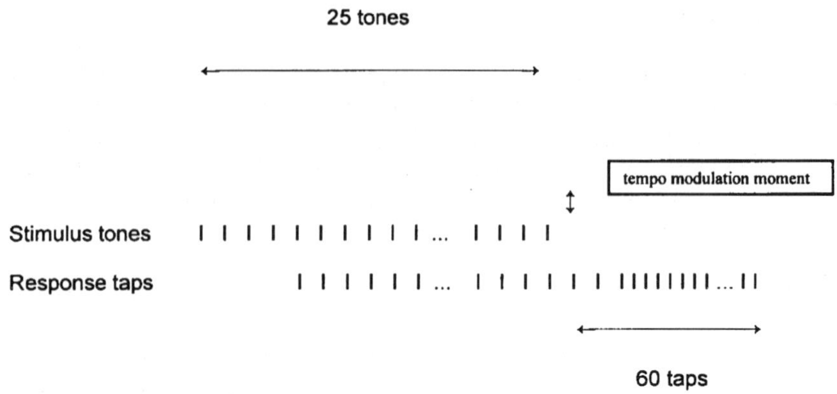

Fig. 1. Schematic representation the of experimental procedure

In the first period, subjects heard a sequence of stimulus tones presented in base tempo. They were asked to synchronize their tapping with the sequence. Stimulus sequences were 25 tones of 100 ms duration at a frequency of 440 Hz. Pauses between tones were 500 ms. Upon termination of the stimulus sequence, subjects were asked to modulate the tempo. They were asked to continue in a newly established tempo. This had to be done without interruption of tapping. Since the moment of tempo modulation appeared unexpectedly (subjects were not aware of the duration of the stimulus sequence), subjects were not instructed to perform the tempo modulation immediately after termination of the stimulus sequence, but rather within a period of 3 to 5 taps. After 60 taps, subjects were asked to stop.

Both speeding up and slowing down directions of tempo modulation were employed. Tempo modulations were of four types: *small*, *large*, *arbitrary*, and *twice*. Before each trial subjects were informed about the category of tempo modulation. The base tempo was 100 beats/minute (one beat lasting 600 ms).

Each subject participated in 40 trials. Four tempo modulation conditions (twice, small, large, arbitrary) by 2 tempo modulation directions (speeding-up or slowing-down) by 5 sessions. Each subject completed five sessions of two blocks, encompassing four trials each. Speeding up was performed in the first block, slowing down in the second block. All four conditions of tempo modulation occurred in random order within a block. Subject completed 8 trials per session. The subjects were asked to avoid making any body movements in order to prevent interference with the tempo modulation task.

2.3 Apparatus

The stimuli were generated and the subjects' tapping was recorded by a computer program running on a PC. The temporal resolution of the time measurements was 250 microseconds. Stimulus tones were generated by the PC's internal sound generator and presented through the PC's loudspeaker. The subject tapped with his index finger on a response key.

3 Results

Motor responses were measured from the moment the stimulus sequence terminated. The temporal intervals between successive taps were evaluated and recorded. In the following data analysis, the first 10 taps, where a new tempo was gradually established, are omitted. The modulated tempo was expressed as a mean beat interval (see Table 1 and Fig.3). Although instructions to modulate tempo *arbitrary*, *small*, *large* theoretically enabled subjects to choose an arbitrary tempo ratio, magnitudes of particular mean beat intervals were characterized by narrow distributions (except in the condition *speed up/arbitrary*).

Table 1. Mean beat intervals (MBI) in ms and standard deviations (SD) of beat intervals for four conditions during tempo modulation in direction speed up and slow down

	Speed up		Slow down	
Condition	MBI	SD	MBI	SD
Arbitrary	353.872	90.524	1007.942	418.670
Twice	331.151	47.008	1178.380	97.499
Small	477.000	353.289	808.624	161.504
Large	239.718	62.439	1759.020	707.506

Detailed analysis, taking into account influences of direction, of condition, and of session on the speed variability of the modulated tempo was done with ANOVA. Direction of tempo modulation (DIR; df=1), condition of tempo modulation (COND; df=3) and session (SESS; df=4) were selected as trial (within subject) factors. Subjects' (1) mean beat intervals and (2) standard deviations were computed over each trial as replicates (df=12). Direction of tempo modulation and tempo modulation condition had main significant effects. Significant interaction occurred between direction and condition of tempo modulation (see Table 2).

The analysis of correspondence between tempo ratios, observed in the set of experimental data and the whole number tempo ratios, was carried out in two steps. The results for each subject and each condition separately

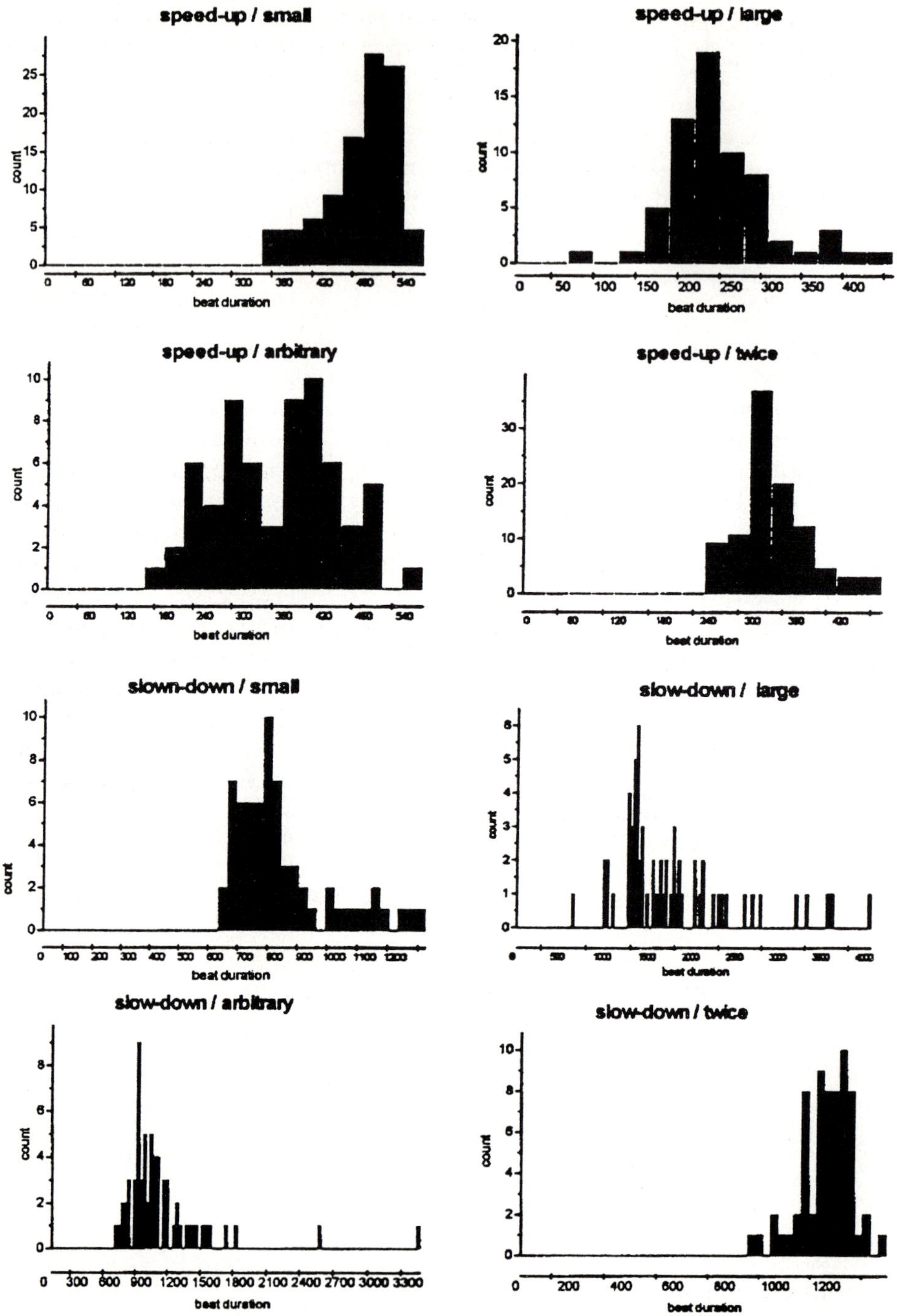

Fig. 2. Histograms of mean beat intervals (in ms) for tempo modulation of both directions. Bar width is 30 ms

Table 2. Degrees of freedom, F ratios and probability levels of within subject effects in repeated measures analyses of variance of mean beat intervals (MBI) and standard deviations (SD). Significant effects are marked with asterisks

		MBI		SD	
Effect	df	F	p	F	p
DIR	1,12	216.262	.000*	64.855	.000*
COND	3,36	19.148	.000*	4.063	.014*
SESS	4,48	.8248	.516	.167	.954
DIR x COND	3,36	37.441	.000*	7.145	.001*
DIR x SESS	4,48	1.616	.185	.092	.984
COND x SESS	12,144	.683	.766	.829	.619
DIRxCONDxSESS	12,144	.715	.734	1.022	.432

were compared with all possible low order whole number values. Paired samples T-test, which analysed the significance of the difference, showed that all data differed significantly from the corresponding whole number ratio values. However, the perceptual difference need not be identical with the statistical difference. Since the former is based mostly on perceptual thresholds, we have to take into account a discrimination threshold from 5 to 8% from the duration of the stimulus. Therefore the best fitting whole number ratio for mean values was found in order to compare relative deviations from these ratios with discrimination thresholds values (see Table 3) Results show that while deviations from the slow down tempo modulations are very small and mostly negative (except in the condition *arbitrary*), deviations for the speed up tempo modulation are larger and of positive direction.

Table 3. The best fitting whole number ratios with relative deviations of observed ratios

	Speed up		Slow down	
Condition	Ratio	Deviation	Ratio	Deviation
Arbitrary	2 : 1	+ 18%	1 : 2	- 16%
Twice	2 : 1	+ 10.4%	1 : 2	- 1.8%
Small	3 : 2	+ 19.3%	3 : 4	+1.1%
Large	3 : 1	+ 19.9%	1 : 3	- 2.3%

4 Discussion

The control condition (double tempo modulation in both directions) gave us important information for reconsidering of the proportional tempo theory.

The task, to modulate a tempo twice, was rather simple compared to the other tasks. The factor of modulation was indeed explicitly defined (1:2). This tempo modulation enabled the subjects to employ additional mental cues like internal counting, which simplified the task. Therefore, we assume, that the results reflect the skill of the performer to create whole number temporal ratio. While slowing down twice was performed exactly (the deviation was only -1.8%), the deviation of speeding up twice reached + 10.4%, which is slightly above the differential threshold. Since all remaining deviations in the speed up conditions also were positive and had magnitudes around 20%, we can hypothesize that there might be certain systematic errors or shifts involved. It indicates that the task of speeding up in this tempo range is performed slightly slower than expected. From results in time psychology we know that various systematic timing errors do exist, for instance an anticipation of 30 ms during synchronization tasks (Fraisse, 1966). Accordingly, the systematic error could have influenced the magnitude of tempo ratios.

5 Conclusion

The goal of our study has not been to put into question the validity of the proportional tempo theory in music, but rather to look for some underlying psychological mechanisms. The results of our study provide evidence in favour of the idea that certain limits might exist in the operational principles of internal clocks with respect to the mentioned tempo proportionality in music. However, strong psychological limits, restricting performers to making sudden tempo changes only in proportions of low order whole number ratios 1:1, 1:2, 2:3, and 3:4, have not been found.

On the other hand, the results convincingly show certain systematic timing behaviors. The vague verbal instructions for tempo modulation allowed random choice of a tempo change, but randomness was not found to occur. In spite of the variance of the magnitude of mean beats intervals, there was an intersubjective agreement in tempo modulation responses under most conditions. It seems that timing in the investigated temporal task could be described in terms of a categorization of tempo differences. This could mean that humans operate within a limited range of tempo change categories. The categorical tempo change is similar to the type of categorization in the production of rhythmical patterns. Fraisse (1978) presented evidence that subjects can operate only with two temporal durations (long and short with ratio 2 : 1). This limit can be exceeded by skilled musicians, who have a capacity to operate with more categories. But, even skilled musicians, in difficult cases, tend to spontaneously simplify rhythmical patterns (Clarke, 1985).

Our research did not show a clear definition of ratios between tempo change categories. There is no confirmation that an average magnitude of the categories lies in the mentioned whole number ratios. Instead, we observed a clear Gestalt effect: speeding up twice became a bit slower than it should be.

We hypothesize that this temporal shift occurred also in the other observed tempo ratios. Confirmation of the hypothesis may be achieved via further research employing a wide range of tempi. Both the observed systematic error and the phenomenon of differential perceptual threshold suggest that the tendency to find exact whole number low order ratios between tempi could be somewhat misleading.

Finally, we have to take into consideration that the observed results were influenced by the speed of the base tempo. If the base tempo is fast, tempo modulations greater than 2 : 1 can be difficult to perform due to a motor limit. If the base tempo is slow, tempo modulations slower than 1 : 2 can cause problems due to damage of rhythmical coherence. Application of a broader range of tempi is necessary. The *non musical* character of the task was probably the second feature which could have influenced the results. It may be expected that tempo modulation in a musical context would limit the choices of tempo change. This could lead to restriction of the number of observed tempo modulation ratios and to a clearer definition of tempo modulation categories.

References

Clarke, E. (1985). Structure and expression in rhythmic performance. In P. Howell, I. Cross, & R. West (Eds.), *Musical structure and cognition.* London: Academic Press.

Ellis, M. (1991). Research note: Thresholds for detecting tempo change. *Psychology of Music*, *19*, 164–169.

Epstein, D. (1979). *Beyond Orpheus.* Cambridge, MA: The MIT Press.

Epstein, D. (1985). Tempo relations: A cross-cultural study. *Music Theory Spectrum*, *7*, 34–70.

Epstein, D. (1995). *Shaping time: Music, the brain, and performance.* New York, NY: Schirmer Books.

Feldman, L., Epstein, D., & Richard, W. (1992). Force dynamics of tempo change in music. *Music Perception*, *10*, 185–204.

Fraisse, P. (1966). L'anticipation de stimulus rythmiques: Vitesse d'établissement et précision de la synchonisation. *L' Année Psychologique*, *66*, 15–36.

Fraisse, P. (1978). Time and rhythm perception. In E. Carterette & M. Friedman (Eds.), *Handbook of perception* (Vol. 8). New York, NY: Academic Press.

Povel, D. (1977). Temporal structure of performed music. *Acta Psychologica*, *41*, 309–320.

Povel, D. (1981). Internal representation of simple temporal patterns. *Journal of Experimental Psychology: Human Perception and Performance*, *7*, 3–18.

Repp, B. (1996). Book review: D. Epstein, Shaping time – music, the brain, and performance. *Music Perception*, *13*, 591–604.

Sundberg, J., & Verillo, V. (1980). On the anatomy of the ritard: A study of timing in music. *The Journal of the Acoustical Society of America*, *68*, 772–779.

A Framework for the Subsymbolic Description of Meter

Dirk Moelants

IPEM, University of Ghent, Blandijnberg 2, B-9000 Ghent, Belgium

Abstract. In this paper the concept of *meter* is investigated, taking into account elements of perception and performance. Considerable problems with the traditional theories lead to a broadened definition of meter as *the temporal framework of music*. A descriptive method for meter based on this definition is proposed, it consists of three main levels: essential beat patterns, metric hierarchies (meter in strict sense) and metric microstructure. The emphasis is on the importance of periodic beats and microstructure and of their combination to reach a complete description of meter. Finally the significance of interpretational elements in the perception of meter is discussed, stressing the need to start from performed music in order to obtain a correct image of the meter in a particular piece.

1 Introduction

In traditional music theory, meter is generally conceived as a regularly recurring pattern of accents determined by the meter indication in the score. This results in a hierarchic structure with two or three levels within the bar and a number of higher levels (hypermeasures) determined by structural elements. The number of higher levels is dependent on the theoretical viewpoint. Kramer (1988), for example, extends the metric structure to complete pieces or movements. He searches for metrically accented events, which are at the higher levels mainly determined by changes of harmony, tonality, timbre and texture. On the other hand, Lerdahl and Jackendoff restrict the number of higher levels to those where a regular alternation of strong and weak beats can be felt by the listener, these authors consider meter as a relatively local phenomenon (Lerdahl & Jackendoff, 1983, p.21).

However, the regular hierarchic description of meter is not sufficient to explain all the complexities that are related to metric structure in everyday musical practice. The strict application of the traditional theory is only possible in simplified examples. When looking at high-quality music, extensions of the strict model are necessary: irregularities at some level are described as *metric deletion* (Lerdahl & Jackendoff, 1983), *elision* (Benjamin, 1984) or *overlap* (Kramer, 1988), complex meters are forced into the model (London, 1995). And what with changing meters or syncopated music in which meter is clearly perceived but not generated in the traditional way? Examples of the latter seem to be carefully avoided in the present literature.

Outside the classical-romantic period, in particular in numerous pre-baroque and 20th century works, the construction of a hierarchic structure is impossible or at least difficult and artificially maintained. Yet, in most cases a metric notation is used and foot-tapping, counting and conducting are still possible. In these cases the question of the definition of meter becomes more critical. In my opinion, it doesn't make sense to restrict the concept of *meter* to a clear hierarchic tree-structure. Meter as it is used in musical practice, entails more essential elements like regular movement, counting and predictability as well as more complex features connected with the specific character of different meters. Some time-points are more important than others, this enables the construction of relations between similar, non-adjacent events, but this mechanism is not restricted to music in which a complex network of superimposed and linearly connected patterns is present. Taking hierarchy as the main concept for defining meter seems an impoverishment. A more complete view on the richness of musical meter should therefore start with a broader definition as *the temporal framework of music*. In this view, meter provides the framework necessary for every type of perceptual organization (Koffka, 1963). The definition retains the basic distinction with surface rhythm: the relations between adjacent elements, linear/horizontal, freely changing and structured by rhythmic grouping. It gives the possibility to construct a more complete view on the phenomenon of musical meter in which the traditional conception is one element of a more elaborated qualitative description and classification of time-points in a piece of music. It also enables us to apply the specifically Western concept of meter to non-Western musical styles. Not surprisingly, Kolinski, proposing a cross-cultural approach to Western musical concepts in the temporal domain came to a similar definition for meter: "organized pulsation functioning as a framework for rhythmic design" (Kolinski, 1973).

Defining meter as the complete temporal framework implies the extension of traditional meter in two directions: The first direction is concerned with the basic counting unit, the second with the subtle (but fairly constant) variations in timing, amplitude and timbre that the performer adds to every class of time-points to give each piece its specific metric character. This extension implies that a purely symbolic description of meter, whether it appears as a time-signature (being too vague from a scientific point of view) or as a multi-level accentual structure (being too rigid to satisfy musicians), is not satisfactory anymore. Instead the information provided by periodic (or nearly periodic) beat-trains and microstructural elements should be combined into a subsymbolic description from which the traditional metric structures (if present) can be derived. This type of description is located between the acoustical (iconic) and the traditional symbolic representation, it can be distinguished from the latter by the fact that it does not purely rely on the knowledge of the reader, but is justified by a causal connection with the acoustical representation and the principles of human perception (Leman, 1993).

2 The Beat-Level

Here we find the basic (non-hierarchic) regularities underlying the temporal structure of music: beats at equal time intervals. This regularity is not absolute. In expressive performance it is always a kind of average that groups time points around a regular *metronome*, moreover it can change its tempo, particularly in cadences or temporarily *freeze*, e.g. on notes held for an unspecified time, often indicated with a fermata sign. Regular beats appear at different intervals, but listeners and performers normally focus on one primary level that serves as a frame of reference (Parncutt, 1994). This level, often called tactus[1], typically falls within the range of *preferred tempo* (Fraisse, 1982; Handel, 1989) i.e. somewhere between 40 and 300 beats per minute with a preference for the tempo around 100 beats per minute. A lot of research has been done to determine the tactus following different strategies (for an overview see Desain & Honing, 1995). The existence of a preferred tempo implies that changes in tempo can have an influence on the perceived temporal structure. According to Povel (1984, pp. 330-331) tempo determines the rate of the temporal grid according to which subjects will tap their tactus. Parncutt (1994) found that a faster tempo leads to a higher rhythmic ambiguity, with a preference for levels at a moderate tempo. Consequently both differences in main tempo between performances and tempo-changes within a performance of a piece of music have their influence on the determination of the tactus and its position relative to other beat-levels. Still other changes can occur in the music: shifts of the primary beat or changes in the importance of less important beat-levels due to changes in (composed) texture or to the choice of accentuation in expressive performance. In this way music keeping the same formal meter can change substantially in its metric character. At the other side the presence of one regular beat, without any clear higher level metric structures, preserves the essentially metric features of predictability and counting and imposes a primitive sense of hierarchy with a distinction between time-points on and off the beat[2]. The preservation of these essential properties creates a sense of meter in music in which only one beat level is

---

[1] In some types of music, like Balkan folk music and the art-music based on it (e.g. several works of Bartók), the tactus is often not exactly equally spaced but appears in long and short versions in a 3:2 ratio. In my view this is the result of an interaction between tactus and microstructure in which the essentially regular tactus is modulated by a strong temporal deviation on certain beats. Not surprisingly in most cases the first beat of the notated bar will be of the *long* type, in accordance with the fact that long values tend to initiate bars (Lee, 1991) and that the first beat of a regular measure tends to be lengthened in performance (Clarke, 1985).

[2] Often the mechanism of subjective accentuation (Vos, 1973) creates a (weak but noticeable) distinction between alternating elements of the tactus, providing an additional hierarchic level for the listener, even if it is not directly present in the performance.

present: It allows e.g. to wave our hand on the regularity of the tactus in Gregorian chant and to dance to the continuous, undifferentiated drone of certain types of contemporary dance music.

We might ask if the presence of a regular beat is an essential characteristic of all music? One of its basic functions is the distinction between *predictable* music and *chaos*. Therefore it is indispensable for ensemble music: a common time measure is the basis for coordination between different voices in polyphony (Zuckerkandl, 1959) but even in monophonic ensemble music the musicians need to be kept together by a common pace, possibly guided by a conductor. Only solo-music seems to escape from the need for at least one beat level. But even in solo-music it is common practice to keep the track of a regular beat: as a rule in music written in traditional notation, and even in improvised music the contemporary western performer tends to follow the pace of an internal beat. For the listener it can be very difficult to recognize and to follow this regularity as it can be thoroughly obscured through the extensive use of changes in tempo, the use of fermatas or the irregular placement of the notes (syncopations, for/afterbeats, cross-rhythms,...). Apart from some contemporary pieces using specific techniques to delete the beat (e.g. extremely long inter-onset intervals, exceeding the perceptual present), only in some primitive societies solo-music that lacks any binding pattern has been reported (Sachs, 1953). Thus we might conclude that in our musical environment the presence of a beat is almost unavoidable, at least for the performance of music, whether it is automatically communicated to the listener is much more doubtful.

Some composers in our century have done considerable efforts to destroy the feeling of a beat, not only in solo but also in ensemble works. Different techniques have been developed to make the perception of any regularity impossible, examples of these techniques are: the combination of different tempi (e.g. Stockhausen: *Zeitmaße*), the superposition of numerous rhythmically independent voices (e.g. Ligeti: *Atmosphères*), superposition of different values of extremely long duration (e.g. Scelsi: *Quattro Pezzi*), or the use of notational systems that lack any reference to traditional meter (e.g. proportionate durational notation in Earl Brown's *Folio and Four Systems*),... (Sabbe, 1994). In most of these examples, however, the individual performers still need to count, to keep the track. Thus, apart from music without any coordinated temporal development, like *sound-scapes*, pieces with a temporal evolution based on cues, or LaMonte Young's notorious conceptual–minimal works, and some solo-pieces (cfr. supra), the realization of music without a perceptually comprehensible temporal framework seems to be restricted to non-human performers: music for automatons (e.g. Nancarrow: *Studies for Player-Piano*), electronic music and other manipulated tape-music, computer music and music in which environmental sounds are used (e.g. Ferrari: *Presque Rien no.1 – Lever du Jour au Bord de la Mer*).

The greater part of our musical environment consists of pieces in which not only the tactus-level is present: other beat-rates with differing saliencies are present and together they form the backbone of the overall metric structure. We find these secondary levels both at a faster rate (defining smaller subdivisions of the main beat they enable the prediction of the time-points of individual events and thus give meaning to the timing deviations in expressive performance) and at a slower rate (producing distinctions between the elements of the basic tactus and thus creating higher level hierarchies of which the regular meter should be the most salient). While the different levels in our musical culture usually work together to form a hierarchic tree structure, this is not a necessity. In the polyrhythmic music of the pygmies of Central Africa for example, beats are present on different levels without giving rise to a classical metric structure. Arom (1985) considers this music as non-metric, mainly because he starts from a very tight definition of meter requiring "equal durations with a regular accentuation". Nevertheless it has a multi-layered temporal framework with at least 3 strictly periodic levels (Arom, 1984): (1) the pulse, the temporal points of reference (often materialized by handclapping); (2) minimal operational values defining the smallest subdivision possible (cfr. time quanta Seifert, Olk, & Schneider, 1995); (3) larger repeating units (*periods*). A simple example that illustrates this system is given in Fig.1. Despite the hierarchic structure there is no obvious sense of meter as we are used to hear, this is mainly due to a lack of regular accentuation[3]. This example shows that a study of the various beat levels is an essential element in the temporal framework of the music. Even if they do not give rise to a meter *in strict sense* they allow a precise description of *metered* music.

3 Meter in Strict Sense

According to Lerdahl and Jackendoff (1983, p.8):

> "Metrical structure expresses the intuition that the events of the piece are related to a regular alternation of strong and weak beats at a number of hierarchical levels."

This is more or less the definition adopted in most theories of musical meter. The main distinction used is that of triple versus duple, at one or different levels. In fact this is already an extension of the common sense notion of meter in which the most salient of these levels, the one usually indicated in the time signature, is regarded as *the meter* of a piece. The introduction of different hierarchic levels allows a more elaborated classification of time-points. In addition to the distinction between on the beat or off the beat and between the strong and the weak beats of the measure, it allows a larger range of variation of metric strength for the different classes of time-points.

[3] It is important to stress that not all accents are metric accents, but that different accentual structures exist independently, for an overview of musical accent see (Parncutt, in press).

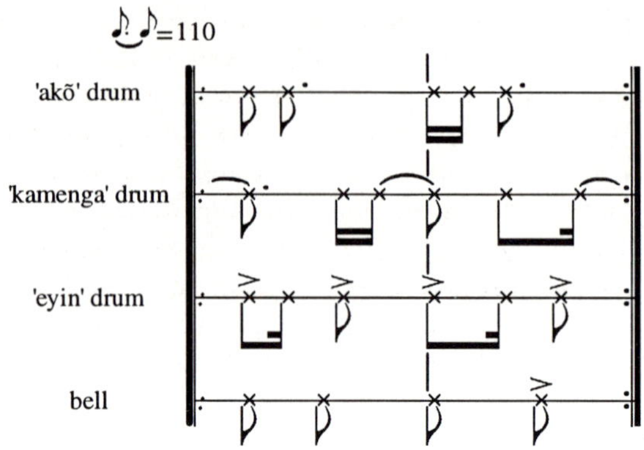

Fig. 1. An example of a rhythmic pattern from Central Africa, adapted from (Arom, 1984). The whole represents one period, the vertical line in the middle divides it in two equal parts each lasting about 0.545 sec, close to preferred tempo and thus serving as tactus or pulse. Within this pulse a 2 against 5 cross rhythm is played bringing the minimal operational value to a 32nd note at about 50 msec

However, as mentioned above, this aspect of meter can not deal with all the phenomena associated with the concept of meter. It gives only an idealized and simplified view on the metric structure of music. Although a careful analysis of the metric hierarchies certainly leads to a rich and interesting view on the musical structure of most of the music of our cultural environment there are too many problems to generalize it toward a general theory of human meter perception. In any case it is an essential part of a meter theory: it represents the main symbolic conceptualization of meter in our language and is hitherto the only method allowing straightforward metric classification. The analysis of meter *in strict sense* has been widely discussed, therefore I confine myself to a few critical remarks on the subject.

First, there is considerable disagreement about the basis for meter perception. While earlier research considered it as a phenomenon primarily derived from the sensory cues provided by the music (Cooper & Meyer, 1960), more recent theories emphasize the importance of internal mental schemata (Palmer & Krumhansl, 1990; Parncutt, 1993). An important argument to believe in the existence of mental schemata is the ease with which the main meter is recognized, even if the cues are ambiguous, interactive or very weak. Second, there is the question whether the perceptual cues for metric structure are already present in the *objective* composition or are primarily communicated through performance cues (cfr. infra). Third, there seems not to be any discussion about the generalization of the simple duple-triple opposition. Most investigations are directly guided towards the detection of binary and ternary structures. But we must be aware of the fact that other basic categorizations of meter are possible. This can be illustrated by theories from the 17th and 18th centuries (Houle, 1987) where other basic distinctions are made: simple versus compound meters, categories based on poetic feet or the distinction between triple (3 x 1) and "inégale" (2 + 1) promoted by Rameau. Although there is no perceptual evidence for these types of classification these alternatives certainly have musical importance, at least regarding the music from this period (if performed in an historically justified way). But as long as these possibilities are not incorporated in psychological studies, their perceptual importance cannot be determined. The broad, flexible framework proposed here could provide a theoretical basis for this kind of research.

4 Metric Microstructure

A description of the number of beats in a unit (on different levels) is not sufficient to describe the metric feeling of a piece. Effects of tempo, articulation and relative duration and amplitude of the beats (microstructure) can lead to a completely different metric feeling. This extension of meter toward microstructure is necessary to explain the difference in metric movement between individual dance movements and other pieces essentially notated in the same meter and tempo that have an obviously distinct metric character. An analysis restricted to the duality strong-weak on each level cannot deal with the metric subtleties of individual pieces and genres (Clynes & Walker, 1982). The application of meter in strict sense will often give rise to a stereotype, an accentual structure that only creates a caricature of the piece's meter.

Particularly in most popular music and in dance (or dance based) music from all ages, a traditional analysis will yield similar, very regular metric structures. In this kind of music the metric analysis is merely restricted to a distinction binary-ternary on one or two levels, the higher levels will almost always be binary (Lerdahl & Jackendoff, 1983, p.101). The central problem of such an analysis seems to be the impossibility to distinguish between different types of binary or ternary structure. For a musician the difference in interpretation between 4/4 and 2/2, 3/4 and 6/8 is obvious within the constraints of the style and related performance practice, while a traditional metric analysis will make no distinction between them. The same is true for the numbers in the denominator of the time signature: at first sight there is no reason to write pieces in 3/8 rather than in 3/4 or 3/2, and the metric analysis will be the same. Nevertheless it is obvious that apart from the quantitative value, each value has a qualitative significance, defining a specific metric character (Zuckerkandl, 1959).

Even if two dance movements have the same time signature, considerable differences in metric feeling remain. First, there is the influence of tempo causing changes of the tactus (cfr. supra) and thus of the perceived time signature. But second, there are also other differences: dance movements with the same time signature and the same (or a similar) tempo have their own specific metric character leading to a completely different interpretation both in musical performance and in dance. Sachs (1953) calculated a table of standard tempos of French baroque dances based on 18th century sources. Here we find 4 different binary dances in the same range of tempo: Allemande (M.M. 120) - Bourrée (M.M. 112–120) - Gavotte (M.M. 97–152) - Rigaudon (M.M. 116–152). Each of these four types of dance can be played at 120 b.p.m. and can have the same time signature, but nevertheless each of them has a distinct metric feeling, mostly expressed in terms of heavy/light, fluency or the upbeat-feeling. Other dance forms require a certain amount of contra-metricity. Although they are notated according to the classical metric practice, their essential character includes elements that contradict the standard meter. This is the case in *composed* dance forms like the saraband and the mazurka as well as in some traditional dance forms. An example of the latter is the Norwegian "gangar" (Blom & Kvifte, 1986), in which the variable meter (6/8 - 3/4) of the music not always coincides with the movements of the related popular dance. However the dance naturally follows the overall metric feeling, the internal movement of the music and is not regarded as particularly syncopated by the performers.

Research on metric microstructure has been done by M. Clynes. He and his colleagues measured the relationship between rhythmic sound patterns and the type of movements which they tend to induce (Clynes & Walker, 1982). They found different types of movement, related to (ethnic and popular) musical styles and to emotional qualities. Another element of Clynes' research is the elaboration of specific *composer's pulses* (Clynes, 1987): fixed patterns of duration and amplitude related to important classical composers. However, perceptual evaluations of the composer's pulses (Repp, 1989, 1990; Thompson, 1989) are not convincing. The main problem seems to be that listeners do not consequently prefer the appropriate *composer's pulse*. Despite this, the empirical research supports the fact that subtle, constant patterns of duration and amplitude have a musical meaning and add artistic value to a performance.

Microstructure appears to be an important issue in performance practice, and it is obviously related to the specific metric character of a particular dance, genre, composer or piece. It is an important factor in the vitality and originality of a composition or traditional tune as it gives its temporal framework an artistic meaning in addition to its structural importance. The main problem with this element of meter is that it is very difficult to investigate: it consists of an interplay of different elements (mainly timing, amplitude and articulation) that do not necessarily point in the same direction, moreover they are often far from obvious for the listener and thus require a detailed investigation at the level of the musical signals. These factors might explain why the study of metric microstructure has been confined to case-studies and the description of general effects (Clarke, 1985), without establishing a detailed classification[4].

5 The Need for the Use of Performed Music in the Study of Meter

An important element in the construction of a model of meter is the impossibility to deduce all this information from the score and thus the need to base the analysis on interpreted music. The score does not contain all the information about time structure (Bengtsson, 1987), the knowledge of the performer is absolutely necessary to obtain an musically acceptable result. Every rhythmic passage is essentially ambivalent: different metric interpretations are always possible, although most of them will sound quite peculiar. In some situations however a composer might have the intention to create a peculiar effect. Therefore it does not seem useful to base a meter-theory on the notated values. Elements purely derived from the notes in the score do not contain reliable information for a universal model, although models based on such elements may generate good results. One of the main problems in this respect is the fact that performers do not base the main part of their metric interpretation on *note-related* elements. The main cues for the choice of a specific metric interpretation are time signatures, barlines and accent marks, these are the means by which the composer informs the performer of the desired metric interpretation (Jones, 1987). Of course these indications will most of the time coincide with the other elements, but this is not a rule, just a question of tradition and taste. The performer uses different expressive means to communicate the meter to the listener, among these intensity accents and the lengthening of the first note of the metric unit (Clarke, 1985) seem to be the most important. From a perceptual view small deviations in the timing of the notes result in a different rhythmic structure that contributes to the salience of particular beat-patterns (Leman, 1995) and thus influences the metric structure.

The expressive possibilities of the performer are very extensive and often very subtle. The complex of interpretational possibilities enable him to impose an *unnatural* interpretation. It is possible, for example, to play an entire sequence syncopated. Syncopation in performance is mainly a way of articulating, immediately grasped by the listener. This particular way of articulating (the clearest form might be the *swelling* of tones towards the metric point of gravity as used by string and wind-players), is used whether the regular beat *crossed* by the syncopation is present or not. Thus the first

[4] The role of timing microstructure has been approached more extensively following the horizontal rhythmic development: focusing on deviations related grouping (e.g. Todd, 1985) or on the relations between different musical dimensions (melody, harmony, phrase,...) and the temporal structure (e.g. Sundberg, Friberg, & Frydén, 1991).

rhythmic figure in a piece can be syncopated, as it is the case in Schumann's *Manfred-Overture* (see Fig.2). This as opposed to theories that first require the construction of a strong meter before syncopations can be perceived, e.g. (Longuet-Higgins & Lee, 1984; Sloboda, 1985). It also explains why writing a

Fig. 2. The first bar of Schumann's *Manfred-Overture* (only the instruments that play in the fragment are notated). The syncopated notation immediately creates tension, without it this motive would be turned into something quite trivial

long sequence of syncopated events (as in Beethoven's *Piano Sonata op.101*, see Fig.3) is not equal to the same passage with a change of time signature. This idea does not contradict the principle of mental schemata for meter perception. The specific way of articulating invokes a metric scheme that does not coincide with the note onsets, this underlines the strength of the scheme. Although the cues are highly ambiguous the appropriate scheme is inferred without any effort.

Fig. 3. Measures 27–31 from Beethoven's *Piano Sonata op.101*, showing the transition from the regular 6/8 movement to the long syncopated passage. Due to the length of the syncopation it is unlikely that it would be fully perceived as syncopated in a mechanical performance. The articulation of the pianist has to support the initial syncopated feeling, to stretch it and thus to avoid the perception of an unsyncopated, rebarred version by the listener

6 Conclusion

In this paper I tried to show that meter in music must be approached very carefully. If we want to come to a complete description of meter it is necessary to base the analysis on performed music and to break the boundaries of a too narrow definition of meter. A symbolic description seems not rich enough to deal with metric diversity. The proposed three-level method can lead to a description of meter in all its aspects and in all its possible appearances. From essential foot-tapping to the emotional qualities of specific meters and from rudimentary temporal frameworks to highly organized structures. This paper provides the theoretical basis for the construction of a subsymbolic description of meter consisting of the different regular beats and their strength (with the tactus as the most salient one) and of microstructural elements interacting with these beat levels. By investigating these elements a symbolic classification in terms of traditional metric structures could be derived, however also other types of symbolic structures could emerge. This objective approach, not from the beginning bounded by the presumptions of traditional music theory allows us to approach the details of human perception and performance.

Acknowledgements

This research was supported by a grant (No. 011D3295) from the special research fund of the research council of the University of Ghent. I am grateful to M. Leman and H. Sabbe for useful comments on earlier versions of this paper.

References

Arom, S. (1984). Structuration du temps dans les musiques d'Afrique centrale: Périodicité, mètre, rythmique et polyrythmie. *Revue de Musicologie, 70*, 5–36.

Arom, S. (1985). *Polyphonies et polyrythmies intrumentales d'afrique centrale (structure et méthodologie).* Paris: Société d'études linguistiques et anthropologiques de France.

Bengtsson, I. (1987). Notation, motion and perception: Some aspects of musical rhythm. In A. Gabrielsson (Ed.), *Action and perception in rhythm and music.* Stockholm: Royal Swedish Academy of Music.

Benjamin, W. (1984). A theory of musical meter. *Music Perception, 1*, 355–413.

Blom, J., & Kvifte, T. (1986). On the problem of inferential ambivalence in musical meter. *Ethnomusicology, 30*, 491–517.

Clarke, E. (1985). Structure and expression in rhythmic performance. In P. Howell, I. Cross, & R. West (Eds.), *Musical structure and cognition.* London: Academic Press.

Clynes, M. (1987). What can a musician learn about music performance from newly discovered microstructure principles (PM and PAS)? In A. Gabrielsson (Ed.), *Action and perception in rhythm and music* (Vol. 39). Stockholm: Royal Swedish Academy of Music.

Clynes, M., & Walker, J. (1982). Neurobiologic functions of rhythm, time, and pulse in music. In M. Clynes (Ed.), *Music, mind and brain: The neuropsychology of music.* New York, NY: Plenum Press.

Cooper, G., & Meyer, L. (1960). *The rhythmic structure of music.* Chicago, IL: University of Chicago Press.

Desain, P., & Honing, H. (1995). Computationeel modelleren van beatinductie. In *Van frictie tot wetenschap — jaarboek 1994–1995.* Amsterdam: Vereniging van Academie-onderzoekers.

Fraisse, P. (1982). Rhythm and tempo. In D. Deutsch (Ed.), *The psycology of music.* New York, NY: Academic Press.

Handel, S. (1989). *Listening. An introduction to the perception of auditory events.* Cambridge, MA: The MIT Press.

Houle, G. (1987). *Meter in music, 1600–1800 — performance, perception, and notation.* Bloomington, Indianapolis: Indiana University Press.

Jones, M. (1987). Perspectives on musical time. In A. Gabrielsson (Ed.), *Action and perception in rhythm and music* (Vol. 39). Stockholm: Royal Swedish Academy of Music.

Koffka, K. (1963). *Principles of Gestalt psychology.* New York, NY: Harcourt, Brace, and World.

Kolinski, M. (1973). A cross-cultural approach to metro-rhythmic patterns. *Ethnomusicology, XVII*, 494–506.

Kramer, J. (1988). *The time of music.* New York, NY: Schirmer Books.

Lee, C. (1991). The perception of metrical structure: Experimental evidence and a model. In P. Howell, I. Cross, & R. West (Eds.), *Representing musical structure.* London: Academic Press.

Leman, M. (1993). Symbolic and subsymbolic description of music. In G. Haus (Ed.), *Music processing.* Madison, WI: A-R Editions.

Leman, M. (1995). *Music and schema theory: Cognitive foundations of systematic musicology.* Berlin, Heidelberg: Springer-Verlag.

Lerdahl, F., & Jackendoff, R. (1983). *A generative theory of tonal music.* Cambridge, MA: The MIT Press.

London, J. (1995). Some examples of complex meters and their implications for models of metric perception. *Music Perception, 13*, 59–77.

Longuet-Higgins, H., & Lee, C. (1984). The rhythmic interpretation of monophonic music. *Music Perception, 1*, 424–441.

Palmer, C., & Krumhansl, C. (1990). Mental representations for musical meter. *Journal of Experimental Psychology: Human Perception and Performance, 16*, 728–741.

Parncutt, R. (1993). Template-matching models of musical pitch and rhythm percpetion. In J. Louhivuori & J. Laaksamo (Eds.), *Proceedings of the First International Conference on Cognitive Musicology* (pp. 1–19). University of Jyväskylä.

Parncutt, R. (1994). A perceptual model of pulse salience and metrical accent in musical rhythms. *Music Perception, 11*, 409–464.

Parncutt, R. (in press). Expressive communication of musical structure in piano performance. In W. Auhagen, B. Gätjen, & K. Niemöller (Eds.), *Systemische Musikwissenschaft — Festschrift Jobst Peter Fricke zum 65. Geburtstag.* Cologne: Musikwissenschaftliches Institut der Universität zu Köln.

Povel, D. (1984). A theoretical framework for rhythm perception. *Psychological Research, 45*, 315–337.

Repp, B. (1989). Expressive microstructure in music: A preliminary perceptual assessment of four composers' "pulses". *Music Perception, 6*, 243–274.

Repp, B. (1990). Composers' pulses: Science or art? *Music Perception, 7*, 423–434.

Sabbe, H. (1994). *Algemene muziekleer voor de 20ste eeuw.* Ghent: Academia Press.

Sachs, C. (1953). *Rhythm and tempo: A study in music history.* New York, NY: W.W. Norton.

Seifert, U., Olk, F., & Schneider, A. (1995). On rhythm perception: Theoretical issues, empirical findings. *Journal of New Music Research, 24*, 164–195.

Sloboda, J. (1985). *The musical mind: The cognitive psychology of music.* Oxford: Clarendon Press.

Sundberg, J., Friberg, A., & Frydén, L. (1991). Common secrets of musicians and listeners: An analysis-by-synthesis study of musical performance. In P. Howell, I. Cross, & R. West (Eds.), *Musical structure and cognition* (pp. 161–197). London: Academic Press.

Thompson, W. (1989). Composer-specific aspects of musical performance: An evaluation of Clynes's theory of pulse for performances of Mozart and Beethoven. *Music Perception*, *7*, 15–42.

Todd, N. (1985). A model of expressive timing in tonal music. *Music Perception*, *3*, 33–58.

Vos, P. (1973). *Waarneming van metrische toonreeksen.* Nijmegen: Stichting Studentenpers.

Zuckerkandl, V. (1959). *The sense of music.* Princeton N.J.: Princeton University Press.

Musical Rhythm: A Formal Model for Determining Local Boundaries, Accents and Metre in a Melodic Surface

Emilios Cambouropoulos

Faculty of Music and Department of Artificial Intelligence, University of Edinburgh, 12 Nicolson square, EH8 9DF Edinburgh, U.K.

Abstract. In this paper a general theory will be introduced that allows the description of a melodic surface – i.e. a score-like representation of a melody – in terms of local grouping, accentuation and metrical structures. Firstly, a formal model will be proposed that detects points of maximum local change that allow a listener to identify local perceptual boundaries in a melodic surface. *The Local Boundary Detection Model (LBDM)* is based on rules that are relating to the Gestalt principles of proximity and similarity. Then, it will be shown that the accentuation structure of a melody may automatically be inferred from the local boundary grouping structure. This is based on the assumption that the phenomenal accents of two contiguous musical events are closely related to the degree by which a local boundary is likely to be perceived between them. Finally, the metrical structure is revealed by matching a hierarchical metrical template onto the accentuation structure. It is suggested that the *Local Boundary Detection Model* presents a more effective method for low-level segmentation in relation to other existing models and it may be incorporated as a supplementary module to more general grouping structure theories. The rhythmic analyses obtained by the methods described herein are tentative, and complementary to higher-level organizational theories.

1 Introduction

Many contemporary theories of rhythm (Cooper & Meyer, 1960; Epstein, 1995; Kramer, 1988; Lerdahl & Jackendoff, 1983; Yeston, 1976) consider rhythm to be the organization/structuring of musical sounds into groups (grouping structure) of more or less salient elements (accentuation structure) that are in constant interplay/interaction with a hierarchy of beats (metrical structure). Metre receives somewhat different treatment in each of these theories and is to a varying extent integrated into the ways rhythm is defined (Moelants, 1997, this book).

For instance, Lerdahl and Jackendoff's definition of rhythm is based on two kinds of structures: namely *grouping structure* that "expresses a hierarchical segmentation of a piece into motives, phrases and sections" and *metrical structure* that "expresses the intuition that the events of a piece are related to a regular alternation of strong and weak beats at a number of hierarchical levels" (Lerdahl & Jackendoff, 1983, p.8). They define three kinds of musical accents: *phenomenal accents* which are due to local intensification such as dynamic stress, high or low register, long notes, harmonic changes and so on, *structural accents* which result from higher-level structural relations such as cadences, and *metrical accents* that correspond to relatively strong beats in a metrical context. Defining a metrical structure is finding a well-formed grid of metrical accents that fits best onto the structure of phenomenal accents: "... the listener's cognitive task is to match the given pattern of phenomenal accentuation as closely as possible to a permissible pattern of metrical accentuation. ... Metrical accent, then, is a mental construct, inferred from but not identical to the patterns of accentuation at the musical surface." (Lerdahl & Jackendoff, 1983, p.18). In their theory, grouping structure is considered to be independent of metrical structure and hence different preference rules are formulated for each: one set of preference rules for the description of groupings and a different independent set for the description of the phenomenal and structural accentuation structure from which metrical structure is inferred (Fig.1a).

The concept that rhythm relates to cognitive grouping of musical events is a Gestalt-based one. The Gestalt principles of perceptual organization are a set of rules-of-thumb that suggest preferential ways of grouping mainly visual events into larger scale schemata. Two of the Gestalt principles state that objects closer together (Proximity principle) or more similar to each other (Similarity principle) tend to be perceived as groups. These principles have been used as a basis for some contemporary theories of musical rhythm. Tenney (1964) discusses the use of the principles of proximity and similarity as a means of providing cohesion and segregation in twentieth century music and, later, Tenney and Polansky (1980) develop a computational system that discovers grouping boundaries in a melodic surface. Musical psychologists (Bregman, 1990; Deutsch, 1982a, 1982b; McAdams, 1984) have experimented and suggested as to how the Gestalt rules may be applied into auditory/musical perception and Deutsch and Feroe (1981) further incorporate such rules in a formal model for representing tonal pitch sequences. The grouping component of Lerdahl and Jackendoff's *Generative Theory of Tonal Music* (Lerdahl & Jackendoff, 1983) is based on the Gestalt theory and an explicit set of rules is thereby described – especially for the low-level grouping boundaries. The formulation of these rules has been supported by the experimental work of I. Deliége (1987).

In the first part of this paper a systematic theory will be described that attempts to define local boundaries in a given melodic surface (see Cambouropoulos, 1996b, for a general representation of pitch intervals). The proposed segmentation model *(Local Boundary Detection Model – LBDM)* will be based on two rules: the *Identity-Change rule* (which is more elementary than the Gestalt principles of proximity and similarity) and the *Proximity rule* (which relates to the Gestalt proximity and similarity principles). The aim has been to develop a formal theory that may suggest all the possible points for local grouping boundaries on a musical surface with various degrees of prominence attached to them rather than a theory that suggests

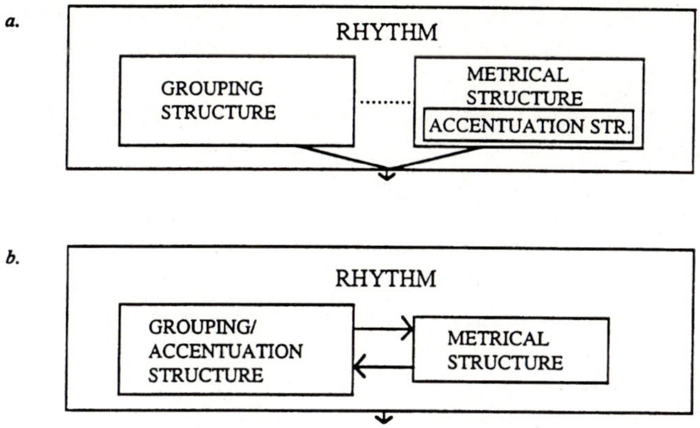

Fig. 1. (a) Lerdahl and Jackendoff's theory of musical rhythm. (b) Proposed model of musical rhythm

some prominent boundaries based on a restricted set of heuristic rules. The discovered boundaries are only seen as potential boundaries as one has to bear in mind that musically interesting groups can be defined only in conjunction with higher-level grouping analysis (parallelism, symmetry, etc.). Low-level grouping boundaries may be coupled with higher-level theories so as to produce *optimum* segmentations.

It will be shown that the formulation of the boundary discovery procedures defined by Lerdahl and Jackendoff (1983) and Tenney and Polansky (1980) have limitations and can be subsumed by the proposed theory. Some examples and counter-examples will be given mainly in relation to the influential formulation of the local detail grouping preference rules by Lerdahl and Jackendoff (1983).

In the second part of the paper it will be maintained that low-level grouping structure and phenomenal accentuation structure are strongly associated; they are actually in a *one-to-one* relation, i.e. if one is defined then the other may automatically be inferred. In other words, if local boundaries for a given melodic surface have been defined then strengths for phenomenal accents may be inferred (the reverse is also possible although not examined in this paper). It is assumed that the phenomenal accents of two contiguous musical events are closely related to the degree by which a local boundary is likely to be perceived between them. A method then is described that mechanically derives accent strengths from the local boundary strengths detected by the *Local Boundary Detection Model.*

The strong link between grouping and accentuation structures is important in that it allows one to develop a model that does not need two separate independent methods for the detection of the local boundaries and the phenomenal accents respectively. In contrast with Lerdahl and Jackendoff's model (Fig.1a) the proposed model directly links phenomenal accentuation structure with grouping structure (Fig.1b). This enables a more economic and efficient formulation of a theory for rhythm.

Once the phenomenal accentuation structure has been defined an attempt can be made to match a well-formed metrical structure to it; this may be possible for a number of hierarchic metric levels of beats or only for one level or possibly for no level at all depending on the kind of music. Metrical structure may be inferred from the accentuation structure but, at the same time, it influences the perception of the accentuation/grouping structure. The interplay between these two kinds of structures is addressed further towards the end of this paper.

In the following sections, formal methods will be described, firstly, for the discovery of local boundaries (low-level grouping structure) in a melodic surface, secondly, for the derivation of the phenomenal accentuation structure from the grouping structure and, lastly, for the selection of a metrical structure that fits best onto the accentuation structure.

2 The Gestalt Principles of Proximity and Similarity in Theories of Musical Rhythm

Some problems in the way the low-level Gestalt principles of perceptual organization have been applied in the organization of temporal musical sequences are briefly discussed below.

The Gestalt principles of proximity and similarity have been applied in both Tenney and Polansky's and Lerdahl and Jackendoff's models in such a way that allow one to interpret them as being different descriptions of the same phenomenon, namely a local maximum in the distance between consecutive musical events for any musical parameter e.g. pitch, start-times, dynamics and so on. Tenney and Polansky state explicitly that the similarity principle – as they define it – actually includes the proximity principle as a special case: "In both, it is the occurrence of a *local maximum in interval magnitudes* which determines clang-initiation" (Tenney & Polansky, 1980, p. 211). Lerdahl and Jackendoff's grouping rules (Lerdahl & Jackendoff, 1983) are defined in such a way that it seems rather plausible that the proximity rules can be subsumed by the change (similarity) rules and the reverse. For example, GPR3a (register rule) states that a greater pitch interval in between smaller neighboring intervals initiates a grouping boundary. This can been seen in two ways: a) that the pitches of the first and last intervals are more *similar* to each other than the pitches of the middle interval or b) that there is a greater *proximity* between the first two pitches – and the last two – rather than between the middle pitches (Handel, 1989, p.198).

In the current paper it will be maintained that although this formalization of the Gestalt principles provides the most important factor for discovering local boundaries a more general approach should account for any change in interval magnitudes. For example, in the following sequence of durations: ♩ ♩ ♩ ♪ ♩ ♩ ♩ a listener easily hears a possible point of segmentation for which neither the Tenney and Polansky nor the Lerdahl and Jackendoff formalisms suggest any boundary. For this reason a different, more elementary rule will be introduced based on the principle of Identity-Change. This issue will be discussed further in the next section and it will be shown that the above example can naturally be accommodated within the proposed model.

The low-level Gestalt principles of proximity and similarity are usually applied on symmetrical non-directional spaces. On applying them to musical temporal spaces, one has to make certain concessions by removing all possible asymmetrical directional properties (e.g. direction of pitch-intervals). There is though one aspect of musical asymmetry that cannot be avoided. This relates to the fact that musical objects are asymmetric objects themselves – even the most simplified homogeneous description of a note distinguishes between its attack and the rest of its body. This asymmetry is reflected in that, for instance, the temporal grouping rules can never give an identical grouping structure to the original and the retrograde form of a melody. It relates to the way that rules of perceptual organization give different grouping boundaries for musical duration sequences and for start-time interval sequences. It will be shown below how the interaction between these duration and start-time interval groupings results in the asymmetric perceptual organization of a sequence of musical events.

We will now attempt to define the Identity-Change rule and the Proximity rule which will form the basis of the *Local Boundary Detection Model.* These rules will be discussed initially for any sequence of two or three objects and then will be applied to longer sequences of musical objects.

3 The *Local Boundary Detection Model*

A formal model that attempts to determine local boundaries in a given melodic surface will now be presented. (For or a detailed description, see Cambouropoulos, 1996a).

3.1 The Identity-Change and Proximity Rules

As we have seen above, the Gestalt principles of proximity and similarity can be interpreted as being different sides of the same coin. In the *Local Boundary Detection Model (LBDM)* an elementary rule will be introduced based on the principle of identity. The Identity-Change rule is more elementary as it can be applied to a minimum of two entities (i.e. two entities can be judged to be identical or not) whereas the Proximity/Similarity rule requires at least three entities (i.e. two entities are closer or more similar than two other entities). This Identity-Change rule, in conjunction with the Proximity rule, forms the basis of the proposed low-level segmentation model.

General Identity-Change Rule: Grouping boundaries may be introduced only between two different entities. Identical entities do not suggest any boundaries between them. This rule is supported by an experiment realized by Garner (1974) wherein an eight-element pattern composed of two different pitch elements, for example XXXOXOOO, is looped indefinitely and listeners are asked to describe the pattern they perceive. Various preferential ways of organization were recorded (there are eight possibilities starting on each element of the sequence) but hardly ever did any listener break a run of same elements.

If the entities compared are intervals (intervals for pitch, start-times, dynamics, etc.) then this rule can be formulated more specifically:

Identity-Change Rule (ICR): Amongst three successive objects boundaries may be introduced on either of the consecutive intervals formed by the objects if these intervals are different. If both intervals are identical no boundary is suggested. When the application of ICR on two consecutive intervals detects a change and suggests a local boundary, this boundary is ambiguous (i.e. the boundary can be placed on either side of the middle object) and each interval receives the same boundary strength value. The second rule (PR) resolves the ambiguity by giving preference to the larger of the two intervals.

Proximity Rule (PR): Amongst three successive objects that form different intervals between them a boundary may be introduced on the larger interval i.e. those two objects will tend to form a group that are closer together (or more similar to each other).

3.2 Applying the ICR and PR Rules on Three Note Sequences

We will assume that for each parametric feature of a musical surface we can construct a sequence of intervals on which the ICR and PR rules may be applied. We will start by presenting the application of the rules to the following parameters: pitch, dynamics, rests and articulation (slurs, staccatti, breath-marks etc. are considered to be expressional rests and are inserted between the notes they mark as normal rests that have a value that is a fraction of the preceding note). The grouping boundaries resulting from the sequence of start-time intervals and durations will be presented at the end of this section.

The relation between two intervals can be of two types: *identity* or *change*. For reasons of asymmetry that will be introduced later on we will depict the change relation in two directional forms: "+" and "-" (Fig.2b and 2c). In the following Figures, dots represent parametric values of musical events and the distances between the dots the interval sizes between these values (Δx, Δy are interval values and are placed at the left-hand side of the interval). In Fig.2a $\Delta x = \Delta y$ and the identity relation is represented by a zero. In Fig.2b

$\Delta x > \Delta y$ and in Fig.2c $\Delta x < \Delta y$, and the change relations are represented by the "+" and "-" signs respectively.

At this stage we will introduce numeric values for the strength of the ICR and PR rules (more research is necessary for the selection of the most appropriate values). A numeric value is given to each interval as indicated below:

ICR: 0 for the identity relation (0 for each interval)
2 for the change relation (1 for each interval)
PR: 0 for the identity relation (0 for each interval)
1 for the change relation (1 for the larger interval)

We get thus the total interval boundary strengths as depicted in Fig.2 (bottom line).

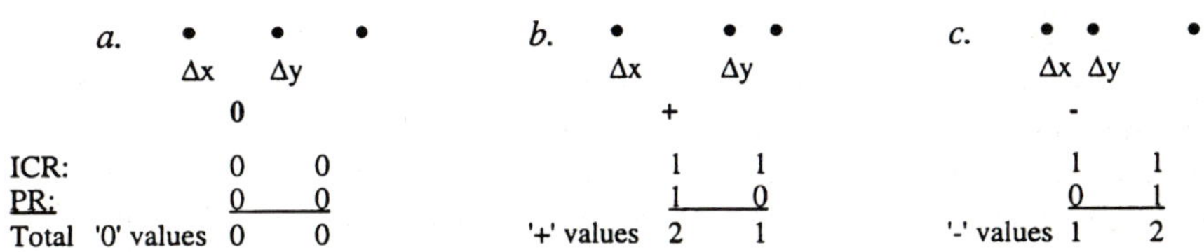

Fig. 2. Boundary strengths (last row) calculated by the use of the ICR and PR rules for three parametric values (e.g. pitch, dynamics etc.) separated by two intervals

We can now examine the duration and start-time interval sequences. The duration of a musical event is an internal attribute of that event whereas start-time intervals are temporal distances between two different successive events. We have thus the application of the ICR and PR rules for the start-time intervals exactly as described above and, additionally, the application of the General ICR for the sequence of durations (numeric strength 2). We now have the following kinds of relations (Fig.3 for two start-time intervals delimited by 3 start-time points (dots) and the two corresponding durations (rectangles) Fig.3. It is now clear that the "+" and "-" change relations are not symmetric. It is not possible to apply the principles of perceptual organisation in the musical temporal domain without introducing local asymmetry.

3.3 Applying the ICR and PR Rules on Longer Melodic Surfaces

For a given parametric interval profile of a musical surface one finds all the kinds of interval relations (0, +, -) that exist between every two successive intervals. If there are 3 or more consecutive "+" or "-" relations (e.g. +++, - - - - -), then only the ones at the ends are considered – the others do not contribute to the numeric strengths. Then, the numeric strengths for each kind of relation are calculated and added for each interval. For a single

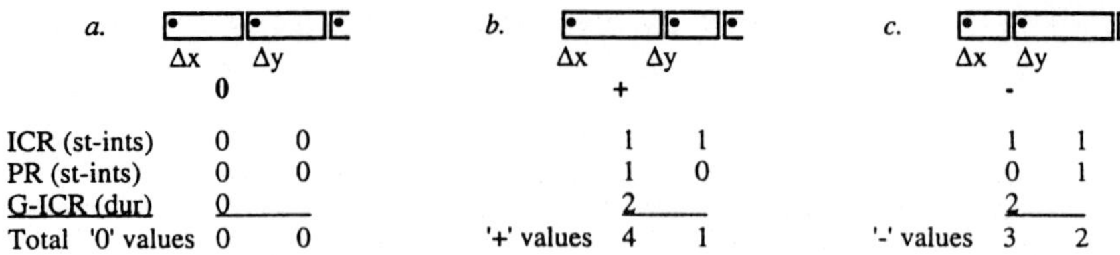

Fig. 3. Boundary strengths (last row) calculated by the use of the ICR and PR rules for three start-time values separated by two start-time intervals and durations

numeric strength sequence the local maxima suggest the most preferable local boundaries (when a local maximum consists of more than one same or almost the same values then an ambiguous boundary is suggested).

In Fig.4 we give a first example of how one can use the ICR and PR rules to calculate the strengths of grouping boundaries for <u>- +</u> sequences. As it happens, almost all of the grouping preference rules of Lerdahl and Jackendoff (1983) (exception: GPR3d (equal note length) and the articulation changes form legato to staccato and the opposite, fall under the <u>*0 + 0*</u> and <u>*0 - 0*</u> combinations), and all the grouping rules suggested by Tenney and Polansky (1980) fall under the <u>- +</u> category of sequences (see Cambouropoulos, 1996a, for the application of the *LBDM* rules to the local detail examples of Lerdahl and Jackendoff's grouping theory). The formulation of the boundary discovery procedures defined by Tenney and Polansky and Lerdahl and Jackendoff are specific instances of the proposed theory. The boundaries in

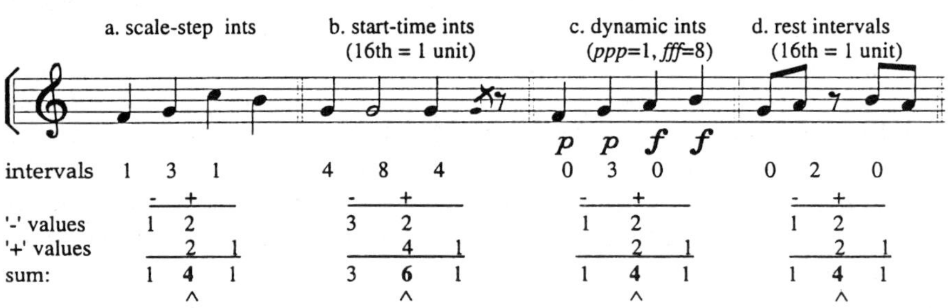

Fig. 4. Examples of boundary strengths (last row) determined by the *LBDM*

the examples of Fig.4 are detected by Tenney and Polansky's and Lerdahl and Jackendoff's methods whereas their models do not suggest any boundaries for the examples in Fig.5. In contrast, the *LBDM* suggests ambiguous boundaries for all the examples of Fig.5 (such ambiguous boundaries may be resolved if higher-level grouping organizational principles are taken into account). The above procedure is realized for every parametric interval pro-

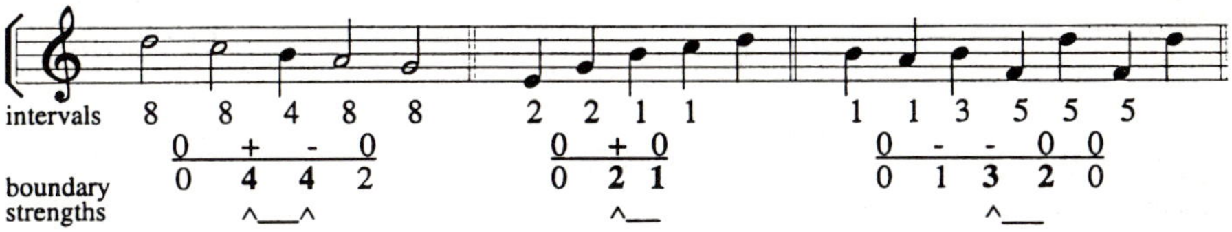

Fig. 5. Examples of boundary strengths (last row) determined by the *LBDM*. These are ambiguous boundaries which may be resolved if higher-level organizational principles are taken into account

file of interest. Then the total sum of all the numeric strength sequences is calculated (weighted or not). The local peaks are the points in a melodic sequence at which boundaries may preferably appear. In Fig.6 the preferred grouping structure is presented for Mozart's opening of the *Symphony in G minor*. The boundary strengths for each parametric interval profile are calculated and then added to produce the total boundary strength sequence A. Sequence B is given by a refined version of *LBDM* which takes in account the degree of difference between two intervals and other factors discussed in Cambouropoulos (1996b). *LBDM* has been successfully applied to many kinds of melodic surfaces – from traditional tonal melodies to contemporary atonal surfaces; see Figs.8-9 and also Cambouropoulos (1996a). This method can be further enriched if, for example, harmonic chord distance or scale-degree tonal distance profiles of the melodic surfaces are incorporated.

4 Phenomenal Accentuation Structure

In this paper it is maintained that local grouping and phenomenal accentuation structures are not independent components of a theory of musical rhythm but that they are in a *one-to-one* relation, i.e. accentuation structure can be derived from the grouping structure and the reverse. If, for instance, one develops an elaborate model of local grouping structure (such as *LBDM*) then, from this, the accentuation structure can automatically be inferred. This hypothesis is fundamentally different from much common practice whereby one set of rules are given for the detection of grouping boundaries and a different set for the determination of accents of musical notes.

The above hypothesis is based on the observation that group boundaries are closely related to the accented/salient events between which they occur. A perceived boundary in a given continuum indicates that the elements that delimit it are more prominent than other events further away. Epstein states:

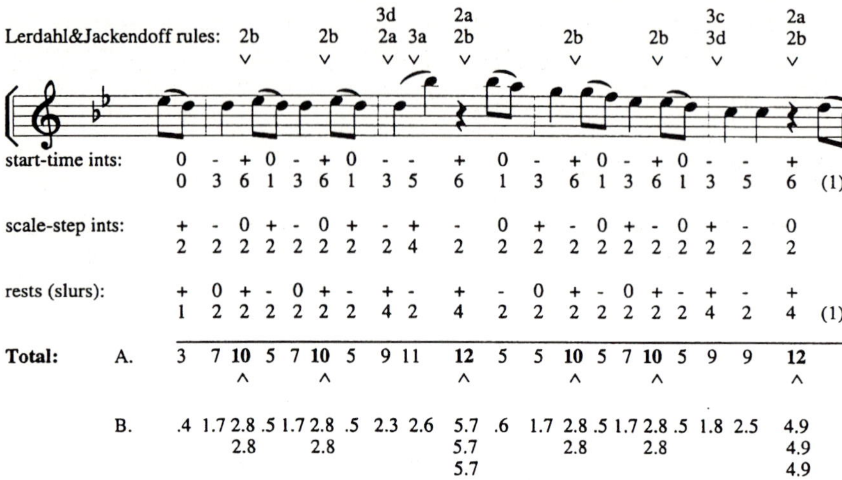

Fig. 6. Low-level grouping structure for the theme of Mozart's *Symphony in G minor*. The boundary strengths sequence A is determined by the *LBDM* whereas sequence B is determined by the refined version of *LBDM* described in Cambouropoulos (1996a)

"Demarcation in effect means emphasis – the emphasis required at that moment when a border of some time segment is to be delineated" (Epstein, 1995, p.24).

In Fig.7 the local boundary strengths are given according to the *Local Boundary Detection Model*. It is hypothesized that *if the boundary strength values are added for every two successive intervals the local accentuation structure of the surface is revealed*. The local maxima in this sequence of accent strengths indicate the elements in the surface that are perceived as being more prominent. In particular, the events delimited by two approximately equal local boundary values (e.g. Fig.7d) are considered to be most salient, i.e. an element that is preceded and followed by a significant boundary indication (ambiguous boundary) tends to be unambiguously highlighted into perception. For the cases where the two events delimiting a boundary receive equal (or almost equal) accent strength values (Fig.7c) there is a general tendency to consider the element that initiates a group as more intense although there are cases where this isn't true (Handel, 1989, Chap.11). As the proposed formal model is considered merely to be complementary to other organizational principles (e.g. metre, parallelism, symmetry, learned structural schemata etc.) these ambiguities are left unresolved at this low level. For example, a given metrical context for the melodic excerpt of Fig.7c may assist in resolving the ambiguity by adding metrical accent to one or the other of the two accented notes. The accentuation structure has been calculated

Fig. 7. Examples of phenomenal accent strengths derived from the LBDM boundary strengths by merely adding every two adjacent boundary strength values

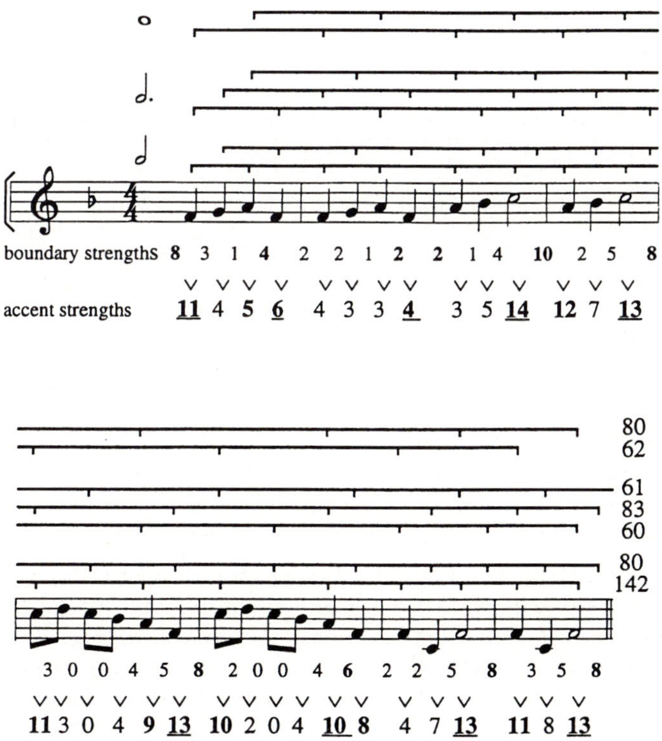

Fig. 8. Accentuation and metric structure for the song *Frère Jacques*

Fig. 9. Accentuation and metric structure for the beginning of J.S.Bach's *Concert for Harpsichord in D minor*

for a variety of melodic surfaces and has produced very reliable results. In Figs.8-9 the accentuation structure is presented for two melodic examples. The local maxima – and the relatively large numeric strengths – indicate the most accented events. Note that most of the strong accents correspond to events that a listener may perceive as most prominent and that the ones that may be considered counter-intuitive (e.g. accent on the 4th quarter-note of *Frère Jacques*) are due to the fact that metrical accents and higher-level principles of organization have not been taken into account (especially for *Frère Jacques*, parallelism/repetition plays a paramount role in the determination of grouping structure).

In the next section it will be shown that the rudimentary phenomenal accentuation structures revealed with the help of the simple mechanism described above may be sufficient for the derivation of the melodic metrical structures – whenever such metrical structures do exist. This further supports the validity of the proposed method for determining accentuation structures.

5 Metrical Structure

Musical time is structured around a cognitive framework of well-formed hierarchically ordered time-points (at least for metric/tonal music). Metrical structure is an abstract system of reference that facilitates the ordering/structuring of sequentially emitted/received musical events (Clarke, 1987).

A metrical structure consists of a number of levels of steady patterns of beats (the beat level at which listeners might tap their foot or clap their hands will be referred to as the tactus). The simplest and most *natural* tactus is when beats are separated by equal time-span units and are delivered at a rate in the neighborhood of 1.7 beats/sec (not much slower than 1 beats/sec, not much faster than 4 beats/sec) (Handel, 1989). It is possible though to have a tactus where beats are separated by non-regular time-span units as in much of the traditional music of the Balkans (e.g. dance songs in 7/8 metre are usually danced/clapped at 1.5:1:1 beat time-span ratios). Time-spans between beats may be further divided into smaller units down to the elementary unit or *fastest pulse* (Seifert, Olk, & Schneider, 1995). Above the tactus beats may be organized into larger measures (usually in regular binary/ternary patterns) and, often, into even larger hypermeasures. In Fig.10 some well-formed metrical structures are presented. It should be noted, though, that some music doesn't have metric structure at all (e.g. much contemporary music) or only a tactus without higher-level metrical hierarchies, e.g. much of African music – see Arom (1991).

A metrical hierarchic grid may be matched onto the accentuation structure of a musical piece (more on template-matching models in Parncutt, 1994). It is asserted that if the grouping/accentuation structure of a piece has been defined then the most appropriate metrical structure may be induced. But, conversely, the metrical structure – once a listener has made a selection – strongly influences and resolves ambiguity in the grouping/accentuation structure. Metrical accents are added onto the accentuation strengths and thus regulate the grouping structure of a piece. Metre is not simply a mental artefact induced from the music but actually has an autonomous psychological existence that is developed within a cultural context and influences actively the way music is performed/perceived (see Clarke, 1985, for an experiment that highlights the influence of different metrical frameworks on the performance of the same melody).

Let's examine now how a metric grid may be matched onto a given accentuation structure. Computational models of the perception of metre – mainly for plain sequences of inter-onset intervals – are described in Lee (1991), Longuet-Higgins and Lee (1982, 1987), Povel and Essens (1985), Rosenthal (1992), and Steedman (1977). In the current model, the total accent strength

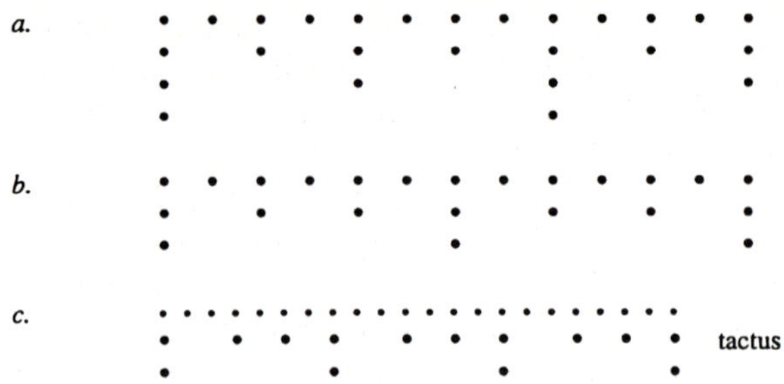

Fig. 10. Examples of well-formed metrical grids

that corresponds to a given metric grid can be calculated by adding the accents of all the events whose inception coincides with the points of the grid. If between different positions/displacements of a metric grid one finds a *significantly* greater total value, then this is considered to be the best fit. If the various placements of a grid receive similar values, then metrical ambiguity is suggested as to that grid.

The two examples presented above (Figs.8-9) are taken from the Western metric/tonal musical tradition, so we would expect that a regular metre of binary/ternary beat patterns would be appropriate (Figs.10a-10b). For both of these examples we consider that the tactus appears at the quarter-note durational value (depends on the tempo). Below is a discussion on the metrical structure of these two melodies.

In Fig.8 we see that at the half-note metric level the total accent strength (indicated at the end of each metric grid) of the binary grid that starts on the first note is much stronger than that of the one that starts on the second quarter-note. This agrees with the metrical perception listeners have and the way metre is indicated on the score. Ternary metrical grids do not suggest any strong preferences (and obviously parallelism considerations would immediately rule them out). Once a binary grid is established, we can examine the next metric level of a whole-note grid. There is no strong preference (there is ambiguity) between the two possible arrangements although the one that starts on the third note is slightly preferred i.e. if articulation and the song word prosody are not taken into account the structure of the piece suggests a gavotte-like metre (bar-lines shifted to the right by two quarter-note beats). Interestingly enough, the prosodic structure of the Greek version of the song adheres to this alternative metrical structure.

The first six bars of Bach's *Concert for Harpsichord in D minor* (Fig.9) are already ambiguous at the tactus; the metrical structure becomes clear only after the seventh bar. The quarter-note beat grid that starts on the first note and the one that starts after an eighth durational value have almost the same total accent strengths (the ambiguity is maintained at the half-note level as well). The first two notes are heard as an upbeat and the listener makes a first selection of a metrical structure that considers the 3rd, 5th and 7th notes as metrically stronger. This assumption is overturned in bar 2 – where the metrical grid is in-phase with the metre indicated on the score – and the beginning of bar 3 is perceived as a suspension. But as more information arrives there is a tendency to shift the metre again and place strong metrical beats on the *suspended* notes. The section that comprises of sixteenth notes is metrically ambiguous. The second half of bar 5 and the first half of bar 6 suggest a metrical structure that conforms with the metric grid that is displaced/shifted by an eighth-durational value. From the second half of bar 6 onwards the metrical structure becomes clear matching the metre indicated in the score. In Fig.9 (top) the melody has been segmented in such a way that the accentuation strength difference in each segment is maximised for the two alternative positions. This metrical analysis seems to correspond to the metrical ambiguity that the composer has intentionally implanted in the melodic surface and that is perceived by the listener.

6 Conclusion

In this paper a formal theory for the low-level rhythmic description of a melodic surface has been presented. The *Local Boundary Detection Model* is based on the *Identity-Change* and *Proximity* rules and detects points of maximum change that allow a listener to identify local boundaries in a melody. This model is more general than the grouping models of both Tenney and Polansky (1980) and Lerdahl and Jackendoff (1983); it can easily be implemented as a computer program and may readily be incorporated as a supplementary module to higher-level theories of rhythmic organization.

It has also been maintained that grouping and accentuation structures are very closely related. Once a grouping structure is defined, the accentuation structure emerges naturally and, from this, the metrical structure may be inferred. It is suggested that the proposed theory is more economic and coherent than most theories of rhythm that treat grouping and accentuation structures as independent components. The evidence presented in this study accounts only for low-level structural features of grouping and accentuation organization. It may be the case that at higher-levels of organization these structures may be partially independent and conflicting. It still is very interesting to see how much is embodied in and can be inferred from a well defined local grouping structure (namely accentuation and metrical structures).

References

Arom, S. (1991). *African polyphony and polyrhythm.* Cambridge: Cambridge University Press.

Bregman, A. (1990). *Auditory scene analysis: The perceptual organization of sound.* Cambridge, MA: The MIT Press.

Cambouropoulos, E. (1996a). A formal theory for the discovery of local boundaries in a melodic surface. In *Proceedings of the Troisièmes Journées d'Informatique Musicale (JIM-96).* Caen: GREYC – Université de Caen.

Cambouropoulos, E. (1996b). A general pitch interval representation: Theory and applications. *Journal of New Music Research, 25,* 231–251.

Clarke, E. (1985). Structure and expression in rhythmic performance. In P. Howell, I. Cross, & R. West (Eds.), *Musical structure and cognition.* London: Academic Press.

Clarke, E. (1987). Levels of structure in the organisation of musical time. *Contemporary Music Review, 2,* 211–238.

Cooper, G., & Meyer, L. (1960). *The rhythmic structure of music.* Chicago, IL: University of Chicago Press.

Deliège, I. (1987). Grouping conditions in listening to music: An approach to Lerdahl and Jackendoff's grouping preference rules. *Music Perception, 4,* 325–359.

Deutsch, D. (1982a). Grouping mechanisms in music. In D. Deutsch (Ed.), *The psychology of music.* New York, NY: Academic Press.

Deutsch, D. (1982b). The processing of pitch combinations. In D. Deutsch (Ed.), *The psychology of music.* New York, NY: Academic Press.

Deutsch, D., & Feroe, J. (1981). The internal representation of pitch sequences in tonal music. *Psychological Review, 88,* 503–22.

Epstein, D. (1995). *Shaping time: Music, the brain, and performance.* New York, NY: Schirmer Books.

Garner, W. (1974). *The processing of information and structure.* Potomac, MD: Erlbaum.

Handel, S. (1989). *Listening. An introduction to the perception of auditory events.* Cambridge, MA: The MIT Press.

Kramer, J. (1988). *The time of music.* New York, NY: Schirmer Books.

Lee, C. (1991). The perception of metrical structure: Experimental evidence and a model. In P. Howell, I. Cross, & R. West (Eds.), *Representing musical structure.* London: Academic Press.

Lerdahl, F., & Jackendoff, R. (1983). *A generative theory of tonal music.* Cambridge, MA: The MIT Press.

Longuet-Higgins, H., & Lee, C. (1982). The perception of musical rhythms. *Perception, 11,* 115–128.

Longuet-Higgins, H., & Lee, C. (1987). The rhythmic interpretation of monophonic music. In H. Longuet-Higgins (Ed.), *Mental processes.* Cambridge, MA: The MIT Press.

McAdams, S. (1984). The auditory image: A metaphor for musical and psychological research on auditory organization. In W. Crozier & A. Chapman (Eds.), *Cognitive processes in the perception of art.* Amsterdam: Elsevier Science Publishers.

Moelants, D. (1997). A framework for the subsymbolic description of meter. In M. Leman (Ed.), *Music, Gestalt, and computing: Studies in systematic musicology.* Berlin, Heidelberg: Springer-Verlag.

Parncutt, R. (1994). Template-matching models of musical pitch and rhythm perception. *Journal of New Music Research, 23*, 145–167.

Povel, D., & Essens, P. (1985). Perception of temporal patterns. *Music Perception, 2*, 411–440.

Rosenthal, D. (1992). Emulation of human rhythm perception. *Computer Music Journal, 16*, 64–76.

Seifert, U., Olk, F., & Schneider, A. (1995). On rhythm perception: Theoretical issues, empirical findings. *Journal of New Music Research, 24*, 164–195.

Steedman, M. (1977). The perception of musical rhythm and metre. *Perception, 6*, 555–569.

Tenney, J. (1964). *Meta + Hodos.* Oakland, CA: Frog Peak Music.

Tenney, J., & Polansky, L. (1980). Temporal Gestalt perception in music. *Journal of Music Theory, 24*, 205–241.

Yeston, M. (1976). *The stratification of musical rhythm.* New Haven, CT: Yale University Press.

Effects of Perceptual Organization and Musical Form on Melodic Expectancies

Carol Krumhansl

Department of Psychology, Cornell University, Ithaca, NY 14853, USA

Abstract. The experiment reported in this chapter examines two aspects of music perception: the role of perceptual organization in generating melodic expectancies, and attentional processing directed at different levels of hierarchical structure in music. The implication-realization model (Narmour, 1990) provides a framework for approaching these questions. It holds that principles of perceptual organization define coherent units within which expectancies operate. The model describes melodic expectancies with five bottom-up principles. The principles are hypothesized to apply not only on the lowest, tone-to-tone level, but also on higher levels between noncontiguous tones. Musical form is described as units at multiple, hierarchically-organized levels. The model holds that when music has similar form, that is, when some characteristics of a pattern are repeated, it draws attention to higher levels. This has the effect of strengthening expectancies operating on higher levels. Thus, musical form is hypothesized to modify the relative strengths of the proposed principles on different levels. An experiment tests these predictions with melodic sequences that contain implications on two levels and exhibit either similar form, *AA'*, or dissimilar form, *AB*. The sequences are followed by continuation tones that listeners judge as to whether they are good continuations. These judgments are tested against a numerical coding of the bottom-up principles of the model. The results support the model's principles at both the low and high levels, and confirm the predicted effects of form.

1 Introduction

According to Koffka (1935, p.437):

> "A melody is a whole, organized in time. ....The earlier notes of the melody have an effect upon the later ones, because they have started a process which demands a definite continuation. A melody ...[is] ... not analogous to beads on a string, but ...[is]... a continuous process.... These events have their own shape which demands a proper continuation."

Numerous studies demonstrate that listeners develop remarkably specific expectancies for pitch patterns in music. Principles of perceptual organization, such as those proposed by the Gestalt psychologists, may guide expectancies along with tonal and harmonic conventions. Carlsen's studies (Carlsen, Divenyi, & Taylor, 1970; Carlsen, 1981; Unyk & Carlsen, 1987) found pitch proximity was one of the major determinants of the continuation tones produced by subjects. Schmuckler (1989, 1990) found expectancies were governed in part by linear and gap-fill patterns as proposed by Meyer (1973).

The bottom-up principles of Narmour's (1990) implication-realization model, which are partially related to Gestalt principles of proximity, similarity, and good continuation, have recently received experimental support. These principles, described in detail below, provided a good account of low-level, that is, tone-to-tone, expectancies. This was tested with melodic fragments ending on open (unclosed) intervals (Krumhansl, 1991, 1995b) as well as simpler two-tone intervals (Cuddy & Lunney, 1995; Krumhansl, 1995a). The primary objective of the present study was to determine whether evidence could also be found for expectancies on higher levels, that is, for expectancies operating over longer temporal spans between noncontiguous tones. If so, are they governed by the same principles of perceptual organization?

The second objective of the present study was to investigate whether similar form affects allocation of attention to different levels of musical organization. In particular, does similarity of form strengthen expectancies on higher levels? Similarity of form varies along a continuum. At one extreme are cases of obvious similarity with exact or near-exact repetition of rhythms, and repeated or transposed melodic patterns. At the other extreme, the similarity may be extremely subtle and obscured by marked differences on the perceptual level. Most psychological studies that have considered the effect of similarity have focused on the clearer cases. These studies demonstrate that repeated or similar melodic and rhythmic patterns influence music perception and memory. Research by Dowling (Dowling, 1978; Dowling & Bartlett, 1981; Bartlett & Dowling, 1980) and Cuddy (Cuddy, Cohen, & Miller, 1979; Smith & Cuddy, 1989) has extensively demonstrated the ability of listeners to perceive the similarity of melodies under transposition; contour, tonality, and meter have all been found to contribute to this ability. Monahan and Carterette (1985) showed the same attributes also influence similarity judgments. At a somewhat more abstract level, Rosner and Meyer (1982, 1986) reported that listeners can classify excepts according to their underlying melodic processes, and Pollard-Gott (1983) showed that listeners perceive the similarity between themes and variations.

The two sequences shown in Fig.1 develop intuitions about musical form on expectancies at different levels of musical structure. According to the implication-realization model (Narmour, 1990), higher levels emerge when strong closure takes place on lower levels. Strong metric emphasis and durational cumulation (which is defined as a pattern of short-to-long duration) are factors that can create closure powerful enough to produce a new, higher level. Tones emphasized in this way are said to *transform* to the higher level. In the examples, the bottom stave shows the music and the upper stave shows the notes that transform to a higher level. In excerpt (a), similarity of form is found beginning at the end of measure 1 and continuing onward (denoted as $A^0 A^1 A^2 ...$). The higher level consists of tones that are emphasized through metric stress and durational cumulation. The effect of similar form highlights the pattern at the higher level and strengthens the descending linear implication. That is, our expectations are strong that the tones G5 - F5 will continue

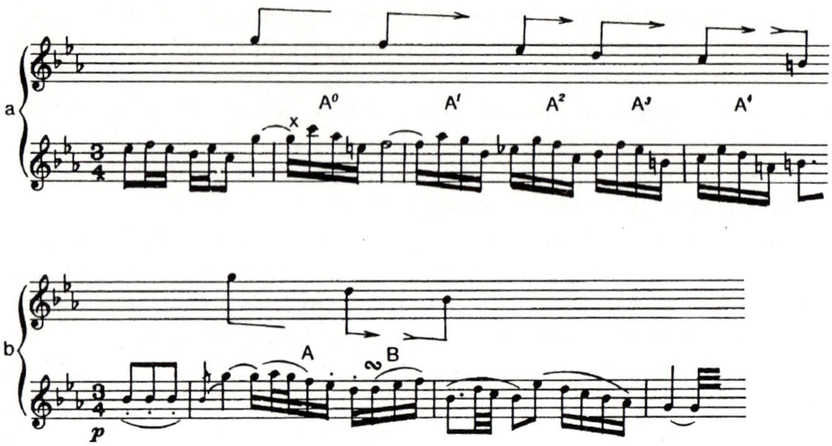

Fig. 1. Two musical excerpts illustrate the concepts of implications on higher levels and similarity of form. Excerpt (a) from Bach, *Sarabande*, *French Suite in C minor*, mm. 1 - 4; excerpt (b) from Mozart, *Piano Trio in B♭*, II (Larghetto), mm. 1 - 3

downward via E♭5 - D5 - C5 - B4. Excerpt (b) exhibits the opposite effect. Dissimilarity of form (denoted *AB*) occurs within measure 1 as a consequence of the rising D5 - D5 - E♭5 - F5 and the ornamental turn on beat 3. The effect of this would seem to be to de-emphasize the pattern at the higher level and weaken the implied descent of the tones G5 - D5, which ultimately continues to the B♭4 at the start of measure 2. Again, the higher level arises primarily because of metric stress and durational cumulation. These intuitions can be characterized more formally by the implication-realization (Narmour, 1990), the relevant parts of which are now summarized.

2 The Bottom-Up Principles of the Implication-Realization Model

The implication-realization model (Narmour, 1990) proposes that expectancy is governed by both bottom-up and top-down processes. Bottom-up processes operate directly on the perceptually available information, and do not depend on prior knowledge. According to the model, five underlying principles govern bottom-up expectancies. (A more detailed presentation of how these principles can be reformulated for psychological test is given in Krumhansl, 1991, 1995b, 1995a, and Cuddy and Lunney, 1995). These principles are based, in part, on the Gestalt concept of perceptual organization, and they can be loosely related to Gestalt principles of similarity, proximity, good continuation, and possibly symmetry. It should be emphasized, however, that the model parameterizes the principles in terms that are specific to music (e.g., interval size and direction). The bottom-up principles are the focus of the present study, but the model holds that top-down processes also influence expectancies. These processes depend on knowledge of the style (*extra-opus* knowledge) and knowledge of the piece (*intra-opus* knowledge). The only top-down factor that will be considered in the present study is the tonal hierarchy of the key of the sequences (Krumhansl & Kessler, 1982).

The implication-realization model describes the cognition of melodies as a succession of points of closure and points of implication. At points of closure, determined by rhythmic, metric, and harmonic stability, expectancies for melodic continuation are weak. At points of implication, when rhythmic, metric, or harmonic instability occurs, expectancies for melodic continuation are strong. The last two tones in a melody at a point of implication form what is called an *implicative interval.* According to the model, the listener develops expectancies for continuation, and these expectancies depend on the size and direction of the implicative interval. The next interval that follows in the melody (the second tone of the implicative interval and the tone that follows) is called a *realized interval.* Besides implicative intervals at the low, tone-to-tone level, implicative intervals may also exist at higher levels. The interval formed by the tones that transform to the higher level also constitutes an implicative interval. It governs expectancies over a longer temporal span. The second tone of the realized interval at the higher level is the tone that metric and durational emphasis transforms to the higher level.

Five bottom-up principles describe the expectancies generated by implicative intervals. It is hypothesized that the same principles apply to implicative intervals at all levels. Reference to Fig.2 may be useful for conceptualizing the principles and understanding how they were coded for data analysis. In the grids, the vertical axis represents the size of the implicative interval (in semitones). The horizontal axis represents the size of the realized interval (in semitones). The realized interval can be either in the same direction as the implicative interval (e.g., up-up), as shown in the right half, or in the opposite direction (e.g., up-down) as shown in the left half. (Both axes could be extended indefinitely, but most intervals in tonal melodies fall within the range shown.) Thus, a point in the grid corresponds to a combination of implicative and realized intervals. If a combination does not satisfy a particular principle it is coded as a 0. The first three principles, registral direction, intervallic difference, and registral return are all-or-none. Therefore, a code of 1 indicates that the principle is satisfied. The last two principles, proximity and closure, are graded in strength with the numerical code indicating the strength.

- The first principle of *registral direction* specifies the implied direction of the realized interval relative to the direction of the implicative interval. The principle states that if the implicative interval is small, defined as five semitones (perfect fourth) or less, then the registral direction is expected

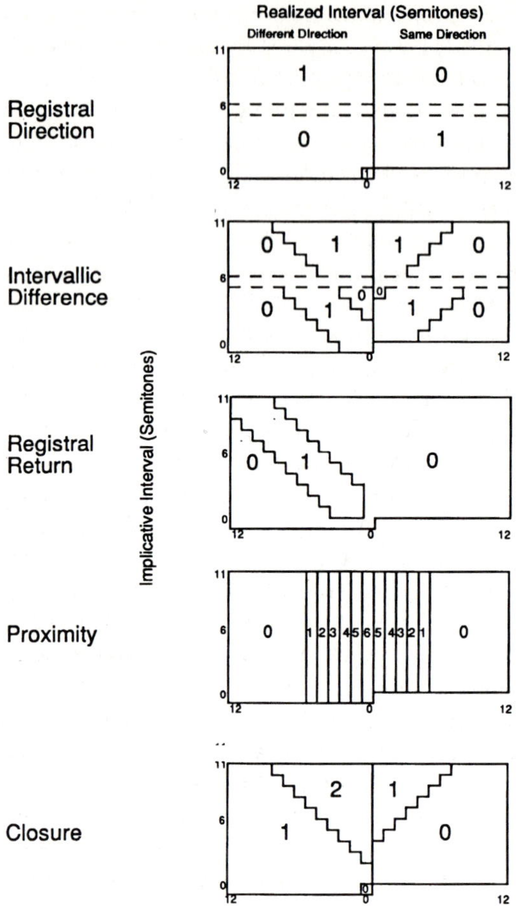

Fig. 2. Shows the coding of the five bottom-up principles of the implication-realization model (Narmour, 1990). The vertical axis represents the size of the implicative interval (in semitones). The horizontal axis represents the size of the realized interval (in semitones) and whether the realized interval is in the same or different direction than the implicative interval. Registral direction concerns the direction of the realized interval relative to the direction of the implicative interval. Intervallic difference concerns the size of the realized interval relative to the size of the implicative interval. Registral return concerns the size of the interval between the first tone of the implicative interval and the second tone of the realized interval. Proximity concerns the size of the realized interval. Closure describes the pairs of implicative and realized intervals that produce closure; it is strongest when registral direction reverses and a large implicative interval is followed by a smaller realized interval.

to continue. In psychological terms, this can be taken to be an example of the Gestalt principle of good continuation. If the implicative interval is large, defined as seven semitones (perfect fifth) or more, then the registral direction is expected to reverse. This might be taken as related to a Gestalt principle of symmetry, because the reversal of registral direction would create a more symmetrical pattern. In the theory, six semitones (augmented fourth/diminished fifth) is considered a threshold and the context determines whether it functions as a small or a large interval.

- The second principle, *intervallic difference*, concerns the implied size of the realized interval relative to the size of the implicative interval. The principle states that if the implicative interval is small, five semitones (perfect fourth) or less, then a similarly sized realized interval is implied, which can be taken to be an application of the Gestalt principle of similarity. The definition of similarly sized depends on whether registral direction changes. If registral direction changes, then similarly sized means that the difference in size between the implicative and realized intervals will be no more than three semitones (minor third). If registral direction does not change, then similarly sized means that the difference in size between the implicative and realized intervals will be no more than four semitones (major third). If the implicative interval is large, seven semitones (perfect fifth) or larger, then a smaller realized interval (by the specified number of semitones) is implied. The case for large implicative intervals would be an application of the Gestalt principle of proximity.
- The next two principles, *registral return* and *proximity*, describe the expected interval size measured relative to the first and second tones of the implicative interval, respectively. Both can be seen as instances of the Gestalt principle of proximity. The principle of registral return governs the size of the interval between the first tone of the implicative interval and the second tone of the realized interval. The principle of registral return is satisfied when the second tone of the realized interval falls within two semitones (major second) of the first tone of the implicative interval. The principle of proximity governs the size of the interval between the second tone of the implicative interval and the second tone of the realized interval (i.e., the size of the realized interval). It is satisfied when the second tone of the realized interval falls within 5 semitones (perfect fourth) of the second tone of the implicative interval. (More simply, proximity holds if the realized interval is five semitones, perfect fourth, or less.) Proximity is considered a graded variable, ranging from maximum for zero semitones (unison) to minimum for five semitones (perfect fourth).
- The final principle, *closure*, specifies the pairs of implicative and realized intervals that produce a sense of closure. A possible confusion stems from the term closure being used in two different senses in the implication-realization model (Narmour, 1990). For consistency with that formulation, however, I will retain both senses. The first is the general meaning of closure, described above, which determines when expectancies for

continuation are weak. The second, referred to here as the principle of closure, specifies the combinations of implicative and realized intervals that produce closure. According to the model, this second sense of closure is strongest when registral direction reverses and a large implicative interval is followed by a smaller realized interval. Closure is moderately strong when either registral direction reverses or a large implicative interval is followed by a smaller realized interval. When neither of these conditions obtains, then the interval pattern is open, that is, unless other factors, such as metrical stress, durational stress, or resolution of dissonance, are operating to produce closure. In Gestalt terms, the registral reversal component of this principle is akin to the principle of symmetry, but the specification of large to small interval size lacks an analogy in Gestalt theory.

Previous research has found supporting evidence for these principles on a tone-to-tone level. Krumhansl (1991, 1995b) used excerpts from songs in three different styles: British folk tunes, atonal songs, and Chinese folk tunes. The excerpts were approximately one and a half phrase, ranging from three to eight measures in length. In all three cases, the judgments made of continuation tones were modeled quite well by the principles of the implication-realization model. This was true independently of style. Moreover, listeners in the experiments varied in their musical training and familiarity with the musical styles. Despite these variations, strong intersubject correlations were found. To investigate whether support for the bottom-up principles would also be found for very minimal melodic contexts, Cuddy and Lunney (1995) and Krumhansl (1995a) used two-tone intervals as contexts. The results supported the model's account of melodic expectancies on the tone-to-tone level. Even though listeners in these experiments varied in the extent of their musical training, strong intersubject agreement was again found. The present experiment tests whether the principles also function at higher levels so that they govern expectancies between noncontiguous tones as described in the next section.

3 Application of the Model to Low- and High-Level Implicative Intervals

Figures 3 and 4 show the eight sequences used in the experiment. E. Narmour wrote them specifically to conform to the relevant aspects of the implication-realization model. The second measure of each sequence ends with an interval that the model considers to be open (unclosed, implicative). This is the interval formed by the last two tones of the second measure. In addition, the sequences were composed such that two tones in each sequence create an implicative interval at a higher level. These appear on the downbeat in the first and second measures. The downbeat in the first measure transforms to a higher level because of the strong durational cumulation (short-to-long duration) of the opening anacruses, or the strong metric stress, or both. The downbeat in the second measure transforms to the higher level because of the durational cumulation, the metric stress, or both. Note that in every case the interval between the end of the first measure and the beginning of the second measure is a relatively large skip; this enhances the metric and durational closure that transforms the downbeat in the second measures to a higher level. I will refer to the implicative intervals on the two levels as low-level and high-level implicative intervals, although additional higher levels may also exist.

Fig. 3. Shows the sequences on which the stimulus materials of the experiment were based. The four sequences presented here are cases of similar form, *AA'*. High-level implicative intervals are formed by the tones on the downbeats in measures 1 and 2; they initiate either scalar or triadic patterns. The registral direction of the high-level implicative interval is, in all sequences, opposite the registral direction of the final, low-level implicative interval.

Fig. 4. Shows the sequences on which the stimulus materials of the experiment were based. These four sequences are cases of dissimilar form, *AB*. High-level implicative intervals are formed by the tones on the downbeats in measures 1 and 2; they initiate either scalar or triadic patterns. The registral direction of the high-level implicative interval is, in all sequences, opposite the registral direction of the final, low-level implicative interval

In every sequence, closure is minimized in both measures after the downbeats. Melodic nonclosure (and thus a sense of openness and continuation) occurs in the absence of cumulative duration (short-to-long duration) and strong metric stress. For maximal openness, registral direction should continue, intervals should be similar, and, if resolution of dissonance occurs, it should do so in a metrically weak position. In these sequences, these factors override the secondary metric stress that would otherwise occur on the second, third, and fourth beats. In addition, stylistically implied harmony and thus implied dissonance frequently weakens the sense of metric closure on these beats. Moreover, octaves and unisons were avoided because of the closural effects of these intervals. Finally, the second measure never ends with a large interval that might interfere with the implications at the higher level. Together, these factors throw into relief the high-level structural tones that occur on the downbeats of each measure.

In all eight sequences, the high-level tones create small implicative intervals (seconds or thirds) that imply continuation of direction and a similarly sized interval. Four of these high-level implicative intervals ascend, and four descend; four of them begin scalar patterns, and four begin triadic patterns. All eight high-level implicative intervals begin on goal notes (scale-steps 1, 3, or 5), and the third tone that continues the scalar or triadic pattern is also a goal note. As far as the low-level implicative intervals are concerned, four end on a nongoal note (scale-step 6, the submediant), two on a mobile tone (scale-step 7, the leading tone), and two on a goal tone (scale-step 1, the tonic). The sequences are similar in length, number of tones, number of registral (contour) changes, mix of melodic steps and skips, and phrase structure. In addition, they all have the same tempo, meter, and key (which is always major).

In testing the quantified predictions of the implication-realization model with the sequences, I distinguish between the principles applied to the low-level implicative intervals and the principles applied to the high-level implicative intervals. The five principles of registral direction, intervallic difference, registral return, proximity, and closure applied to the low-level implicative intervals will be denoted LRD, LID, LRR, LPR, and LCL, respectively; applied to the high-level implicative interval, they will be denoted HRD, HID, HRR, HPR, and HCL, respectively. Table 1 shows the coding of these principles, plus the tonal hierarchy, for the sequences on top of Figs.3 and 4. This illustrates by concrete example which principles are satisfied by different possible continuation tones for these two sequences. The first question of interest is whether the perceptual data supports both low- and high-level implications. In other words, can evidence be found that listeners' expectancies are influenced by both low- and high-level implicative intervals? As described more fully later, the second question of interest is whether the predicted effects of form influence the relative weights of the principles at low- and high-levels. First, however, an issue of a somewhat technical nature arises, namely, how the particular sequences used in the experiment affect the predicted quantitative results.

4 The Contextual Distribution of the Principles

The specific quantitative predictions of the bottom-up principles of the implication-realization model will depend on the particular sequences to which they are applied. I refer to this dependency as the *contextual distribution* of the principles. This dependency limits the degree to which a single test of the model can be generalized to other sequences, and has implications for how the data for any particular set of sequences are analyzed. The core issue is that the relationships between the principles change across different

Table 1. Predictions of implication-realization model for two example sequences: (a) Sequence 1: similar form *AA'*, (b) Sequence 5: dissimilar form *AB*

Tone	LRD	LID	LRR	LPR	LCL	HRD	HID	HRR	HPR	HCL	Tonal Hierarchy
G3	1	0	0	0	0	0	0	0	0	1	3.48
A3	1	0	0	0	0	0	0	0	0	1	4.38
Bb3	1	0	0	0	0	0	0	0	0	1	4.09
C4	1	1	0	1	0	0	0	0	0	1	5.19
D4	1	1	0	3	0	0	0	0	0	1	3.66
E4	1	1	0	5	0	0	1	1	1	1	2.88
F4	0	1	0	6	1	0	1	1	2	1	6.35
G4	0	1	1	4	1	0	1	1	4	1	3.48
A4	0	1	1	2	1	0	0	0	6	2	4.38
Bb4	0	0	0	1	1	1	1	0	5	0	4.09
C5	0	0	0	0	1	1	1	0	3	0	5.19
D5	0	0	0	0	1	1	1	0	1	0	3.66
E5	0	0	0	0	1	1	1	0	0	0	2.88
F5	0	0	0	0	1	1	0	0	0	0	6.35
G5	0	0	0	0	1	1	0	0	0	0	3.48

Tone	LRD	LID	LRR	LPR	LCL	HRD	HID	HRR	HPR	HCL	Tonal Hierarchy
G3	1	0	0	0	0	0	0	0	0	1	3.48
A3	1	0	0	0	0	0	0	0	0	1	4.38
Bb3	1	0	0	0	0	0	0	0	0	1	4.09
C4	1	1	0	1	0	0	0	0	0	1	5.19
D4	1	1	0	3	0	0	0	0	0	1	3.66
E4	1	1	0	5	0	0	1	1	1	1	2.88
F4	0	1	0	6	1	0	1	1	2	1	6.35
G4	0	1	1	4	1	0	1	1	4	1	3.48
A4	0	1	1	2	1	0	0	0	6	2	4.38
Bb4	0	0	0	1	1	1	1	0	5	0	4.09
C5	0	0	0	0	1	1	1	0	3	0	5.19
D5	0	0	0	0	1	1	1	0	1	0	3.66
E5	0	0	0	0	1	1	1	0	0	0	2.88
F5	0	0	0	0	1	1	0	0	0	0	6.35
G5	0	0	0	0	1	1	0	0	0	0	3.48

sets of sequences. For some sequences, two principles may make very similar predictions whereas, for other sequences, the same principles may make very dissimilar predictions. The pattern of relationships between the principles is especially relevant to the predicted effects of form described in the next section. It should be noted that this problem is distinct from the nonindependence of the principles in general; that problem can be addressed empirically only with a comprehensive set of implicative and realized intervals as was done by Krumhansl (1995a).

A concrete example illustrating the contextual distribution of the principles is how registral direction and closure apply to the sequences in the experiment. For small implicative intervals, the principle of registral direction specifies that registral direction will continue, whereas the principle of closure specifies that registral direction will reverse. In the eight sequences used in the experiment all final low-level implicative intervals are small, being either one or two semitones (minor or major second) in size. Consequently, exactly those tones that satisfy low-level registral direction (LRD) will violate low-level closure (LCL), and vice versa. In other words, a perfect negative correlation holds between these two principles for this set of sequences. If large low-level implicative intervals had been present instead, then the two principles would make positively correlated predictions; tones reversing registral direction would satisfy low-level registral direction (LRD) and produce either a moderate or strong degree of low-level closure (LCL). The same logic applies to implicative intervals on higher levels. For these sequences, the implicative intervals on the higher level are classified as small. Hence, high-level registral direction (HRD) and high-level closure (HCL) are negatively correlated, although not perfectly so because some of the cases are strongly closed and others only moderately closed for the somewhat larger implicative intervals at the higher level.

Another example illustrating the contextual distribution of the principles, and thus the correlations between their predictions, is the relation between low-level registral direction (LRD) and high-level registral direction (HRD). In the sequences used in the experiment, the low-level implicative interval is always in the direction opposite the high-level implicative interval. In the first sequence, for instance, the low-level implicative interval is a descending G4 - F4, whereas the high-level implicative interval is an ascending F4 - A4. This property of the set of sequences means that low-level registral direction (LRD) is negatively correlated with high-level registral direction (HRD). Recall the observation made in the last paragraph concerning the inverse relationship between registral direction and closure for small implicative intervals. This inverse relationship implies that the predictions for low-level closure (LCL) will correlate negatively with those for high-level closure (HCL) because the implicative intervals at both low- and high-levels are small. These interrelationships will be summarized in a table as the effects of form on low- and high-level implications are described.

5 Effects of Similar Form on Low- and High-Level Implications

I turn now to the predicted effects of form, keeping in mind the problem of the contextual distribution of the principles. The four sequences in Fig.3 were designed to be very clear instances of similar form. They are called *similar form* sequences and are denoted *AA'*. In each of these examples, the rhythmic pattern of the first measure is repeated in the second measure, and the melodic pattern of the first measure is repeated at a different pitch level (transposed) in the second measure. The four sequences of Fig.4 do not have this property. In these, both the rhythmic and melodic patterns of the first and second measures contrast quite strongly. Although it is difficult to assess precisely the degree of contrast, it is clearly greater in these sequences than in the similar form sequences. Thus, they will be called 'dissimilar form' sequences and denoted *AB*. The weights for the various predictor variables are hypothesized to depend on form; the predictions are summarized in Table 2.

I will discuss these predictions in a different order from the table because they fall into two different conceptual groups. The first conceptual group is governed by the intuition that similar form sequences are more open (unclosed) on the higher level than sequences with dissimilar form. This intuition carries with it predictions about the direction of the realized interval on the higher level. Specifically, it makes predictions for the principle of registral direction (HRD), that the direction will continue, and the inversely correlated (for these sequences) principle of closure (HCL), that closure will be weak. The second conceptual group is governed by the intuition that attention is focused more on the higher-level implications in similar form sequences than in dissimilar form sequences. This intuition carries with it predictions about the size of the realized interval. Specifically, it makes predictions for the principle of intervallic difference (HID), that the realized interval on the higher level will be similar in size to the implicative interval, and the principle of proximity (HPR), that it will be a small interval. As will be discussed, the principle of registral reversal (HRR) is ambiguous regarding predicted effects of form because it makes predictions about the direction and the size of the realized interval that are incompatible.

First, concerning the direction of the realized interval, sequences with similar form, *AA'*, are predicted to be more open (unclosed) than sequences with dissimilar form, *AB*. In other words, listeners more strongly expect that the registral direction established at the higher level will continue when the form is similar than when it is dissimilar. This makes predictions for the weights of both registral direction and closure. High-level registral direction (HRD) should be stronger for *AA'* sequences than for *AB* sequences and, conversely, high-level closure (HCL) should be stronger for *AB* sequences than for *AA'* sequences. Because of the inverse relationship between low- and high-level registral direction in these sequences, it follows that low-level

Table 2. Predicted Effects of Similar Form (*AA'*) versus Dissimilar Form (*AB*)

Principle	Prediction	Reason
HRD	*AA'* > *AB*	High-level implication for registral direction to continue (unclosed) stronger for *AA'* than for *AB*
LRD	*AA'* < *AB*	Inverse relationship between registral direction at low-and high-levels
HID	*AA'* > *AB*	High-level implication for interval size stronger for *AA'* than for *AB*
LID	*AA'* < *AB*	Low-level implication for interval size stronger for *AB* than for *AA'*
HRR	*AA'* < *AB*	High-level implication for registral direction to reverse (closed) stronger for *AB* than for *AA'*
LRR	*AA'* > *AB*	Inverse relationship between closure at low- and high-levels
		or
HRR	*AA'* > *AB*	High-level implication for interval size stronger for *AA'* than for *AB*
LRR	*AA'* < *AB*	Low-level implication for interval size stronger for *AB* than for *AA'*
HPR	*AA'* > *AB*	High-level implication for interval size stronger for *AA'* than for *AB*
LPR	*AA'* < *AB*	Low-level implication for interval size stronger for *AB* than for *AA'*
HCL	*AA'* < *AB*	High-level implication for registral direction to reverse (closed) stronger for *AB* than for *AA'*
LCL	*AA'* > *AB*	Inverse relationship between closure at low- and high-levels

registral direction (LRD) should be stronger for *AB* sequences than for *AA'* sequences. And, because of the inverse relationship between low- and high-level closure in these sequences, low-level closure (LCL) should be stronger for *AA'* sequences than for *AB* sequences.

Second, concerning the size of the realized interval, the high-level implications are predicted to be stronger for sequences with similar form, *AA'*, than for sequences with dissimilar form, *AB*. This is based on the intuition that the similar form perceptually emphasizes the high-level patterns. In contrast, dissimilar form is hypothesized to emphasize the low-level patterns. For the principles governing the size of the realized intervals, this makes the following predictions. High-level intervallic difference (HID) should be stronger for *AA'* sequences than for *AB* sequences, whereas low-level intervallic difference (LID) should be stronger for *AB* sequences than for *AA'* sequences. Similarly, high-level proximity (HPR) should be stronger for *AA'* sequences than for *AB* sequences, and low-level proximity (LPR) should be stronger for *AB* sequences than for *AA'* sequences.

These two intuitions concerning the effect of similar form, however, make competing predictions for the fifth, remaining principle of registral return. That principle specifies both the direction (reversal) and size (similarity) of the realized interval. Thus, it might fall either into the first group of predictions that concern the direction of the interval, or the second group of predictions that concern its size. If the predictions for direction predominate, then high-level registral return (HRR) should be stronger for *AB* sequences than for *AA'* sequences because registral return produces closure on the high-level. The inverse relationship between low- and high-level registral return would, in this case, predict that low-level registral return (LRR) should be stronger for *AA'* sequences than for *AB* sequences. If, on the other hand, the predictions for size predominate, then the opposite predictions follow. That is, because the similar form emphasizes the size implications of the high-level implicative interval, high-level registral return (HRR) should be stronger for *AA'* sequences than for *AB* sequences. And, because dissimilar form emphasizes the size implications of the low-level implicative interval, low-level registral return (LRR) should be stronger for *AB* sequences than for *AA'* sequences.

The experiment tested the rather intricate set of predictions shown in Table 2. In the experiment, the *AA'* and *AB* sequences were followed by a set of continuation tones that were the diatonic scale tones in a two-octave range. The listeners judged the degree to which these tones were good, natural continuations of the contexts. For each sequence, the five bottom-up principles applied to the low- and high-level implicative intervals generated ten predictor variables. That is, any particular continuation tone was coded for whether it satisfied low- and high-level registral return (LRR and HRR), low- and high-level intervallic difference (LID and HRR), low- and high-level registral return (LRR and HRR), low- and high-level proximity (LPR and HPR), and low- and high-level closure (LCL and HCL). The analysis of the judgments tested whether these variables accurately predict the judgments, and whether their weights show the predicted effects of similar versus dissimilar form.

6 Method

6.1 Subjects

Sixteen listeners were recruited from the Cornell University community. They were paid $5.00 for participating in an experimental session that lasted slightly more than one hour. Eight of the listeners met the criteria for musical training, which was that they had received at least six years of instruction on a musical instrument or voice, and that they had been involved in some musical activity within the last year. On average, this group had a total of 14.5 years of instrumental and vocal instruction. The other eight listeners did not meet the criteria for musical training. On average, they had received 3.0 years of musical instruction; none was currently involved in musical activities.

6.2 Apparatus

Stimuli were programmed on a MacIntosh IIcx computer using the MAX software. The computer was connected through an Apple MIDI Interface to a Yamaha TX816 FM (frequency modulation) tone generator. The stimuli were presented with a Yamaha Power Amplifier (P2150) and a Yamaha 1204 MC series mixing console through a single JBL Model 4312A loudspeaker at a comfortable listening level. Listeners recorded their responses using the mouse of the computer.

6.3 Stimulus materials

The stimulus sequences were played in a piano timbre at a moderate tempo at a comfortable listening level. The metrical structure of the sequences was emphasized by a subtle increase in amplitude on the downbeat of each measure. Each trial consisted of the sequence followed by a single continuation tone sounded on the downbeat of the third measure. The eight blocks of trials, which corresponded with the eight sequences, were presented in different, randomly determined orders to the subjects. Each block consisted of 18 trials, the first three of which were practice trials with randomly selected continuation tones. These three practice trials were followed by 15 experimental trials in which the fragment was followed by 15 different continuation tones. In all cases, the continuation tones were in the diatonic scale of the key of the fragments: F, G, A, B♭, C, D, and E. The range of the continuation tones was approximately centered around the final tone of the sequences. It was G3 - G5 for the blocks of trials based on sequences 1, 2, 5, and 6, and C4 - C6 for the blocks of trials based on sequences 3, 4, 7, and 8.

6.4 Procedure

Listeners, who were tested individually, were given the instructions in written form. The instructions described how listeners were to use the three controls exhibited on the computer screen. They were told to click on the button at the top of the display to start each trial. After listening to the trial, they were to adjust the vertical position of a slider to indicate whether the last tone was a good, natural continuation of the music; the top of the slider was labeled *very good* and the bottom of the slider was labeled *very bad*. After adjusting the slider, they were to click on the button at the bottom, which signaled the computer to record the response and prepare the next trial. The experimental session began with 8 practice trials to permit listeners to become familiar with the procedure and to ask any questions that they might have about the instructions.

7 Results

Excluding the practice trials, each subject made a response to each of 15 continuation tones in each of the 8 blocks of trials for a total of 120 responses. The position to which subjects moved the slider to make their response on each trial was coded by the computer on a scale from 0 to 127. To assess the degree of intersubject agreement, each subject's responses were correlated with those of every other subject. Each of the 120 intersubject correlations was significant (at $p < .01$). The correlation between the average data for the musicians and nonmusicians was also significant (at $p < .0001$). Because of the strong agreement between listeners and between the groups of listeners, the following analyses consider the data for listeners independently of their level of musical training. (Other analyses, which looked for effects of musical training, did not find any consistent or interpretable differences and so they will not be reported.)

The first step in testing the model used the data for the eight two-measure long contexts, *AA'* and *AB*, averaged over listeners. Low-level closure (LCL) was not included in the model because of its perfect negative correlation with low-level registral direction (LRD). To these nine predictor variables was added a tenth variable, the tonal hierarchy of the key of F major using the values from Krumhansl and Kessler (1982). The multiple regression analysis found an $R(10,109) = .85$, $p < .0001$. The same model was also fit to the data for the 16 individual subjects. Fifteen of the 16 multiple correlations were significant (at $p < .005$), and the one remaining multiple correlation approached significance ($p = .12$). Thus, the model with all factors included provided an excellent fit to the data, both for the group and for the individual listeners. The overall fit of the model was also good for both the *AA'* and *AB* sequences when they were analyzed separately. The resulting multiple correlation values were both $R(10,49) = .87$, $p < .0001$.

These regression analyses have a relatively large number of predictor variables with a complex pattern of intercorrelations, so the regression weights were difficult to interpret. Consequently, the effects of the predictor variables were analyzed in separate regression tests. The series of regression analyses tested the relative influences of the five pairs of low- and high-level principles for the *AA'* and *AB* sequences. The results are summarized in Table 3, which should be compared to the predictions shown in Table 2. All the multiple correlations shown were significant (at $p < .001$), as were the contributions of the individual predictor variables (at $p < .01$) with the three exceptions involving registral return noted below.

Table 3. Obtained Effects of Similar Form (*AA'*) versus Dissimilar Form (*AB*)[1]

Principle	Prediction	Weight for *AA'*		Weight for *AB*	Significance of Difference
		AA': R(2,57)=.58		*AB*: R(2,57)=.56	
HRD	*AA'* > *AB*	b=-.50	>	b=-.67	t(15)=2.50, p=.03
LRD	*AA'* < *AB*	b=-.76	<	b=-.57	t(15)=-3.33, p<.01
		AA': R(2,57)=.72		*AB*: R(2,57)=.66	
HID	*AA'* > *AB*	b=.55	>	b=.39	t(15)= 3.57, p<.01
LID	*AA'* < *AB*	b=.38	<	b=.53	t(15)=-4.70, p<.01
		AA': R(2,57)=.52		*AB*: R(2,57)=.50	
HRR	*AA'* < *AB*[3]	b=.18[2]	<	b=.33[2]	t(15)=-2.30, p=.04
LRR	*AA'* > *AB*[3]	b=.41	>	b=.22[2]	t(15)=4.61, p<.01
		AA': R(2,57)=.77		*AB*: R(2,57)=.79	
HPR	*AA'* > *AB*	b=.62	>	b=.54	t(15)=2.29, p=.04
LPR	*AA'* < *AB*	b=.31	<	b=.55	t(15)=-3.37, p<.01
		AA': R(2,57)=.58		*AB*: R(2,57)=.54	
HCL	*AA'* < *AB*	b=.44	<	b=.59	t(15)=-2.49, p=.02
LCL	*AA'* > *AB*	b=.67	>	b=.46	t(15)=3.26, p<.01

[1]All multiple correlations shown were significant at p<.0001.
[2]These normalized regression weights were not significant.
[3]These are the predictions if high-level implications for registral direction predominate over high-level implications for interval size.

Consider the first multiple regression in which the data for the *AA'* sequences and the *AB* sequences were analyzed with the model consisting of low- and high-level registral direction (LRD and HRD). The multiple correlations shown in the table for *AA'* and *AB* were both significant, as were both predictor variables. The standardized regression weights, the b values, are shown in the table. The negative values indicate that listeners generally tended to give higher ratings to continuation tones that reversed the direction of high-level implicative intervals (giving closure). As predicted, this effect was weaker for *AA'* sequences than for *AB* sequences; the b values for high-level registral direction (HRD) were larger (less negative) for *AA'* than for *AB* sequences. Again as predicted, the difference was reversed for low-level registral direction (LRD). To test whether these b values for *AA'* and *AB* sequences were significantly different, the same analysis was done for each subject and the resulting b values were analyzed. The statistic t, shown at the right, indicated that the differences in weights were significant.

Separate multiple regressions examined the remaining four pairs of principles. The second regression model used low- and high-level intervallic difference (LID and HID) as predictor variables. Again, significant results were obtained. As predicted, the effect of high-level intervallic difference was larger for the *AA'* sequences than for *AB* sequences, and the effect of low-level intervallic difference was larger for the *AB* sequences than for *AA'* sequences. These results support the prediction that attention to the higher level is stronger in the *AA'* sequences, generating stronger expectations for interval size on the higher level and weaker expectations for interval size on the lower level.

The next multiple regression used low- and high-level registral return (LRR and HRR). Although the multiple correlations were both significant, three of the individual predictor variables were not: high-level registral return (HRR) in the *AA'* analysis, and both low- and high- level registral return (LRR and HRR) in the *AB* analysis. These nonsignificant weights may result from the incompatible predictions for registral return of *AA'* sequences. For these, openness at the higher level would predict continuation of direction, whereas attention to interval size on the higher level would predict a return to the pitch region of the first tone of the implicative interval. What is more important, although the weights for these principles were not consistently significant, their differences were. The direction of the differences suggests that the high-level implications for the direction of the realized interval predominate over those for the size of the interval.

The next multiple regression using low- and high-level proximity (LPR and HPR) produced consistently significant results. As predicted, high-level proximity was weighted more heavily for *AA'* sequences than for *AB* sequences; the reverse was found for low-level proximity. Thus, attention to the higher level in the *AA'* sequences strengthened the expectation for a relatively small realized interval at the higher level and weakened the expectation for a relatively small realized interval at the lower level.

The final regression using low- and high-level closure (LCL and HCL) also produced consistently significant results. As predicted, high-level closure was weighted more heavily for *AB* sequences than for *AA'* sequences; the reverse was found for low-level closure. In summary, the analyses strongly confirmed the predicted effects of form on low- and high-level implicative intervals. The detailed predictions and the reasons for them were summarized in Table 2. This pattern was confirmed by the results shown in Table 3; every predicted difference in the weights was observed.

8 Discussion

The first objective of the experiment was to determine whether implications at higher levels influence listeners' expectancies for melodic continuations. Toward this end, the sequences presented to listeners contained tones that, according to the implication-realization model (Narmour, 1990), formed implicative intervals at a higher level. Predictor variables were coded to represent the model's predictions for both the low- and the high-level implicative intervals. These, together with the tonal hierarchy of the key of the sequences, provided an excellent account of the listeners' ratings of continuation tones presented after the melodic contexts. Thus, the present experiment demonstrates that the principles function not only at the lowest, tone-to-tone level but also at higher levels governing expectancies between noncontiguous tones.

The present study was limited to two hierarchical levels. This raises the question of whether there are psychological limits to the number of levels that can simultaneously induce expectancies. The limit may be a function of the depth of embedding, owing to the cognitive load imposed by the complex interactions between principles on different levels. This problem was apparent even with the two-leveled sequences investigated here. Alternatively, or in addition, memory may limit the span of time over which implications can be active. The sequences here were relatively short, well within the span of the perceptual present (Fraisse, 1982). Intuitively, it would seem that implications would be weakened over longer time spans, although this is a matter that cannot be addressed without further experimentation.

The results replicate and extend earlier results (Krumhansl, 1991, 1995b, 1995a; Cuddy & Lunney, 1995) supporting the bottom-up principles of the implication-realization model as formulated by Narmour (1990). The consistent support for the model found across different contexts, different musical styles, and different listener backgrounds in these studies bears emphasis. It suggests that the principles describe basic aspects of musical experience that are quite general over listeners and musical styles. This would be expected in so far as the principles can be related to general principles of perceptual organization that do not require extensive training or enculturation. As described earlier, the principles can be partially related to Gestalt principles, such as proximity, similarity, good continuation, and possibly symmetry. Thus, the support found for the model may result from general psychological processes, such as those described by Bregman (1990), organizing auditory information into coherent, well-formed patterns.

It should be recalled, however, that the implication-realization model is formulated in specifically musical terms. It makes precise claims concerning the interval size that distinguishes between small and large implicative intervals, and between proximal and nonproximal tones. Detailed tests of these specifications were made in the experiment by Krumhansl (1995a), which presented implicative intervals over a two-octave range, each followed by comparison tones over a two-octave range. That study confirmed the model's division between small and large intervals at the interval of a tritone. The division at the tritone may be related to this interval's relatively rare appearance in melodies (Krumhansl, 1990), and its weak tonal implications, see, for example, (Cuddy & Badertscher, 1987; Krumhansl, 1995a). However, the results of the Krumhansl (1995a) study suggested the principles of proximity and registral return might better be characterized as simple linear functions of interval size, and found no independent support for the principles of closure and intervallic difference. At present, it appears that the detailed parameterization of the model cannot be justified fully in psychological terms.

The second objective of the experiment was to investigate the effect of form. Half of the sequences, the *AA'* sequences, had similar melodic and rhythmic patterns in the first and second measures. The remaining sequences, the *AB* sequences, had dissimilar melodic and rhythmic patterns in the first and second measures. The predicted effects of this manipulation were twofold. First, similar form sequences, *AA'*, were expected to be more open, more strongly implying continuation of the direction of the pattern at the higher level, than dissimilar form sequences, *AB*. Second, by perceptually emphasizing the higher level, the similar form sequences, *AA'*, were expected to show stronger effects of the high-level implications for interval size than the dissimilar form sequences, *AB*.

To translate these general predictions into specific predictions for the weights of the bottom-up principles, I introduced the concept of the contextual distribution of the principles. The predictor variables of the model intercorrelate in a way that depends on the particular sequences to which the model is applied. The contextual distribution of the principles in this application produced strong intercorrelations (some of which were positive and some of which were negative) among four of the predictor variables because of the characteristics of the sequences: low- and high-level registral direction (LRD and HRD), and low- and high-level closure (LCL and HCL). Once these intercorrelations were taken into account, it was possible to make detailed predictions concerning form's effects on the weights of the predictor variables. The full pattern of predictions was confirmed by the experiment. In all cases, the data confirmed the predicted differences in weights between *AA'* and *AB* sequences.

Thus, the present study also extends previous findings by demonstrating effects of form on melodic expectancies. The first prediction concerning form was that when musical form is similar, listeners expect the pattern to be more open on the higher level than when musical form is dissimilar. The results were consistent with this prediction suggesting that, once a pattern is established and then repeated, listeners expect that the pattern will continue.

Underlying this result is the ability of listeners, demonstrated in numerous studies cited earlier, to perceive melodic and rhythmic similarity under transposition and other transformations. The second prediction concerning form was that when musical form is similar, high-level implications for interval size are stronger than when musical form is dissimilar. The results were also consistent with this prediction. Three closely related psychological explanations may account for these results. First, the repeated melodic and rhythmic patterns may shift attention to the higher level. Second, the repeated melodic and rhythmic patterns may facilitate encoding and remembering the pattern in its entirety. Sequences with similar form would correspond to more efficient representations in coding models that have been proposed (Simon & Sumner, 1968; Deutsch & Feroe, 1981; Deutsch, 1980; Boltz & Jones, 1986; Boltz, Marshburn, Jones, & Johnson, 1985). Third, information about the beginning of the sequence may be purged from memory once a new pattern is introduced. These three processes would all account for the results obtained, and further studies may be able to disentangle their relative influences.

9 Conclusion

I close with a general remark about real-time processing of music and the central role played by expectancies in listening to music. According to Meyer (1956), "...expectation is always ahead of the music, creating a background of diffuse tension against which particular delays articulate the affective curve and create meaning." How can this be brought into correspondence with the rather simple principles governing expectations that emerge from psychological studies? I will use the implication-realization model (Narmour, 1990) as a framework for this discussion, although the general point would apply equally to modified or alternative models of this general sort. The model proposes a small number of perceptual principles that can be made active by the on-going pattern in the music, in fact, just five basic principles: registral direction, intervallic difference, registral return, proximity, and closure. How is this consistent with the richness and variety of our musical experience?

The present study suggests a partial answer to this is that expectations are generated simultaneously on a number of different levels. This creates a constantly shifting pattern of expectations because of the complex interactions between principles on different levels, sometimes compatible and sometimes incompatible. Moreover, the present study suggests that musical form may shift the listener's attention to one level of structure or another, dynamically weighting some expectations more heavily than others. In addition, principles of melodic expectations undoubtedly have parallels in other dimensions, such as rhythm, timbre, dynamics, and so on, which would combine in complex ways with melodic expectations. Experienced listeners also bring to bear knowledge of music written in a style similar to the piece, for example, tonal hierarchies, harmonic progressions, and hierarchies of metrical stress (what are called extra-opus style factors in the implication-realization model). Finally, listeners' expectation may be influenced by knowledge of the particular musical composition (called intra-opus style in the model). For example, distinctive motivic and rhythmic patterns heard in the piece may generate expectations for their repetition or variation.

The situation is represented schematically in Fig.5. As a piece of music progresses, a variety of psychological processes are latent. Included among them are extra-opus style and intra-opus style, as just described. In addition, the implication-realization model hypothesizes the existence of five general bottom-up processes that do not depend on knowledge of either extra-opus or intra-opus style. These are embodied in the five principles at the low-level: registral direction, intervallic difference, registral return, proximity, and closure. In addition, the principles may apply to implicative intervals at higher levels. For simplicity, these principles are represented at just one higher level in the Fig.4. As time, represented horizontally, progresses, different sources of expectancy will be made active at different points in time.

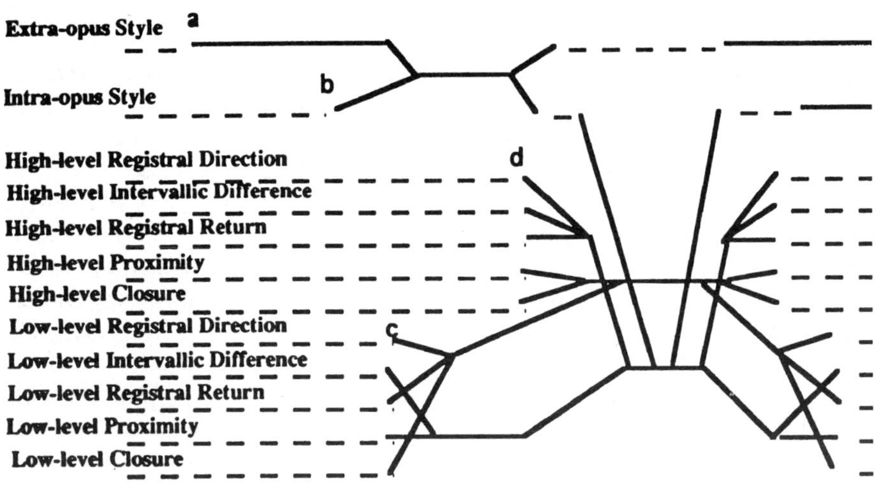

Fig. 5. Schematically illustrates the variety of psychological processes that may be engaged by a piece of music. Represented are extra-opus style, intra-opus style, and the five bottom-up principles of the implication-realization model the low level and one higher level: registral direction, intervallic difference, registral return, proximity, and closure. Time is represented along the horizontal axis. An activated psychological process is indicated by a solid line; convergences between the different sources of expectancy is indicated by convergence between the lines

As shown schematically in the Fig.5, suppose that the tonality of the music is perceived relatively early (at point a), engaging expectancies based on typical tonal and harmonic patterns of the key of the piece. This is represented by the transition between the dotted and solid line for extra-opus style. Then (at point b), a motivic pattern is heard that is similar to a previously heard pattern, generating expectancies that coincide to some degree with the expectancies from extra-opus sources, as indicated by the convergence between the solid lines for extra-opus and intra-opus style. Perhaps just after this moment (at point c), metrical, durational, and harmonic instability establish a low-level implicative interval. This interval engages the bottom-up processes as specified by the implication-realization model. In this hypothetical example, the size of the particular implicative interval is such that low-level closure, registral return, and registral direction generate similar expectancies for continuation, and these are different from those predicted by proximity and intervallic difference. Perhaps at a somewhat later time (at point d), an implicative interval at a higher level is established, engaging expectancies that exhibit their own patterns of expectancies that coincides to some degree with expectancies from the low-level implicative interval, as indicated by the convergence between low- and high-level principles. As indicated, these expectancies may also coincide to some degree with expectancies generated from extra- and intra-opus style knowledge. These complex interactions between factors would then continue as the sequence progressed.

Although this example is hypothetical, it is largely consistent with the kinds of processes that have been examined by psychological experimentation. Gestalt principles, such as proximity, similarity, and good continuation play a fundamental role in the organization of auditory information into coherent well-formed units. It is not surprising to find that they also play a role in determining expectations. Music capitalizes on the interactions possible between these principles, which are especially complex when a number of different levels of patterning are present. This produces a dynamic, constantly changing engagement of cognitive processes by the music as it unfolds in time. Similar processes operate on metrical, harmonic, and rhythmic aspects of the music along with the melodic aspect investigated here, and similar complexity may be found there as well. Thus, although principles of perceptual organization appear on the surface to be somewhat fixed and mechanical, in music they are engaged in such a way as to generate a continually changing pattern of cognitive expectancies commensurate with the richness and variety of our musical experience.

Acknowledgement

This article was begun while the author was a Fellow at the Center for Advanced Study in the Behavioral Sciences, supported by a grant from the National Science Foundation (SES-9022192) and a Sabbatical Leave Award from the James Keenan Cattell Fund. E. Narmour contributed extensively to the research. He composed the stimulus sequences, provided the analyses of the sequences shown in the figures, prepared the figures, and consulted extensively on all aspects of the project. E. Scheirer wrote the computer program and conducted the experimental sessions.

References

Bartlett, J., & Dowling, W. (1980). The recognition of transposed melodies: A key-distance effect in developmental perspective. *Journal of Experimental Psychology: Human Perception and Performance, 6*, 501–515.

Boltz, M., & Jones, M. (1986). Does rule recursion make melodies easier to reproduce? If not, what does? *Cognitive Psychology, 18*, 389–431.

Boltz, M., Marshburn, E., Jones, M., & Johnson, W. (1985). Serial-pattern structure and temporal-order recognition. *Perception and Psychophysics, 37*, 209–217.

Bregman, A. (1990). *Auditory scene analysis: The perceptual organization of sound.* Cambridge, MA: The MIT Press.

Carlsen, J. (1981). Some factors which influence melodic expectancy. *Psychomusicology, 1*, 12–29.

Carlsen, J., Divenyi, P., & Taylor, J. (1970). A preliminary study of perceptual expectancy in melodic configurations. *Council for Research in Music Education Bulletin, 22*, 4–12.

Cuddy, L., & Badertscher, B. (1987). Recovery of the tonal hierarchy: Some comparisons across age and levels of musical experience. *Perception and Psychophysics, 41*, 609–620.

Cuddy, L., Cohen, A., & Miller, J. (1979). Melody recognition: The experimental application of rules. *Canadian Journal of Psychology, 33*, 148–157.

Cuddy, L., & Lunney, C. (1995). Expectancies generated by melodic intervals: Perceptual judgments of melodic continuity. *Perception and Psychophysics, 57*, 451–462.

Deutsch, D. (1980). The processing of structured and unstructured tonal sequences. *Perception and Psychophysics, 28*, 381–389.

Deutsch, D., & Feroe, J. (1981). The internal representation of pitch sequences in tonal music. *Psychological Review, 88*, 503–22.

Dowling, W. (1978). Scale and contour: Two components of a theory of memory for melodies. *Psychological Review, 85*, 341–354.

Dowling, W., & Bartlett, J. (1981). The importance of interval information in long-term memory for melodies. *Psychomusicology, 1*, 30–49.

Fraisse, P. (1982). Rhythm and tempo. In D. Deutsch (Ed.), *The psychology of music.* New York, NY: Academic Press.

Koffka, K. (1935). *Principles of Gestalt psychology.* New York, NY: Harcourt, Brace, and Company.

Krumhansl, C. (1990). Tonal hierarchies and rare intervals in music cognition. *Music Perception, 7*, 309–324.

Krumhansl, C. (1991). Melodic structure: Theoretical and empirical descriptions. In J. Sundberg, L. Nord, & R. Carlson (Eds.), *Music, language, speech, and brain.* London: MacMillan.

Krumhansl, C. (1995a). Effects of musical context on similarity and expectancy. *Systematische Musikwissenschaft*, *3*, 211–250.

Krumhansl, C. (1995b). Music psychology and music theory: Problems and prospects. *Music Theory Spectrum*, *17*, 53–90.

Krumhansl, C., & Kessler, E. (1982). Tracing the dynamic changes in perceived tonal organization in a spatial representation of musical keys. *Psychological Review*, *89*, 334–368.

Meyer, L. (1956). *Emotion and meaning in music.* Chicago, IL: University of Chicago Press.

Meyer, L. (1973). *Explaining music.* Berkeley, CA: University of California Press.

Monahan, C., & Carterette, E. (1985). Pitch and durations as determinants of musical space. *Music Perception*, *3*, 1–32.

Narmour, E. (1990). *The analysis and cognition of basic melodic structures: The implication-realization model.* Chicago, IL: Chicago University Press.

Pollard-Gott, L. (1983). Emergence of thematic concepts in repeated listening to music. *Cognitive Psychology*, *15*, 66–94.

Rosner, B., & Meyer, L. (1982). Melodic processes and the perception of music. In D. Deutsch (Ed.), *The psychology of music.* New York, NY: Academic Press.

Rosner, B., & Meyer, L. (1986). The perceptual roles of melodic process, contour, and form. *Music Perception*, *4*, 1–39.

Schmuckler, M. (1989). Expectation in music: Investigation of melodic and harmonic processes. *Music Perception*, *7*, 109–149.

Schmuckler, M. (1990). The performance of global expectations. *Psychomusicology*, *9*, 122–147.

Simon, H., & Sumner, R. (1968). Pattern in music. In B. Kleinmuntz (Ed.), *Formal representation of human judgement.* New York, NY: Wiley.

Smith, K., & Cuddy, L. (1989). Effects of metric and harmonic rhythm on the detection of pitch alterations in melodic sequences. *Journal of Experimental Psychology: Human Perception and Performance*, *15*, 457–471.

Unyk, A., & Carlsen, J. (1987). The influence of expectancy on melodic perception. *Psychomusicology*, *7*, 3–23.

Continuations as Completions: Studying Melodic Expectation in the Creative Microdomain *Seek Well*

Steve Larson

School of Music, University of Oregon, 97403-1225 Eugene, OR, USA

Abstract. Listening to music is a creative process in which we shape the sounds we hear into expressive meanings tempered by our nature and experience. And when experienced listeners of tonal music are engaged in that process, they are constantly making predictions about what will happen next. Because of this central connection between meaning and expectation, exploring melodic expectation allows us to listen to how the musical mind works. The purpose of this paper is to describe a context for studying some aspects of melodic expectation: a *creative microdomain* called *Seek Well*. In this domain, one is given a monophonic melodic beginning (in notes of equal durations) and asked to finish it by adding any number of notes (of equal duration). Reference to familiar melodies is discouraged so that responses may represent shared intuitions about tonal music. This chapter describes two experiments that gathered data via *Seek Well*. The results challenge central assumptions of previous studies of melodic expectation, illustrate the value of *Seek Well* as a method of gathering data, and raise additional interesting questions.

1 Introduction

Expectation plays a central role in the understanding and enjoyment of music. And a wide variety of methods have been used for studying musical expectations.

Some experimental approaches severely restrict the contexts in which subjects produce continuations. J. Carlsen and his collaborators have asked subjects to sing continuations of two-note beginnings (Carlsen, Divenyi, & Taylor, 1970; Carlsen, 1981; Unyk & Carlsen, 1987). W. Lake has also asked subjects to sing continuations of two-note beginnings, but for each two-note beginning, he first established a major-key context (Lake, 1987). In the experiments of both Carlsen and Lake, subjects sang continuations that added more than one new note, but researchers analyzed only the first added tone. While these studies have added to our knowledge of melodic expectation, some musicians and scholars (e.g. Schmuckler, 1990) have suggested that such severely restricted contexts limit the applicability of the findings of such studies. And restricting analysis to just the first added tone seems not only to ignore a significant part of the response, but also to assume – without persuasive justification – that only that tone is important in describing expectations.

Experiments that ask listeners to judge how well a probe tone or chord fits an established context may also be regarded as asking listeners to rate the degree to which that tone or chord is expected in that context. C. Krumhansl and her collaborators have asked subjects to provide numerical ratings of such *fit* (some of these studies are discussed in Krumhansl, 1990). One of these studies (Krumhansl, 1995) uses the probe-tone technique to gather data to test predictions of the bottom-up component of Narmour's implication-realization model of melodic expectancy (Narmour, 1990). J. Bharucha and his collaborators have measured the amount of time it takes subjects to make judgments about the intonation of chords, relating the response time to the relationship between the chord judged and the chord that preceded it (Bharucha & Stoeckig, 1986, 1987; Tekman & Bharucha, 1992). While the contexts for these experiments are often larger and more like real music, subjects are asked to judge limited materials that they do not produce. Other experimental approaches use excerpts of real music. M. Schmuckler (Schmuckler, 1989, 1990) has asked subjects to judge or produce continuations to excerpts of previously composed music. Krumhansl (1995) cites a number of other studies of musical expectancy that rely on excerpts of composed music. Although the use of previously composed examples brings us closer to actual musical practice, it also has drawbacks. Because the experimenter does not control specific details, questions may arise about the significance of those details or about assumptions related to the specific musical style of the excerpt chosen.

The method of exploring melodic expectation proposed in this paper – a creative microdomain called *Seek Well* – finds a middleground between the atomistic oversimplification of studying only the first added tone to a two-note beginning and the confusing complexity of (and hidden factors in) comparing judgments of listeners to entire, stylistically-specific excerpts. *Seek Well* is based on four assumptions. The first assumption is that listeners' expectations are not merely for melodic continuations, but in fact for entire melodic completions. The second assumption is that such completions match one another note-for-note often enough to justify viewing those expectations as based upon cognitive processes that may be described as rule-driven. The third assumption is that hypotheses about such rule-driven processes can best be tested by implementing them in a computer program that will be judged successful if it can generate completions that match the completions of a pool of subjects as often as those of any subject in that pool do. The fourth assumption is that expectation is best studied and explained within a domain that is not only well enough defined to allow such tests, but also complex enough to retain some of the sophistication of real musical behavior.

These assumptions themselves suggest four testable hypotheses:

1. Entire responses by different subjects will match one another note-for-note at levels significantly beyond chance and often enough to produce a small set of shared responses.

2. Grouping responses solely on the basis of the first added tone – as is often done with other experimental methods – confuses data by making different responses appear similar and by making similar responses appear different.
3. Key-determination is central to the process of generating melodic expectations and relevant to classifying and understanding responses; and
4. The completions that match note-for-note, as well as their distribution, so well represent intuitions shared by experienced musicians that similar results may be produced under varied conditions.

In what follows, I will summarize the idea of *creative microdomains*, describe a specific new microdomain called *Seek Well*, and describe two experiments that support the hypotheses mention above.

2 Creative Microdomains for Studying Cognition

In *Fluid Concepts and Creative Analogies* (Hofstadter & the Fluid Analogies Research Group, 1995), D. Hofstadter and his colleagues describe work in cognitive science that studies creativity in human cognition by designing and experimenting with *creative microdomains* and by building computer models of creative thought in those domains. A *creative microdomain* is a limited context in which one is asked to make choices – in essence, a game. To date, a number have been designed, these include: *Seek Whence*, which extrapolates sequences of integers; *Copycat*, which creates analogies between strings of letters; *Jumbo*, which creates word-like anagrams of letter strings without consulting a dictionary; *Tabletop*, which creates analogies between collections of items found on a dining table; and *Letter Spirit*, which creates a whole typeface alphabet from a single *seed* letter. Computer programs have been created that solve problems in each of these domains. *Fluid Concepts and Creative Analogies* demonstrates persuasively that each of these domains provides real insight into the fundamental mechanisms of thought.

Because those mechanisms are so fundamental, we should expect them to be relevant to music cognition. In fact, Hofstadter observes that issues arising in these microdomains are central to the art of melody-making. He mentions issues of the hierarchical nature of musical structure, the influence of context on function, the implicit polyphony of compound melody, and the importance of pattern in expectation and musical meaning.

3 Seek Well

The purpose of *Seek Well* is to find a musical continuation to a given melody. And a guiding principle behind my work with *Seek Well* is that the crux of discovering good melodies lies in the high-level perception of underlying patterns. *Seek Well* was developed in consultation with Hofstadter, and he and I have begun to design a computer program that will play *Seek Well*.

Since the domain is a musical one, its most immediate payoff should be in music cognition, especially the study of creativity and expressive meaning in music. That is, I expect *Seek Well* to tell us some things about how experienced listeners of tonal music think about music and how they think in music.

By *tonal music*, I mean much of the concert music and popular music of the last couple of centuries in western Europe and America – music based on functional triadic tonality. I have played *Seek Well* enough to feel that it captures something that experienced tonal listeners share. It might even be regarded as a miniature *Turing test* for musical intelligence in tonal music. This restriction to *common-practice* tonal music is an admittedly artificial one. Defined this way, tonal music is only a small percentage of the music of the world and from a small percentage of its history. But this is the musical system that I know best, and therefore seems the best place for me to start. Nevertheless, since the focus is on the nature of musical hearing and musical creativity, I would not be surprised if *Seek Well* should tell us something about human creativity that might ultimately be applicable to understanding music of other periods and cultures.

By *how we think about music*, I mean how we use words or images to respond to musical experiences.

By *how we think in music* I mean how we hear notes (that is, pitches and durations instead of words or images) to respond to musical experiences. One example of *thinking in music* occurs when a bit of melody stops unexpectedly. In such a situation, especially if we hear that bit of melody as a musical *question*, we may *auralize* a musical *answer*.

Nevertheless, it would be a mistake to regard *Seek Well* as a purely musical domain. Problem solving in *Seek Well* should inform us about creativity and problem solving in general. Comparisons between *Seek Well* and other domains should inform us about the degree to which aspects of creativity are domain-specific.

Please complete the following melodic beginning:

Fig. 1. A sample problem in *Seek Well*

Figure 1 offers a sample problem in *Seek Well*. Players are given a simple monophonic melodic beginning. The beginning may include any number of notes, but they must all be of equal durations. Players are then asked to sing a completion (of any length). Instructions discourage reference to familiar melodies so that continuations may represent shared intuitions about tonal music rather than play on knowledge of specific pieces. The only restriction is that completions represent *thinking in music*, rather than *thinking about music*. For instance, if hearing the first three notes leads one to sing C-A-G-E, that would be in the spirit of *Seek Well*. But to get the answer C-A-G-E by spelling the name of a twentieth-century composer would not be in the spirit of *Seek Well*.

What is the best simple melodic conclusion for this three-note beginning? Notice that *best* and *simple* are not defined here. There are two reasons for this. First, I think that we can go a long way without explicitly agreeing on what these terms mean – in fact, I think that it may be better to rely on problems in *Seek Well* to provide the context that determines the meaning of these terms. Second, I think Seek Well itself can teach us something about how we understand these concepts.

The two experiments described in what follows use *Seek Well* to gather data and test the hypotheses mentioned above.

4 Experiment One

The first experiment described in this paper was performed at the University of Oregon in February 1997. Using the method of *Seek Well*, data were gathered to test the hypotheses listed above.

4.1 Method

Subjects: Sixty-five musicians were recruited from the University of Oregon School of Music. All were music majors or music faculty. All were unpaid volunteers for a session that lasted less than three minutes.

Stimulus: For sopranos and altos, the three tones shown in Fig.1 were presented. For tenors and basses, the same tones were presented an octave lower. Tones of one second in duration and with no breaks between them were synthesized in general MIDI Acoustic Grand Piano sound through a Roland SC-7 Sound Module and recorded onto a cassette tape using a Sony Audio-Video Control Center (Model STR-AV200) and an HX Pro Stereo Cassette Deck (Model TC-WR670). For the experiment, the cassette tape was played back on a JVC KD-25 Stereo Cassette Deck at a comfortable listening level.

Procedure: Subjects were given the following instructions:

> Thank you for participating in this experiment. Its purpose is to gather information about how musicians like yourself expect melodies to be completed. I believe that this will help us understand how musicians *think in*

> *tonal music* and it may help me to be a better music teacher. Please think of this as a game. Give the response you think most other musicians will give when given the following instructions.
> You will hear a short melodic beginning of long notes in equal durations. Then you'll sing back that beginning, adding one or more notes of the same duration to complete the melody in the way you think best projects your sense of *the way tonal music goes*. So you should sing the simplest or most natural conclusion you can imagine. Singing a familiar melody is discouraged so that responses will represent shared intuitions about tonal music.
> I will post the responses later (without identifying who sang which ones). Please don't discuss this experiment with anyone else until after the responses are posted.

After an opportunity to ask questions, subjects were tested individually. They heard the stimulus and then sang a response. I notated the response and then sang or played it back for them to confirm that I had correctly notated their response.

4.2 Results

Most subjects sang their completions with little or no hesitation. The sixty-five subjects produced twenty-six unique responses. Figure 2 lists every response that was sung by more than one subject. The eight responses listed in Fig.2 represent forty-seven of the subjects (72%). These responses are listed, left-to-right and top-to-bottom, in the order of their popularity. Each response is numbered (but notice that ♯6, ♯7, and ♯8 tied at two responses each). Then the number of subjects who sang that response is expressed as a percentage of the forty-seven represented in Fig.2. The first response was sung by fourteen subjects (30%). Furthermore, those responses that suggest the key of C are listed in the left-hand column and those that suggest the key of F are listed in the right-hand column (subject protocols confirm that this division corresponds to their own sense of the key implied).

4.3 Discussion

The first hypothesis mentioned above is that entire responses by different subjects will match one another note-for-note at levels significantly above chance and often enough to produce a small set of shared responses. Those experimental methods cited above that collect or analyze only the first added tone of a continuation appear to assume the opposite.

Two basic sets of reasons are given or implied for focusing on just the first added tone. The first set is based on assumptions about how such continuations are produced. According to one of these assumptions, each new continuation tone is predicted one-at-a-time by consulting some internalized table of probabilities or by following *bottom-up* principles of note-to-note

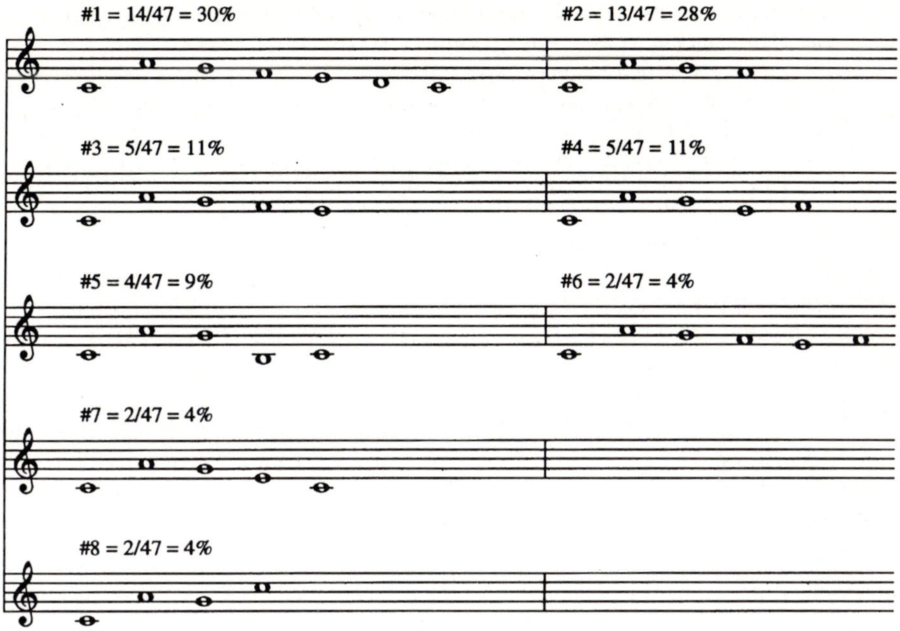

Fig. 2. Shared responses, Experiment One.

expectancies. The addition of one new continuation tone creates a new beginning. This new beginning then leads to the prediction of a single new tone, again by consulting some internalized table of probabilities or bottom-up principles. This assumption leads to two assertions: (a) that each new continuation tone is less constrained by the original melodic beginning, and (b) that each new continuation tone is thus of less interest. The second set of reasons for focusing on just the first added tone is based on assumptions about how such continuations should be explained. According to one of these assumptions, the purpose of these studies is restricted to gathering data and describing them statistically. If each new continuation tone multiplies the number of possible continuations, then the number of possibilities explodes exponentially. And if we allow that continuations may be of various lengths, then comparing complete continuations can appear dauntingly complex. Limiting analysis to just the first added tones thus responds to assumptions both about subjects' expectations and about researchers' purposes.

The data in Fig.2 seem to contradict both of these assumptions and to support the hypothesis that entire responses by different subjects will match one another note-for-note at levels significantly beyond chance and often enough to produce a small set of shared responses.

The first assumption, that longer responses are less constrained by the stimulus, would lead us to expect longer responses to match less often. This seems right; with more notes, there would seem to be more opportunities to disagree. But in Fig.2, the most popular response is also the longest. The shortest response is the least popular. Even for the whole data set (all sixty-five subjects), the correlation between response length and number of matches is negative (r = -0.186).

The second assumption, that considering entire responses will lead to an exponential explosion of possibilities, is also contradicted by the data. If one considers only the first added tone, one must include all subject responses. But in *Seek Well*, *outliers* are eliminated in a principled way. To ensure that data reflect shared tendencies, responses that do not match note-for-note with at least one other response are omitted. Thus, although their number is small (28%), some responses were justifiably omitted from Fig.2. The number of responses in Fig.2, despite their length, is only eight. In fact, if we look at all the responses (as must be done in considering only the first added tone), we find that there are also eight different first added tones. Clearly, we have no exponential explosion – even if we consider all twenty-six unique responses.

The second hypothesis mentioned above is that grouping responses solely on the basis of the first added tone confuses data by making different responses appear similar and by making similar responses appear different. Again, Fig.2 appears to support this hypothesis. If we consider only the first added tones (the tones after the initial C-A-G in Fig.2), then the four responses in Fig.2 whose first added tone is F are lumped together. But these responses include two that suggest the key of C and two that suggest the key of F, making these very different classes of continuations seem inappropriately similar. On the other hand, the responses in the right-hand column are genuinely similar. They all suggest the key of F and they all end on F (in fact, the longer ones may be heard as embellishments of the shortest of these, created by hearing in the same key, but interpreting the meter differently – more on meter below). Since they have different first added tones, grouping responses on the basis of the first added tone would again confuse the data, here by making these essentially similar responses appear different.

By the way, when I have played this game informally with others, I have sometimes asked people to come up with several continuations for this beginning. Most people sing continuations in either C or F, but not both. This suggests that once a key determination is made, it is difficult to dislodge (Butler, 1989). Furthermore, most people who hear continuations in F report that they heard the cue in a way that was different metrically – they either heard the first note as an upbeat in duple meter or as a downbeat in triple meter. This suggests that perception of meter and key are inter-related (Kidd, 1984). Note that the imposition of equal durations on the melodies has not eliminated rhythm and meter from our considerations.

Thus, these data support the third of the hypotheses mentioned above, that key-determination is a central part of the process of generating melodic expectations and relevant to classifying and understanding responses (Larson, 1993).

5 Experiment Two

The second experiment described in this paper was performed at the 1996 national meeting of the Society for Music Theory. Using a variation of *Seek Well*, data were gathered to test the hypotheses listed above.

5.1 Method

Subjects: Sixty-five musicians were recruited from those attending the Society for Music Theory conference. The experiment was advertised as a contest in which the winner would be awarded a bottle of wine (a Staton Hills 1994 Oregon Pinot Noir). The experiment was administered at a poster session presenting work of the Music Cognition Group of Society for Music Theory.

Stimuli: Instead of recorded sound, musicians were given melodic beginnings (cues) in notation and asked to notate their completions. Furthermore, participants were asked to assume that they heard the cues in the context of C major.

Procedure: The cues are presented in Fig.3. The instructions were presented on the contest blank as follows:

> *Instructions:* Write one melodic continuation (adding one or more whole notes) for each of the following beginnings. Assume that each melody is heard in a context in which C is the tonic. Assume that each contestant has been asked to write continuations they hear, continuations that best represent the tendencies of common-practice tonality. Please write the continuation that you think will appear most often on other contestants' pages. Do not discuss your answers with anyone who will subsequently participate in the contest!
> *Scoring:* You score a point every time another contestant submits a continuation that is identical to yours.
> *Prize:* The winner will receive a free bottle of wine. If there is a tie, there will be a drawing between the top contestants.

5.2 Results

The last of the cues listed in Fig.3 is the same as the cue used in Experiment One. In response to that cue, the sixty-two subjects (as compared with sixty-five in Experiment One) produced twenty-six unique responses (the same number as in Experiment One). The left-hand column of Fig.4 lists every response that was notated by more than one subject. Each response is numbered (but notice that ♯1 and ♯2 tied, ♯4 and ♯5 tied, and ♯7 and ♯8 tied). Then the number of subjects who sang that response is expressed as a percentage

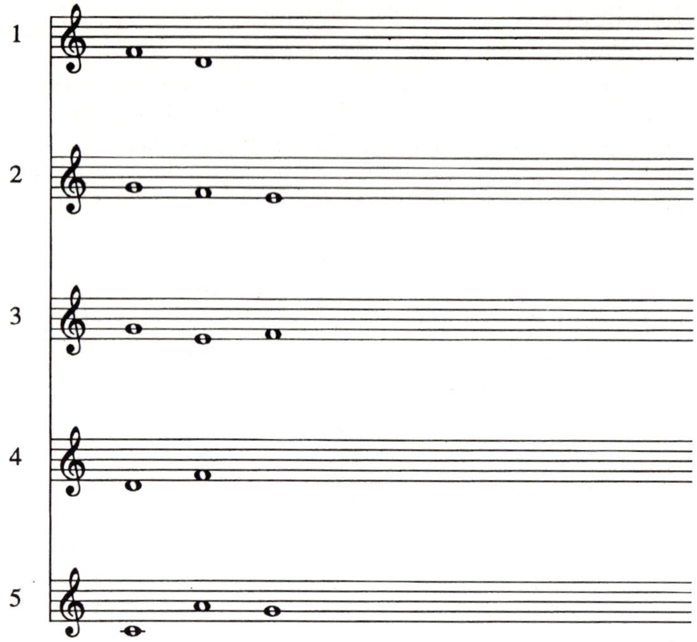

Fig. 3. Cues in Experiment Two

of the forty-six subjects represented by the first seven responses (because the eighth response suggests the key of F, the two subjects represented by that response are not included in calculating the percentages of subjects singing shared responses). For comparison, the right-hand column of Fig.4 lists all the responses sung by subjects in Experiment One that suggest C major (as well as the response that matches the F major completion). Excluding the last response, the five C-major responses listed in the right-hand column represent twenty-seven subjects.

5.3 Discussion

The left-hand column of Fig.4 shows the completions notated by professional music theorists instructed to *auralize* the cues as if in C major (Experiment Two). The right-hand column of Fig.4 shows the responses sung by musicians whose completions suggest they heard the cues as if in C major (Experiment One). When we consider the thousands of possible continuations, the agreement on just these seven responses is striking. And the similarity of their distributions is astounding. (Even if we consider the distributions of only these responses, the correlation is high: r = .738. If we consider all possible diatonic responses of up to four added tones in the same vocal range as those listed – there are 4680 – the correlation is very high.)

Thus, these data offer additional support for the three hypotheses discussed above. They also support the fourth hypothesis: that the completions that match note-for-note, as well as their distribution, so well represent intuitions shared by experienced musicians that similar results may be produced under varied conditions.

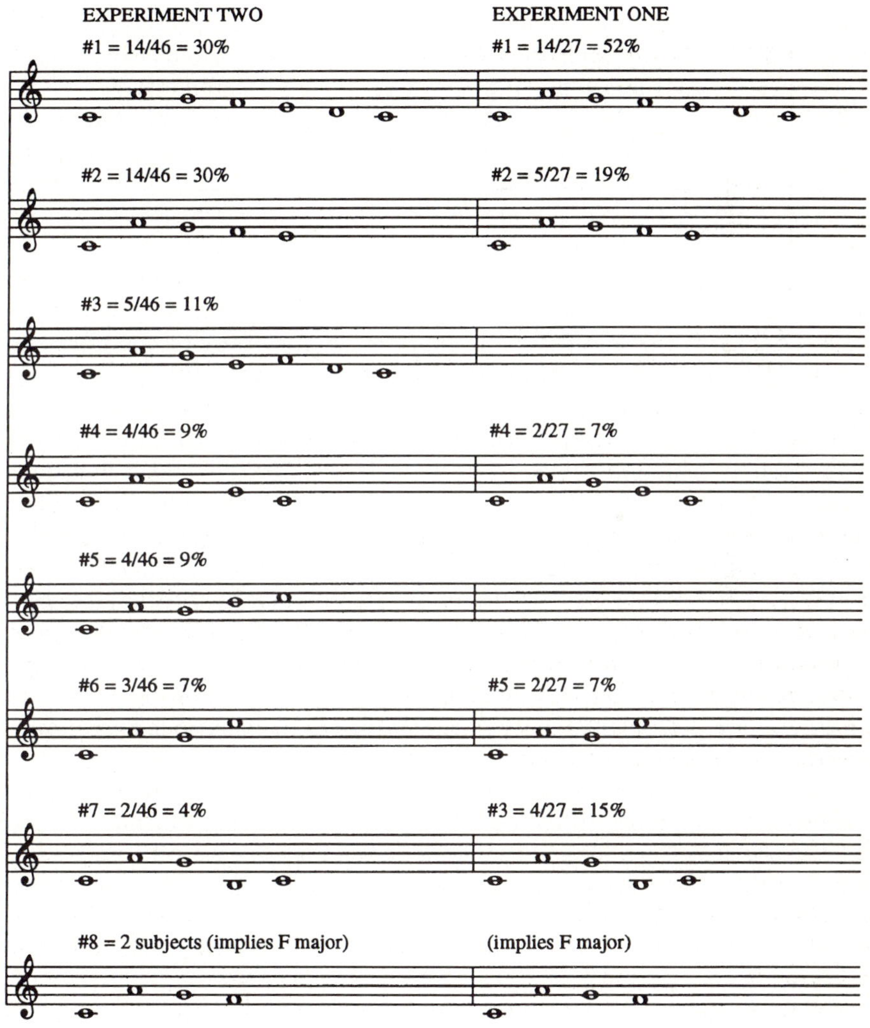

Fig. 4. Shared responses, both experiments

6 Conclusion

In these experiments, most subjects sang a response that agreed note-for-note with another subject's entire response – and the total number of different entire responses was small. In these experiments, collecting and analyzing only the first added tones would have confused the data (making similar responses look different and making different responses look similar). And in these experiments, key determination and melodic continuation appear intertwined. These findings suggest that we should regard the melodic expectations of subjects (as represented by their sung or notated continuations) as expectations not so much for continuations, but as expectations for completions.

These findings also suggest that *Seek Well* is an excellent domain for studying such completions. It does not stipulate or establish the key. It allows for beginnings or completions of any length. Yet it is restricted to single notes of equal durations. The balance appears to allow for the richness of complete musical responses within well-controlled experimental conditions.

These findings also suggest that good explanations of melodic expectation should be able to generate entire completions, rather than merely to rate the degree to which a first added tone is expected. Some of my recent work suggests that this is possible. I have created a simple computer program called Next Generation, based on the ideas of melodic completion described in Larson (1994). Next Generation can match note-for-note the entire continuations sung by subjects in a previous experiment (Lake, 1987) – and it can do so as often as (or more often than) do those of any of those subjects (Larson, 1995). However, that experiment (like Experiment Two above) established the key before presenting cues. Next Generation must also be given the key and it considers only the last two notes of a cue in creating predictions. Nevertheless, the success of Next Generation suggests that explanations capable of generating entire completions may be possible.

Finally, these findings raise further questions. If we consider entire responses, can we detect differences in the nature or distributions of responses that reflect experience with tonal music? That reflect cultural or other differences? Are there variations on the game that do alter the nature or distributions of responses? What about listeners from other cultures? Under what circumstances can different melodic beginnings suggest different styles? What additional hypotheses about melodic expectation can be tested with *Seek Well*?

References

Bharucha, J., & Stoeckig, K. (1986). Reaction time and musical expectancy: Priming of chords. *Journal of Experimental Psychology: Human Perception and Performance, 12*, 403–410.

Bharucha, J., & Stoeckig, K. (1987). Priming of chords: Spreading activation or overlapping frequency spectra? *Perception and Psychophysics, 41*, 519–524.

Butler, D. (1989). Describing the perception of tonality in music: A proposal for a theory of intervallic rivalry. *Music Perception, 6*, 219–241.

Carlsen, J. (1981). Some factors which influence melodic expectancy. *Psychomusicology, 1*, 12–29.

Carlsen, J., Divenyi, P., & Taylor, J. (1970). A preliminary study of perceptual expectancy in melodic configurations. *Council for Research in Music Education Bulletin, 22*, 4–12.

Hofstadter, D., & the Fluid Analogies Research Group. (1995). *Fluid concepts and creative analogies: Computer models of the fundamental mechanisms of thought.* New York, NY: Basic Books.

Kidd, G. (1984). *The perception of tonal relationships in music.* Unpublished doctoral dissertation, Ohio State University.

Krumhansl, C. (1990). *Cognitive foundations of musical pitch.* New York, NY: Oxford University Press.

Krumhansl, C. (1995). Music psychology and music theory: Problems and prospects. *Music Theory Spectrum, 17*, 53–90.

Lake, W. (1987). *Melodic perception and cognition: The influence of tonality.* Unpublished doctoral dissertation, University of Michigan.

Larson, S. (1993). *Computer models of melodic implication and key determination in tonal music* (CRCC Technical Report No. 77). Bloomington, IN: Center for Research on Concepts and Cognition.

Larson, S. (1994). Musical forces, step collections, tonal pitch space, and melodic expectation. In *Proceedings of the Third International Conference on Music Perception and Cognition* (pp. 227–229). Liège: ESCOM.

Larson, S. (1995). *"Musical forces" and melodic expectations: Comparing computer models with human subjects.* (Unpublished manuscript)

Narmour, E. (1990). *The analysis and cognition of basic melodic structures: The implication-realization model.* Chicago, IL: Chicago University Press.

Schmuckler, M. (1989). Expectation in music: Investigation of melodic and harmonic processes. *Music Perception, 7*, 109–149.

Schmuckler, M. (1990). The performance of global expectations. *Psychomusicology, 9*, 122–147.

Tekman, H., & Bharucha, J. (1992). Time course of chord priming. *Perception and Psychophysics, 51*, 33–39.

Unyk, A., & Carlsen, J. (1987). The influence of expectancy on melodic perception. *Psychomusicology*, *7*, 3–23.

IV

From Timbre to Texture

Optimizing Self-Organizing Timbre Maps: Two Approaches

Petri Toiviainen

Department of Musicology, University of Jyväskylä, PL 35 Jyväskylä, Finland

Abstract. The effect of using different auditory images and distance metrics on the final configuration of a self-organized timbre map is examined by comparing distance matrices obtained from simulations with a similarity rating matrix, obtained using the same set of stimuli as in the simulations. Two approaches are described. In the static approach, each stimulus is represented as a single multi-component vector. Gradient images, which are intended to represent idealizations of physiological gradient maps in the auditory pathway, are constructed. The optimal auditory image and distance metric, with respect to the similarity rating data, are searched using the gradient method. In the dynamic approach, each input stimulus is represented by a set of spectral vectors.The response patterns are constructed through temporal integration. The onset portions of the stimuli are emphasized using a scaling procedure which operates qualitatively in the same way as lateral inhibition.

1 Introduction

There have been a number of studies aiming at extracting the most salient acoustic attributes affecting the perception of timbre (Iverson & Krumhansl, 1993). A method widely used in these studies is similarity rating: subjects are asked to rate, on a given scale, the similarity of all possible pairs in the set of stimuli. Multidimensional scaling is then used to map the tones into a low-dimensional timbre space. By examining multidimensional scaling maps it has been found that the brightness of a tone is one of the main contributors to the perception of timbre. Dynamic attributes, mostly in the onset portion of a tone, have also been shown to have a considerable influence on the perception of timbre.

A timbre space can also be constructed from a set of acoustical signals by means of connectionist models; for this purpose, the Kohonen self-organizing map (Kohonen, 1995) has been used by a number of researchers (Cosi, De Poli, & Lauzzana, 1994; De Poli, Prandoni, & Tonella, 1993; Feiten & Günzel, 1994; Toiviainen, Kaipainen, & Louhivuori, 1995; Toiviainen, 1996).

There is evidence that the human auditory system encodes frequency transitions (Steiger & Bregman, 1981). In the cochlear nucleus, cells responsive to onsets and offsets have been found (Pickles, 1982). Shamma et al (1992) found that the distribution of responses encodes the locally averaged gradient of the acoustic spectrum.

During the self-organizing process of the self-organizing map, the input vectors are compared with the synaptic vectors of the network using a distance metric. The prevalent choice for distance metric in the self-organizing map simulations is the Euclidean metric. It, however, does not necessarily correspond to how the auditory system compares stimuli.

The goals of the present study were (1) to examine the effect of different auditory images and distance metrics on the final configuration of a self-organized timbre map; and (2) to explore whether the self-organizing map is able to project the used auditory images into two dimensions while maintaining the metrical relations between the images. The main method was that of calculating correlation coefficient values between distance matrices obtained from the simulations and a similarity rating matrix obtained using the same set of tones. The overall aim was to find the auditory image and distance metric which would yield the highest correlation with the similarity rating data.

This paper describes two approaches, subsequently referred to as the static and the dynamic approach. In the static approach, each input stimulus is presented to the self-organizing map as a single multicomponent vector. The onset of the stimuli is emphasized using an exponential mapping of time. Spectral images are combined with gradient images, i.e., images which are supposed to represent qualitatively responses of neurons sensitive to frequency and amplitude modulation as well as neurons encoding the spectral gradient. In the dynamic approach, each input stimulus is represented by a set of spectral vectors, each of which represents the instantaneous spectrum of the stimulus at a given time. The response patterns are constructed using temporal integration on the self-organizing map; the emphasizing of onsets is carried out using a scaling procedure which operates qualitatively in the same way as a nonlinear activation function.

2 Materials

Synthetic stimuli were used instead of acoustic instrumental sounds in order to acquire a rich spectrum of timbres with an optimally economical set of variables and to keep the number of stimuli reasonable.

The 27 tones used in the experiments were produced by additive synthesis. Each tone had an equal sampling rate (20000 Hz), number of partials (16), and frequency of partials (fundamental frequency 440 Hz; a harmonic overtone series). The variable parameters were chosen so that they would correspond to the most salient acoustic attributes found in previous studies (Iverson & Krumhansl, 1993).

These parameters, and their ranges of variation in the stimulus set, were:

- brightness (ratio of maximum amplitudes of successive partials; 0.3-0.7);
- attack time (the time interval for the first partial to reach maximum amplitude; 0-350 ms);

- attack asynchronity (difference of attack times between 1st and 16th partial; 0-100 ms);
- sustain time (the time interval for the first partial to have maximum amplitude; 0-500 ms);
- decay time (the time interval, during which the amplitude of the first partial reduces to half; 50-100 ms); and
- time scaling (ratio of sustain and decay times of successive partials; 0.5-1.0).

Each tone ended with a 50 ms release period; that is, period of linear decay of amplitude. A mathematical description of the synthesis algorithm as well as the values of the parameters used in the creation of each tone in the stimulus set can be found in Toiviainen et al. (1995).

The values of the parameters were chosen so that they produced a wide range of different tones, including tones resembling those of piano, of strings, brass, and woodwind instruments. After the stimuli had been produced, their amplitudes were carefully adjusted by hand so as to make each tone identical in loudness.

3 Similarity Rating Experiment

Nine subjects participated in the similarity rating experiment. All of them have at least moderate skill in some musical instrument. During the experiment subjects heard each possible pair of tones, presented in random order. They were asked to rate the similarity of each pair on a scale between zero and twelve. In this study, the similarity rating matrix obtained by averaging the similarity rating matrices of each subject was used.

4 Static Approach

A more detailed description of this approach can be found in (Toiviainen, 1996).

4.1 Auditory Images

The timbre maps were constructed by a two-stage process. First, the auditory images were computed using an auditory model. Second, the obtained auditory images were fed into a self-organizing map which then projected them onto a two-dimensional map.

To obtain auditory images, the stimuli were preprocessed, using the peripheral part of an auditory model by Van Immerseel and Martens (1992), modified by Leman (1994) for musical purposes. The output of the model is a 20-component vector, updated every 0.4 ms; each component represents the probability of neural firing since the last update.

The Minkowski distance of two real-valued vectors, $\mathbf{p} \equiv (\xi_1, \ldots, \xi_n)$ and $\mathbf{q} \equiv (\eta_1, \ldots, \eta_n)$, is defined by

$$||\mathbf{p} - \mathbf{q}||_\lambda \equiv \left[\sum_{i=1}^{n} |\xi_i - \eta_i|^\lambda \right]^{1/\lambda} . \tag{1}$$

By varying the value of the Minkowski exponent λ, a wide range of different metrics can be achieved.The spectral image of a tone, having a duration T, can be expressed as a scalar-valued function on the subset $[f_0, f_1] \times [0, T]$ of the frequency-time-plane:

$$S : s(f,t), f \in [f_0, f_1], t \in [0, T]. \tag{2}$$

Here f_0 and f_1 denote the minimum and maximum of the frequencies included in the image, respectively; $s(f,t)$ is obtained by temporally integrating the output of the auditory model:

$$s(f,t) = \int_0^t p(f,t') \frac{1}{\tau} e^{-t'/\tau} dt'. \tag{3}$$

Here $p(f,t)$ denotes the neural firing probability of a neuron with a best frequency f, within a time interval $[t - dt, t]$; τ denotes the time constant.

The onset of the spectral image is emphasized by defining an exponential mapping of time:

$$\tau_k(t) \equiv \begin{cases} t & , \quad k = 0 \\ T \dfrac{e^{kt} - 1}{e^{kT} - 1}, & k > 0 \end{cases} \tag{4}$$

and a mapping on the frequency-time plane

$$\sigma_k : \sigma_k(f,t) = s(f, \tau_k(t)). \tag{5}$$

The exponential parameter k defines the degree of emphasis on the onset; subsequently, it is referred to as the emphasis parameter. The onset-emphasized spectral image, Σ_k, is now defined to be

$$\Sigma_k : \sigma_k(f,t), f \in [f_0, f_1], t \in [0, T]. \tag{6}$$

The physiological gradient maps are modelled by means of directional derivatives. For a given direction vector $\mathbf{e} = (e_f, e_t)$ on the frequency-time-plane, the rate of change of function σ_k along the direction of $\mathbf{e}$ is expressed by its directional derivative, defined by

$$\gamma_k^{\mathbf{e}} \equiv \frac{\partial \sigma_k}{\partial \mathbf{e}} \equiv \mathbf{e} \cdot \nabla \sigma_k \equiv e_f \frac{\partial \sigma_k}{\partial f} + e_t \frac{\partial \sigma_k}{\partial t}. \tag{7}$$

For a given $\mathbf{e}$ and k, the corresponding gradient image, $\gamma_k^{\mathbf{e}}$, is now defined to be

$$\Gamma_k^{\mathbf{e}} : \gamma_k^{\mathbf{e}}(f,t), f \in [f_0, f_1], t \in [0,T]. \tag{8}$$

An illustration of gradient images is presented in Fig.1. According to Ru and Shamma (1997), several features of the auditory spectral pattern are analyzed and mapped out along the isofrequency axis of the auditory cortex. He suggests that neurons with asymmetric response fields around their best frequencies are capable of encoding the local skewness of the spectrum. The f-gradient image in Fig.1b can be regarded to represent the response of a group of such neurons. Ru and Shamma (1997) further states that many cells on the auditory cortex have been found to have impulse response functions such that they are capable of encoding the local dynamics of the spectrum around their best frequencies. The t-gradient image in Fig.1c can be regarded to represent the response of a group of such neurons, more specifically, neurons which are responsive to onsets around their best frequency. Further, the c1- and c2-gradient images (Fig.1d-e) stand for the response of a group of neurons with both asymmetric response fields and sensitivity to modulation. Ru and Shamma suggest that the auditory spectrum, including the aforementioned features of it, is represented in the auditory cortex with multiple degrees of frequency and time resolution. This is supported by many perceptual invariants such as the ability to recognize melodies despite large changes in tempo, or to perceive music through short duration interruptions in the sound stream. This kind of multiresolution representation is, however, not implemented in this model.

For a set $\{\mathbf{e}_1, \mathbf{e}_2, \ldots, \mathbf{e}_n\}$ of direction vectors on the frequency-time plane, a composite auditory image can be constructed as a vector-valued function, defined by

$$\begin{aligned}\mathbf{\Phi} : \varphi(f,t) \equiv (\alpha_s \sigma_k(f,t), \alpha_1 \gamma_k^{\mathbf{e}_1}(f,t), \ldots, \alpha_n \gamma_k^{\mathbf{e}_n}(f,t)),\\ f \in [f_0, f_1], t \in [0,T].\end{aligned} \tag{9}$$

The α's are the weighting factors of the corresponding images.The Minkowski distance of two (vector-valued) auditory images, $\mathbf{\Phi}_1$ and $\mathbf{\Phi}_2$, is defined, analogously to Equation 1, by

$$|||\mathbf{\Phi}_1 - \mathbf{\Phi}_2||| \equiv \left[\iint ||\varphi_1(f,t) - \varphi_2(f,t)||_\lambda^\lambda \, df \, dt\right]^{1/\lambda}, \tag{10}$$

where $||\varphi_1(f,t) - \varphi_2(f,t)||_\lambda$ is the Minkowski norm of vector $\varphi_1(f,t) - \varphi_2(f,t)$, and the integration is carried out over $[f_0, f_1] \times [0,T]$.

4.2 Metrics on the Self-Organizing Map

To obtain distance matrices from the self-organizing map simulations, two different distance metrics were used. In the response focus metric (RFM),

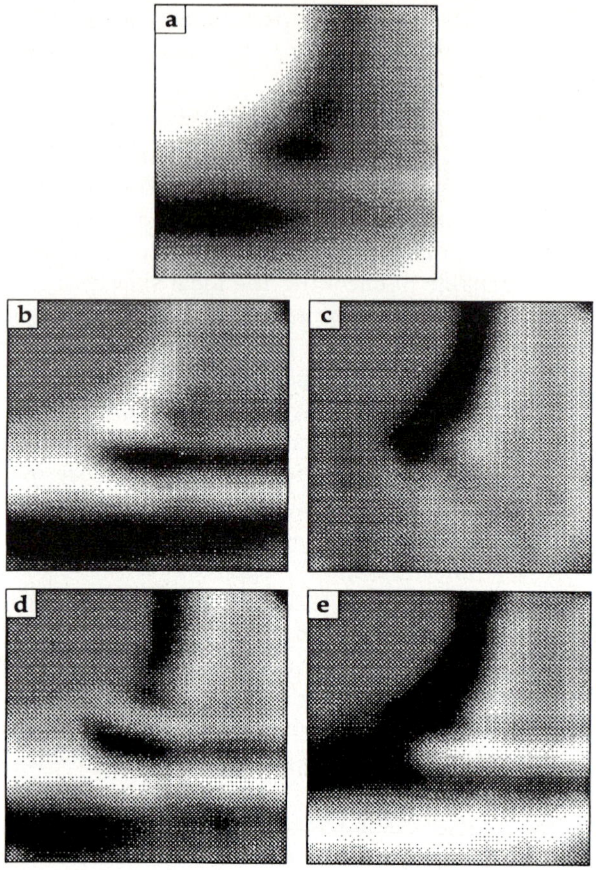

Fig. 1. An illustration of spectral and gradient images of tone 24 of the set of stimuli used (Toiviainen et al., 1995). In each of the subfigures the horizontal axis stands for time and the vertical axis for frequency. (a) The spectral image; the gradient images for direction vectors, (b) $\mathbf{e} = (0, 1)$ (parallel to the f-axis), (c) $\mathbf{e} = (1, 0)$ (parallel to the t-axis), (d) $\mathbf{e} = (1/\sqrt{2}, 1/\sqrt{2})$ (45 degrees counterclockwise from the t-axis), and (e) $\mathbf{e} = (1/\sqrt{2}, -1/\sqrt{2})$ (45 degrees clockwise from the t-axis). Dark areas denote high values. The gradient images b-e are subsequently referred to as the f-, t-, c1-, and c2-gradient, respectively. From Toiviainen (1996), republished with permission of Swets & Zeitlinger

the locus of response to each input vector is defined to be the position of the neuron whose weight vector is closest to the input vector in question. In the centroid-of-activation metric (CAM), the locus of response is defined to be the centroid of the activation pattern. The distance between two responses is then defined as being the Euclidean distance between the loci of response. The distance metric between the auditory images will subsequently be referred to as the AIM.

4.3 Results

Auditory Images. To begin with, the correlation between the similarity rating and the auditory image distance matrix, denoted by $corr(\mathbf{D}, \mathbf{S})$, was calculated using the spectral images only, with parameter values $\lambda = 2$ (Euclidean metric), $k = 0$ (no emphasizing of onsets). This yielded the correlation value $corr(\mathbf{D}, \mathbf{S}) = 0.677$.

Next, λ and k were varied between $1 \leq \lambda \leq 10$ and $0 \leq k \leq 0.48$, with steps of 1 and 0.02, respectively. The maximum correlation value, obtained with $\lambda = 1$ and $k = 0.20$, was $corr(\mathbf{D}, \mathbf{S}) = 0.868$.

The significance of gradient images in timbre perception was studied by constructing combined auditory images using the spectral image and the four gradient images presented in Fig.1. The gradient ascent method was used to find the combination of parameter values $\mathbf{p}^* = (\lambda^*, k^*, \alpha_s^*, \alpha_f^*, \alpha_t^*, \alpha_{c1}^*, \alpha_{c2}^*)$ which would yield the maximum value of $corr(\mathbf{D}, \mathbf{S})$. Since in the calculation of the Pearson correlation coefficient the variances of the two variables are normalized, multiplying the α's with a constant does not alter the correlation. Therefore, α_s was fixed to value 1. The search for the maximum was started with zero values for the other α's. In a series of calculations, the initial values for λ and k were chosen from the neighborhood of the point $\lambda = 1$, $k = 0.20$. The maximum value achieved was $corr(\mathbf{D}, \mathbf{S}) = 0.882$; this was obtained with the parameter values $\lambda = 0.690$, $k = 0.21$, $\alpha_f = 1.510$, $\alpha_t = 0.010$, $\alpha_{c1} = 0.262$, and $\alpha_{c2} = 0.000$. As a result of adding gradient images and allowing the Minkowski exponent to change continuously, a slight increase in correlation was thus achieved, the main contributors to this being the parameters λ and α_f.

Simulations with the Self-Organizing Map. Using various sets of auditory images as input, a series of simulations with the self-organizing map was carried out; the dimension of the input vectors was 400 (20x20), unless otherwise indicated. During the training procedure, the Minkowski metric was used. In each simulation, a 12 x 12 Kohonen network was trained for 50000 cycles, using a constant learning rate of 0.05. In the beginning, the radius of the topological neighborhood was 6, and it was linearly decreased after every cycle so as to reach zero at the end of the training. From the obtained timbre maps, distance matrices were calculated using both the RFM and the

CAM. These matrices were then compared with the similarity rating data by calculating the Pearson correlation.

First, the self-organizing map was trained with auditory images which did not include any gradient images. The correlation values obtained from simulations with two different sets of auditory images are presented in Fig.2, bars a-b.

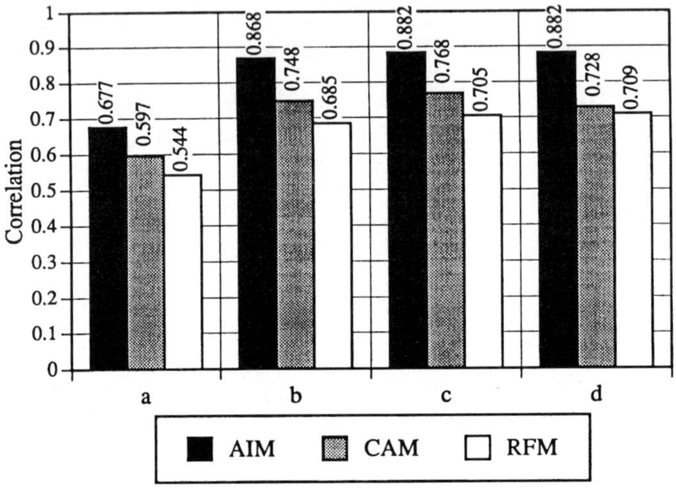

Fig. 2. Correlations obtained from four simulations using AIM, CAM, and RFM. a) $\lambda = 2$, $k = 0$, no gradient images; b) $\lambda = 1$, $k = 0.20$, no gradient images; c) spectral and f-gradient image, one net; d) spectral and f-gradient image, three nets

In both simulations, the AIM correlation was the highest, followed by the CAM and RFM correlations, in that order. This might indicate that it is not possible to reduce timbre to two dimensions without compromising the metrical relationships between the input stimuli. Second, the post-Kohonen correlation values are high when the AIM correlation is high, and vice versa; this shows that the chosen preprocessing strategy is an essential factor in this kind of modelling. Last, the CAM correlations are regularly higher than the respective RFM correlations; taking into account the whole activation patterns of the self-organizing map seems, thus, to yield a better metric than paying attention only to the unit with the highest response.

At the next stage, the self-organizing map was trained with sets of auditory images which included the spectral and the f-gradient image of each tone. The values of the parameters were $\lambda = 1$, $k = 0.21$, $\alpha_f = 1.510$, and $\alpha_t = \alpha_{c1} = \alpha_{c2} = 0$. The last three gradient images were ignored to reduce computational load. Two different network architectures were used. First, a single self-organizing map was trained with auditory images consisting of a spectral and an f-gradient image; the components of the latter were multiplied by 1.510. In the second approach, two self-organizing maps were first trained, one with the spectral images and the other with the f-gradient images. For each input stimulus, the coordinates of the centroids of activation on these maps, which thus formed a 4-dimensional vector, were used as input to a third self-organizing map.

The CAM and RFM correlation values, obtained using these combined auditory images, are presented in Fig.2, bars c-d. Also in these simulations the CAM and RFM correlations were lower than the respective AIM correlations. High AIM correlations tended to imply high CAM and RFM correlations, and vice versa. Using the one-network architecture yielded the highest CAM correlation value, 0.768; the highest RFM correlation value, 0.709, was obtained using the three-network architecture.

5 Dynamic Approach

In the static approach described above, each input stimulus was presented to the self-organizing map as one multicomponent vector: each component represented the spectral energy within a given range of frequency and time. In this paragraph, an alternative approach is presented, where each input stimulus is represented by a set of spectral vectors, each of which represents the instantaneous spectrum of the stimulus at a given time. A similar approach has been proposed by Feiten and Günzel (1994). The activation patterns on the self-organizing map are constructed using temporal integration; the emphasizing of onsets is carried out using a scaling procedure which operates qualitatively in the same way as lateral inhibition. Since the gradient images used in the static approach failed to increase the correlation significantly, they were not included in this study.

5.1 Auditory Images

The temporally integrated output of the auditory model (see Equation 3) was sampled every 10 ms, starting at 10 ms after the onset of each input stimulus. Since the length of the input stimuli was 500 ms, this yielded 50 spectral vectors per each stimulus, or a total of 27 x 50 = 1350 input vectors. A self-organizing map of 20 x 20 units was trained with these input vectors.

5.2 Timbre Maps

To calculate the responses of the self-organizing map to each stimulus, the input activation of neuron i at time t, $a_i(t)$, was defined by

$$a_i(t) = \frac{d_{max} - ||\mathbf{v}(t) - \mathbf{w}_i||}{d_{max} - d_{min}}, \quad t = 0, 1, 2, \ldots, 49, \tag{11}$$

where $\mathbf{v}(t)$ denotes the input vector at time t, $\mathbf{w}_i$ the weight vector of neuron i, $||\cdot||$ the Minkowski norm, and d_{max} and d_{min} the maximum and minimum Minkowski distances of weight vectors from the input vector, e.g.:

$$d_{max} = \max_i \{||\mathbf{v}(t) - \mathbf{w}_i||\} . \tag{12}$$

The input activation values are, thus, scaled so that they lie between 0 and 1. The integrated response of neuron i at time t, $A_i(t)$, was given by

$$A_i(t) = \begin{cases} a_i(0) & , \quad t = 0 \\ A_i(t-1) + (1-\varepsilon)^t a_i(t) & , \quad 0 < t \leq 49 \end{cases} \tag{13}$$

where ε is the inhibition parameter. When $\varepsilon = 0$, subsequent input activation patterns are simply summed up to obtain the integrated response pattern. When $\varepsilon > 0$, the gain factor $(1-\varepsilon)^t$ decreases exponentially with time. The effect of the inhibition parameter on the time evolution of the pattern of integrated responses is illustrated in Fig.3. When there is no inhibition, the pattern of integrated response keeps evolving throughout the duration of the stimulus. When inhibition is increased to $\varepsilon = 0.16$, the pattern of integrated response does not change significantly after about 120 ms; the final response pattern is thus mainly determined by the onset period of the tone.

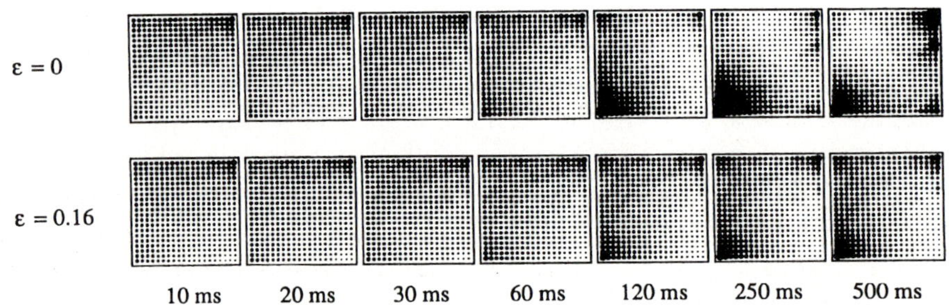

Fig. 3. The time evolution of the pattern of integrated response to tone 24 of the stimulus set. The response pattern is shown at 10 ms, 20 ms, 30 ms, 60 ms, 120 ms, 250 ms, and 500 ms after the onset of the tone for two values of inhibition, $\varepsilon = 0$ and $\varepsilon = 0.16$. In each subfigure, each circle represents the instantaneous integrated response of one neuron, its radius being proportional to the latter

5.3 Results

The distance between the network's responses to two stimuli, j and k, was defined to be the city-block distance of the two integrated activation patterns

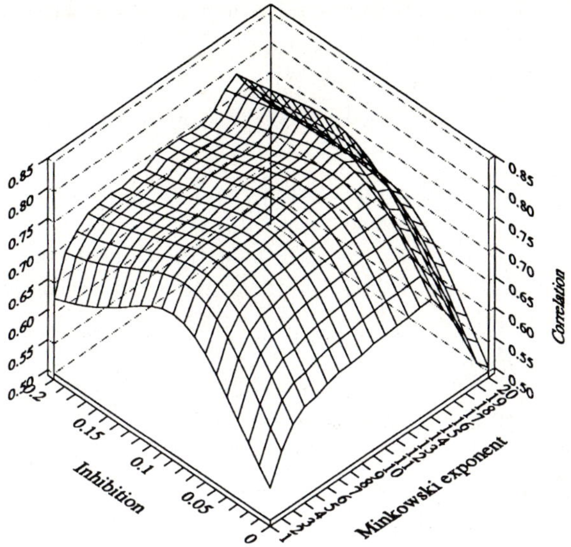

Fig. 4. The values of $corr(\mathbf{D}, \mathbf{S})$ as a function of the inhibition parameter ε and the Minkowski exponent λ

at the end of the stimuli, i.e.,

$$d_{jk} = \sum_{i=1}^{N} \left| A_i^j(49) - A_i^k(49) \right| , \tag{14}$$

where $A_i^j(t)$ is the integrated response of neuron i to stimulus j at time t and N is the number of units in the self-organizing map. The distance matrix thus obtained was compared with the similarity rating matrix by means of the correlation coefficient.

In the simulations, values of the Minkowski exponent between $1 \leq \lambda \leq 20$, with steps of 1, were used to train the self-organizing map and calculate the input activations. The integrated responses were calculated using values of the inhibition parameter between $0 \leq \varepsilon \leq 0.2$, with steps of 0.01.

Figure 4 depicts the thus obtained values of $corr(\mathbf{D}, \mathbf{S})$. The graph shows that the highest values are obtained when $\varepsilon \approx 0.1$. Moreover, high values of λ tend to yield high correlations, although this effect is not significant when $\lambda > 5$. The maximum correlation, $corr(\mathbf{D}, \mathbf{S}) = 0.817$, was obtained using the values $\lambda = 17$ and $\varepsilon = 0.11$. This is significantly higher than the value $corr(\mathbf{D}, \mathbf{S}) = 0.602$, obtained using Euclidean metric ($\lambda = 2$) and no inhibition. In order to study the time development of the integrated responses, the average rate of change of the latter was calculated according to

$$\Delta(t) = \frac{1}{M} \sum_{j=1}^{M} \left\| \frac{\mathbf{A}^j(t)}{\|\mathbf{A}^j(t)\|_1} - \frac{\mathbf{A}^j(t-1)}{\|\mathbf{A}^j(t-1)\|_1} \right\|_1 , \qquad (15)$$

where M is the number of input stimuli, and $\mathbf{A}^j(t) = (A_1^j(t), \ldots, A_N^j(t))$ is the vector of integrated response to stimulus j at time t. Fig.5 displays, on a logarithmic scale, the values of $\Delta(t)$ as a function of time for Minkowski exponent $\lambda = 5$ and two values of inhibition: $\varepsilon = 0$ (no inhibition) and $\varepsilon = 0.08$ (optimal inhibition). As can be expected, increasing the amount of inhibition prevents the integrated responses from developing at the end of stimuli, thus causing the final integrated response to be mainly determined by the onset period. For $\varepsilon = 0$, the minimum of $\Delta(t)$ is 0.0037 at $t = 450ms$; for $\varepsilon = 0.08$, the rate of change falls below that value as early as at $t = 210ms$.

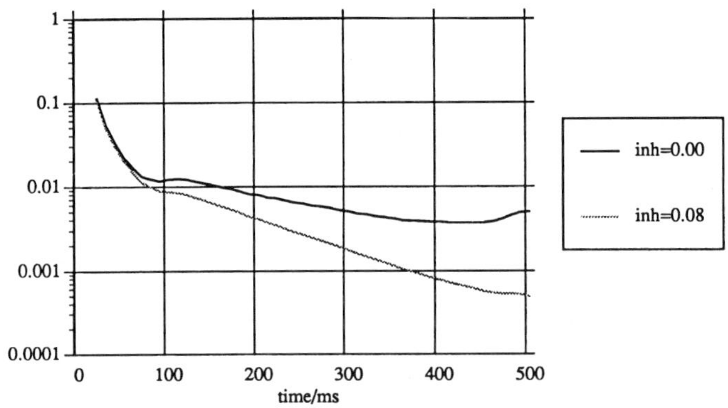

Fig. 5. The rate of change of integrated responses, averaged over the set of input stimuli, for two values of inhibition, $\varepsilon = 0$ and $\varepsilon = 0.08$. The value of the Minkowski exponent is $\lambda = 5$

6 Conclusions

The role of frequency transition and gradient maps in timbre perception has not yet been carefully examined. This study failed to clarify this question: although the gradient images, defined earlier, can be regarded as idealizations of the aforementioned maps, their frequency and time resolutions are certainly insufficient for extracting quick transitions present in the onset portions of musical tones. Further study is needed in order to elucidate this problem.

The results obtained from both the static and the dynamic approach support the assumption that, in order to construct timbre maps whose configurations would correspond to the mental representations of timbres of given tones, it is necessary to emphasize the onset portion of the tones to a proper degree. In the static approach, this was carried out on the preprocessing stage using an exponential scaling of time, whereas the dynamic approach utilized temporal integration on the self-organizing map, with a scaling procedure.

The optimal distance metrics found in the two approaches were significantly dissimilar. In the static approach, highest correlations were obtained using values of the Minkowski exponent of the order $\lambda \approx 1$. In the dynamic approach, however, the values $\lambda > 5$ yielded the best correlations; the highest correlation was obtained using a value as high as $\lambda = 17$. In other words, the static approach seems to work best when the mutual distance of two vectors is mainly determined by the number of components for which the vectors differ significantly, whereas the dynamic approach favours a metric where the distance is mainly determined by the largest difference present in any component. The reason for this anomaly remains to be explained.

Appendix

Sound Examples (CD-track 30)

Example #1. The 27 tone stimuli used in the experiments.

References

Cosi, P., De Poli, G., & Lauzzana, G. (1994). Auditory modelling and self-organizing neural networks for timbre classification. *Journal of New Music Research*, *23*, 71–98.

De Poli, G., Prandoni, P., & Tonella, P. (1993). Timbre clustering by self-organizing neural networks. In G. Haus & I. Pighi (Eds.), *Proceedings of the Xth Colloquio di Informatica Musicale.* Milan: AIMI.

Feiten, B., & Günzel, S. (1994). Automatic indexing of a sound database using self-organizing neural nets. *Computer Music Journal*, *18*, 53–65.

Iverson, P., & Krumhansl, C. (1993). Isolating the dynamic attributes of musical timbre. *The Journal of the Acoustical Society of America*, *94*, 2595–2603.

Kohonen, T. (1995). *Self-organizing maps.* Berlin, Heidelberg: Springer-Verlag.

Leman, M. (1994). Schema-based tone center recognition of musical signals. *Journal of New Music Research*, *23*, 169–204.

Pickles, J. (1982). *An introduction to the physiology of hearing.* London: Academic Press.

Ru, P., & Shamma, S. (1997). Representation of musical timbre in the auditory cortex. *Journal of New Music Research*, *26*.

Shamma, S., Vranic, S., & Wiser, P. (1992). Spectral gradient columns in primary auditory cortex: Physiological and psychoacoustical correlates. *Advances in the Biosciences*, *83*, 397–404.

Steiger, H., & Bregman, A. (1981). Capturing frequency components of glided tones: Frequency separation, orientation, and alignment. *Perception and Psychophysics*, *30*, 425–435.

Toiviainen, P. (1996). Optimizing auditory images and distance metrics for self-organizing timbre maps. *Journal of New Music Research*, *25*, 1–30.

Toiviainen, P., Kaipainen, M., & Louhivuori, J. (1995). Musical timbre: Similarity ratings correlate with computational feature space distances. *Journal of New Music Research*, *24*, 282–298.

Van Immerseel, L., & Martens, J. (1992). Pitch and voiced/unvoiced determination with an auditory model. *The Journal of the Acoustical Society of America*, *91*, 3511–3526.

Towards a More General Understanding of the Nasality Phenomenon

Milan Rusko

Institute of Control Theory and Robotics, Slovak Academy of Sciences,
Dúbravská cesta 9, Bratislava, Slovak Republic

Abstract. Nasality, nasal sound, nasalized sound – these expressions are used in phonetics and speech research, in singing theory, and in musical acoustics. In this paper we try to find a classification for several different types of nasality and to show links between the meaning of the expression *nasality* in speech, singing, and musical acoustics. A number of spectral measurements, listening experiments and discussions were undertaken to find relationships for nasality in these three fields.

1 Introduction

Nasality is one of the first sound phenomena human beings use in vocal communication in the early stages of their life. Notice that sounds like "na na", "ne ne" and similar, are some of the first more complex sounds or first syllables pronounced by small children. Some of the nasal consonants are also contained in the word expressing "mother" in many languages (English "mummy", Slovak "mama", Hungarian "anyu", etc.).

As the nasality is important from many points of view, extensive research has been done to study the complex acoustical principals of generating nasalized sounds in human speech (Abramson, Nye, Henderson, & Marshall, 1981). Undoubtedly one of the most comprehensive studies was done by G. Fant in his work on the acoustics of speech (Fant, 1960/1964).

According to the basic phonetical definition, nasalized sound is a sound created with a contribution from the nasal tract. This definition is of course either insufficient or even not suitable at all for nasalized sounds in human singing or for sounds of musical instruments with *nasal* timbre. The aim of this paper is not only to explain the phonetical model of nasalization, but also to discuss the possibility for the definition of a unified model of the nasalized sound, which would include sounds of speech, singing and music. We would also like to define a *nasality measure* for the future, which would allow us to evaluate the nasality quantitatively. This will be based on acoustical theory of nasalization as well as on the listening experiments which helped us to understand the cognitive aspects of nasality.

Another viewpoint, that should not be neglected, is the aesthetic one. The speakers often use the nasalization (frequently together with higher pitch) to produce a *funny* voice. In male singing the *buffo tenore* and *basso buffo* with nasal timbre are used to perform comical roles. In music the nasal sound of a bassoon is often used to express comical atmosphere.

Nasalization can also make the sound more aggressive, attack-like. The high degree of nasalization is contained in a speaker's voice expressing strong stress or even anger (closed teeth, deeper breathing and wide-open nostrils).

Speaking about the sounds of musical instruments, the situation is not so clear, as there is no nasal tract contributing to the sound production. Nevertheless, the similarities in the timbre of nasalized sounds of human speech and singing, and in that of musical instruments, reflected in the expression the nasal timbre of the instrument, are worth investigating. The conclusions of the acoustical analyses of several instruments and listening experiments are included in this paper.

2 Nasalization in Speech

2.1 Nasals

The nasal consonants are voiced sounds produced when the glottal wave excites an open nasal cavity and closed oral cavity. The signal waveform of nasals is similar to that of vowels, but generally is weaker in energy due to the limited ability of the nasal cavity to radiate a sound. In forming a nasal, a complete closure is made toward the front of the vocal tract, either with the lips (labial or front nasal /m/ as in "more"), with the tongue resting on the gum ridge (alveolar or mid-nasal /n/ as in "noon"), or by the tongue pressing on the soft or hard palate (velar or back-nasal /η/ as in "sing"). The velum is opened widely to allow for sound propagation through the nasal cavity (Deller, Proakis, & Hansen, 1993). As the timbre of the sound (or corresponding formant structure) changes according to the location of the closure, we will take into account three other phonemes prealveolar /n/, /n,/, /ã/, which can be found for instance in French. The first of them is produced with the tip of the tongue touching the hard palate close to the teeth (prealveolar /n/ in "an" (year)) and is very similar to the English /n/, which differs only in using smaller volume of the mouth cavity as a shunting resonator and has the closure more far from the teeth. In this study we will neglect this small difference and will consider these two phonemes to be identical.

In the second one the closure is created by the central part of the tongue lifted to the boundary between the hard and soft palates (alveopalatal /n,/ in "champagne").

An interesting statement concerning the third variation of this phoneme, corresponding with our listening experiments can be found in an early work by Kempelen (1791/1990): "/ã/, pronounced by the French in the words "enlever" or "enfin" is considered to sound most nasal. Under careful investigation we find out the opposite: in spite of the fact that the nose is open, the quantity of the sound radiated through it is smaller than for the other sounds." Kempelen thinks it is caused by the fact that the mouth tract is not fully closed and a vowel is coarticulated with the nasal. He continues :

"...We have mentioned that during the articulation of vowels the nose tract must be closed. If it is not, the vowel sounds unclear and it is recognized by an ear to be nasal in the place where it generally should not be, so that it sounds very expressive (disturbing)."

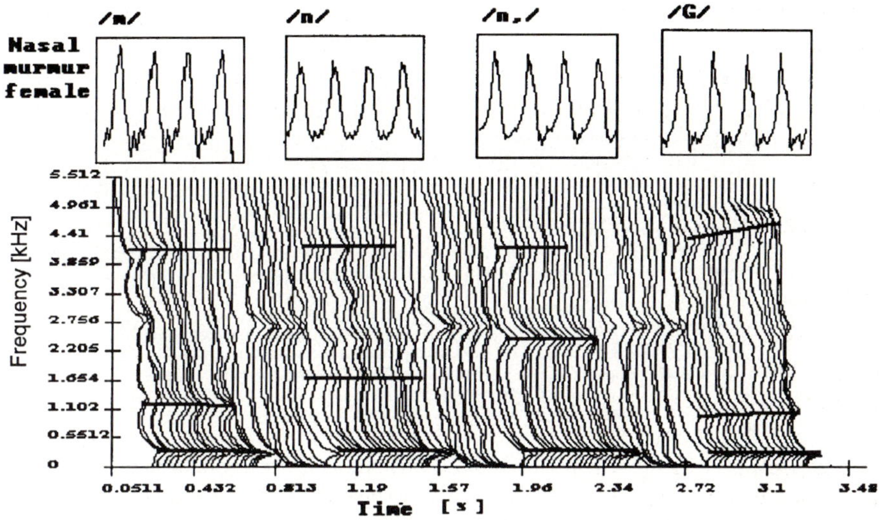

Fig. 1. Waterfall graph showing formant structure of nasals in a female voice

Acoustical Features of Nasals

- closed mouth cavity acoustically coupled with pharyngeal and open nasal cavities ⇒ trapping energy at certain frequencies (⇒ anti-resonances or spectral nulls)
- a low formant F1 = 250Hz dominates the spectrum
- F2 is weak resulting from the anti-resonance of the closed oral cavity
- F3 occurs near 2200 Hz
- Spectral zero in the frequency response is inversely proportional to the length of the constricted oral cavity. (750 to 1250 Hz for /m/, 1450-2200 Hz for /n/, and above 3000 Hz for/η/
- the phoneme /ã/ belongs to the sounds which will be discussed later in Sect.2.2

Since the human auditory system is only partially able to perceptually resolve spectral nulls, discrimination of nasals based on the place of articulation is normally cued by formant transitions in adjacent sounds (Deller et al., 1993, p.136). A special case can be found in some languages, where the whole word is pronounced with the mouth closed (such as in the interjection "mhm" in Slovak, expressing agreement or "?m?m", expressing disagreement. Two types of glottal excitation are used in "?mhm" – vowel-like in the first half and whisper-like in the second. Note that these phenomena are typical for *brumendo* singing too. When nasal is followed by vowel, the change in formant structure is abrupt for /m/ and /n/, but gradual for /n,/.

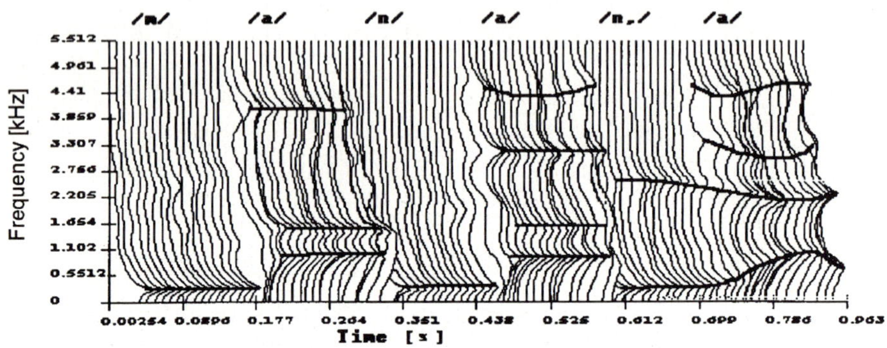

Fig. 2. Formant changes in the syllables consisting of a nasal and the /a/ vowel in a female voice. (The presence of the both of these dynamic changes contributes to the higher level of the perceived nasality)

2.2 Nasalized Phonemes

Although nasals are the only phonemes that use only the nostrils to radiate sound, some phonemes become nasalized if they precede or follow a nasal, primarily in vowels, when the velum begins to drop in anticipation of the ensuing consonant. So the last phoneme mentioned in 2.1. can be looked upon as a nasalized version of the vowel /a/.

It should be mentioned that if "m" or "n" are pronounced alone, they are always followed by a neutral vowel (the closed lips or the closing formed by tongue and hard palate open, and give birth to a neutral vowel explosion) to make the pronunciation louder.

Acoustical Features of Nasalized Vowels:

- F1 is weakened, broader F1 bandwidths, less peaked (than those without nasal coupling)
- the degree of the effect of nasalization on the spectrum depends on the amount of coupling between the two cavities
- additional formants appear between the ones typical for the vowel, the most important one being higher than F1 and lying at 1000 Hz

2.3 Closed Nose Nasalization

A way of producing nasalized speech sounds, when the nostrils are closed, and only the lips are used for radiation, but the velum leaves the entrance to the nasal cavities open – and so they are shunting the vocal tract – we call *closed nose nasalization*. This type of nasalization impacts upon not only vowels, but mainly upon nasals, as these cannot be pronounced properly, and also other voiced consonants – and results in the well known effect of *full nose*. As the functioning of the lip opening is not affected, the amplitudes of radiated sound can be comparable to that without nasalization. This happens in some special situations (e.g. when a speaker is trying not to show his identity during a telephone conversation, or when an actor is performing some comical role, etc.) as the speaker closes his nostrils with his fingers. In listening experiments we found that sounds with this type of nasalization are considered to be *most nasal*.

Acoustical Characterization of the Sounds with Closed Nose Nasality:

- F1 is weakened. If it is still recognized, its frequency is shifted higher
- A part of the energy is shifted to the higher frequency range (to the formants at 2800 to 3500 Hz)
- The narrower the constriction created by the velum at the entrance to the mouth cavity the stronger the influence of the closed nasal cavities (and the more *nasal* the timbre). This is the effect of higher acoustic coupling of the two systems

2.4 Fully Closed Nasalization

This type of nasalization is obtained when both the nostrils and the mouth are closed, but glottal excitation works (until the whole volume of the nasal cavity and mouth cavity (with cheeks) is filled with the air). The radiation through the walls of the vocal tract is very small and so is the energy of fully closed nasalized sounds. These sounds are non-speech sounds used in vocal communication very rarely.

3 Nasality in Singing

Nasality in singing can be defined as using the nasal cavities as resonators to amplify the produced sound. It helps the voice to be more effective without using more energy for the sound production. Our opinion is this is a result of the fact that nasalized sounds have more of their energy in a higher frequency range (a so-called *higher formant* in singing lies in a range 2800 to 3200 Hz). The radiation characteristics of the vocal tract is more unidirectional for these frequencies, and so the voice radiation is more effective in its direction to the audience (in comparison to the velvety and round voices using lower formant

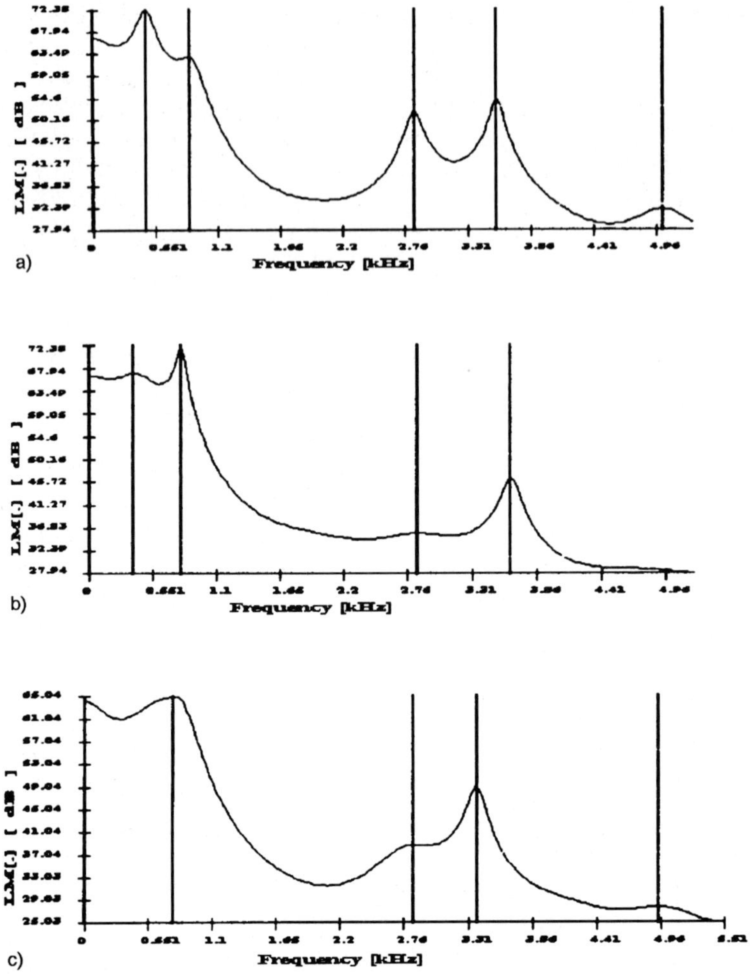

Fig. 3. Linear prediction smoothed spectra showing formant position of (a) normal, (b) nasalized and (c) closed nose nasalized vowel /o/ in male speech. The first formant is mostly attenuated in (c).

resonance with its central frequency around 600 Hz). Nasalized singing is also very hygienic, as the glottal cords need not to be fully excited. These could also be some of the reasons why Arab muezzins use nasalized singing.

Using nasality for a special purpose can be found in pop-music in some countries (e.g. Slovakia). The singers, in order to get timbre more similar to their American ideals, use – when singing in Slovak – American pronunciation for many of the phonemes and a much higher degree of nasalization, which is typical for American speech. (Note there is a national and geographical aesthetic norm for the level of nasality in speech and singing.)

3.1 Brumendo

The differences between normal singing and brumendo, in which all the voice energy is radiated through the nostrils, are evident from the graphs of their long time average spectra. The spectrum of brumendo is much more similar to that of nasals and there is no contribution from vowels with a high energy content, which is essential in normal singing spectrum. The similarity of the brumendo sound to that of musical instruments is used in folk music. For instance, in the Slovak village of Jedl'ové Kostolany singers use brumendo to imitate the sound of the bagpipe drone.

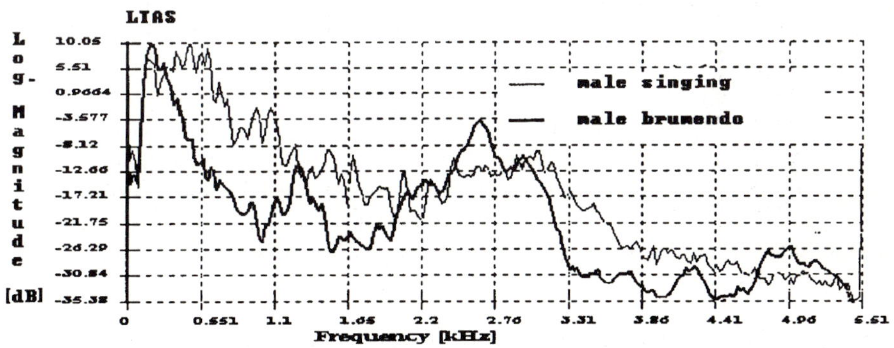

Fig. 4. Long time average spectra of the same song sung in normal voice (thin line) and brumendo (thick line)

4 Nasality in the Timbre of Musical Instruments

Several authors write about nasality in the sound of musical instruments, but the reasons and manifestation of its presence are not the same in different cases and it seems the musical acousticians still have only a vague notion of what nasality generally is. For instance Špelda (1978) says: "The appearance of the 7th, 11th, and 13th harmonics is one of the reasons for the nasal timbre of the saxophone sound." In another place he says, "The insufficient volume of the resonator is the reason for the nasal property of the timbre of viola tones." or "If the resonance maximum is in the range around 1200 Hz – 1600 Hz (so called nasal formant), there is a danger that the tones of the instrument get an unpleasant nasal timbre."

There is no satisfactory definition of *nasal timbre* for the sounds of musical instruments. When taking acoustic features of nasality in speech as a base for the definition, we must take into account several different types of nasality in phonemes, and recognize at least four classes of nasality. We also must be aware of the fact that the sound production mechanism in instruments is different from that in speech. So the boundaries between classes will not be clear and a spectrum of a sound can have the *mixed* properties of two or more classes. One of the interviewed people (musically and technically *uneducated*) defined nasal sound as a sound which has to pass through some barrier and is damped at the same time. In fact he gave a true picture of what is going on in nasality in speech. In instruments – a barrier in a sound wave propagation path represents a kind of filtering (which can be obtained using resonators with smaller volumes that work like acoustical filters with its pass-band covering only higher harmonics). Highly damped areas are for instance sounds of stringed instruments with a skin membrane used instead of a wooden front resonance plate (such as a banjo). Instruments played *con sordino* fully fit this model too. Oboes, which seem to sound extremely nasal according to the answers of our interview group, also use a highly damped double reed for sound generation. Listening experiments, which consisted of listening to several sound samples from speech, singing and the tones of musical instruments, subjective evaluation of their perceived nasality and discussions on the definition of the nasality have indicated that a whole complex of phenomena is represented by the expression *nasality* in the understanding of the interviewed people.

The presence of some dynamic acoustic processes, that have their equivalents in speech, (such as tongue and/or velum movements during sound production – like in /n,/ /a/ (see Fig.2), or the neutral vowel explosion in the isolated /m/) increase the subjective feeling of nasality in the sound. A similar increase in the perceived level of nasality was observed by us in a synthesized sound when frequency or phase modulation was applied.

The closer the pitch of the instrument sound to that of its phonetic template known in speech, the stronger the feeling of nasality. The timbre of tones of a higher pitch is often compared to singing rather than to speech sounds for this reason. This fact indicates that a *nasality measure*, if defined, should be normalized to pitch. One possible way could be to use a bark scale instead of a linear frequency scale.

Table 1. Classification according to some acoustic and aesthetic relations between nasal sounds

Nasalization Class	Phonemes	Singing	Musical Instruments
m-type	/m/, /n/, /n,/, /η/	brumendo	wooden string instruments (cello, viola, etc.)
vowel-type	nasalized vowels a, e, i, o u and @ in normal speech	tenore buffo basso buffo, female voices using nose with noise cavity resonance	violin "con sordino", fiddle, clarinets (saxophones, hornpipe, etc.), trumpets, organ register "nasal"
closed nose	vowels and voiced consonants pronounced with closed nose	singing with nose (rarely used) closed to reach an extreme -ly comical timbre	oboes (schalmei, zurma, piffero, bassoon, French horn, etc.), muted trumpets, stringed instruments with skin membrane(banjo) and/or resonators with smaller volume (musical bow, sitar, trumscheidt)
fully closed	a sound which can be produced with both nostrils and lips closed (not a typical speech sound)	not used in singing	famfrnoch (Czech folk instrument: a pot with a skin membrane, and a horsetail for excitation)

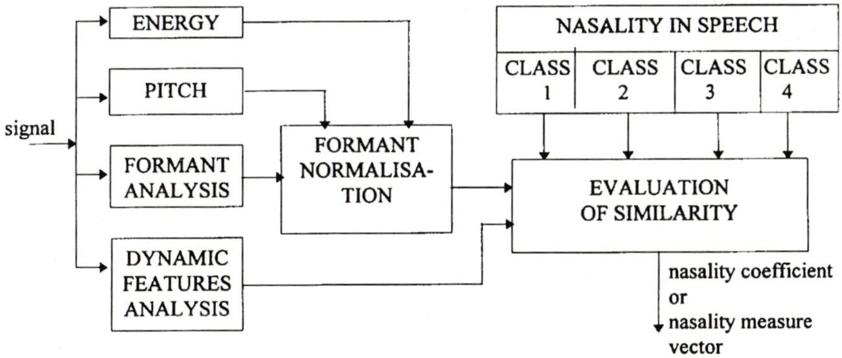

Fig. 5. Our recommended procedure as a possible model for computing the nasality coefficient or the nasality measure vector

5 Conclusion

In ending, we should state that the sound which is evaluated as most nasal does not belong to the first class (together with nasals) but to the third class (in speech called "sounds with closed nose nasalization"). In the sound of musical instruments the level and the type of nasality is not constant and changes dynamically in the process of duration of the tone. The perception of nasality is not based only on the spectral properties of the sound, but also depends a lot on the similarity of the dynamic changes (transients, onset, dynamic spectral changes) to their aesthetic templates in speech, and the similarity of the formant structure to that of the vowels. The exact and technically satisfying general definition of nasality will take more time, more measurements, and more discussions amongst musicologists, phoneticians and acousticians, and the same can be said about the possibility of measuring the amount of nasality in sound. Nevertheless, this problem is of considerable interest for people working in the fields of speech recognition, speaker identification, sound synthesis, the acoustics of musical instruments etc., and the research is worth continuing.

Acknowledgements

We would like to thank voice pedagogist O.Šimová (Personal communication) and musicologist L.Kačic (Personal communication) for numerous fruitful discussions on the topic. This research was funded by The Grant Agency for Science in Slovakia, grant No.2301896.

References

Abramson, A., Nye, P., Henderson, J., & Marshall, C. (1981). Vowel height and the perception of consonantal nasality. *The Journal of the Acoustical Society of America, 70*, 329–339.

Deller, J., Proakis, J., & Hansen, J. (1993). *Discrete-time processing of speech signals.* New York, NY: Macmillan.

Fant, G. (1964). *Acoustic theory of speech production.* Moscow: Mouton. (Original work published 1960; Russian Translation)

Kempelen, W. (1990). *Mechanismus der menslichen Sprache nebst der Bechreibung seiner sprechenden Maschine.* Bratislava: Tatran. (Original work published 1791; Slovak Translation)

Špelda, A. (1978). *Hudebnó akustika.* Praha: SPN.

Karl Erich Schumann's Principles of Timbre as a Helpful Tool in Stream Segregation Research

Christoph Reuter

University of Cologne, Albertus-Magnus-Platz, D-50931 Cologne, Germany

Abstract. The following text deals with timbre-dependent stream segregation. Combining the ideas of Gestalt principles, auditory scene analysis and Schumann's principles of timbre, the following hypotheses are proposed and examined: (i) Alternating timbres with matching formant positions are perceived for the most part as one sole melodic stream. (ii) Alternating timbres with non-matching formant positions are perceived for the most part as two interwoven melodic streams. (iii) Alternating timbres which have on the one hand formants and on the other hand fluctuations as primary attributes are perceived for the most part as two interwoven melodic streams. (iv) Alternating timbres of which the main attributes are fluctuations are perceived for the most part as one sole melodic stream. In this experiment based on Wessel (1985), hypotheses 2 and 3 were fully confirmed, whereas hypotheses 1 and 4 could not be fully confirmed. Possible causes for this are discussed: the invariable melodic contour of the stimuli as well as the tempo in which they were played; furthermore, the mental situation of the subjects at the time of the experiment.

1 Wertheimer and Schumann: Gestalt Principles and Principles of Timbre

In the 1920's, two major discoveries in the area of psychology and timbre perception were made.

On the one hand, it was M. Wertheimer who was the first to describe the laws of Gestalt perception in 1923 (Wertheimer, 1923). His investigations served as the basic foundation for explaining the observed phenomena of stream segregation, continuity illusion, or, auditory scene analysis, starting with the investigations of the trill threshold made by Miller and Heise (1950), and then continuing with the work done by Dowling and Harwood (1986), van Noorden (1977), Huron (1989), Wessel (1985), Tougas and Bregman (1985, 1990), Bregman et al. (1990), Bregman (1990), up to McAdams (1982), McAdams and Bregman (1985), among many others.

On the other hand, K.E. Schumann (Schumann, 1929) propounded the laws of perception of wind instrument timbres in 1929, following thorough investigations. His observations served as the basic foundation for explaining the production of wind instrument timbres and their typical formant areas (the pulse forming process), as well as partial masking and the blending together of orchestral instrument timbres (Fricke, 1975, 1993; Voigt, 1975; Auhagen, 1987; Reuter, 1995, 1996).

Schumann was a pupil of C. Stumpf and a fellow researcher of W. Köhler. His work, written in several volumes (now known under the name "Physik der Klangfarben"), was strongly influenced by both of them. Today, only the second volume exists in a few copies; it deals with the positions and behavior of the formants of woodwind instruments. [1]

Schumann discovers four principles of timbre:

1. The principle of *formant areas* ("Formantstreckengesetz"): Formants of musical instruments are fixed areas of the spectrum which are independent of the fundamental. Partials located in this areas have exceptionally strong amplitudes, so that the timbre is influenced mainly by these partials (Schumann, 1929, pp.15-18,98,100)
2. The principle of *formant shifting* ("Formantverschiebungsgesetz"): With increasing dynamics the strongest amplitude of the partial in the formant area is shifted to a partial of higher order in the same formant area (Schumann, 1929, pp.15-18,98,100)
3. The principle of *formant skipping* ("Sprunggesetz"): This principle is a subcase of Principle No. 2.: with very loud dynamics the strongest amplitude of the first (or lowest) formant area is shifted to a partial in the second (higher) formant area, skipping over the partials between these areas (Schumann, 1929, pp.98,100)
4. The principle of *formant intervals* ("Formantintervallgesetz"): The partial with the strongest amplitude located in the first (or lowest) formant area and the partial with the strongest amplitude located in the second (or higher) formant area result in an interval, which is typical for the respective instrument (Schumann, 1929, pp.100,131,208)

This last principle is questionable to a certain extent, since it contradicts Principle no. 2. Furthermore, one obtains different formant intervals when placing the recording microphone in different positions, due to the comb filter effect.

However, Principles 1, 2 and 3 were confirmed particularly for woodwind instruments (especially for reed- and double-reed-instruments) (Fricke, 1993; Reuter, 1995). With some restrictions they seem to be also valid for brass instrument timbres, whereas in string instrument timbres it is not possible to recognize unequivocal formant areas that could be generalized (Reuter, 1996, pp.221,228,230).

Using electronically or digitally produced pulses, one can synthesize authentic brass and woodwind instrument timbres behaving according to the above described principles of timbre (the so-called pulse forming process, described in Fricke, 1993, Reuter, 1995). These synthesized timbres sound so natural that one cannot distinguish between the real wind instrument and the synthesized sound.

[1] One of these copies can be found in the World Wide Web under the following address: http://www.uni-koeln.de/phil-fak/muwi/publ/schumann/schumann.html.

In order to investigate stream segregation due to timbral differences alone, it would be fitting to try to combine both of the above described schools of research in one experiment (Gestalt principles and principles of timbre). To begin with, since such an experiment can be seen as representing a new attempt at applying the methodology known from Wessel (1985), that experiment is presented and its values as well as its shortcomings are discussed in the following section.

2 Sharpness and Attack Transients: Stream Segregation and Timbre Space

Very few investigations have been carried forth until now in the area of stream segregation solely due to differences in modulation or timbre (Bregman et al., 1990; Bregman, 1990; Dowling & Harwood, 1986; Huron, 1989; Iverson, in press; McAdams, 1982; McAdams & Bregman, 1985; Tougas & Bregman, 1985). D. Wessel's 1979 experiment (Wessel, 1985) is one of the best known and most often cited.

Wessel used sounds from the experiments of Grey (1977) and the manipulated sounds of Grey and Gordon (1978) in order to emulate Grey's timbre space in two dimensions (as to timbre space, see also Reuter, 1995, p.177ff). In Wessel's new timbre space, the two axes describe both sharpness (or brightness) and velocity of attack (Wessel, 1985, pp.684-685). Always using two different sounds extracted from this timbre space, Wessel constructed a constantly recurring melodic pattern consisting in two ascending fourths (b-e'-a'), and in which both timbres (say, *A* and *B*) always alternate in succession (i.e.: *A-B-A-B*...). When both sounds were far apart from each other on the sharpness axis, many subjects heard two distinct melodic streams which crossed paths, whereas with a short distance in sharpness between both timbres only one melody was heard (Wessel, 1985, p.650). Wessel explained this effect by referring to the difference in the distribution of spectral energy between the two timbres (Wessel, 1985, p.651). Nevertheless, due to the fact that these sampled sounds were transposed, the spectral peaks were also shifted along with the frequency, whereas, in a true musical context, the spectral peaks always stay in place while the frequency is shifted up and down. Anyone can easily observe an example for timbre *unnaturally* shifting along with the frequency: just play an Long Play record at 45 *rpm* instead of 33 *rpm* (this corresponds roughly to transposing the fundamental up a fourth). This shows us that, in fact, the phenomenon of melodic fission that Wessel observed is actually quite unexpected, considering the conditions under which his experiment took place. Moreover, he discloses neither the number of subjects that took part nor the number of subjects who actually heard two melodies depending on the distance in timbral space. He implicitly admits that this effect did not appear in every individual case ("...and for many listeners two interwoven descending lines are formed, each with its own timbral identity." (Wessel, 1985, p.540).

3 Fluctuations and Formant Positions: Stream Segregation and Schumann's Principles of Timbre

Due to the above cited reasons, my new version of the Wessel experiment does not transpose the individual probes. All of the pitches employed below are the original pitches as played in the samples. Furthermore, the following experiment does not concentrate as much on the sharpness of the sounds, but rather gives special emphasis to the positions of the formants and to the presence of fluctuations in the sounds. This is done because timbre can be described much more thoroughly by using the formant concept than by using that of sharpness (Kendall & Carterette, 1993).

If we compare the formant areas of woodwind and brass instruments that have been measured up until now, the result is that some pairs of instruments have formant areas that are practically identical (for instance bassoon and horn), whereas other pairs of instruments possess totally diverging formant areas (for instance bassoon and trumpet). Moreover, there are instruments for which no characteristic formants can be found, but only fluctuations (for bowed string instruments, see Reuter, 1995, pp.149,154, 1996, pp.221-230).

Following the results presented in my thesis (Reuter, 1996) concerning the distinguishability of instruments playing in unison, the hitherto hopelessly elaborate area of instrumental combinations can be now summarized in four key statements:

1. Timbres of which the formant areas mainly diverge from one another can be distinguished well from one another when playing in unison (comparable to the results of the experiments of Sandell, 1995, p.239).
2. Timbres of which the formant areas mostly correspond with one another are hard or impossible to tell apart from each other when playing in unison. They blend homogeneously.
3. Timbres that are mainly characterized by fluctuations cannot be distinguished from one another when playing in unison: they blend homogeneously.
4. When instruments that are mainly characterized by fluctuations play together with instruments whose main timbral characteristic is formants, then the difference in sound pressure level between the two groups or types determines if they are perceived separately or not.

3.1 Hypotheses

If we transfer this approach from simultaneously sounding timbres to successively alternating ones, we can propose the following hypotheses:

1. For alternating timbres with corresponding formant positions (for instance, bassoon and horn), it could be expected that, due to their similar timbral characteristics, they would be heard as one sole melodic stream.

2. For alternating timbres with non-corresponding formant positions (for instance, bassoon and trumpet), it could be expected that, due to their differing timbral characteristics, they would be heard mostly as two interwoven melodic streams.
3. For alternating timbres with formants on the one hand and fluctuations on the other (for instance, the bassoon and the cello section), it could be expected that, due to diverging timbral characteristics, mostly two interwoven melodic streams would be heard.
4. For alternating timbres that are mostly characterized by fluctuations (for instance the violin section and the cello section), it could be expected that, due to similar timbral characteristics, mostly one sole melodic stream would be heard.

3.2 Experiment

In order to test the validity of these hypotheses, the following experiment was undertaken: 30 subjects were presented with short melodies (composed of the pitches g-b-d'). These melodies were either played by one sole instrument or section, or the timbres alternated from one tone to the next, see Fig.1. After each of these examples, the subjects were to decide whether they had heard one sole melody (pattern 1) or whether they had heard two interwoven streams (pattern 2 or 3). In the case that they head heard two interwoven streams, then they were to decide whether the melody that started on the beat was in the foreground (pattern 2, darkly colored notes onbeat) or if it was the pattern that started offbeat (pattern 3, darkly colored notes offbeat).

Fig. 1. One stream (pattern 1) or two interwoven streams (pattern 2 and 3)

The timbres used here (the bassoon, the horn, the trumpet, the violin section (6 violins) and the cello section (3 cellos) were played alternately at the same sound pressure level in the following combinations (see Appendix):

- Instruments with corresponding formant positions (bassoon and valve-horn): bn/cor: bassoon (first) and valve horn (second). cor/bn: valve horn (first) and bassoon (second).
- Instruments with non-corresponding formant positions (bassoon and trumpet): bn/tp: bassoon (first) and trumpet (second). tp/bn: trumpet (first) and bassoon (second).
- Instruments with formants and with fluctuations (bassoon and celli): bn/vc: bassoon (first) and celli (second). vc/bn: celli (first) and bassoon (second).
- Instruments with fluctuations (violins and celli): vl/vc: violins (first) and celli (second). vc/vl: celli (first) and violins (second).

Additionally, melodies were presented in which only one instrument was playing. The tempo of the samples was 3 tones/sec in the first part of the experiment, and 6 tones/sec in the second part. During the course of the ca. 20-min.-long hearing experiment, each combination was presented three times in random succession. In other words: with 30 subjects taking part, each combination was judged a total of 90 times. Each musical example (ca. 2 sec.) was followed by a pause of 8 sec. in which the subjects were to place the mark on the questionnaire (forced choice) corresponding to what they had perceived. As preparation for the experiment, before each session patterns 1, 2 and 3 were presented using pure (sine) tones with identical (pattern 1) or alternating (patterns 2 and 3) sound pressure levels in order give the subjects an explanation of the streaming phenomena that they were about to hear. The participants were encouraged to adjust the volume to a level that they found comfortable.

3.3 Results

The evaluation of the questionnaires brought the following results:

- In the case of alternating instruments with corresponding formant positions (bassoon and horn, Table 1), the subject's assessment varied between one (pattern 1) and two melodic streams (pattern 2 or 3). Compared with the other results (see below), the judgments here seem to have been made quite randomly, according to where the subjects focused their attention.
- When instruments with non-corresponding formant positions alternated within one melody (bassoon and trumpet, Table 2), this led to a clear perception of two interwoven streams.
- Two interwoven streams were also heard when a formant-characterized instrument alternated with timbres characterized by fluctuations (bassoon and celli, Table 3)

- When two timbres alternated which were both characterized by fluctuations (Table 4), this resulted once more in a rather random distribution of the listeners' judgments that is hard to interpret. The perception of the melodic streams seemed once more to follow their momentary focusing of attention
- Melodies consisting of only one timbre (the reference stimuli, Table 5) were heard in almost all cases as one melodic stream.

Table 1. Alternating timbres with corresponding formant positions (1st line, bassoon [bn] and valve horn [cor]) that lead to the perception of one or more melodic streams (1st column)

3 tones/sec	bn/cor	cor/bn	6 tones/sec	bn/cor	cor/bn
1 stream	46	34	1 stream	31	38
2 streams (on beat)	43	20	2 streams (on beat)	50	19
2 streams (off-beat)	1	23	2 streams (off-beat)	9	33

Table 2. Alternating timbres with non-corresponding formant positions (1st line, bassoon [bn] and trumpet [tp]) that lead to the perception of one or more melodic streams (1st column)

3 tones/sec	bn/tp	tp/bn	6 tones/sec	bn/tp	tp/bn
1 stream	1	0	1 stream	2	1
2 streams (on beat)	11	76	2 streams (on beat)	19	76
2 streams (off-beat)	78	14	2 streams (off-beat)	69	13

Table 3. Alternating timbres with formants and fluctuations (1st line, bassoon [bn] and celli [vc]) that lead to the perception of one or more melodic streams (1st column)

3 tones/sec	bn/vc	vc/bn	6 tones/sec	bn/vc	vc/bn
1 stream	3	0	1 stream	1	2
2 streams (on beat)	8	77	2 streams (on beat)	18	71
2 streams (off-beat)	79	13	2 streams (off-beat)	71	17

Finally, by examining which combinations tend to lead rather to an on-beat melodic perception (pattern 2) and which ones lead rather to an anacrustic, off-beat melodic perception (pattern 3), we find that, once the melody has

Table 4. Alternating timbres with fluctuations (1st line, violins [vl] and celli [vc]) that lead to the perception of one or more melodic streams (1st column)

3 tones/sec	vl/vc	vc/vl
1 stream	40	17
2 streams (on beat)	24	26
2 streams (off-beat)	26	47

6 tones/sec	vl/vc	vc/vl
1 stream	36	24
2 streams (on beat)	40	45
2 streams (off-beat)	14	21

Table 5. Individual timbres (1st line, bassoon [bn], trumpet [tp], valve horn [cor], violins [vl] and celli [vc]) that lead to the perception of one or more melodic streams (1st column)

3 tones/sec	bn	tp	cor	vl	vc
1 stream	89	82	86	70	76
2 streams (on beat)	1	8	4	18	8
2 streams (off-beat)	0	0	0	2	6

6 tones/sec	bn	tp	cor	vl	vc
1 stream	81	87	89	69	80
2 streams (on beat)	7	3	1	10	9
2 streams (off-beat)	2	0	0	11	1

been split, the timbre which is characterized by a higher formant becomes the melodic stream which, in the combination of bassoon and trumpet, comes to the foreground — or, in the case of the combination of bassoon and celli, the timbre that has the most pronounced fluctuations comes to the foreground. To put this more generally and in Gestalt terms, this means that, following the division of the melody into two streams due to timbre, the brighter timbre becomes the Gestalt or theme whereas the timbre which is less bright becomes the background, i.e. is perceived as an accompaniment.

4 Discussion

The hypothesis: "in the case of timbres with differing formant positions two melodic streams are perceived" was fully confirmed (Hypothesis no. 2). The same is true for melodies in which timbres characterized by formants alternate with timbres characterized by fluctuations (Hypothesis no. 3). The supporting hypotheses: "in the case of alternating timbres with corresponding timbral characteristics (Hypothesis no. 1: the same formant position, Hypothesis no. 4: fluctuations of the same intensity) only one melodic stream is heard" could neither be disproved nor satisfactorily confirmed. Thus, the question remains open: why did the judgments of alternating timbres with corresponding characteristics vary so widely?

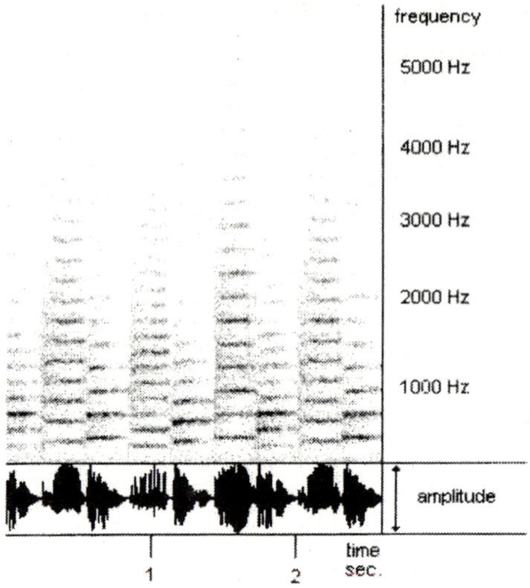

Fig. 2. When alternating bn - tp - bn - tp ... the melody of the trumpet is more prominent. Sample rate=44.1 kHz, Resolution=16 bit, FFT=2048 points, FFT-Window=11 msec

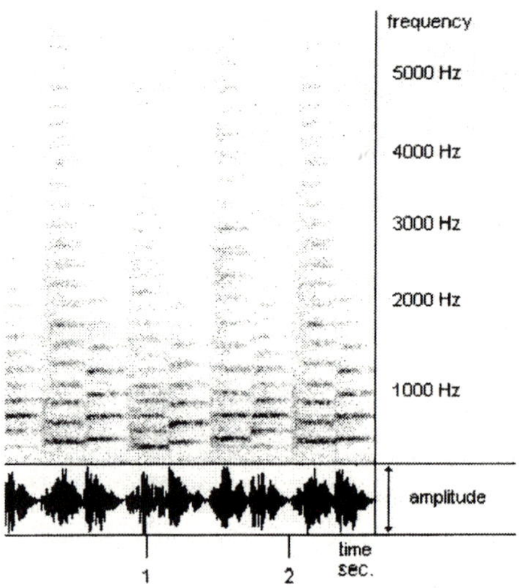

Fig. 3. When alternating bn - vc - bn - vc ... the melody of the cello section is more prominent. Sample rate=44.1 kHz, Resolution=16 bit, FFT=2048 points, FFT-Window=11 msec

An answer might be found in the investigations of van Noorden (1977): using pure-tone intervals played at different tempi, he was able to show that tempo and interval size determine from what point on a melody is heard as one or more interwoven streams and from when on perception can be influenced by the focusing of attention (van Noorden, 1977, p.1041).

The tempi (3 viz. 6 tones/sec) and interval sizes (fifths, major and minor thirds = 7, 3 and 4 half tones) used in the experiment presented here all lie in an area in which, as van Noorden found, streaming depends on the focusing of attention.

We could leave it at that and say that it is due to the tempo that the streaming judgments in the case of corresponding timbre characteristics were made rather randomly. If we bear in mind, though, that in almost all cases the melodies were heard as one sole stream when they were played by only one instrument (see above), then we must suppose that other factors contributed decisively to the subjects' undecidedness.

One of these factors could be, for example, the fact that some of the subjects judged the melodies supposing that the experiment was testing their analytical hearing acuity, and thus focused their attention more on the ambiguous examples (patterns 2 and 3). Although before each series of tests it was expressly emphasized that musical abilities per se were by no means meant to be tested (as opposed to the situation in, say, an ear training test), it would surely not be far-fetched to suppose that the 30 subjects, being musicologists, musicians and sound technicians, also wanted to show how well they could hear.

A further reason might be found in the experimental setup itself: on the one hand, only three patterns of perception were given to which the subjects could relate their melodic judgments (patterns 1-3). Perhaps other modes of perception could have been possible, for ex. a shifting of the melodic streams after each third or fourth note, or that all G's, B's and D's each form their own melodic stream, etc. This should be taken into account when repeating the experiment.

As to the melody itself: due to the alternating change of timbre in an unchanging melody, the points at which the streams (when they were heard) split apart were always at the same places. Thus, after the first stream segregated melodies, the subjects could have undergone a certain learning effect which then also could have been transferred to the melodies with corresponding timbres.

5 Conclusion

Apart from these critical remarks, the results in the examples with contrasting timbres are still quite encouraging. The results for alternating timbres with similar features (corresponding formant positions or only fluctuations) are not distinct enough to obtain a plausible interpretation in the sense of

Schumann's principles of timbre. Nevertheless, it has been demonstrated that differing formant positions lead to the streaming effect, where the instrument with the higher main formant is heard in the foreground. A similar phenomenon can also be observed when timbres with formants alternate with timbres with fluctuations.

Acknowledgement

I wish to thank Stanley Hanks for the English translation.

Appendix

Sound Examples (CD-tracks 31-38)

Each sound is played twice:

- Instruments with corresponding formant positions (bassoon and valve horn):
  Example #1: bn/cor: bassoon (first) and valve horn (second).
  Example #2: cor/bn: valve horn (first) and bassoon (second).

- Instruments with non-corresponding formant positions (bassoon and trumpet):
  Example #3: bn/tp: bassoon (first) and trumpet (second).
  Example #4: tp/bn: trumpet (first) and bassoon (second).

- Instruments with formants and with fluctuations (bassoon and celli):
  Example #5: bn/vc: bassoon (first) and celli (second).
  Example #6: vc/bn: celli (first) and bassoon (second).

- Instruments with fluctuations (violins and celli):
  Example #7: vl/vc: violins (first) and celli (second).
  Example #8: vc/vl: celli (first) and violins (second).

References

Auhagen, W. (1987). Dreiecksimpulsfolgen als Modell der Anregungsfunktion von Blasinstrumenten. *Fortschritte der Akustik, 13*, 709–712.

Bregman, A. (1990). *Auditory scene analysis: The perceptual organization of sound.* Cambridge, MA: The MIT Press.

Bregman, A., Levitan, R., & Liao, C. (1990). Fusion of auditory components: Effects of the frequency of amplitude modulation. *Perception and Psychophysics, 47*, 68–73.

Dowling, W., & Harwood, D. (1986). *Music cognition.* New York, NY: Academic Press.

Fricke, J. (1975). Formantbildende Impulsfolgen bei Blasinstrumenten. In *Fortschritte der Akustik, 4. DAGA '75* (pp. 407–411). Braunschweig.

Fricke, J. (1993). Die Wechselwirkung von Mensch und Instrument im Zusammenspiel von Physik und Psychologie. In B. Enders (Ed.), *Neue Musiktechnologie, Vortärge und Berichte vom Klangart-Kongress 1991 an der Univ. Osnabrück.* Mainz: Schott.

Grey, J. (1977). Multidimensional perceptual scaling of musical timbres. *The Journal of the Acoustical Society of America, 61*, 1270–1277.

Grey, J., & Gordon, J. (1978). Perceptual effects of spectral modifications on musical timbres. *The Journal of the Acoustical Society of America, 63*, 1493–1500.

Huron, D. (1989). Voice denumerability in polyphonic music of homogeneous timbres. *Music Perception, 6*, 361–382.

Iverson, P. (in press). Auditory stream segregation by musical timbre: effects of static and dynamic acoustic attributes. *Journal of Experimental Psychology: Human Perception and Performance.*

Kendall, R., & Carterette, E. (1993). Verbal attributes of simultaneous wind instrument timbres: I. von Bismarck's adjectives. *Music Perception, 10*, 445–467.

McAdams, S. (1982). Spectral fusion and the creation of auditory images. In M. Clynes (Ed.), *Music, mind and brain: A neuropsychology of music.* New York, NY: Plenum Press.

McAdams, S., & Bregman, A. (1985). Hearing musical streams. In C. Roads & J. Strawn (Eds.), *Foundations of computer music.* Cambridge, MA: The MIT Press.

Miller, G., & Heise, G. (1950). The trill threshold. *The Journal of the Acoustical Society of America, 22*, 637–638.

Reuter, C. (1995). *Der Einschwingvorgang nichtperkussiver Musikinstrumente.* Frankfurt/M.: P. Lang.

Reuter, C. (1996). *Die auditive Diskrimination von Orchesterinstrumenten.* Frankfurt/M.: P. Lang.

Sandell, G. (1995). Roles for spectral centroid and other factors in determining "blended" instruments pairings in orchestration. *Music Perception, 13*, 209–246.

Schumann, K. (1929). *Physik der Klangfarben.* Berlin: Habilschr.

Tougas, Y., & Bregman, A. (1985). Crossing of auditory streams. *Journal of Experimental Psychology: Human Perception and Performance, 11*, 788–798.

Tougas, Y., & Bregman, A. (1990). Auditory streaming and the continuity illusion. *Perception and Psychophysics, 47*, 121–126.

van Noorden, L. (1977). Minimum differences of level and frequency for perceptual fission of tone sequences ABAB. *The Journal of the Acoustical Society of America, 61*, 1041–1045.

Voigt, W. (1975). Untersuchungen zur Formantbildung in Klängen von Fagott und Dulzianen. In *Kölner Beiträge zur Musikforschung 80.* Regensburg: Bosse.

Wertheimer, M. (1923). Untersuchungen zur Lehre von der Gestalt. *Psychologische Forschung, 4*, 301–350.

Wessel, D. (1985). Timbre space as a musical control structure. In C. Roads & J. Strawn (Eds.), *Foundations of computer music.* Cambridge, MA: The MIT Press.

Cross-Synthesis Using Interverted Principal Harmonic Sub-Spaces

Thierry Rochebois and Gérard Charbonneau

Université Paris Sud, Institut d'Electronique Fondamentale, Bâtiment 220, F-91405 Orsay Cedex, France

Abstract. Nowadays, analysis and synthesis processes allow a wide variety of timbral effects. One of the most popular effects is known as cross-synthesis. Cross-synthesis consists in combining timbral parameters from two distinct sounds to obtain a new *hybrid* sound. This paper introduces the use of principal component analysis (PCA) in an analysis/synthesis process and shows its flexibility for the creation of hybrid sounds.

1 Introduction

Timbre or tone color is often defined as the property that distinguishes two tones having the same pitch and loudness. The perceived timbre has a multidimensional characteristic (Grey, 1975). Its physical counterpart is still hard to define. We can state with A. Schoenberg that: "The evaluation of tone color ("Klangfarbe"), the second dimension of tone, is thus in a still much less cultivated, much less organized state than is the aesthetic evaluation of these last-named harmonies."(Schoenberg, 1922/1978).

Cross-synthesis involves an analysis process, the interversion of parameters, and the synthesis of *hybrid* sounds. The results of such a process allow the listener to understand the relationships between the physical parameters extracted by the analysis and perceptual dimensions of timbre. Cross-synthesis is often performed using a set of filters controlled by the spectral envelope of a sound. Figure 1 shows a typical analog implementation of this method:

- A bank of band pass filters and a set of level detectors perform a spectral analysis of a *formantic tone*
- The *instrument tone* is decomposed into a set of band limited signals by another bank of band pass filters
- These band limited signals are mixed together according to the spectral analysis performed on the *formantic tone*

This method allows for instance to make acoustical instruments sound like singing voices. However, this kind of synthesis is limited, because it only permits to duplicate formantic features of a sound to another. Thus, this kind of cross-synthesis allows the listener to link the physical concept of *formants* to the perceptual concept of voice articulation and color.

On the other hand, multidimensional scaling techniques are already known as efficient tools for perceptual timbre study (Grey, 1975). These methods, and especially principal component analysis, have also been envisioned for efficient multiple wavetable synthesis (Beauchamp & Horner, 1995; Stapleton & Bass, 1988).

This paper presents a new method for creating hybrid sounds. This process uses the principal component analysis to extract valuable spectral and temporal features. We present in this paper the two steps involved in the method, namely, the sound analysis using the principal component analysis (PCA), and the sound synthesis using interverted reduced data. The results of this cross-synthesis process allow the listener to link the mathematical concept of *principal harmonic sub-space* and *trajectory* to the perceptual concept of *tone palette* and *tone dynamics*. This link lets us foresee that this analysis/synthesis process can be a valuable basis of a cognitive model of timbre.

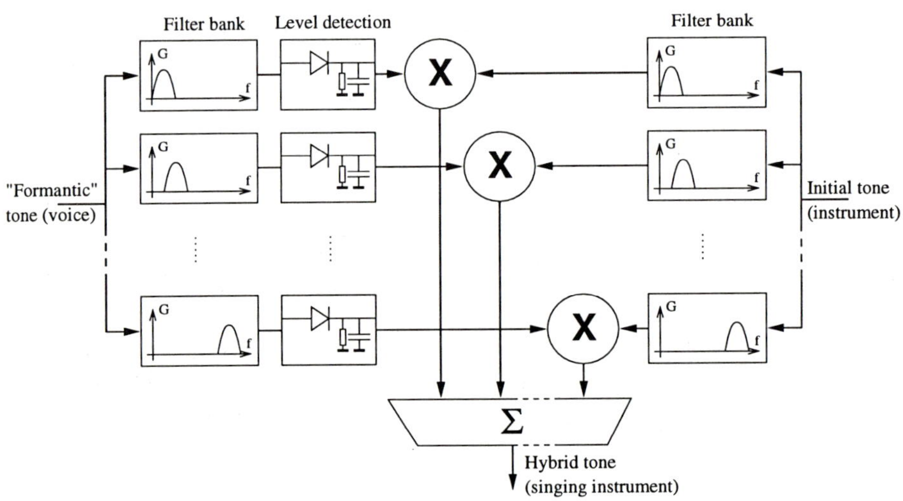

Fig. 1. Formantic Cross Synthesis (vocoder)

2 Sound Analysis

2.1 Spectral Analysis

We only consider tones having a defined pitch. In this context, a spectral analysis consists of computing the harmonic amplitudes and frequencies of a pseudo periodic signal versus time. Precisely, we restrict our study to strictly harmonic tones: the frequency of any component is the exact multiple of the fundamental frequency at any time, nevertheless this frequency may vary (allowing vibrato, glissando,..etc..).

The analysis method we use consists in interpolating a Fourier transform (McAulay & Quatieri, 1986). The principle of this method is described on Fig.2. This allows to extract all the components of a sound, whether they are harmonic or not. Applied to pseudo periodic tones, this technique gives very good results, even with sounds having large frequency variations.

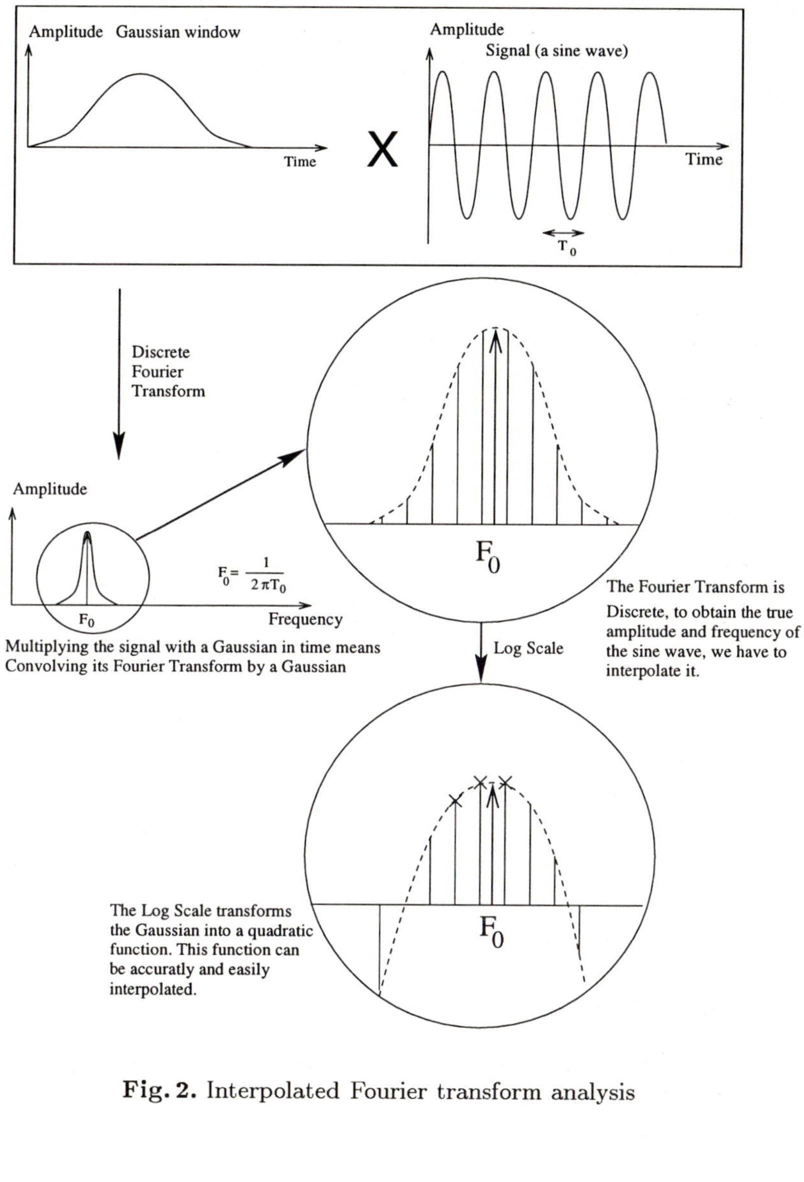

Fig. 2. Interpolated Fourier transform analysis

2.2 Principal Component Analysis

The harmonic spectral analysis leads to a large amount of data. Thus it is useful to reduce this data, while eliminating redundancies. For this purpose, we used the PCA method applied to the amplitudes of harmonics.

To understand the effects of this technique in our context, let us think in terms of basis of functions. Imagine that you want to draw a trajectory representing the spectral evolution of a sound. The first axis will represent the amplitude of the first harmonic, the second axis will represent the amplitude of the second harmonic and so on... The problem is that one needs about 50 harmonics to synthesize correctly a sound. So, you will have to draw the trajectory in a 50 dimension space! The set of harmonic components S_n is the basis for this *harmonic* space. The amplitudes $A_n(t)$ are the coordinates of the harmonic spectrum at time t in the harmonic space.

PCA optimally chooses a sub-space from the original 50 dimension harmonic space. This is very similar to choosing the best point of view for drawing an object: you choose the best way to represent a three dimensional object on a two dimensional plane. Applied to our problem, PCA gives us a low dimension *drawing* of the original 50 dimensional trajectory. A three dimensional data reduction is often adequate for usual musical sounds (this assertion can be validated by calculating the error caused by the data reduction). Both the trajectory and the definition of the optimal sub-space contain relevant data. The interpretation of both the harmonic sub-space and the curve representing the dynamic spectrum will allow us to build up a kind of multiple wavetable synthesis.

2.3 Principal Harmonic Vectors

The PCA method gives a new basis P_n of the original sound space. Each vector P_n is expressed as a linear combination of vectors S_n. In brief, the signal is no longer represented as a sum of numerous sinusoids S_n , but rather by a sum of more elaborate functions defined by the harmonic vectors P_n. Harmonic vectors P_n are ordered according to their importance for the represented sound. We build up the harmonic sub-space of each sound using the first 3 harmonic vectors P_1, P_2, P_3, which we will call the principal harmonic vectors. P_1 has an obvious significance: it represents the mean harmonic spectrum of the sound. P_2 and P_3 represent the *spectral directions* in which the spectrum evolves. The principal harmonic sub-space of a sound is static, it corresponds to its spectral behavior. Thus, we can guess that the principal harmonic sub-space *tries to model* the static structure that generated the sound i.e. the instrument.

2.4 Trajectories in the Principal Harmonic Sub-Spaces

A point in the harmonic sub-space corresponds to the harmonic spectrum of the sound at time t. Taking into account the previous remarks, it is possible to interpret the trajectory of the harmonic spectrum in this sub-space. Let $V_1(t)$, $V_2(t)$, $V_3(t)$, be the coordinates of the harmonic spectrum on axes P_1, P_2, and P_3. The coordinate $V_1(t)$ corresponds to the amplitude of the projection on principal axis P_1 of the harmonic spectrum. Because axis P_1 is related to the mean harmonic spectrum of the sound, $V_1(t)$ just about represents the sound amplitude. Coordinates $V_2(t)$ and $V_3(t)$ correspond to the deviation between the current harmonic spectrum and the mean harmonic spectrum. Thus, $V_2(t)$ and $V_3(t)$ are related to the spectral dynamics. The trajectory of the sound is dynamic and corresponds to the spectral variations of the sound. Thus, we can guess that the trajectory corresponds to the way the instrument was played.

2.5 Analysis Example

- Figure 3 represents the trajectory of the flute tone formerly used by Grey (1975). We can see the three parts of the sounds and the associated timbral variations. On the attack and release parts, the most important variation concerns the first axis i.e. the amplitude variations. On the sustain part, the most important variation concerns the second axis: during this sustain, the amplitude of the sound is steady while it's spectral content changes.
- Figure 4a represents the first 3 principal harmonic vectors that corresponds to the principal harmonic vectors P_1, P_2, and P_3.
- Figure 4b represents the coordinates $V_1(t)$, $V_2(t)$, and $V_3(t)$ as enveloppes. The dotted lines correspond to the error caused by the data reduction.

Thus, the data reduction is valuable for analysing and studying timbre, we will see how it can be used in a synthesis context.

3 Cross-Synthesis

The principal harmonic sub-space analysis leads to a wide variety of synthesis possibilities: sub-spaces can be understood as a kind of *tone palette* while trajectories can be understood as a kind of *drawing*. Cross-synthesis will illustrate this concept. With the PCA method, original data have been reduced, and the remaining spectral data consist of the seven following curves:

- The three principal spectra P_1, P_2, and P_3.
- The trajectory of the harmonic spectrum (coordinates $V_1(t)$, $V_2(t)$, and $V_3(t)$).
- The fundamental frequency $f_1(t)$.

Let us take two very simplistic tones, having a constant spectrum and an amplitude envelope. The PCA method would extract the amplitude envelope as $V_1(t)$, and the spectrum as P_1 for each tone. As they describe spectral

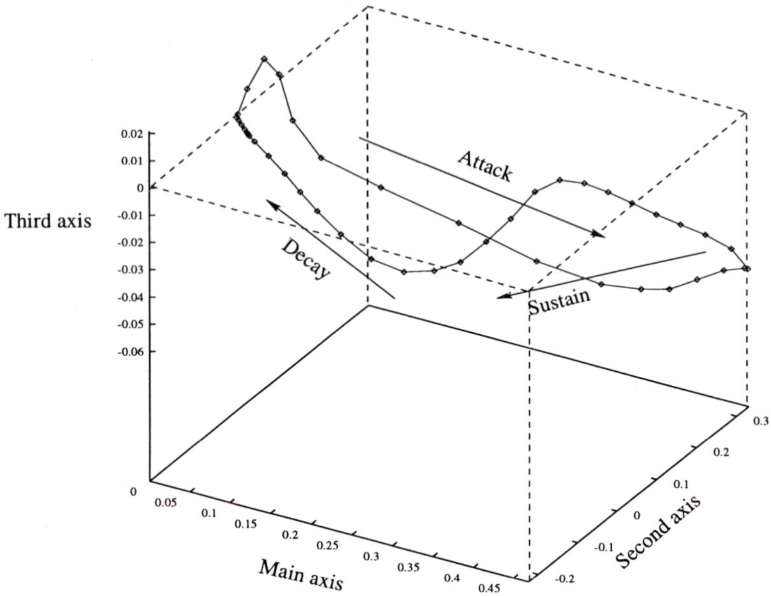

Fig. 3. Flute tone trajectory

variations, $V_2(t)$ and $V_3(t)$ would be equal to 0. In this simplistic case, it is easy to imagine a method for generating hybrid sounds: one only has to intervert the two amplitude envelopes. For example, it is possible to use the spectrum of an organ, and then to apply it the amplitude envelope of a piano. The resulting sound is a *struck* organ tone.

The previous method can only be applied to sounds having constant spectra. It is possible to extend it by using reduced spectral data which generalize the steady spectrum and amplitude envelope concept. For building up an hybrid sound pair, one only needs to intervert the principal harmonic vectors P_1, P_2, P_3 of two sounds (Fig.5). Each tone will present the dynamic behavior of one sound and the spectral features of the other. This process has been used successfully with many sounds played on acoustical instruments. For our preliminary experiments we used the McGill University sound database CDROM (Opolko & Wapnik, 1991). This database includes about all the instruments of the orchestra, each instrument being played with various styles and over its full scale.

In most of the cases the cross-synthesis leads to naturally sounding tones. For example, the dynamic behavior of a flute with the spectral contents of an oboe leads to a *blown* oboe sound. By contrast, the cross-synthesis of very different sounds leads to rather synthetic hybrid tones. For instance, the dynamic behavior of a piano crossed with a cello spectrum sounds like

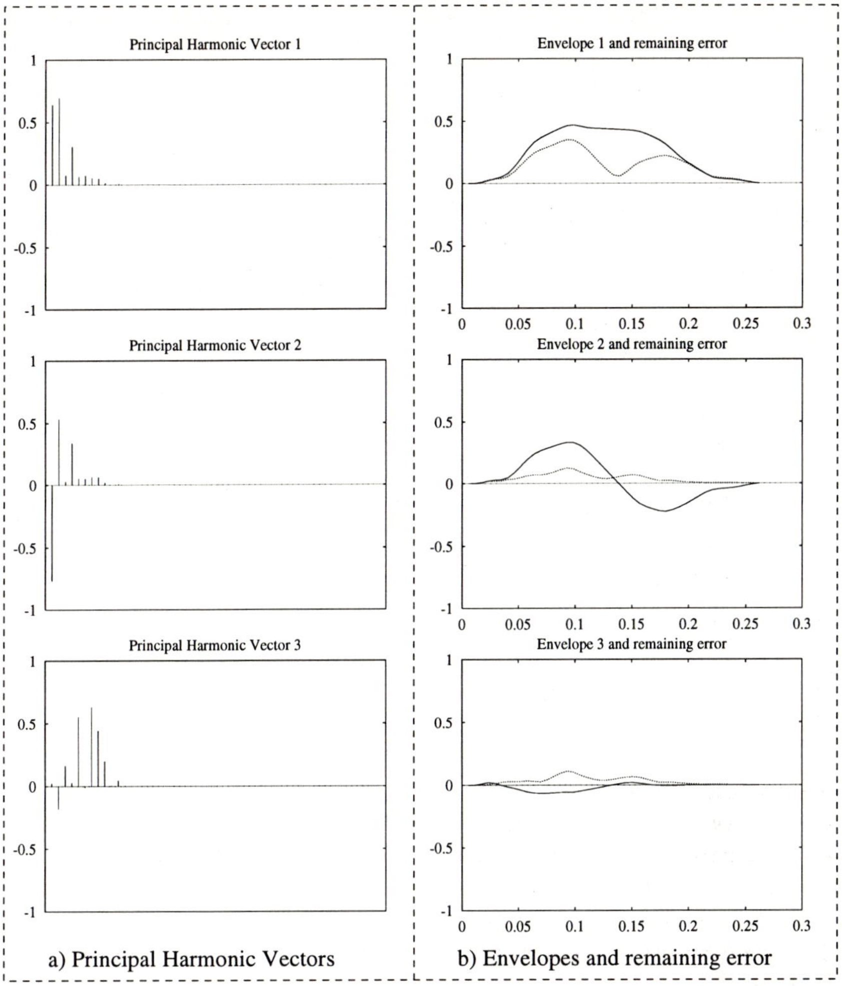

Fig. 4. Flute tone reduced data

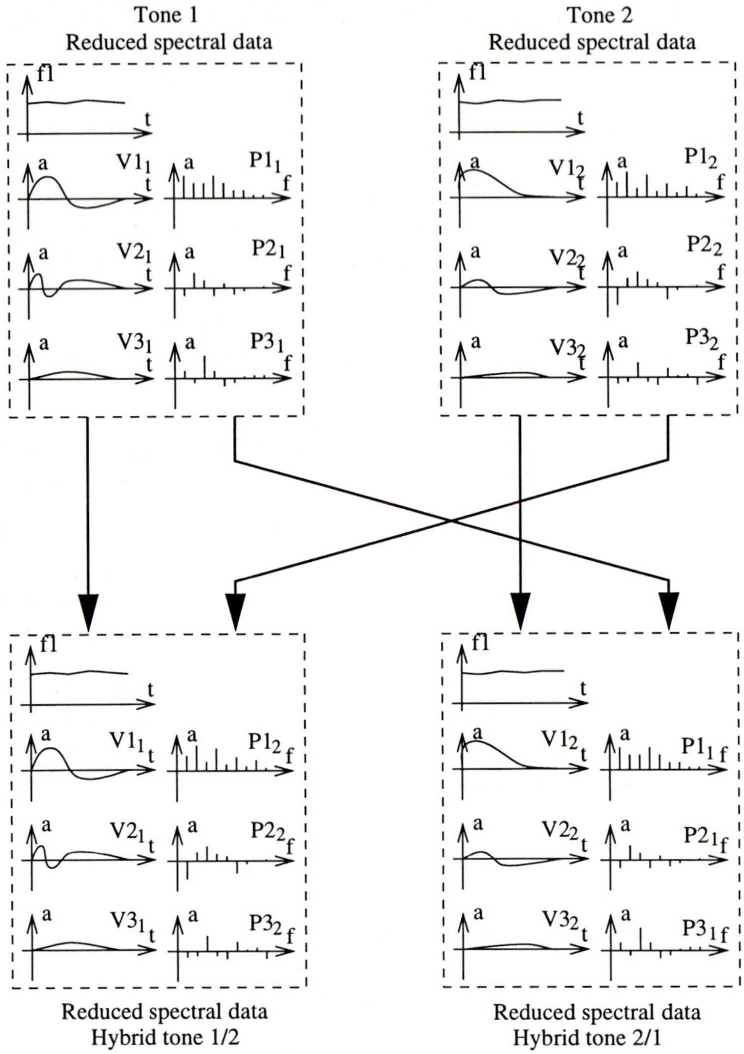

Fig. 5. Cross-synthesis

a harpsichord rather than like a cello played pizzicato. Nevertheless, those strange tones are often interesting.

4 Conclusion

The cross-synthesis described here, which is obtained by interverting principal harmonic sub-spaces is still at an experimental level. However, cross-synthesis is just an example of what can be done using principal harmonic sub-spaces. As sub-spaces can be compared to a painter's palette and trajectories can be compared to drawings, the design of flexible synthesis software and music notation would make it convenient for music composing. Compared to the *vectorial synthesis* introduced in 1986 by Sequential Circuits (Roads, 1987) (that consists in dynamically mixing four oscillators to obtain time varying timbres), this method is based on consistent wavetable sets extracted from real instruments. Many technical improvements are under development:

– the use of complex valued amplitudes will allow to take account of phase variations and so will be valuable to model inharmonic tones.
– a dedicated analysis/synthesis process must take account of noisy sounds (Serra & Smith, 1990) (hammer noise of the piano, blow of the flutist...).

Additionally, the use of this synthesis process, the development of fast additive synthesis algorithms (Depalle & Rodet, 1995), and the availability of sound synthesis dedicated DSP (Deforeit & Heckroth, 1995) will allow the design of real-time synthesizers having expressive timbral variations.

Appendix

Sound Examples (CD-tracks 39-44)

– Sound exapmple #1: Grey's Flute Tone. This sound's trajectory is shown on Fig.3 and reduced data set is shown on Fig.4. This sound is presented four times. The first time with its original length. The second, third and fourth time, it is time-stretched by factors 2, 4 and 8. These time-stretched sounds were obtained by using the reduced data set. These sounds allow the listener to understand the relationships between timbre evolution and the data represented on Fig. 3 and Fig.4.

The next examples are cross synthesis examples, based on four sounds exerpted from the McGill University sound database CDROM.

– Sound exapmple #2: First the original sounds are presented:
  - flute
  - oboe

  - cello
  - piano
- Sound exapmple #3: The first hybridization is presented:
  - flute
  - oboe
  - first hybrid tone: flute sub-space, oboe trajectory
  - second hybrid tone: oboe sub-space, flute trajectory
- Sound exapmple #4: The second hybridization is presented:
  - cello
  - piano
  - first hybrid tone: cello sub-space, piano trajectory
  - second hybrid tone: piano sub-space, cello trajectory
- Sound exapmple #5: The third hybridization is presented:
  - flute
  - piano
  - first hybrid tone: flute sub-space, piano trajectory
  - second hybrid tone: piano sub-space, flute trajectory
- Sound exapmple #6: The fourth hybridization is presented:
  - oboe
  - cello
  - first hybrid tone: oboe sub-space, cello trajectory
  - second hybrid tone: cello sub-space, oboe trajectory

References

Beauchamp, J., & Horner, A. (1995). Wavetable interpolation synthesis based on time-variant spectral analysis of musical sounds. *Audio Engineering Society Preprint, Feb 1995.*

Deforeit, C., & Heckroth, J. (1995). A music synthesizer architecture which integrates a specialized DSP core and a 16 bit microprocessor on a single chip. *Audio Engineering Society Preprint, 3958(E3), Feb 1995.*

Depalle, P., & Rodet, X. (1995). *Sound synthesis process, US patent number 5,401,897, Mar 1995.*

Grey, J. (1975). An exploration of musical timbre. *CCRMA Report, STAN-M-2.*

McAulay, R., & Quatieri, T. (1986). Speech analysis/synthesis based on a sinusoidal representation. *IEEE Transactions on Acoustics, Speech and Signal Processing, ASSP-34*, 744–754.

Opolko, F., & Wapnik, J. (1991). *Classical sounds.* McGill University Master Samples. (CD-ROM, Vol. 1, McGill University Master Samples)

Roads, C. (1987). Prophet VS: User's report. *Computer Music Journal, 11*, 74–75.

Schoenberg, A. (1978). *Theory of harmony.* London, Boston: Faber and Faber. (Original work published 1922)

Serra, X., & Smith, J. (1990). Spectral modeling synthesis: A sound analysis/synthesis system based on a deterministic plus stochastic decomposition. *Computer Music Journal, 14*, 12–24.

Stapleton, J., & Bass, S. (1988). Synthesis of musical tones based on the Karhunen-Loève transform. *IEEE Transactions on Acoustics, Speech and Signal Processing, 36*, 305–319.

Gestalt Phenomena in Musical Texture

Dalia Cohen[1] and Shlomo Dubnov[2]

[1] 13 Rashba St., 92264 Jerusalem, Israel
[2] The Hebrew University of Jerusalem, Department of Musicology, Mount Scopus, 91905 Jerusalem, Israel

Abstract. We attempt to examine textural phenomena and their significance in music, by comparing and contrasting texture with timbre and *learned schemes*. The latter are the main subject of musical theories and relate to organizations of parameters (particularly intervals) in a quantitative, measurable form not suitable for describing texture. We extend the definition of texture to cover principles of organization that need qualitative or statistical tools in order to be described. The following points will be addressed in this study: (i) Principles for classification of texture. We suggest textural schemes pertaining to registers of all the parameters, to contours, and to operations. (ii) Combinations of textural schemes with learned schemes. (iii) The relationship to the stylistic ideal. Our assumption is that all of the schemes are selected so as to fit the stylistic ideal, and most of them (even the *learned* ones) are not arbitrary and are subject to psychoacoustic and cognitive constraints.

1 Introduction

In the present paper we attempt to understand the manifestations and significance of texture in music in contrast to other parameters, bearing in mind Gestalt principles and the stylistic ideal [1], which varies from culture to culture and from era to era. Most of Western tonal music theory relates to learned schemes that coalesce on the cognitive level (scales, chords, certain forms, and so on), whereas texture is considered subsidiary – a factor that "cannot exist independently" (Levy, 1982, p.403). Very little direct attention has been paid to the meaning of texture in Western tonal music, and the available studies on the subject (Levy, 1982; Lorince Jr, 1966; Ratner, 1980; Rothgeb, 1977) address mainly the parameter of pitch.

In the music theory of non-Western cultures, texture (in the sense used here) is often an integral part of the learned schemes. Moreover, Lomax (1968, 1977), in his extensive study that compared 400 musical cultures in order to

[1] The stylistic ideal, which can be thought of as a message that shapes the style, has been called by various names such as *ideal*, *ideology*, and *poiesis*. In our opinion, the stylistic ideal can be characterized by four main variables: (1) connection/lack of connection with the outside world; (2) expression of calmness vs. excitement; (3) focusing on the moment vs. the overall structure; (4) clear or blurred units and directionality.

expose possible links between sociocultural and musical characteristics, relates in fact to the manifestations of timbre and texture (without mentioning them explicitly) rather than to interval systems. In twentieth-century music, especially when it lacks learned schemes and is based on timbres and textures only, texture is addressed not only by researchers (Lansky, 1974; Goldstein, 1974; Czink, 1987; Scolnic, 1993), but by the composers themselves (Varèse, 1967; Stockhausen, 1957; Boulez, 1971; Ligeti, 1993).

There are even studies dealing with the formulation of textural phenomena on the acoustic level in electronic and computer music (Smalley, 1986; Czink, 1987). Nevertheless, no conceptual framework has been laid out for understanding texture in all its manifestations, and certain questions arise:

- Is it at all possible to develop a suitable definition for texture that would encompass all its manifestations?
- Can one speak of Gestalt phenomena, that is, the existence of units or schemes, in texture?
- What is the role of texture in relation to the other parameters in various styles?
- Is texture of significance for the stylistic ideal? In what way?

2 Proposal for Characterizing Textural Phenomena by Comparison

2.1 Comparison of Texture with Timbre and Learned Schemes

Texture, as we define it, can be viewed as a superparameter that is related to various combinations of other parameters. In the most general sense, textural phenomena are familiar to us from nature as psychoacoustic factors that evoke various emotional associations.

Music in a given style contains two kinds of information: one having to do with learned schemes and the other with textural information. The relationship between the two is one of the characteristics of the style.

Just as the most general definition of timbre is based on negation of the other parameters, here, too, we define texture as a principle of organization that is not derived from learned schemes that have coalesced on the cognitive level and that differ from culture to culture. Positively speaking, in the most general sense we can define texture as the way of distributing the sound (of defined or undefined pitches) in the dimensions of frequency, time, and intensity. This definition also holds true for timbre, but texture is divisible and refers to longer durations than does timbre. Timbre can be defined even for a duration of approximately 20 milliseconds and is indivisible when we hear it, despite its extreme complexity. Texture, on the other hand, requires some separability along the frequency or time axis; a single-note event has no texture.

Another property of texture is that it is not well defined quantitatively. By contrast, in most pre-twentieth-century styles the parameters (intervals, rhythms, and meters) are not only defined quantitatively but also belong to systems of learned schemes that serve as building blocks for the musical organization of the particular style.

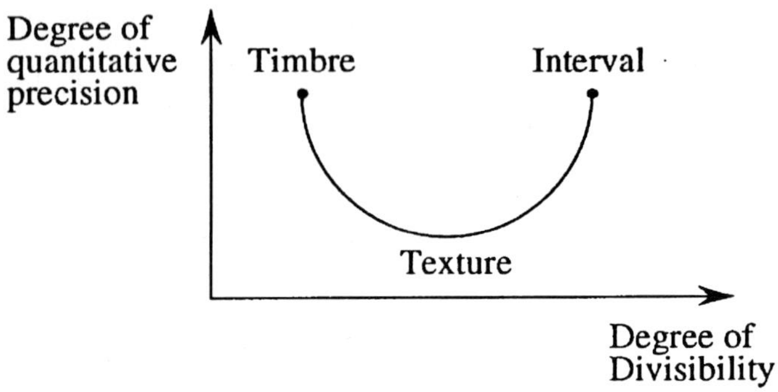

Fig. 1. Texture is contrasted both with intervals and with timbre with respect to the degree of quantitative precision. It is situated between them as far as degree of divisibility is concerned

Difference in learned schemes refers to quantitative difference. For example, the major scale differs from the minor scale in the sizes of three intervals, whereas in texture the important factor is the relative relationship of greater than, less than, or equal to. It refers to the direction of change and not its exact magnitude. We have termed such a relationship a *type*. Type relationships can appear in many forms. For example, a *metric type* is poetic meter, which, unlike musical meter, is not well-defined quantitatively; *type of intonation* characterizes the scale skeleton of Arab singing in practice (Cohen, 1969). Each musical realization of texture is a manifestation of a certain *type*, and conversely, each *type* has many realizations, with the typical ones having high probability of occurrence. Thus, one might speak about texture either as a probability distribution P that underlies a certain type, or look at a typical musical example as a representative of that type. We can measure the difference/similarity by using probability separability measures $D(P_1, P_2)$, which measure the *distance* between distribution functions P_1 and P_2. Such measures are widely used in the pattern recognition literature, and we will not discuss them here.

2.2 Analogy to Speech and Bird Calls

Comparisons and analogies between music and speech have been offered from various perspectives. Here we wish to propose an additional analogy between

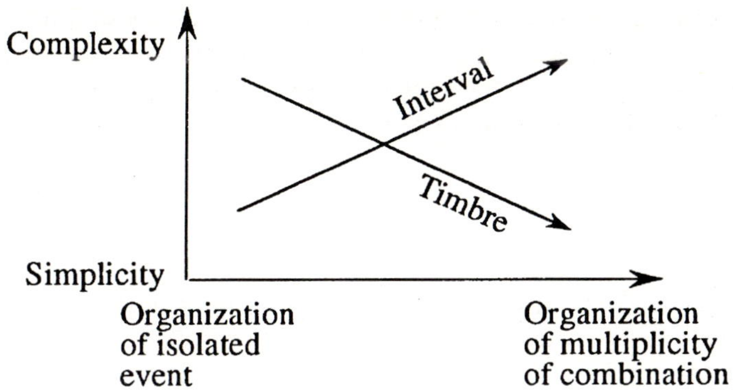

Fig. 2. Intermediate states between texture and interval are arranged along the definability axis of difference/similarity. One pole represents a system of coherent, well defined intervals, whereas the other represents textures defined by *type* relationships

two complementary pairs: learned schemes/texture in music, and lexical stratum/prosody in speech.

The stratum of prosody, or the *musical factors* in speech – called *intonation* in linguistics (Bolinger, 1972) – has two functions: (1) to represent the laws of syntax; and (2) to convey the speaker's attitude to the listener and to what he or she is saying – that is, to convey emotion. Texture in music has the same two functions: (1) *syntactic* – to clarify or blur the musical structure based on learned schemes; (2) to convey emotion, that is, to characterize the musical expression in terms of extra-musical factors. From the standpoint of the psychoacoustic parameters, the musical factors in speech are represented by the texture, that is, the distribution of loudness, duration, and pitch (without reference to precise quantitative magnitudes). Like the texture in music that may or may not concur with the learned schemes, the musical factors in speech may or may not be correlated with the message contained in the lexical stratum[2]. They can also appear as an independent message, as in various nonverbal exclamations. Many experiments have been done in this field (Fónagy & Magdis, 1972), and the findings of these experiments indicate an interesting regularity in the expression of emotion by means of the prosodic level of speech. This regularity has served as a basis for formulating the laws of Palestrina counterpoint, in which the stylistic ideal calls for calm (Cohen, 1971). Moreover, in comprehensive studies of bird calls, we have compared their calls to the musical features of speech, although they have no lexical

[2] Try saying "I'd like to see you" twice – the first time emphasizing the semantic content, the second time emphasizing the opposite meaning, in other words, "go to hell!"

stratum. Interestingly enough, the universal principles of tension and relaxation that determine the laws of Palestrina counterpoint (and not only them, of course) also apply to bird calls in different situations (Cohen, 1983) (for more details, see Fig.9).

2.3 The Contribution of Musical Timbre, Interval, and Texture to the Static and Dynamic Poles

One of the characteristics of a piece of music is its relationship to the time axis, which extends between two poles: one representing the sense of *flow* and directionality that accompanies a series of musical events (typical of Western tonal music), the other representing the sense of focusing on the present moment. Aspects of music that are related to these poles have been given various names: vertical/horizontal; inside time/outside of time and space/time (Czink, 1987); material/structure (Cage, 1966); local/global; momentariness/overall directionality and complexity (Cohen, 1994). Here we shall call this pair *static/dynamic*. Although these features of music with respect to time cannot appear independently, relations between them vary enormously, reflecting one of the important characteristics of a given stylistic ideal. Indeed, the importance of focusing on the moment has been stated explicitly both in non-Western cultures (Chou, 1970) and by twentieth-century composers. We should also note that many twentieth-century compositions are based, consciously or otherwise, on the compositional principles of non-Western music (Chou, 1974). With regard to the possibilities for simple or complex organization in the short or long term, the most prominent representative of the static pole is timbre, which is also significant as an isolated event. In contrast, the most prominent representative of the dynamic pole is the interval (under certain conditions), whose significance stems mainly from its context and which serves as a foundation for many learned schemes that contribute to directional flow.

Particularly noteworthy is the contribution to the static pole of the *borderline states* (as we shall see below) between texture and the other parameters. Texture can be placed between timbre and intervals in terms of the static/dynamic and other aspects, as we shall see below (Fig.4).

3 Borderline States

Due to psychoacoustic constraints (which determine thresholds for our ability to make distinctions), there are borderline states for various musical parameters and texture. In these states, either the parameters cannot be distinguished quantitatively, so that they approach texture; or there is textural change but its direction cannot be identified (producing a sensation of random change). Alternatively, there may be no differentiation between simultaneous textural events. In all these states, the degree to which events can

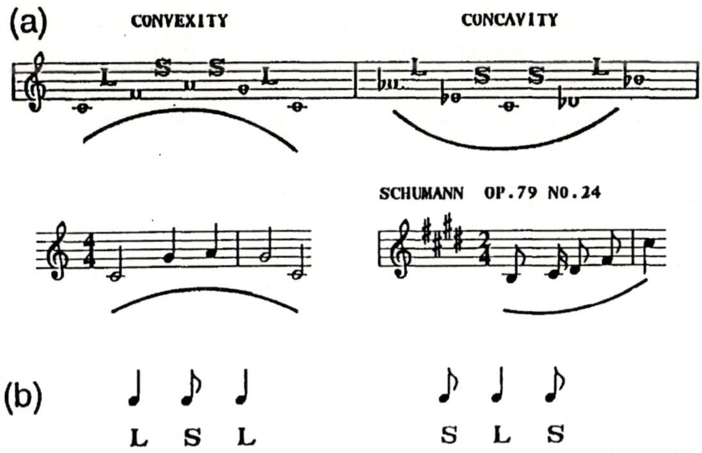

Fig. 3. Inverse relationship between interval and timbre: interval (simple as an isolated event) allows for the most complex organization, while timbre (highly complex as an isolated event) does not allow for a clear and complex organization

be defined is small. There are many levels of borderline states. In the most general terms we will distinguish between relatively stable situations and the various borderline states.

3.1 Borderline States between Texture and Timbre

In borderline states between texture and timbre, on the one hand we cannot detect the texture due to psychoacoustic constraints, and on the other hand we are able to sense the changes that occur within the timbre. The latter occurs in continuous sounds or notes: either when the attack phase is relatively long and the changes in it are meaningful to the listener, as in artistic singing (Rapoport, 1997, this book), or when significant continuous changes in timbre take place throughout the the envelope, as with a bell, or other idiophonic instruments. Note that in ancient China when one spoke of the meaning of a single note, one was referring to the constantly changing idiophonic sound. Texture can thus be regarded as an expansion of timbre, and conversely, timbre can be viewed as a borderline case of texture, with a gray area between them (Fig.4). There are even situations in which an event is perceived both as texture and as timbre – for example, a chord whose components we can identify, is defined by the texture (chords that are open or closed or in different inversions) as well as by the timbre (major or minor, or even the timbre of a particular inversion, in addition to the timbre of the instrument on which the chord is played).

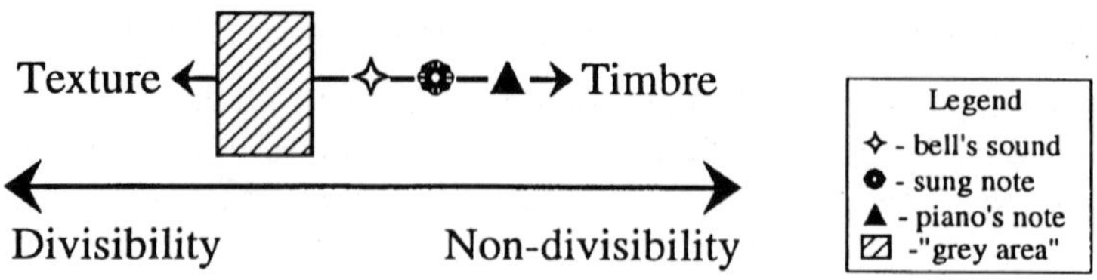

Fig. 4. Examples of borderline states between timbre and texture, when we sense changes in timbre

3.2 Borderline States between Texture and other Parameters

As between timbre and texture, there are also borderline states between texture and intervals and between texture and rhythm. These are the situations in which small or rapid changes occur in pitch or duration, and psychoacoustic constraints prevent clear identification of the pitches, the intervals, or the rhythm. Situations in which intervals turn into texture are intentional in many non-Western musical cultures, as well as in contemporary music.

4 Classification and Schemes of Texture

Three main factors may produce cognitive units: (1) a scheme or Gestalt principle that fuses a collection of events into a unit; (2) a repetition of the collection (A, A, A); (3) a substantial difference of the collection from the surroundings. Learned schemes serve as a unit-forming factor on various levels. With respect to the texture, too, as we shall see, there are some *natural* schemes that combine to form a single unit. Here we shall try to *classify texture by the nature of the events and the manner in which they are distributed.* The distribution will be classified as follows: (1) static distribution – by the range of occurrence of each parameter, based on segregation into separate horizontal layers and the range of scatter in each layer; (2) dynamic distribution – by the curves of change of the different parameters in each layer and between layers, with or without concurrence between the curves. Thus, texture may appear in a single layer with regard to all parameters or in many; without change along the time axis or with a great deal of change, producing numerous musical units of varying degrees of definability.

The nature of the events will be classified as follows: (1) by their stability or definability, from events made up of separate, well-defined elements to events in various borderline states (described above), distinguishing between different degrees of randomness; (2) by connection or lack of connection between the texture and learned schemes. Below we describe the various classifications and their meanings.

4.1 Classification of Textural Events by their Stability

This classification is by the degree of separability or inseparability (borderline states) and the degree of definability of pitch. One can characterize these types using computer sound synthesis and algorithmic composition methods. Below is a selection of types of sound at various levels of indefinability:

- White noise: This is a perceptually indivisible sound undefined in all ways. It can be produced by random number generators.
- Continuous sounds composed of continuous-spectrum segments in which the sensation of pitch is not well-defined, like the *fricative* phonemes in speech, still belong more to the timbre side of the border. With a computer they are produced by filtering the white input noise.
- Sounds composed of a set of discrete frequencies, such as idiophonic instruments with continuous sound whose timbre gradually changes. These *complex* sounds are an example of the borderline case between texture and timbre. Such sounds, characterized by inharmonic spectra, are easily simulated by FM synthesis. The separability is sensed due to the lack of harmonic relations between the partials (Dubnov, Tishby, & Cohen, 1997).
- Timbres with well-defined pitch. We classify these according to the degree of their divisibility. Before, we distinguished between sound events with indivisible timbre and sound events in which we perceive changes, such that they are in a borderline area with respect to texture. Moreover, it is interesting that even timbres, which are considered indivisible, can be classified with respect to divisibility. This classification relates to the *concurrence/nonconcurrence* of its overtones (Dubnov et al., 1997). The non concurrent case approaches the state of texture within timbre. There are various methods of synthesizing pitched sounds, and the sense of texture is achieved by applying a random fluctuation to the harmonics. Applying a random modulation to the frequencies of the sinusoidal components in an additive synthesis method or randomizing the input pulses in *granular* synthesis creates smooth transitions along the continuum of pitch-texture-noise as randomness increases.
- In the case of very unclear interval construction, as in some of the music pieces by Ligeti, we arrive at a border between interval and texture typical of much modern music. In some limiting cases these textures can turn either into noise or into a well-defined pitch. In computer music, many algorithmic composition methods are actually concerned with producing textures. Thus, although many of Ligeti's works are written in conventional notation, one should regard it as sound-texture for purposes of analysis.

4.2 Categories of Schemes According to Manner of Distribution

1. Register Schemes (Ranges of Occurrence).
   These may be related to various parameters that can be arranged on a

scale: tempo, pitch, and intensity. The parameters are classified in accordance with their absolute range and the degree of scatter, described in terms of the three main quantities: low, medium, and high. In many cases the medium size is a *normative optimum*. A focus on the optimum (the medium) as opposed to the two extremes reflects the stylistic ideal. Such a classification system is commonly used in various cultures for tempo, tessitura, degree of melismatism, and so on. For example, the chants of Tibetan monks have a one-part layer (monophony) with extremely low scatter in the low range of pitch. In Palestrina's polyphonic music, whose ideal is calmness, all the indices – the scatter and absolute range of each part and between the parts – are *medium*.

2. Curves of Change (Contour) Along the Time Axis.
   The changes in the various parameters can be represented by six basic types of curves: level, zigzag (including sudden change), ascending, descending, convex, and concave, with concurrence or nonconcurrence between parameters (Fig.5).

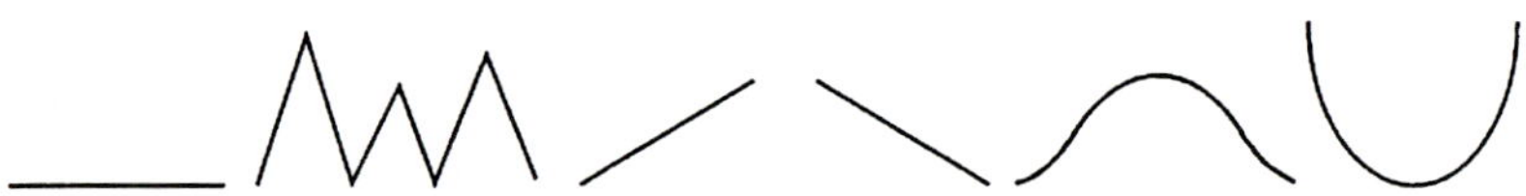

1 2 3 4 5 6

Fig. 5. The six basic curves (contours) of the various parameters, which define a certain part of the texture scheme

These curves may appear in various combinations and on various levels. For example, a curve of intensification may be described as being composed of a chain of convex curves. As we shall see, these types of curves are significant for musical expression in an interesting way. Some scholars refer to melodic curves on the immediate level as *gestures*, which are sometimes considered to be rhetorical means (Ratner, 1980). In the previous paragraph we defined stable texture primitives and used probability functions to describe them. Gradual changes in texture could be described as slow *modulation* acting on the parameters of the probability function.

(a) The flat curve (successive repetitions):
    Its meaning varies in accordance with the number of repetitions, the size of the repeated unit, and the precision of the repetition. Multiple repetitions of small units may serve as background for emphasizing events at another level or for creating tension due to uncertainty regarding the rest of the progression. This tension is limited by the

boredom threshold (Cohen, 1971). In contrast, a single repetition (two occurrences) of a directional unit based on any scheme intensifies the clear directionality. Hence, it is not surprising that the rules of Palestrina counterpoint prohibit various kinds of repetition, whereas multiple repetitions of small units are extremely common in Baroque music, and the symmetrical periodic phrases (one repetition) are so typical of the Classical period.

The 2^n scheme is an expansion of the simple symmetry obtained by a single repetition of a directional unit (two occurrences). In the Classical period, symmetry, which is considered to express maximum clarity, generally includes a series of repetitions, each of which contains its predecessor within it, so that a 2^n series is obtained:

$$1+1+2+4+8+16$$

This series can also be obtained from a repeated division of the whole into two parts. The 2^n scheme was common already in Far Eastern (especially ancient Chinese) music, and currently appears mainly in Gamelan (orchestral) music in Indonesia (Fig.6). Note that the repetitions are not exact; they may include various transformations.

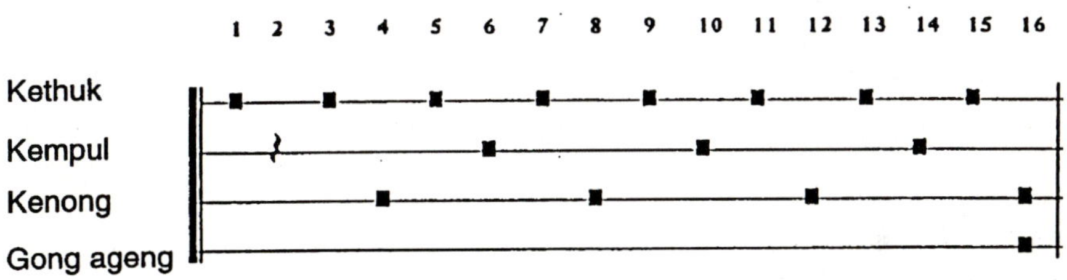

Fig. 6. *Gongan*, a rhythmic pattern in Gamelan music (in Indonesia), has 2^n beats (up to 256) performed by four idiophonic instruments

(b) Sudden change or zigzag curve:
Sudden changes lead to a sense of drama or to other emotions. Indeed, Palestrina counterpoint restricts the magnitude of changes – permitted intervals, ambitus, and range of rhythmic values, and combinations thereof. In Beethoven, on the other hand, there are sudden marked silences, changes of register, volume, and rhythmic values, even within a single theme. Sudden changes are rare in non-Western music, which is characterized by graduality.

(c) The ascending curve (gradual intensification):
This curve, which is common in large units in many non-Western cultures, is expressed through various parameters: ambitus, density, loudness, pitch register, and the degree of definability. The principle of the ascending curve has also been given special names, emphasizing

the fact that it is a natural process (*alap-jor-jahala* in Indian music; *jo-ha-kiu* in Japanese music, levels of *irama* [density] in Gamelan). It also exists, to a large extent, in twentieth-century Western music. A good example of applying the principle to all parameters, with a sudden interruption at the zenith of intensification followed by a short coda, is Charles Ives' The Housatonic at Stockbridge, from *Three Places in New England* (1903-1914); and the fourth of Webern's *Sechs Stücke für Orchester* (Op. 6) (1928).

(d) The descending curve:
This curve, of course, always appears as the second part of a convex curve. The unique aspect of this progression is that it appears at the beginning of a piece, i.e., when it begins at the maximum point. It appears often in twentieth-century music; however, it can also be found in tonal music. A striking example of its appearance in tonal music is the opening of the *Mass in B minor* by J. S. Bach. With respect to loudness only, we find Chopin's *Prelude in C minor* (Op. 28, No. 20) to be a good example (Fig.7). It is almost never found outside the West.

(e) The convex and concave curves:
In the extreme case, convexity refers to a gradual rise to a single peak, followed by a gradual decline. It allows for maximum predictability as to the next step in the process, and it helps consolidate the units. This curve prevails in folk tunes all over the world (Nettl, 1977; Huron, 1996), except for melodies meant to arouse specific emotions such as dirges. In art music, it is prominent in the vocal works of Palestrina, whose ideal, as stated above, was tranquility (Fig.8d). Here the convexity is expressed on various levels. At the level of the phrase, it appears in the parameters of pitch and duration. It also makes an appearance in other parameters, such as loudness, density, or ambitus, thereby helping to establish a sense of unit on various levels (Fig.8e). The convex curve makes a covert appearance in most works of Western tonal music, in the soprano line (according to Schenkerian analysis). Unlike the convex curve, the concave curve does not allow prediction of the duration of the process, because the curve has no upper limit. Concave curves appear in styles that seek to arouse emotions; for example, at the immediate level of performance of Rig Veda chants (Fig.8c), which are full of an explicit tension, and at various levels in Romantic music. On the other hand, the rules of Palestrina counterpoint forbid it.

Both the convex curve and the 2^n scheme are common schemes that represent types of Gestalt rules that contribute to the formation of a single directional unit. A comparison of the two is quite interesting. Both are extremely efficient methods of achieving predictability, but one (convexity) emphasizes the cohesion of the unit, whereas the other (2^n) emphasizes the divisibility and formation of levels in musical organization. Without

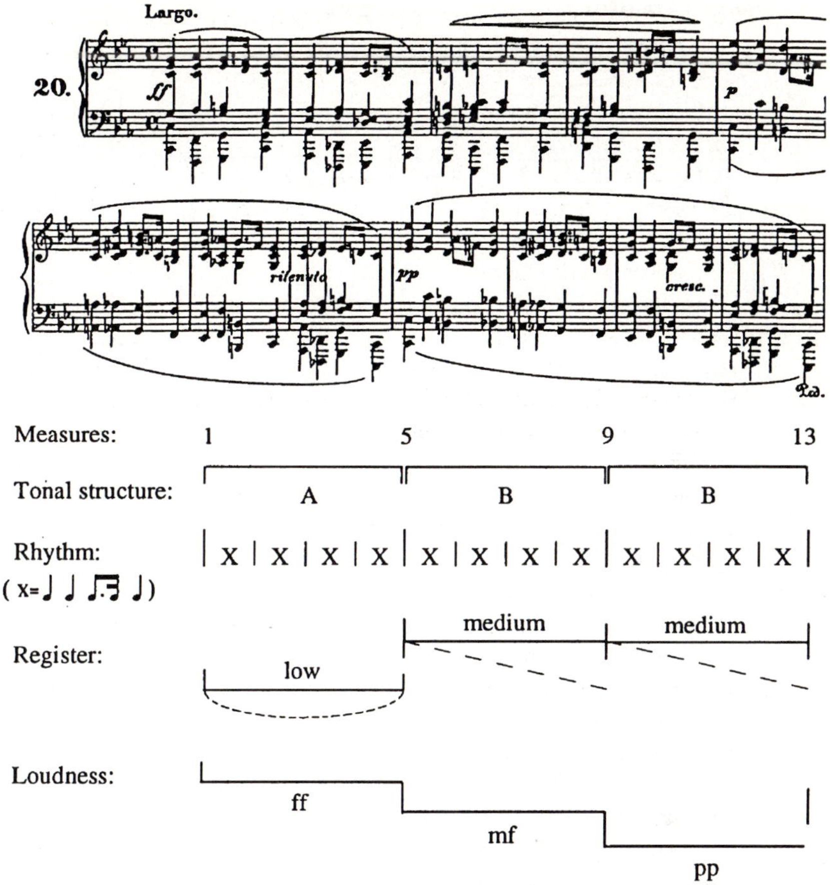

Fig. 7. Contours in Chopin's prelude: horizontal for rhythm (and in simultaneous density of chords, each one containing 5-7 notes); ascending for register; descending for loudness

going into detail, one can see that the convex/2^n curves correspond to stable/self-similar *modulation functions*.

3. Hierarchical Units of Texture – from the Static to the Dynamic Distribution.

   The curves of change over time can be seen as an organization of the static distribution on a higher hierarchical level. Accordingly, a particular static texture can be constructed by repetitive or random placement of basic elements or figures. Once this process has been held constant for a certain amount of time, we have a stationary phenomenon whose distribution can be defined. In practice, the stationariness is rarely observed for long time periods, and the parameters of the distribution vary slowly. Above we defined several types of such curves of change, which are basic

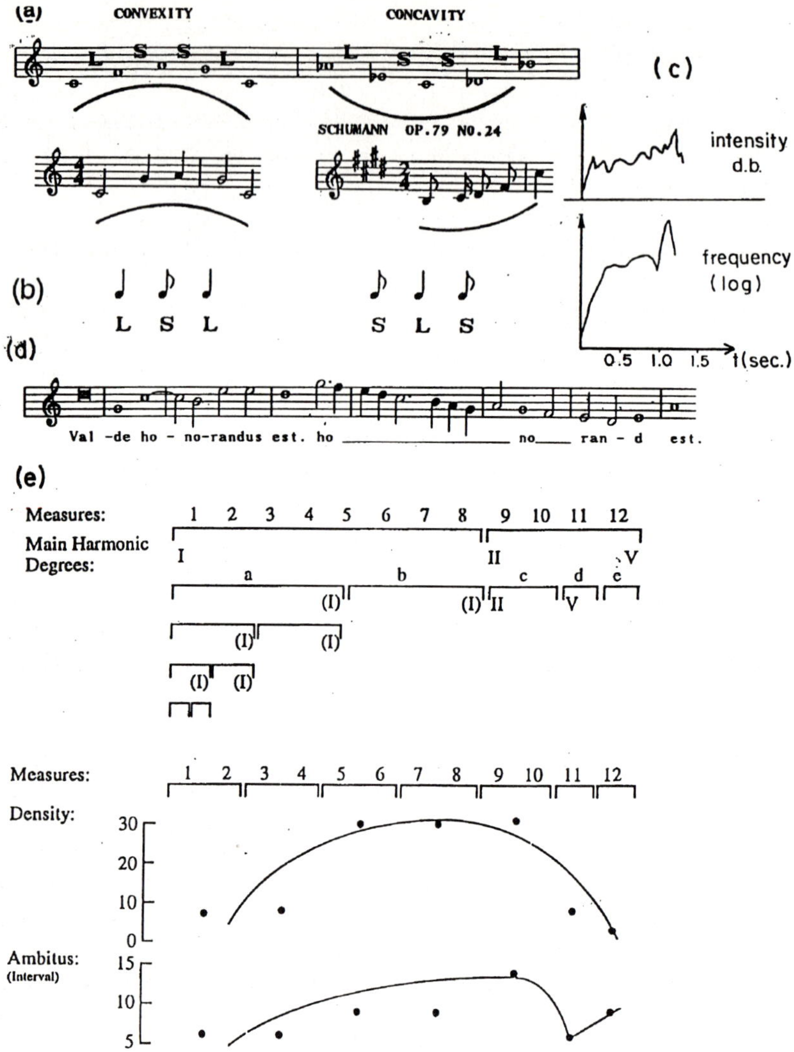

Fig. 8. Convex and concave curves: I. On the immediate level: (a) for the interval (s=relatively small; L=Large); (b) for the duration; (c) concavity for pitch and intensity in micro-motives in the Rig Veda chants. II. On the phrase level: (d) convexity for the interval and duration in the soprano line, taken from a motet by Palestrina; (e) convexity for density and ambitus (for each couple of bars) in a 2^n phrase opening Mozart's *Sonata in C* (K. 545)

ways of controlling and operating on the static texture properties. In the literature, the curves of change are also known as control functions, break point functions, gestures, and so on. Once the basic hierarchy of stationary distribution, characterized by a vector of parameters and curves of change acting and modifying these parameters, has been established, one obtains a basic unit of dynamic texture. This process can be repeated several times: the basic dynamic units can be juxtaposed in some stationary manner, or, on a higher level, the process of combining the units into texture can be varied in time, too.

4. Transformation and Symmetry

   (a) Transformation:
   An expansion of the single repetition is obtained from the phenomenon of transformation, which is a repetition of a unit with an operation on different parameters in different schemes at various levels of organization. Interestingly, the operations can be grouped into five categories whose principles govern general cognitive operations: *contrast*, which has a wide range of manifestations (Cohen, 1994); *expansion and contraction* of various parameters; *shifts* in cyclical systems along the time or pitch axis; *fusion and segregation*; and *equivalence*. We can thus see transformations as principles of textural organization that link units.

   (b) Symmetry:
   One frequently hears today of principles of symmetry that are common to inanimate objects in nature, living organisms, and even works of art [3]. Without going into detail, we shall merely state that in music symmetry is expressed both in schemes and in operations (Cohen, in press). In the present paper we have stressed the two symmetrical textural schemes – the convex curve and 2^n – as representing Gestalt principles that fuse a collection of events into a unit. Similarly, we mentioned *natural* operations that represent symmetrical organization in the texture. Finally, there is the common symmetrical organization expressed as ABA. Grouping together all the different phenomena mentioned here, some of them extremely hidden, within the category of *symmetry* is a relatively new idea that reinforces the significance of forms of textural organization in music from another point of view.

Thus, texture appears in a wide range of forms, even without taking into account the possible transformations: all six kinds of curves may appear either in one stratum or in several strata, in a large or small range of change within each stratum and between the strata, in different locations within the absolute ranges of the various parameters, and with concurrence or nonconcurrence between them.

[3] Note the existence of an international association for symmetry in nature and in man-made products.

5 States of Concurrence/Nonconcurrence

5.1 The General Significance of the Phenomena

These states refer to various kinds of combinations of simultaneous events. For example, two simultaneous parameters may change either in the same direction (concurrence) or in opposite directions (*nonconcurrence*, such as an increase in pitch with a decrease in intensity). Complete concurrence is obtained when an event that is emphasized in terms of all parameters (peak in pitch, intensity, and duration) appears on the stressed beat in the measure. In the case of simultaneous schemes, their borders may be identical (concurrence) or not (nonconcurrence); in the latter case the result is uncertainty about the musical unit (which is determined by the combination of schemes). The distinction between states of concurrence and nonconcurrence appears to be extremely fundamental; further confirmation of this distinction can be learned from its significance in bird calls in expressing tension and relaxation, as we mentioned above (Fig.9).

Nonconcurrence can occur in an enormous variety of ways, and even within the timbre of a single note such as concurrence or nonconcurrence between the phases of the overtones (Dubnov et al., 1997), and it adds to complexity and uncertainty of various kinds that emerge in most musical analyses. Thus, concurrence/nonconcurrence can be viewed as another characteristic of the texture.

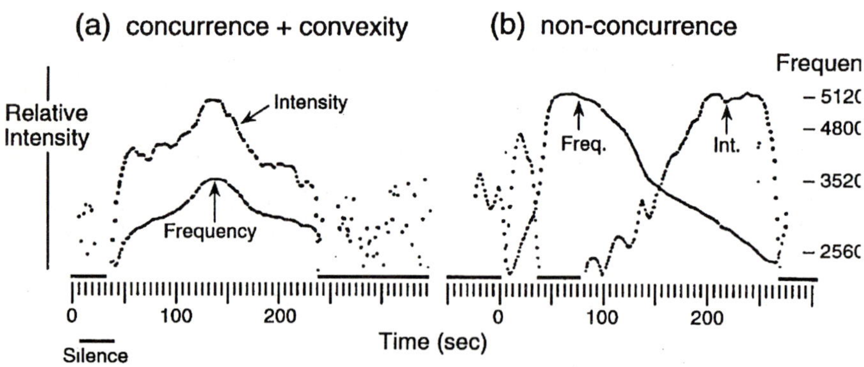

Fig. 9. Concurrence/nonconcurrence between pitch and intensity curves in *syllables* of bird calls (babblers): (a) in calm situations; (b) in tense situations

5.2 Texture and Learned Schemes, with Concurrence or Nonconcurrence

In Western tonal music, every learned scheme may be realized in a variety of textures, and the texture may be concurrent or nonconcurrent with the learned schemes. In a situation of concurrence, textural changes follow changes in the learned schemes, thereby highlighting the structure based on the schemes. In a situation of nonconcurrence the texture may blur the structure. In twentieth-century music, the texture is often not only unconnected to learned schemes but may also be realized in a variety of intervals or durations while retaining the *type* as expressed in its specific notation. The relationship between texture and learned schemes with respect to their role and importance differs from style to style.

In Palestrina counterpoint there were severe restrictions on states of nonconcurrence (Cohen, 1971). In the Classical sonata harmonic exchanges generally occur at the beginning of a measure and the various themes appear in various textures (concurrence) such that the texture supports the tonal organization (Lorince Jr, 1966). Of course, there are exceptions, and states of nonconcurrence are particularly conspicuous by their rarity. In Bach's works, in contrast, long sections are marked conspicuously by uniform texture, with an abundance of states of nonconcurrence of various sorts, some of which are quite latent (Fig.10).

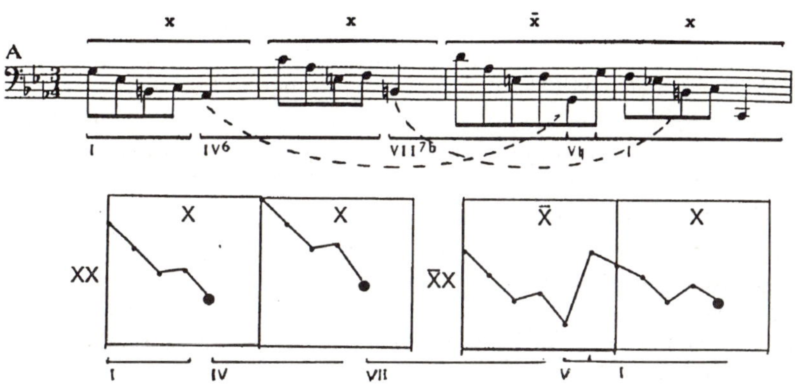

Fig. 10. Nonconcurrence between boundaries of contour units (textural schemes – x) and chordal units (learned schemes) at the beginning of the Saraband from Bach's *Fifth Suite for Cello Solo*

6 Conclusion

In conclusion, we have attempted to clarify textural phenomena, taking into account Gestalt principles and the stylistic ideal. We have shown here that texture, which is also familiar to us from the world outside of music, is a complementary contrast to both timbre and the learned schemes (which are not necessarily arbitrary and reach their peak in the organization of intervals; this issue is not discussed here). Schemes, which represent the rules of Gestalt, may also relate to texture and are expressed not only in categories defining the manner of distribution of the parameters and their combinations with concurrence or nonconcurrence, but also in categories of operations that, though unfamiliar from nature, represent (*natural*) principles of our cognitive activity. Textural schemes also contribute, though sometimes latently, to the organization of Western tonal music. Below is a brief summary of our answers to the questions raised:

- In the broadest sense, our definition of texture is suitable for describing many musical phenomena that we presented in the paper. In special cases, such as in borderline states and due to psychoacoustic constraints, some of the definitions are irrelevant.
- There are indeed texture-based Gestalt phenomena. First of all, we should note that what are known as Gestalt phenomena, which have been formulated (Huron, 1996) and corroborated empirically (Deutsch & Feroe, 1981), are textural ones (according to our definition here). Furthermore, one should bear in mind the conspicuous symmetrical schemes, including the convex curve and 2^n, and the basic operation of the various transformations and the contours and registers of the various parameters.
- Texture may be regarded as a *superparameter* that suggests three types of relationships with the parameters measured and their learned schemes: it can support the organization based on a scheme, it can blur it, or it can have gray areas. Moreover, it can serve as a major factor in organization and can be realized in various ways, so as to assume the function of the learned schemes.
- By definition, texture, unlike learned schemes, has direct, inherent extramusical meanings and can therefore contribute directly to states of calmness/excitement; certainty/uncertainty; and types of momentary/overall and clear/suspensive directionality. All of these are characteristics of the stylistic ideal. The learned schemes, unlike texture, allow for an extremely complex hierarchical overall organization, and their contribution (not explained here) to characteristics of the stylistic ideal is more indirect. In any given style one *chooses*, whether consciously or unconsciously, to emphasize learned schemes or textural organization, types of learned schemes, and types of textural schemes (including kinds of uncertainty), all of which are vital to the definition of the style.

It is interesting to note also that texture, as an intermediate property between timbre and interval organization, provides an important link between electroacoustic music, and *stylistic* music based on learned schemes. In many contemporary compositions that involve technology, a computer is used to provide timbres/sounds that extend the instrumental gestures; and vice versa, the pitch organization is built so as to reflect or augment textural properties of a sound.

The research naturally calls for more investigation, but we strongly believe that texture should already be regarded as an indispensable factor in musical analysis of all styles.

References

Bolinger, D. (Ed.). (1972). *Intonation: Selected readings.* Harmondsworth, Middlesex: Penguin Education.

Boulez, P. (1971). *Boulez on music today.* London: Faber and Faber.

Cage, J. (1966). Rhythm etc. In G. Kepes (Ed.), *Module symmetry proportion.* London: Studio Vista.

Chou, C. (1970). Single tone as musical entities: An approach to structured deviations in tonal characteristics. In *American Society of University Composers, Proceedings* (Vol. 3).

Chou, C. (1974). Asian music and Western composition. In J. Vinton (Ed.), *Dictionary of 20th century music.* London: Thames and Hudson.

Cohen, D. (1969). Patterns and frameworks of intonation. *Journal of Music Theory, 13*, 66–92.

Cohen, D. (1971). Palestrina counterpoint: A musical expression of unexcited speech. *Journal of Music Theory, 15*, 68–111.

Cohen, D. (1983). Birdcalls and the rules of Palestrina counterpoint: Towards the discovery of universal qualities in vocal expression. *Israel Studies in Musicology, 3*, 96–123.

Cohen, D. (1994). Directionality and complexity in music. *Musikometrika, 6*, 27–77.

Cohen, D. (in press). Symmetry in music. *Culture and Science, 8.*

Czink, A. (1987). *The morphology of musical sound.* (manuscript)

Deutsch, D., & Feroe, J. (1981). The internal representation of pitch sequences in tonal music. *Psychological Review, 88*, 503–22.

Dubnov, S., Tishby, N., & Cohen, D. (1997). Polyspectra and measures of sound timbre and texture. *Journal of New Music Research.*

Fónagy, I., & Magdis, K. (1972). Emotional patterns in intonation and music. In D. Bolinger (Ed.), *Intonation: Selected readings* (pp. 286–312). Harmondsworth, Middlesex: Penguin Education.

Goldstein, M. (1974). Sound texture. In J. Vinton (Ed.), *Dictionary of 20th century music.* London: Thames and Hudson.

Huron. (1996). On the kinematics of melodic contour: Deceleration, declination, and arch-trajectories in vocal phrases. In *Proceedings of the 4th ICMPC.* Montreal, Canada.

Lansky, P. (1974). Texture. In J. Vinton (Ed.), *Dictionary of 20th century music.* London: Thames and Hudson.

Levy, J. (1982). Texture as a sign in class and early romantic music. *Jams, 35*, 402–531.

Ligeti, G. (1993). States, events, transformation. *Perspectives of New Music, 31*, 164–171.

Lomax, A. (1968). *Folk song style and culture.* Washington, DC: American Association for the Advancement of Science.

Lomax, A. (1977). Universals in song. *The World of Music, 19*, 117–129.

Lorince Jr, F. (1966). *A study of musical texture in relation to sonata-form as evidenced in selected keyboard sonatas from C.P.E. Bach through Beethoven.* Unpublished doctoral dissertation, Univ. of Rochester.

Nettl, B. (1977). On the question of universals. *The World of Music, 19*, 2–7.

Rapoport, E. (1997). Singing, mind and brain – unit pulse, rhythm, emotion and expression. In M. Leman (Ed.), *Music, Gestalt, and computing: Studies in cognitive and systematic musicology.* Berlin, Heidelberg: Springer-Verlag.

Ratner, L. (1980). Texture, a rhetorical element in Beethoven's quartets. *Israel Studies in Musicology, 2*, 51–62.

Rothgeb, J. (1977). Design as a key to structure in tonal music. In M. Yeston (Ed.), *Readings in Schenker analysis and other approaches* (pp. 72–93). New Haven, CT: Yale University Press.

Scolnic, V. (1993). *Pitch organization in aleatoric counterpoint in Lutoslawski's music of the sixties?* Unpublished doctoral dissertation, Hebrew University of Jerusalem.

Smalley, D. (1986). Spectro-morphology and structuring processes. In S. Emmerson (Ed.), *The language of electroacoustic music.* London: Macmillan.

Stockhausen, K. (1957). How time passes... *Die Reihe, 3*, 10–40. (English edition)

Varèse, E. (1967). The liberation of sound, excerpts from lectures of 1936–1962, ed. by Ch. Chou. In *Contemporary composers on contemporary music.* New York, NY.

V

From Musical Expression to Interactive Computer Systems

Technology of Interpretation and Expressive Pulses

Andranick Tanguiane

FernUniversität, Feithstr. 140, D-58084 Hagen, Germany

Abstract. Musical performance is considered from the points of view of interpretation and nuance, the former requiring mainly analytical, the latter technical skills. Interpretation is understood as finding a structure in a musical text by its segmentation and determining the relation between Gestalt segments. Nuancing is understood as a means of showing the segments by intonation patterns. A system of rules has been developed which aim to do segmentation and the First Movement of Vivaldi's *Violin Concerto in G major* (Op.3 No.3) is taken as an example of how to use the analysis-based method for the interpretation. The interpretation has been realized both by a computer and by a violin pupil appropriately instructed.

1 Introduction

Following the studies in Tanguiane (1992a, 1992b, 1996) we develop an approach to interpretation by (i) finding a structure in a score by segmentation and (ii) showing this structure while playing. The background segmentation can be either rule-based, or heuristics-based, or both (Lerdahl & Jackendoff, 1983; Schaffer, 1995). It is assumed that a segment in a score is shown by an *intonation wave*, or *intonation envelope* in dynamics and timing (Todd, 1985; Clarke, 1988). To be perceived, it should have clear borders and distinguishable phases of intensification, culmination, and relaxation.

A musical structure, however, may have several levels of segmentation. At the lowest level, the melodic line may be divided into motives which are to be considered as the smallest meaningful Gestalt segments, each having a unique stress (Thomassen, 1982; Drake & Palmer, 1993). The intonation envelope is then determined by dynamics and tempo deviations from given nominal values, according to a certain pattern (Parncutt, 1994). The deviations from the nominal values for every motive are regulated by Gestalt patterns of the superior segmentation level, corresponding to phrases, each having a unique intonation peak said to be the *target*. Such a hierarchical grouping of segments with unique *centers of gravity* has to be realized up to the level of musical form. The intonation pattern hierarchy is similar to Clynes' (1983) expressive pulses, understood as a hierarchy of patterns of regular deviations from nominal time and dynamics values.

To provide experimental evidence to our arguments, we consider the First Movement of Vivaldi's *Violin Concerto in G major*(Op.3 No.3). Its interpretation has been realized by an eight-year old violin pupil taught to distinguish phrases by dynamics, timing, and articulation and by a computer program with dynamics and timing specifications derived from the segment structure of the piece.

2 Terminology

We consider the performance of music as an activity which aims to create some special time arrangement of acoustical events. This means that events constitute time patterns whose transformations determine transformations of musical time. In this process, repetitive rhythmic patterns may be used as reference marks for varying tempo. The acoustical events are multi-attributed in that they have a number of pitches, timbre combinations, durations, etc. Each attribute can furthermore be developed independently from the others, which implies a rather complex arrangement of musical time.

In our music teaching experiments, we have tried to conceive of interpretation as an activity which aims to realize these temporal processes in a logically consistent way. In addition, an *emotional* performance has been considered starting from this advanced point.

In what follows, we shall refer to three levels of segments which are distinguished with regard to their length, and the type of their *centers of gravity* as shown by their *intonation*. The intonation is formed by variation in pitch, dynamics, timing, and articulation. It is a means to show temporal connectedness and tension between acoustical events or to oppose and rank the groups of events.

The execution of nuances aims at either enhancing the pitch contour or giving it a certain contrast. This can be done in several alternative ways which is the task of interpretation. In this context, it is straightforward to make a distinction between the concepts of *motive*, *phrase*, and *period*.

- A *motive* is conceived of as the shortest meaningful melodic segment with a single intonation *stress*. It is analogous to an indivisible speech expression which may contain several words but only one stress. In poetry and vocal music, a motive corresponds to a line, or a part of a line which can be singled out.
- A *phrase* is a union of several motives with a single intonation *target*. In poetry/vocal music a phrase usually corresponds to a text line whose ending is the intonation target.
- A *period* is a union of several phrases with a single intonational *culmination* and a single intonation *resolution*. In poetry and vocal music a period corresponds to a rhyme with a rhythmically and intonationally equilibrated pattern.

Note that a motive is characterized by a single-peaked intonation envelope, while a phrase has a multiple-peak envelope dominated by the target. A period is characterized by an envelope with several waves but a unique culmination. Generally speaking, a hierarchical organization of information is rather universal. For instance, a broadcasting wave is a carrier of the acoustical wave which is its envelope. In turn the acoustical wave is a carrier of spectral information, the spectrum is a carrier of formants, etc. Similarly, an intonation wave is a carrier of information of a higher structural level which is determined by relationships between intonational peaks, peaks of groups of peaks, etc, thus determining the time envelope of musical form. Since we consider musical performance as an art of time arrangement, the intonation waves are regarded as carriers of time marks whose multidimensional structure constitute the meaning of music. The accent, target, culmination, and resolution are all hierarchically organized *gravitation points* of the musical drive.

3 First movement of Vivaldi's *Violin Concerto* Op.3 No.3

The way in which dynamics, timing, and articulation (timbre) are used can be illustrated in a performance of the First Movement of Vivaldi's *Violin Concerto in G major* (Op.3 No.3) (Figs. 1-2). The piece is a *must* in violin teaching. Like most of Vivaldi's music it contains a lot of repetitions in several appearances: literal repeats of certain phrases, sequences, repeated arpeggio and harmonic figurations, etc. What is the meaning of all these repetitions and how should they be performed?

From our standpoint, the repetitions determine a specific time space with its own dynamic density and hierarchically organized *centers of gravity*. Repetitive segments are carriers of time relationships of a higher level which may be understood as a rhythm on rhythms. In percussion music, the multilevel pulse is presented directly, often not being masked by melody and rhythmic variations. The performance task is therefore twofold, (i) to show elements, and (ii) to show the time structure built from these elements.

To realize this purpose one has to distinguish between phrases of different types, providing a certain dynamic, rhythmic, and articulation contrast while not forgetting about the wholeness of the piece. The latter means that we limit the diversity of nuances in order to show that certain musical elements belong to certain basic types.

We shall not analyse the piece. Instead, we give a description of three types of segments used throughout the piece. The segments are associated with certain recognizable execution techniques.

- The theme (measures 1–2). The theme has to be played *forte* with the intonation peak at the 5–7th beats. This means that in spite of the powerful first note, the force should be somewhat reserved. Rhythmically, the first bar of the theme should be felt as a measure-long "Auftakt" with no accents which break the intonation wave. This postponed accent of the theme provides a characteristic rhythmical contrast with other phrases which have much shorter wave patterns. Every entry of the theme breaks the fractional motion and makes an effect of a stop in the throughout development.

- Harmonic figurations (first time in measures 3–7). The second segment's first note is the last eight of the 2nd measure. Such a segmentation follows from the subsequent repetition of the same segment. Therefore, the last note of the theme has to be separated from the first note of the next segment by a clear bow stop. The fractional structure of the segment should be shown by metrical accents of eights with the target at the B. The next segment is a literal repetition which should be played softer, in order to provide the effect of an echo. This principle of playing literal repetitions as echo should be sustained throughout the whole piece. In contrast to that, developing repetitions, e.g., by transpositions are performed with increasing tension. For instance, it can be seen in the next (third) segment. The third segment of the phrase is an ascending sequential figuration with a cadence. It is as long as the two preceding segments together and has the target at the 6th beat (from the beginning of the phrase). Thus the second phrase has structure $1+1+2$ which should be shown in performance.
- Transient sequential phrases (first time in measures 7–11). Each short phrase of this type has one measure long intonation envelope. It is definitively shorter than that of the theme and lacks internal repetitiveness inherent in the second phrase. Being arranged into a sequence of ascending transpositions, these short phrases resemble the last segment of the previous item in augmentation. They can be understood as a means of increasing tension; therefore, they should be performed *crescendo* with a light acceleration which determine a the intonation pattern of the second level. It is suggested to distinguish this (third) type of segments by playing each of them less motorically, with relative *rubato* within the segments. However, at the upper level these segments should be performed rhythmically to enhance the effect of sequential repetitiveness of an ascending stair. (Note the contrast in intonation envelopes.)

4 Computer Simulation

The computer simulation has enabled a specification of the nuances and principles of their applications. It seems that our methodology can be used in a playback animation option to music score editors, in a way which is similar to experiments by Honing (1990) and Mazzola (1994).

We have applied both micro and macro-level rules. Micro-level rules deal with motives and short phrases where performance is based mainly on metrical cues which have been used to sustain metrically strong notes up to 110% of the nominal values (more suspension sounds as exaggeration), and enhancing them with dynamical stress +3 dB. At the macro-level (phrases longer than 1 measure) we have applied the following rules:

1. Long ascending sequences of transposed patterns are performed with gradually increasing tempo up to 105% of the initial value, with increased dynamics, with the goal being the last metrically accentuated note.
2. Descending sequences of transpositions are performed *diminuendo*. with a tempo decrease.
3. A literal repetition is performed with abrupt *piano* as echo and with no tempo change.
4. Motive and phrase boundaries are marked by a tempo decrease. The higher the segment stands in the segment hierarchy, the more salient the tempo variation. Consequently, the ending has the greatest final retard (in our experiments 60% of initial tempo).
5. Each entry of the theme is marked by the tempo reset to its initial value.

The computer experiments show that long intonation waves need less salient envelopes than short intonation waves. It looks as if the perception becomes more refined over larger data windows. We can draw an analogy to the Fourier transform where a lower time resolution enables one to obtain a finer frequency resolution.

As a general remark, after having applied the rules mentioned, the computer playback has been no longer mechanical and, after adjustments, has become quasi-emotional. The segmentation has been done manually, but it can be done semi-automatically by known segmentation rules (Lerdahl & Jackendoff, 1983). In our experiments we have used only rhythmical cues specified in detail in Tanguiane (1993, Chapter 6).

However, we are far from thinking that our or similar rule-based system can run automatically. In most cases manual adjustments are highly desirable. However, we are almost sure that the number of basic execution rules is not numerous and most of them can be implemented in a ready-to-use routines. An interesting problem is to specify nuances with a few parameters which could be adjusted by a potential user or selected from a menu. For example, in our experiments the tempo curve has been evaluated from the segment length (longer segments require less tempo increase if the tempo increase should not be perceived). In this case the parameter is: Timing animation (not perceptible, with tempo change parameter calculated automatically), or Tempo change (with a quantitative adjustable parameter).

5 Teaching Experience

The approach described above has been used in teaching an eight year old violin pupil to interpret the Vivaldi piece. The pupil has been taught that making an interpretation is like building a toy with *Lego* using a construction set of elementary bricks of different shape and color. The *shape* corresponds to phrases of certain resemblance, and *color* is the execution technique.

The idea of intonation envelopes has been explained both verbally and by playing. The three different types of performance of the three different types of Gestalt segments have been distinguished. Next, the rules have been applied to different pieces of music in order to illustrate their general meaning.

One of the difficulties in children's teaching turned out to be the performance of long nuances, such as homogeneous crescendo with tempo increases. Another difficulty turned out to be the feeling of initial tempo (for tempo reset at the theme entries). The ensemble playing with the teacher providing the piano accompaniment has been used to fix the main ideas of interpretation. The exercises have been aimed at both getting the desired dynamics and timing over phrases, and at acquiring a feeling of the musical form in general. Besides improving technical skills, and learning the text, a pupil has been faced to the problem of achieving a clearly specified sound. The exercises have become more creative, more purpose-oriented, and more motivated. As a side effect, the efforts to get the nuances required have had a positive effect on the sound quality in general. As a result, a meaningful interpretation of the piece (rather large and complex for a young musician) has been achieved.

6 Revision of Clynes' Model

Another conclusion relates to Clynes' idea of expressive pulses. Recall that, according to Clynes (1983), an expressive pulse is understood as a hierarchy of patterns of regular deviations from nominal time and dynamics values. In addition, these patterns are related to codes of emotions (Clynes, 1977). Clynes has suggested that a particular expressive pulse can be fitted to the style of a given composer thus becoming its emotional Gestalt. Such a view has been criticized by Repp (1989, 1990) and Thompson (1989) whose objections were later commented on by Clynes (1995).

In contrast to Clynes' approach, however, we do not restrict ourselves exclusively to bar segmentation. Instead, we assume that the hierarchical pulse is essentially context-dependent and determined by meaningful segmentation and intonation. Clynes has determined pulses of several great composers, but he has not found the pulse of Vivaldi. From the standpoint of our model, it is explained by the fact that the styles of most classical composers are characterized by a rather rigid metric structure and, consequently, regular segmentation patterns. Therefore, it is possible to find prevailing pulses of these composers regardless of their particular compositions. This approach fails, however, when the composer is less restricted to the metric structure, which is the case of Vivaldi's music with irregular number of repetitions at different levels. In particular, the length of sequences of transpositions in Vivaldi's music is often motivated by harmonic reasons rather than by the idea of rhythmic symmetry. In this case the pulse is determined by the phrase structure of the given piece rather than by the *composer's style*. We therefore conclude that the specific pulse of a composer results from using stable phrasing patterns as an indirect characteristic of style. This feature is inherent in music of certain but not all composers. Therefore, we believe that expressive pulses should be considered in a *context-dependent* way.

7 Conclusion

In this paper we have argued that interpretation can be understood as a process of segmentation and determination of accents within the segments. Musical form is characterized by a hierarchy of intonation waves in dynamics and timing and this hierarchy determines the performance structure as a system of successive targets. Gestalts of different types (e.g., relating to the theme, or to a transient passage) have distinctly different intonation patterns. Dynamics and timing are used consistently to show tension and relaxation (e.g., *crescendo* with *accelerando*).

Computer experiments have enabled us to judge about the sufficiency or insufficiency of a set of performance rules. These rules have been also used in teaching. They form a basis for free performance, not for restricting emotional freedom. Rational explanations make the interpretation task well defined, and the training becomes more purpose-oriented. Based on this approach, the idea of Clynes' expressive pulses has been revised and related to the segment structure rather than to the metric structure of a piece.

Appendix

Sound Examples (CD-tracks 45-48)

The sound examples have been made with the Roland Orchestra M-OC1 expansion module and Kurzweil digital piano RG 200. Both have been MIDI-interfaced to an IBM-compatible computer. The Roland module has five solo violin voices which have been used interchangeably to simulate the articulation variety. The Kurzweil piano (voice Grand Piano 1) has been used for the accompaniment to both the computer and human violin performance. The following sound examples enable comparison between computer unanimated and animated performances, and a violin pupil performance.

- Sound Example #1: The theme of Vivaldi's five Roland M-OC1 synthesized violin voices used in the sequel, in the order of increasing sound tension:
  Voice # 214 (Solo Vn 1)
  Voice # 218 (Stereo Vn)
  Voice # 216 (Chorus Vn)
  Voice # 215 (Solo Vn 2)
  Voice # 217 (Agitato Vn)
- Sound Example #2: Computer performance with no nuances by Roland M-OC1 voice # 214 (Solo Vn 1) and Kurzweil RG 200 voice Grand Piano 1.
- Sound Example #3: Computer performance with dynamics changes, tempo deviations, and articulation variety by using the five violin voices listed.

– Sound Example #4: Violin part: M. Tangian, 8 years old. Piano part: A. Tangian (Tanguiane). The recording has been realized at the Fern Universität, ZFS-studios on July 31, 1996,

Fig. 1. First movement of Vivaldi's *Violin Concerto* Op.3 No.3, part 1

Fig. 2. First movement of Vivaldi's *Violin Concerto* Op.3 No.3, part 2

References

Clarke, E. (1988). Generative principles in music performance. In J. Sloboda (Ed.), *Generative processes in music: The psychology of performance, improvisation and composition.* Oxford: Science Publications.

Clynes, M. (1977). *Sentics: A touch of emotions.* San Francisco, CA: Double Day Anchor.

Clynes, M. (1983). Expressive microstructure in music, linked to living qualities. In J. Sundberg (Ed.), *Studies of musical performance* (Vol. 39). Stockholm: Royal Swedish Academy of Music.

Clynes, M. (1995). Microstructural musical linguistics: Composer's pulses are liked most by the best musicians. *Cognition, 55*, 269–310.

Drake, C., & Palmer, C. (1993). Accent structures in music performance. *Music Perception, 10*, 343–378.

Honing, H. (1990). POCO: An environment for analysis, modifying, and generating expression in music. In *Proceedings International Computer Music Conference (ICMC-1990).* San Francisco: Computer Music Association.

Lerdahl, F., & Jackendoff, R. (1983). *A generative theory of tonal music.* Cambridge, MA: The MIT Press.

Mazzola, G. (1994). Musical interpretation via vector fields: A new human workstation on the NEXTSTEP environment. In *Proceedings 1993 Music Acoustics Conference* (pp. 112–115). Stockholm: Royal Swedish Academy of Music.

Parncutt, R. (1994). A perceptual model of pulse salience and metrical accent in musical rhythms. *Music Perception, 11*, 409–464.

Repp, B. (1989). Expressive microstructure in music: A preliminary perceptual assessment of four composers' "pulses". *Music Perception, 6*, 243–274.

Repp, B. (1990). Patterns of expressive timing in performances of a Beethoven minuet by nineteen famous pianists. *The Journal of the Acoustical Society of America, 93*, 622–641.

Schaffer, H. (1995). Musical performance as interpretation. *Psychology of Music, 19*, 3–21.

Tanguiane, A. (1992a). Analytical approach to performance interpretation. In A. Marsden & A. Pople (Eds.), *Computer representations and models in music.* London: Academic Press.

Tanguiane, A. (1992b). Analyzing three interpretations of the same piece of music. In R. Dalmonte & M. Baroni (Eds.), *Secondo convegno Europeo di analisi Musicale* (pp. 21–40). Trento: Università degli Studi di Trento.

Tanguiane, A. (1993). *Artificial perception and music recognition.* Berlin, Heidelberg: Springer-Verlag.

Tanguiane, A. (1996). *Interpretation of music and text by multilevel segmentation.* (Forthcoming)

Thomassen, M. (1982). Melodic accent: Experiments and a tentative model. *The Journal of the Acoustical Society of America*, *71*, 1596–1605.

Thompson, W. (1989). Composer-specific aspects of musical performance: An evaluation of Clynes's theory of pulse for performances of Mozart and Beethoven. *Music Perception*, *7*, 15–42.

Todd, N. (1985). A model for expressive timing in music. *Music Perception*, *3*, 33–58.

Intonational Protention in the Performance of Melodic Octaves on the Violin

Janina Fyk

Fryderyk Chopin Academy of Music, Music Acoustics Laboratory,
Okólnik 2, 00-368 Warsaw, Poland

Abstract. The purpose of this study was to examine whether, in violin performance, frequency changes occur in the tone preceding a given tone which might represent an announcement of the direction of that tone. In order to examine this possibility, melodic octaves were performed on the violin upwards from B4 (494 Hz) and downwards from B5 (988 Hz) on the A and E strings with and without position shift. Altogether 20 octaves performed upwards and 20 downwards were recorded monophonically on tape at a speed of 38.1 cm/s. A method of measuring the frequency of the fundamental partials of the violin tones was applied which took into account the succeeding stages of the tone. Specially prepared computer programs were used for this purpose. Particular attention was focused on frequency changes in the final stage of the first tone of the melodic octave. There was strong evidence to suggest that already in the last state of a given tone an announcement of a pitch change in the following tone is issued. Frequency modification in the last state of a tone which is in agreement with the movement direction of the following tone suggests that the performer incorporates protention (anticipation) in the process of shaping subsequent pitches in order to obtain homogeneous sound unity.

1 Introduction

Music is an art which is actualized in time. According to Husserl (1966/1989), a melody, as well as other elements of a piece of music, can be regarded as an object which is spread out over time. In the perception of a melody, tones which have just been heard represent a perceptive *now*; yet at the same time, they are constantly *made present* in the course of the melody. This constant flow of *making present* creates a unity between what is past and what is to come, and raises expectations concerning future musical events. Therefore Ingarden (1973, p.90) maintains that, "a listener perceiving the current phase of a melody does not admittedly actually hear the following phase of the melody, yet s/he feels the directional drift in which the melody is flowing". This formation of the time sequence in a piece of music has particular significance with respect to intonation in violin playing. Since a melody represents a perceptual entirety, the attainment of so-called pure intonation on the violin – whether at the level of the tone, the interval or the musical phrase – is possible only when the intention of the performer and the movement of the fingers go hand in hand with an awareness of the musical context. Only in this case do conditions exist which are conducive to the shaping and control of intonation which is in keeping with the total playing event. In addition, as the Polish cellist, K. Wilkomirski aptly puts it (Wilkomirski, 1965, p.90): "Imagination reveals *ideal* tone pictures, imagined but nevertheless precise, supple and free from imperfections and inaccuracies. The ear perceives real tangible tone phenomena, which are assessed and critically analysed on the basis of the auditory imagination" (p.97). It is likely that, as far as violin movement skill is concerned, one of the effects of directed anticipation of oncoming musical events will be a quicker and more precise movement of the fingers in performing succeeding tones of a melody. The question can be asked, therefore, whether the pitch of a given tone of a melody is already modified towards the end of its execution, as a result of the fact that it precedes this particular tone of the melody and not any other. Do pitch changes take place in a tone currently being performed as a result of the fact that a position shift related to the accomplishment of the pitch of the following tone is about to take place? To answer this question, this study aims to examine whether, in violin performance, frequency changes occur in the final state of the tone preceding a subsequent tone which might represent an announcement of the direction of that tone.

To address the above mentioned question, the author has concentrated on the most basic and elementary level of melodic intonation, and for the purpose of this study, the musical material has been restricted to two tones which make up a melodic octave. This interval satisfactorily illustrates melodic movement and the idea that each individual part affects every other. Even at this level, the manifestation of Gestalt is apparent, and indeed this study is related to the psychological and philosophical foundations of Gestalt theory (Köhler, 1929) with special emphasis on the functional approach and Husserl's phenomenological theory of time (see Husserl, 1966/1989; Eichert, Schmidt, & Seifert, 1997, this book).

It was hypothesized that *a frequency change occurs in the final state of the first octave tone which announces the direction of the second octave tone.* Based on Husserl's concept of *protention* this phenomenon has been called *intonational protention.* Intonational protention denotes the performer's awareness of the musical context closest to the tone presently being performed. This leads to an earlier reaction to the action which is about to take place, i.e., a change in pitch in the following tone of the interval. Thus, protentional frequency change represents an earlier reaction while change in pitch in the following tone signifies action. In other words, the hypothesis set out to examine the performer's reaction to the action, and for this reason special attention is paid to the connection and interaction between mental anticipation and behavioral (pitch change) levels.

Finally, it must be emphasized that although the problem of the present research is restricted to the intonational aspect of octave performance on the violin, there may be a possible connection with both temporal and neuropsychological aspects of perception and performance (Fraisse, 1966, 1982, 1987; Scheid & Eccles, 1975; Todd, 1994, 1995).

2 Procedure

Two Polish violinists of international standing performed the melodic octaves B4–B5 (494 Hz – 988 Hz) and B5–B4 (988 Hz – 494 Hz), five times each. The tones making up the octaves were performed on the A and E strings in the first position and on the A string with a position shift. Vibrato was not applied. The octaves were recorded in Studio S1 at the Fr. Chopin Academy of Music in Warsaw. The musical material was recorded monophonically on tape, at a speed of 38.1 cm/sec.

A tone frequency measuring method was applied which took into account three states of the violin tone envelope: attack, sustain and decay. On the basis of results of previous experiments (Fyk, 1985, 1987) two stages were distinguished in the first state (attack): The physical beginning and the perceptual beginning. In the second state (sustain) again two stages were distinguished: The quasi-steady stage from the point of view of frequency only and another from the point of view of both frequency and amplitude. The third and final state (decay) was also the fifth stage of the tone. Individual stages made up time continua within a given state but there was no continuity between separate states. The discontinuity between the states was not fixed but depended on the duration and amplitude envelope of the tone played. This is represented schematically in Fig. 1.

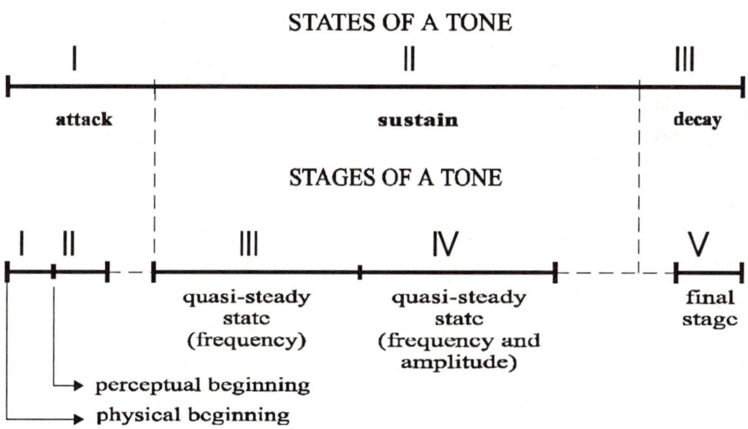

Fig. 1. Segmentation of the violin tone into three characteristic states and five stages

The above mentioned discontinuity between particular states of the tone, which can be observed in the diagram in the form of broken lines, allowed us to delineate more clearly particular tone states. In this way we can avoid obscuring information concerning the frequencies of given states of the tone. It is worth noting that as the amplitude envelope of the first octave tone was longer and the second one shorter, there was some differentiation in discontinuity between states of two octave tones.

The physical beginning of the tone accounted for 5-10 periods of the emitted tone. The next ten periods of the tone were taken up by the perceptual beginning. Each of the two subsequent stages (the second state or sustain state of the tone) accounted for about 160 periods of tone B4 and 320 periods of B5. The final (fifth) stage accounted for the last 10-15 periods of the tone.

The frequency in succeeding stages of the tone corresponded to the frequencies of the average period calculated on the basis of the measurement of duration of a given number of periods of the fundamental partial. The duration of the periods was measured with an accuracy of up to 1 ms. Specially prepared computer programs were used for this purpose. The violin tones recorded on the magnetic tape were fed directly from the tape-recorder into the measuring system. Individual tones were introduced into the computer memory as quasi-stationary signals by means of 8-bit analog-to-digital converter. During the period measurements a sampling rate of 50 kHz was applied. The frequency of the tone in the quasi-steady state was taken as representative for a given octave tone.

3 Analysis of the Results

Altogether 20 octaves were performed upwards and 20 downwards. For both performers, the average duration of both tones of the interval was similar and represented about 670 ms. in the case of Violinist 1 and 700 ms. for Violinist 2. The loudness of the tones corresponded to mezzo forte. The amplitude of the first tone grew gradually for 125 ms (mean value) until stability was achieved. During the last 50 ms. of the first tone the amplitude diminished only slightly, as a result of which the time necessary to achieve amplitude stability to the second tone was short and did not exceed 50 ms. (mean value). In the last stage of the second tone, the amplitude gradually diminished.

As a result of segmentation of the tones of the melodic octave into states and stages, it was possible to calculate frequency deviations from the quasi-steady state tone frequency in succeeding stages of the tone. The main emphasis was placed on frequency changes in the quasi-steady state and the final state (the Vth stage of the first tone of the octave).

3.1 The Frequency of the First Tone in the Quasi-Steady State

Noticeable stabilization of the tone with regard to frequency is characteristic of this state of the tone. The maximum frequency difference between stages III and IV of the tone, which together make up the quasi-steady state, amounted to 10 cents for octave tones performed without position shift (both tones in the first position) and 22 cents for octave tones with position shift. The tone frequency stability in the quasi-steady state (the standard frequency deviation in stages III and IV of the tone was taken as a measure of this) was greatest for tones performed in the first position (standard deviation was equal to 2 cents) as opposed to standard deviation less than or equal to 4 cents for tones performed with position shift. In the case of 2 tones, where the frequency difference was +/– 10 cents, it can be assumed that the performers were still correcting the pitch of these tones at this state of the tone.

3.2 The Frequency of the First Tone in the Final State

The maximum frequency difference between the quasi-steady state and the final stage of the tone amounted to 29 cents in the case of a tone performed in the first position. Frequency stabilization in this state of the tone was great for tones performed in the first position (standard deviation = 14 cents) than for tones performed with position shift (standard deviation = 52 cents). In the final state of the first tone, for octaves performed upwards, 16 out of 19 tones performed had a higher frequency than that occurring in the quasi-steady state: by an average of 6 cents when the second tone was performed in the same position and by an average of 48 cents when the second tone was performed in a different position. The number of performances of the first tone in the octave analysed containing a frequency change in the final stage of the tone together with the direction and magnitude of that change are presented in Table 1.

Table 1. Number of frequency changes in the final state of the first octave tone. "+" is an increase in frequency in relation to mean frequency in the quasi-steady state "–" is a decrease in frequency in relation to mean frequency in the quasi-steady state

Octave tones	Strings	Number of performances	Number of performances and mean change of frequency in the 5th stage of the first tone (in cents)	
B4–B5	A & E	10	7 + 16	3 – 2
B4–B5	A	9	9 + 48	
B5–B4	E & A	10	2 + 11	8 – 19
B5–B4	A	10		10 – 46

As the data contained in Table 1 indicate, in 34 out of 39 of the octaves analysed, frequency changes in the final state of the first tone announce a change in movement direction in the second tone of the octave. These changes were protentional in character (using Husserl's term) because the performers were aware in advance of the changes in tone pitch which were about to take place, although these changes would only find expression in a future perceptual experience. The protential character of the frequency changes in the final stage of the first tone of the octave, whose pitch tendency is in agreement with the movement direction of the next tone, is presented in Fig 2.

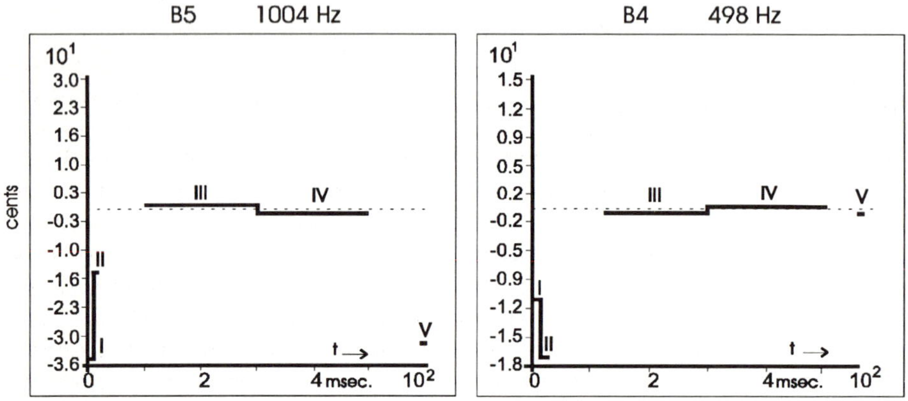

Fig. 2. Protentional changes of frequency in the last stage of the first tone of octave B5–B4. Both tones were played in the first position. The dotted line corresponds to the frequency in the quasi-steady stage

The effect illustrated above provides evidence of the performer's previous awareness of the musical context closest to the tone. This in turn led to performer reflex reaction to the action which was about to take place.

4 Conclusions

Pitch change in the second tone of the melodic octave was preceded by a slight frequency modification in the last state of the first tone of the octave. Translating this into motor movement, we can say that a slight change in the hand position at the end of the execution of the first tone of the octave was a prediction of the hand shift which is about to take place in order to attain the pitch of the second tone of the octave. Thus, the results of the experiments which we carried out indicate that the direction of the movement of the second tone of the interval has an influence on frequency modification in the last state of the first tone of the octave. This modification leads to a slight frequency increase in the final stage of the first tone when the second tone is higher, and to a slight decrease when the second tone is lower. In this way, during the last stage of the first tone, the performer *prepares himself* for a frequency change in the following tone. Consequently, the last state of the tone is an announcement of a frequency change in the next tone. Frequency modification in the last state of a tone, which is in accordance with the direction of the following tone, is therefore a result of the performer's incorporation of anticipation, i.e., protention in order to obtain homogeneous sound unity. Using Meyer's (1956) words, we can say that such frequency modification is dependent on the "specificity of the original expectation" (p. 26).

Although Husserl's protention has philosophical foundations, from the point of view of the performer's action, intonational protention manifests itself as a kind of systematic kinesthetic synchronization, similar in character to premature motor responses, e.g. tap of the forefinger to periodic sounds (Fraisse, 1966, 1982, 1987). As Fraisse (1982) reported: "In synchronization the response is produced at the same time as the appearance of the stimulus. A similar behavior is possible only if the motor command is anticipated in regard to the moment when the stimulus is produced" (p. 154). For this reason, intonational protention is a true reflection of the holistic process in performance, i.e., the interaction of the performer's musical intention and kinaesthetic synchronization with what Godøy calls "actual properties of musical substance" (Godøy, 1994). It is worth noting that intonational protention enhances the Gestalt quality of the interval as a figure composed of two tones. Indeed, intonational protention could be explained in terms of Godøy's triangular model of image, action and sound interactions, which integrates the main aspects of musical imagery in the conceptual scheme of thinking shapes (Godøy, 1997, this book). One possible explanation of intonational protention is provided by Pribram's holonomic theory of the mind and perception according to which neural responses are represented in the brain cortex by *holographic* images of the stimuli. This theory is particularly useful especially with respect to feedforward and feedback operations (Pribram, 1971; Eichert et al., 1997, this book).

In the case of intonational protention as with all phenomena, the conditions conducive to its occurrence may be more or less favorable. It is therefore necessary to carry out further research into this phenomenon, which will take into account other conditions relevant to the performance of melodic intervals. For example, when playing the violin it is possible to perform melodic intervals in the same position, except that the lower tone is played on the lower string but with a *higher* finger, and the higher tone on the next string up but with a *lower* finger. It would also be useful to examine the performance variant in which the lower tone of the interval is performed on the lower string but in a higher position, and the higher tone on the next string up but in a lower position. In the case of melody performance, protentional frequency changes may also be influenced by psychoacoustic factors such as tone duration, the relative loudness of the melody tones or tempo and phrase structure. It is not out of the question that these changes may, to some degree, be associated with motion at the level of the limbs or the whole body, as has been suggested in Todd's two-component dynamic model of the perception of rhythm (Todd, 1994, 1995).

Such a wide ranging form of research would allow us to establish the most favorable conditions for the appearance of the phenomenon called by the author of this work *intonational protention*. Despite the limitations of this work, on the basis of the results obtained and on the basis of the results obtained and the conclusions drawn, a picture of intonation emerges which is dynamic with respect to the performer's actions in combination with direct protention. Protentional frequency changes in violin tone provide evidence therefore that intonation in violin playing is a dynamic process, consciously directed and controlled by the player.

Acknowledgment

This research was supported by a grant P 107 022 07 from the State Committee for Scientific Research.

References

Eichert, R., Schmidt, L., & Seifert, U. (1997). Logic, Gestalt theory, and neural computation in research on auditory perceptual organization. In M. Leman (Ed.), *Music, Gestalt, and computing: Studies in cognitive systematic musicology.* Berlin, Heidelberg: Springer-Verlag.

Fraisse, P. (1966). L'anticipation de stimulus rythmiques: Vitesse d'établissement et précision de la synchonisation. *L' Année Psychologique, 66*, 15–36.

Fraisse, P. (1982). Rhythm and tempo. In D. Deutsch (Ed.), *The psychology of music.* New York, NY: Academic Press.

Fraisse, P. (1987). A historical approach to rhythm as perception. In A. Gabrielsson (Ed.), *Action and perception in rhythm and music* (Vol. 55). Stockholm: Royal Svedish Academy of Music.

Fyk, J. (1985). Vocal pitch-matching ability in children as a function of sound duration. *Bulletin of the Council for Research in Music Education, 85*, 76–89.

Fyk, J. (1987). Duration of tones required for satisfactory precision of pitch-matching. *Bulletin of the Council for Research in Music Education, 91*, 38–44.

Godøy, R. (1994). Shapes and spaces in musical thought. In I. Deliège (Ed.), *Proceedings of the 3rd International Conference for Music Perception and Cognition* (pp. 177–178). Université de Liège.

Godøy, R. (1997). Knowledge in music theory by shapes of musical objects and sound-producing actions. In M. Leman (Ed.), *Music, Gestalt, and computing: Studies in cognitive and systematic musicology.* Berlin, Heidelberg: Springer-Verlag.

Husserl, E. (1989). *Wyklady z fenomenologii wewnetrznej swiadomosci czasu. trnsl. Janusz Sidorek.* Warsaw: PWN-Edition. (Original work published 1966)

Ingarden, R. (1973). *A musical work and its identity.* Cracow: PWM-Edition. (In Polish)

Köhler, W. (1929). *Gestaltpyschology.* New York, NY: Liveright.

Meyer, L. (1956). *Emotion and meaning in music.* Chicago, IL: University of Chicago Press.

Pribram, K. (1971). *Languages of the brain.* Englewood Cliffs, NJ: Prentice-Hall.

Scheid, P., & Eccles, J. C. (1975). Music and speech: Artistic functions of the brain. *Psychology of Music, 3*, 21–35.

Todd, N. (1994). Multi-scale analysis of expressive signals: Recovery of structure and motion. In A. Friberg, J. Iwarsson, E. Jansson, & J. Sundberg (Eds.), *Proceedings of the Stockholm Music Acoustics Conference, 1993* (pp. 146–149). Taberg: Royal Swedish Academy of Music.

Todd, N. (1995). The kinematics of musical expression. *The Journal of the Acoustical Society of America, 97*, 1940–1949.

Wilkomirski, K. (1965). *The cello technique and problems of performance.* Cracow: PWM-Edition. (In Polish)

Sonological Analysis of Clarinet Expressivity

Sergio Canazza, Giovanni De Poli, Stefano Rinaldin and Alvise Vidolin

CSC-DEI, University of Padova, Via Gradenigo 6a, I-35131 Padova, Italy

Abstract. Seven different performances of a fragment of W. A. Mozart's *Concert for Clarinet* (K622) were recorded. Each performance was inspired giving the player a sensorial type adjective in order to suggest various kinds of *expressive intentions*. Certain particularly significant sonological parameters were then measured. These were linked to the time factor (correlated to variations in the amplitude and duration) and frequency (variations in the timbre). The principle aim was to identify which physical parameters, and how many of them, were subject to modifications when the expressive intention of the performer was varied. The sonological analysis shows how a performer changes the physical parameters of frequency and time in function of expressive performance of a score. The evaluation of a synthesized performance, based of the values of the parameters measured, confirms the validity of the procedure.

1 Introduction

A musical score is not an exact description of the sound to be produced but is, rather, the written representation of the cognitive parameters as imagined by the composer. A literal interpretation of the score would produce a non musical or *mechanical* performance. It is common practise, on the part of composers, to use some conventional annotations in order to better explain the sense of a musical phrase and, on the part of the performer, to take certain liberties in the interpretation of the score. This gives rise to a *normal* or scholastic interpretation which aims to be musically acceptable while paying the maximum respect to the score. However, no musician plays the same piece in the same way on every single occasion. Each performance depends on the performer's emotional state at that particular moment as well as his/her hypothetical dialogue with other musicians. Thus, alongside the performer's various expressive intentions of this *normal* interpretation, there are also some deviations depending entirely on subjective artistic choices.

Computers provide suitable conceptual and operative tools for a better understanding of the musician's performance by means of modelling. The present study starts from sonological analysis of a performed score. The aim is to build a model of the expressive deviations for use in a computer performance. Apart from the research by B. Repp (1990, 1992, 1994, 1995) and N. Todd (1985, 1992, 1995), not much attention has been paid to this latter aspect in that it was thought to be a purely artistic one and, as such, could not be analyzed.

2 Description of the Experiment

A phrase (Fig.1) taken from W.A. Mozart's *Clarinet Concert in E Major* (K622) was performed by a professional clarinet player. His performance

Fig. 1. First Movement of the *Clarinet Concert in E Major* (K622) by W.A. Mozart

was inspired by the following adjectives: *light*, *heavy*, *soft*, *hard*, *bright*, and *dark*. The scholastic performance, described by the adjective *normal*, was also added to these and used as a standard measure of comparison in the sonological analysis of the various interpretations. Unconventional adjectives, in the musical field, were deliberately chosen in order to give the performer the greatest possible liberty of expression. The recordings were carried out in three cycles, each cycle consisting of the seven different interpretations. The clarinettist then choosed an interpretation for the seven adjectives that, in his opinion, corresponded best to the proposed adjectives. He thereby tried to minimize the influence that the order of execution might have had on the performer. The recordings were made at the CSC of Padua University in a monophonic digital form at 16 bits and 44100 Hz. The most representative sonological parameters were then measured. They fall into the domain of time (correlated to amplitude and duration) and frequency (variations in timbre).

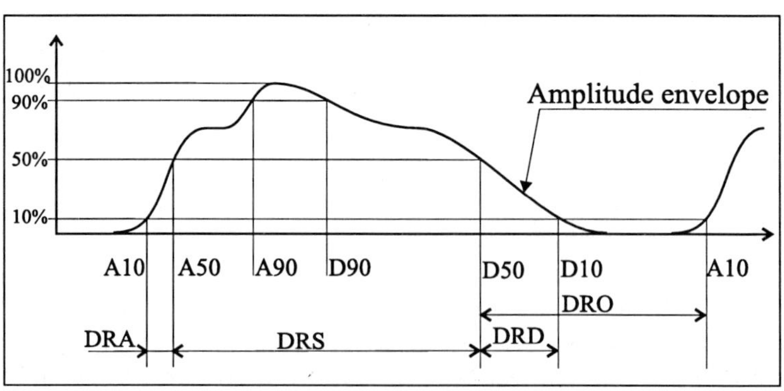

Fig. 2. The characteristic parameters of a note's amplitude envelope

Table 1. Initial instant deviations. Differences with respect to ideal and normal interpretations. Times are given in ms

	Normal	Bright	Dark	Hard	Soft	Heavy	Light
II - IInom	32	14	-1	41	-7	27	7
II - IImean	16	-2	-17	24	-22	9	-8

3 Time Analysis

The time analysis starts, especially for an instrument such as the clarinet, with measuring and modeling the amplitude envelopes of the notes. There is no standard procedure to measure the envelopes. Figure 2 shows the parameters that are taken into consideration in relation to the envelope. To calculate the attack, sustain and decay of the notes in the various performances, the time instants corresponding to the notes' envelopes at the 10%, 50% and 90% threshold of the maximum rising amplitude ($A10$, $A50$, $A90$) and decaying amplitude ($D10$, $D50$, $D90$) were measured.

From the measures of amplitude envelope, the note's initial instant II is computed as $A10$. In legato notes, II was considered to be the instant of the minimum amplitude of the envelope between the two notes. The metre of the pieces was also analyzed, measuring, in particular, the following values: tempo MM of the piece (Fig. 3), the duration of each beat and half beat, the deviation of the initial instant from the *nominal* interpretation, whereby all the notes began at the metric instant. With reference to this, Table 1 shows the mean (averaged on the $N = 14$ notes) deviations of the initial instants, both from the nominal interpretation (II - $IInom$) and from the average of the seven normalized performances (II - $IImean$).

Some other parameters of great expressive relevance were also measured. One parameter that is very useful in describing the degree of staccato used in a performance is the off-time duration DRO (Friberg, Fryden, Bodin, & Sundberg, 1991). The DRO of the n-th note is here computed as $DRO(n) = A10(n+1) - D50(n)$, see Fig. 2. It is possible to calculate the DRO_M (mean off-time duration) as follows:

$$DRO_M = \frac{1}{N}\sum_{n=1}^{N} DRO(n,x)\,\frac{MM(x)}{MM(nor)}$$

where $MM(nor)$ represents the tempo of the *normal* performance, $MM(x)$ the tempo of the actual x-th performance, while N is the total number of notes in the performance. The DRO_M value is therefore normalized with respect to the normal performance and averaged with respect to the 14 notes. Figure 4 shows the values of this parameter in relation to the seven performances. The $D50$ values were used because they turned out to be the most significant ones, after having examined the amplitude envelopes. The envelopes of many notes have, in fact, a characteristic trend until, approximately, half-way along the decay. Then the slopes flatten out. In this sense, the values $D90$ and $D10$ would be of little use in determining the end of the previous note. The variations in these parameters have been studied not only from the point of view of the different expressive intentions, but also in terms of the notes of the score which has, in itself, its own intrinsic expressivity. Some notes, are in fact, little influenced by the musician's intentions in that these are bound by physical limits (i.e. brief time) and/or structural factors such as, for example notes at the beginning or at the end of a melodic phrase or rhythmic group (Drake & Palmer, 1993).

Figures 3 and 4 show the average values of the most significant temporal parameters measured for the various performances. The attack duration was calculated as $DRA = A50 - A10$. The value was then averaged with respect to the 1, 9, 11 and 13 notes, which represent the combination that gave the lowest value to the variance. Thus, the average attack duration was obtained (DRA_M). The decay duration is $DRD = D10 - D50$. The notes with the lowest value of variance, and therefore used to calculate the average decay duration (DRD_M) were, in this case, the 8, 10 and 14 ones. The sustain duration is $DRS = D50 - A50$.

To characterize the sustain of each performance, the parameter DRS_M was calculated. First of all the average duration $DRS(v)$ of the notes with the same nominal value v was calculated, after this the various $DRS(v)$ values, normalized to the eight note value, were averaged in order to obtain the DRS_M.

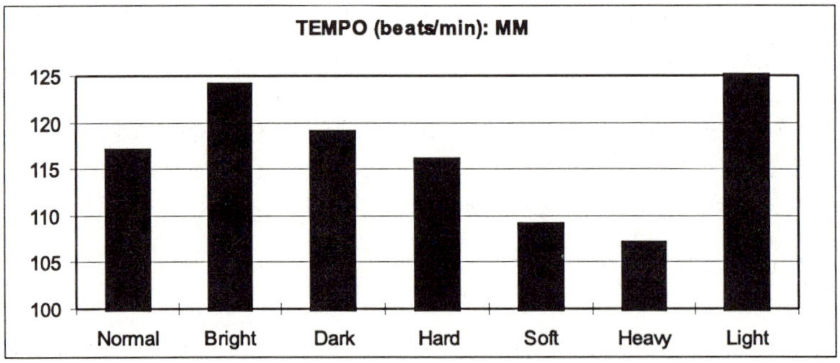

Fig. 3. Value of tempo *MM* in the seven performances

From the measurements of the envelopes, it was also possible to deduce how the clarinettist performed the various interpretations, giving each a different *emphasis* (ER) in terms of the relationship between the amplitude of

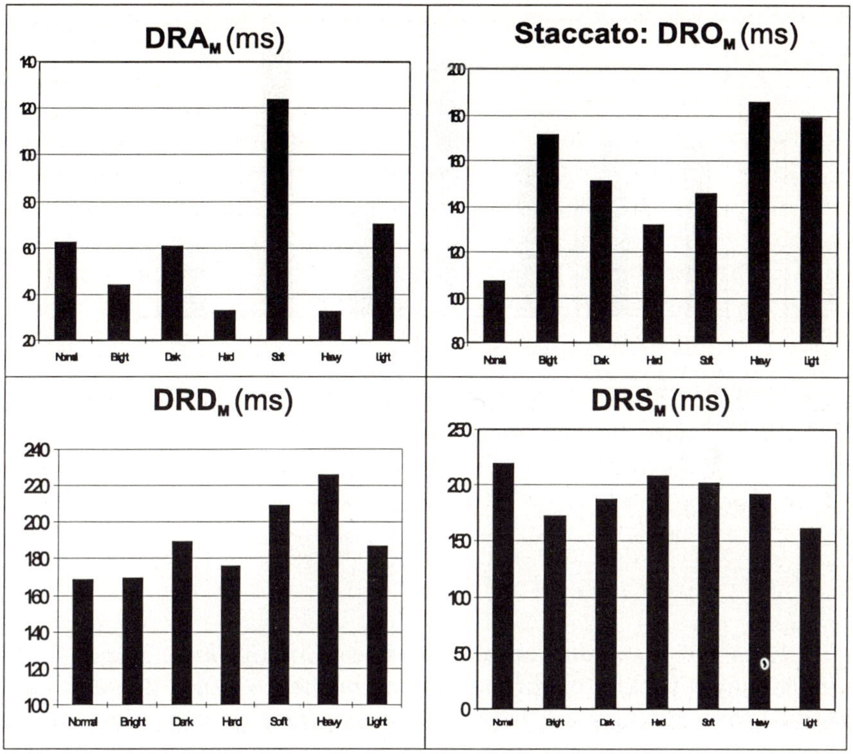

Fig. 4. Values of some temporal parameters in the seven performances

the main accent and the average accent. Table 2 shows the values for this parameter for each performance.

4 Frequency Analysis

The analysis in the domain of frequency then allowed for the identification of the variations in the timbre. The time-varying amplitude of each harmonics was obtained by making use of a heterodyne filter to break up the input audio signal into its sinusoidal components (Moorer, 1977). From this analysis, an attempt was made to try to deduce the main parameters that could describe the characteristics of the observed spectra. The timbre seems to depend on three parameters linked, respectively, to the brightness of the sound, the attack time and spectral irregularity (Krimphoff, McAdams, & Winsberg, 1994). The brightness dimension can be quantitatively represented by the

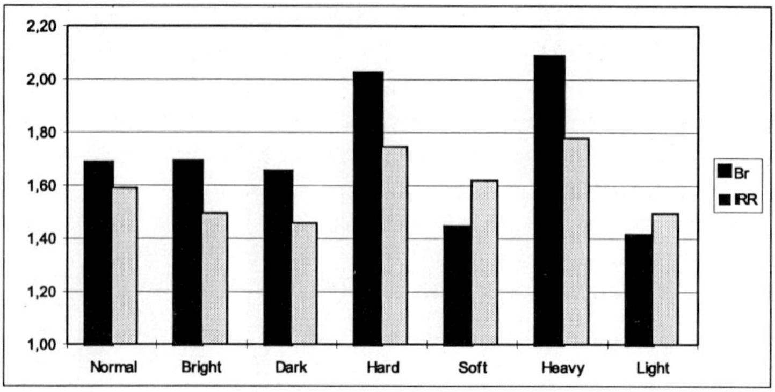

Fig. 5. Brightness Br and Irregularity of the spectrum IRR calculated in the various interpretations

spectral centroid, as follows:

$$\mathrm{Br} = \frac{\sum_k kA_k}{\sum_k A_k}$$

where A_k is the maximum amplitude (linear) of the k-th harmonic. The other significant timbric parameter defined in the frequency domain, can be quantified from the spectral irregularity index (IRR), which indicates the distance that the harmonics move away from the smoothed version of the spectrum. This was calculated by taking the amplitude average, in decibels, of three consecutive harmonics:

$$\mathrm{IRR} = \log\left(20\sum_{k=2}^{M-1}\left|\log A_k - \frac{\log A_{k+1} + \log A_k + \log A_{k-1}}{3}\right|\right)$$

where M is the number of significant harmonics.

It was seen that the most significant variation in the trend of the various harmonics took place at the initial phase of the note, that is, from the beginning and continued until the first harmonic reached its maximum value. However, during certain interpretations (Dark and Soft), the harmonics were delayed with respect to the fundamental one. The initial instant of the various harmonics varied with the order of the harmonic itself. The higher the harmonic, the greater the delay with respect to the fundamental one. This was roughly proportional to the duration of the note. The third harmonic

Table 2. Emphasis of the seven interpretations (Amplitude of the main accent/Average accent)

	Normal	Bright	Dark	Hard	Soft	Heavy	Light
Emphasis	1.58	1.06	1.62	1.78	1.5	1.91	1.21

was used as an indicator of the overall delay with respect to the fundamental one. This delay (Partial Delay = PD) too was measured, using a threshold value of 10% on the k-th harmonic: $PD(x) = A10(x,k) - A10(x,1)$, where x is the interpretation and k represents the order of the reference harmonic. In our case $k = 3$. Figures 5 and 6 show the trend of these parameters according to the intention of the performer.

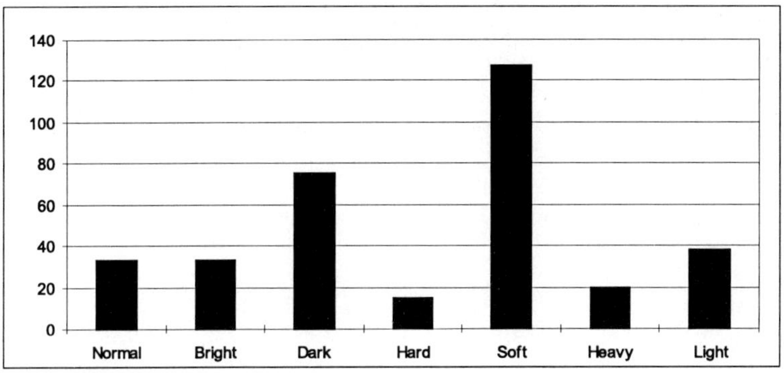

Fig. 6. Third harmonic Delay with respect to the fundamental harmonic (PD), calculated for the various interpretations

5 Analysis validation

The expressivity of the clarinet is strictly linked to the instrumental structure. The most relevant sonological parameters have been measured in function of the typical constraints of that instrument. With the aim of validating the analysis, a set of synthesized sounds were made. Cluster analysis, developed on perceptual tests, demonstrated that a fixed waveform synthesis gave difficulties in the recognition of expressive intentions for listeners not trained in electronic music. Therefore we decided to use physical model synthesis. In this way it was possible to apply the results of the sonological analysis to the synthesis – in order to obtain performances closer to the ones of a real clarinet. The use of a physical model allows the control of the performance not only note-by-note, but also at the phrase level; this is essential, for example, for the legato in which it is necessary to control the amplitude envelope of the complete phrase. In a physical synthesis model, the control is obtained naturally by modelling the pressure envelope of the air introduced inside the instrument. Some characteristic envelopes of the real instrument, typical for different expressive intention, were pointed out in order to develop pressure models to be employed in the physical model synthesis of the clarinet. We have used the analysis-by-synthesis method, in order to check the consistency of the approximations made. A set of synthesized phrases of the fragment shown in Fig.1 were performed, using a physical model of clarinet (Smith, 1986; Rocchesso & Turra, 1993). Each synthesis was related to a different expressive intention (Bright, Dark, Hard, Light, Soft, Normal, Heavy). Perceptual tests, developed on these synthesis, pointed out that musicians can correctly recognize the expressivity of the performances. Factor analysis showed how listeners were able to map the stimuli on a bi-dimensional space, in correspondence with their expressive intentions (Canazza, De Poli, & Vidolin, 1997, this book).

It seems fair to describe the various interpretations (Table 3) according to some parameters that have been the subject of this particular analysis. These measurements agreed well enough with the intentions that had been

Table 3. Expressive intentions described according to some parameters measured

Stimuli	Low	High
Bright	DRA, DRS, DRD	MM, DRO
Dark	ER, IRR	PD
Hard	DRO, DRA, PD	$DRS, BR, IRR, InitInstantDelay$
Soft	BR, MM	DRA, DRS, PD
Heavy	PD, MM	BR, IRR, DRD, DRO
Light	DRS, BR	MM, DRO

verbally described by the musician in an interview after the performances:

- Bright: fast, vivacious interpretation. A virtuoso performance. Sforzato first note.
- Dark: closer to a *normal* interpretation. Played almost as if it were warm. The clarinettist blew more warm air using an open throat.
- Hard: the notes were held for their entire value and well separated. An attempt to obtain a flat envelope around the note.
- Soft: the passage from one note to the next effected without using tongues (Strawn, 1986).
- Heavy: like the dark one, but slower.
- Light: abbreviating the value of the note but without changing the attack time. Fast performance, with soft attack and shortened holding.

6 Conclusions

The results of the sonological analysis seem to confirm the hypothesis that various performances of the same musical piece have their own sonological characteristics that can, in reality, be measured. The measurements agreed with the intentions that had been verbally described by the musician in an interview after the performances. A set of synthesis was made with the aim of evaluating the relevance of the chosen parameters. Fixed waveform and physical modeling synthesis, based on the values of the parameters measured confirmed the consistency of the data obtained. A factor analysis carried out on the data from a perceptive test, based on these synthesis, also showed, in fact, how listeners effectively correlated the pieces to the relative expressive intentions.

Appendix

Audio Examples (CD-tracks 49-62)

- Sound Examples #1-7: Mozart's *Concert for Clarinet* (K622): original recordings. Each performance was inspired giving the player a sensorial type adjective in order to suggest various kinds of *expressive intentions*. Performances according to the following adjectives:
  - Bright
  - Dark
  - Hard
  - Light
  - Soft
  - Normal
  - Heavy

- Sound Examples #8-14: Mozart's *Concert for Clarinet* (K622): Physical Model Synthesis. Physical modeling synthesis, based on the values of the parameters measured in the performances. Synthesized sounds according to the performance parameters of the following adjectives:
  - Bright
  - Dark
  - Hard
  - Light
  - Soft
  - Normal
  - Heavy

References

Canazza, S., De Poli, G., & Vidolin, A. (1997). Perceptual analysis of the musical expressive intention in a clarinet performance. In M. Leman (Ed.), *Music, Gestalt, and computing: Studies in cognitive and systematic musicology.* Berlin, Heidelberg: Springer-Verlag.

Drake, C., & Palmer, C. (1993). Accent structures in music performance. *Music Perception, 10*, 343–378.

Friberg, A., Fryden, L., Bodin, L., & Sundberg, J. (1991). Performance rules for computer-controlled contemporary keybord music. *Computer Music Journal, 15*, 49–55.

Krimphoff, J., McAdams, S., & Winsberg, S. (1994). Caracterisation du timbre des sons complexes II: Analysis acoustiques et quantificata psychophysics. *Journal de Physique, C5*, 625–628.

Moorer, J. (1977). Signal processing aspects of computer music: A survey. *Proceedings of the IEEE, 65*, 1108–1132.

Repp, B. (1990). Patterns of expressive timing in performances of a Beethoven minuet by nineteen famous pianists. *The Journal of the Acoustical Society of America, 93*, 622–641.

Repp, B. (1992). Diversity and commonality in music performance: An analysis of timing microstructure in Schumann's "Traumerei". *The Journal of the Acoustical Society of America, 95*, 2546–2566.

Repp, B. (1994). Relational invariance of expressive microstructure across global changes in music performance: An exploratory study. *Psychological Research, 56*, 269–284.

Repp, B. (1995). Quantitative effects of global tempo on expressive timing in music performance: Some perceptual evidence. *Music Perception, 13*, 39–57.

Rocchesso, D., & Turra, F. (1993). A generalized excitation for real-time sound synthesis by physical models. In *Proceedings of the Stockholm Music Acoustics Conference* (pp. 584–588). Stockholm.

Smith, J. (1986). Efficient simulation of the read-bore and bow-string mechanism. In *Proceedings of the International Computer Music Conference (ICMC-86)* (pp. 275–280). CMA.

Strawn, J. (1986). Orchestral instrument: Analysis of performed transitions. *Journal of the Audio Engineering Society, 34*, 867–880.

Todd, N. (1985). A model of expressive timing in tonal music. *Music Perception, 3*, 33–58.

Todd, N. (1992). The dynamics of dynamics: A model of musical expression. *The Journal of Acoustical Society of America, 91*, 3540–3550.

Todd, N. (1995). The kynematics of musical expression. *The Journal of Acoustical Society of America, 9*, 1940–1949.

Perceptual Analysis of the Musical Expressive Intention in a Clarinet Performance

Sergio Canazza, Giovanni De Poli and Alvise Vidolin

CSC-DEI, University of Padova, Via Gradenigo 6a, I-35131 Padova, Italy

Abstract. Usually, in the spoken language, a speaker can give different meanings to his words by introducing variations in the prosodic factors of the sentence. In a musical performance, the player can also introduce different expressive intentions within, naturally, the limitations of the score. Attention here, is turned towards the analysis how these intentions are communicated to the listeners. Perceptive tests were used to determine how some listeners' impressions arranged the musical pieces heard within a hypothetical n-dimensional space. Factor Analysis, Multi-Dimensional Scaling and Cluster Analysis applied to the subjects' replies verified that groups of listeners with different cultural preparation, could recognize the performer's intentions. It was then possible to reduce the size of the problem, in that the subjects had created a space into which the pieces were set, so that these could be subdivided into a reduced number of clusters. The synthesis of some performances and the successive perceptive analysis of the same, verified the results emerging from this study.

1 Introduction

Theoretical studies on musical performance and interpretation carried out according to rigorous scientific principles are a relatively recent development, and one of the biggest difficulties lies in the player's inability to explain his/her interpretation clearly, in theoretical terms. At the same time, there is also some difficulty in gathering musical data by the physical measurement of the overall sonorous result of a performance. Moreover, the study of musical performance and interpretation raises problems which go beyond the specific area of taking measurements, running into the wider field of non-verbal expressive communication that involves various aspects of human sensory perception. Musical performance, in the western tradition, is based on the score. This is a document of graphic symbols by means of which the composer's musical idea is communicated to the listener, even when the two belong to different historical periods. The player usually takes a certain amount of liberty in his/her interpretation of the score in order to better express the sense of a musical phrase. A performance that is stylistically correct and played paying the greatest respect to the style of the score, is here termed *normal*. However, the same score can be played with different degrees of expressivity, depending on the interpreter's *expressive intention*, with the result that each performance is different, closely correlated to subjective artistic choices. This paper presents a perceptive analysis with the successive model of expressive deviations in the interpretation of a musical phrase performed with different expressive intentions. In this context, *expressive intention* is taken to mean how a musician's inspiration varied according to certain adjectives that had been given before each performance. This aspect of a performance has not been widely studied, in that it has always been considered a pre-eminently artistic one and could not, therefore, be analyzed. With the advent of computers however, new methods of analysis have been developed and, therefore, the possibility of verifying, by synthesis, the validity of the analytical data or performance model formulated on the theoretical plane. Consequently, this field of investigation is stimulating ever greater interest not only from the scientific and cognitive point of view, but also from the applicative one both in terms of composing music and, more generally, in multimedia systems. More attention is, in fact, being given to this argument and some interesting studies have been carried out, particularly, by Gabrielsson (1973, 1993, 1995), Nakamura (1987), Namba, Kuwano, Hatoh and Kato (1991), Seashore (1938), Senju and Ohgushi (1987) and Sundberg and co-workers (Friberg, Fryden, Bodin, & Sundberg, 1991).

2 Method

Some musical performances, linked to certain adjectives that had been chosen to suggest and stimulate different interpretations of a musical phrase, were recorded. The adjectives chosen are not commonly used in the field of music and to this end, sensorial type adjectives were chosen in that this type seemed more suitable for sonological interpretation. Emotional type adjectives were, for the present, avoided in that it seemed more opportune to limit the overall semantics under examination. Thus, the choice of adjectives fell on the following: light, heavy, soft, hard, bright and dark and as such, each had its opposite (soft vs. hard) in order to deliberately induce contrasting performances on the part of the musician. A normal performance, as described above, was also recorded. Seven different interpretations of a fragment of Mozart's *Concert for Clarinet* (K622), were performed by a professional clarinet player and recorded in monophonic digital form at 16 bits and 44100 Hz at the CSC, Padua University. Initially the focus was on the judgement dimensions used by two groups of subjects called in to listen to the various interpretations of the same musical piece. A factor analysis was applied to this measure of perceptive order. The analysis was used in order to see how the listeners organized the pieces in their own minds, and how many dimensions (*performance types*) the subject were able to distinguish. The data were analyzed in yet another way, by transposing the matrix of judgments and conducting another factor analysis, in order to examine how the individual performances ranked on semantic-space as defined by the adjectives, from the factor score. A MultiDimensional Scaling (MDS) analysis was used to verify how much the various musical interpretations could, effectively, be distinguished one from the other. A Cluster Analysis was also used in order to see if there was any similarity between the two listening groups and based on the cumulative data emerging from the tests carried out on the two groups of subjects. The results deduced from the statistical analysis were, finally, used to try to construct a model that represented a musically expressive space. The validity of this space was verified by constructing some musical synthesis and analyzing them. The results showed that the performer's intentions and the listeners' impressions, in general, agreed.

3 Experiment 1

Two tests were carried out, respectively, on a group of twelve musicians, graduated at the Pollini Conservatory in Padua, and a group of twelve subjects who had no specific musical preparation. Both groups were asked to describe the piece of music heard along a graduated scale of the six adjectives mentioned above. An agreement index was calculated (Robinson's A) to verify the consistency of the replies. This coefficient indicates the degree of agreement expressed by the 12 trained musicians regarding the distribution of the variable performances:

$$\bar{A} = 1 - \frac{\sum_i \sum_j (x_{ij} - \mathbf{x}_i)^2}{\sum_i \sum_j (x_{ij} - \mathbf{x})^2} = 1 - \frac{DiscordanceMeasure}{\max(PossibleDiscordance)} = 1 - \frac{D}{\max(D)}$$

where x_{ij} is the value attributed to the adjective i by the subject j; $\mathbf{x}_i$ is the average of the values given to adjective i, while $\mathbf{x}$ is the overall average. There was a surprising degree of agreement between the subjects, despite the highly subjective nature of the question, reaching a value of 0.592. From the MDS it was seen that the listeners had, effectively, arranged the pieces separately in their own minds and had tended to place them at an equal distance from the normally played piece. The Euclidean distances deduced from the replies are shown in the triangular matrix in Table 1.

Table 1. Matrix of the Euclidean distances

	Bright	Dark	Hard	Light	Soft	Normal	Heavy
Bright	0						
Dark	2.66	0					
Hard	3.17	2.06	0				
Light	0.85	1.57	2.66	0			
Soft	2.66	2.16	4.01	1.69	0		
Normal	2.06	0.85	2.16	0.85	1.93	0	
Heavy	2.66	0.85	1.67	1.67	2.66	0.85	0

From the results of the factor analysis applied to the data from both tests, it was clear that the subjects had placed the performances along only three axes. The first three factors, in fact, explained 85.2% of the total variance and were the only ones having eigenvalues greater than 1. Table 2 shows the percentage variance explained by the factors and their respective eigenvalues. The factor loadings, after Varimax rotation, are listed in Table 3. Varimax rotation was used in order to simplify the factors' interpretation.

Table 2. Eigenvalues and variance explained by the factors. The first three factors explained 85.2% of the total variance and were the only ones having eigenvalues greater than 1. Thus it is sufficient to consider only the first three factors

Factor	Eigenvalue	Percentage of variance	Percentage
1	3.33	47.5	47.5
2	1.48	21.1	68.7
3	1.16	16.5	85.2
4	0.73	10.4	95.6
5	0.31	4.4	100

Table 3. Factor loadings after the Varimax rotation. Factor loadings of first and third factor are plotted in Fig.1

	Factor 1	Factor 2	Factor 3
Bright	-0.89	0.29	-0.2
Dark	0.25	-0.94	0.17
Hard	0.57	0.01	-0.61
Light	-0.58	0.63	0.51
Soft	-0.56	0.1	0.88
Normal	0.17	0.83	0.45
Heavy	0.83	0.56	-0.16

The three dimensional space so obtained represents a model of expressivity, by which the subjects arranged the pieces in their own minds. Figure 1 shows the arrangement of the pieces along two of these axes. The former seemed to be closely correlated to the tempo factor, while the third factor was connected to the notes' attack time. The normal vs. dark were placed at the extreme ends of the second factor and appeared to depend on the score's own expressivity.

From the factor analysis limited to the data obtained from the test with non musicians, on the other hand, the pieces were subdivided along one single judgement dimension. A dominant factor (60.2% of the total variance) has been observed which tended to divide the pieces into two groups only, separating the soft, light and dark from the rest. Even the dendrogram deduced from the cluster analysis (see Fig.2) showed the clear distinction between

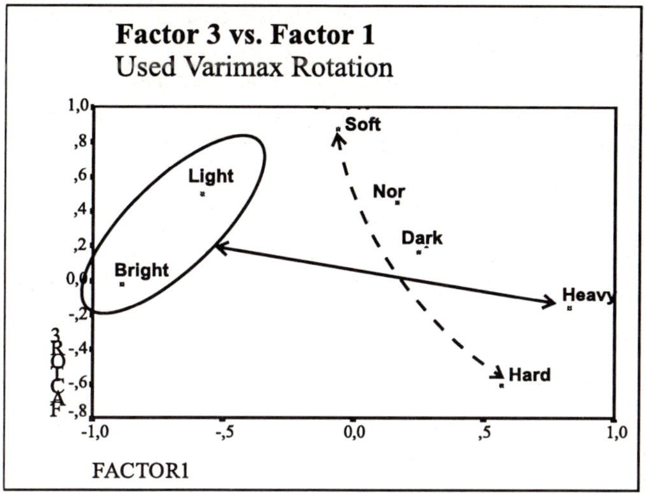

Fig. 1. Graph of factor loadings deduced from factor analysis. Two expressive types can be observed: the former (bright-light vs. heavy) seemed to be closely correlated to the 'speed of performance' factor (Tempo), while the third factor (soft vs. hard) was connected to the notes' attack time

the group of trained musicians (subj.1 - subj.12) which was highly cohesive, or rather, a small distance index and the group of non musicians (subj.13 - subj.24) which showed a greater variance in the judgements given.

4 Experiment 2

A new test carried out on the same musical performances but giving the two groups of subjects a wider semantic choice, allowed for a more detailed study of the listeners' judgement categories. The seventeen new adjectives, again of a sensorial nature, were chosen so as to offer the subjects a exhaustive sampling of a semantic space and here, the list of chosen adjectives did not include their opposites. This new list of adjectives (which did not contain those in the previous list) included the following: "nero" (black), "greve" (oppressive), "grave" (serious), "tetro" (dismal), "massiccio" (massive), "rigido" (rigid), "soffice" (mellow), "tenero" (tender), "dolce" (sweet), "aereo" (airy), "lieve" (gentle), "spumeggiante" (effervescent), "vaporoso" (vaporous), "fresco" (fresh), "brusco" (abrupt), "netto" (sharp). Once again, the experiment consisted of two tests, the first using 12 subjects trained at the Conservatory, eight of whom had not taken part in the first experiment, and twelve non musicians, none of whom had taken part in the first experiment. The factor analysis of the performance confirmed the same two factors which had emerged from the previous experiment. From the factor analysis carried out using the adjectives as the variables, it was seen that the subjects

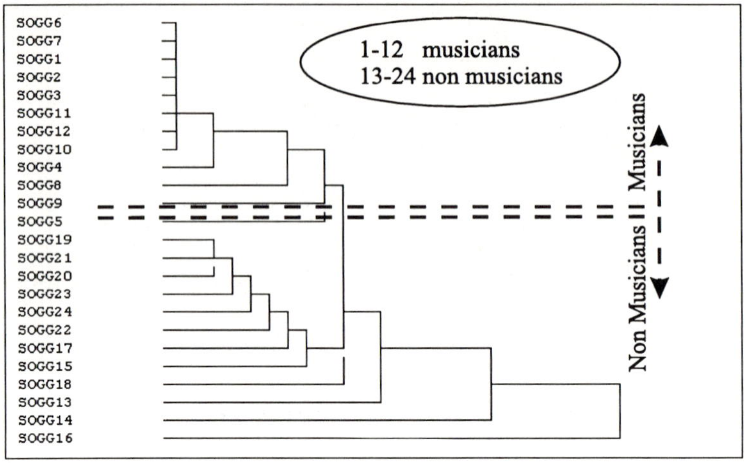

Fig. 2. Cluster analysis dendrogram using the centroid method. The ascissa represents the distance index between the subjects. It can be noticed that distance among musicians' judgment is lower than the distance among musically untrained subjects

clearly recognized the performer's intentions. Figure 3 shows the graph of the factor loadings and, by means of the factor scores, it was possible to insert the performances into the factor space deduced from the adjectives (Repp, 1990).

5 Experiment 3

The analysis-by-synthesis method was used to verify the validity of the results obtained. Therefore, seven interpretations of a fragment of G.F. Händel's *Sonata for Alto and Basso Continuo* were synthesized with the program CSOUND (Vercoe, 1993). Given that the factor analysis of the results of the test carried out on the original pieces of music showed that these were arranged in such a way as to agree with some values of the sonological variables, the factor loadings, opportunely scaled, were used as the values for these parameters. Thus, the same expressive type of stimuli deduced by the listeners and constructed on the basis of one of Händel's scores, were obtained. At the same time, these were far enough away from each other in that the fragments taken from Mozart's *Concert for Clarinet* had shown that these latter had factor loadings which were different from each other. With reference to the secondary parameters, or rather, those not clearly picked up by the listeners, values deduced from sonological measurements of the fragment of Mozart's work were used. The first factor, in particular, was used to determine the values of the Tempo and Climax Emphasis (ER = principle/average accent) parameters, while the second one was used to establish

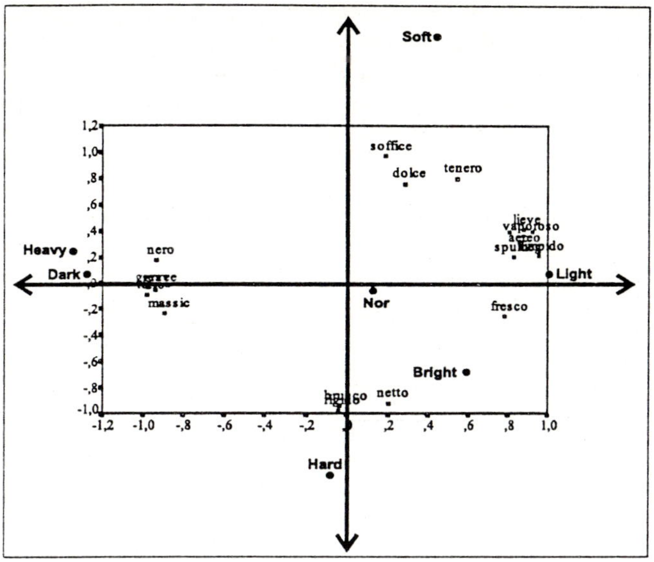

Fig. 3. Factor analysis on the adjectives. The first factor explains 49.6% of the total variance, the second 28.5%. Performances, placed in the plane using factor scores as coordinates, are strongly correlated with the corresponding adjective

the notes' attack time (*DRA*, in ms) (Canazza, De Poli, Rinaldin, & Vidolin, 1997, this book). Figure 4 shows the values of the main parameters used to differentiate the synthesis. The relative distances of the pieces were deduced by means of MultiDimensional Scaling carried out on the cumulative data obtained from Experiment 1.

The validity of the synthesis was checked by subjecting them to the perceptive experiments carried out on pieces by Mozart. More precisely, two listening tests were repeated for each experiment and, again, the subjects were divided into two groups, one group consisting of trained musicians and the other of non musicians. The analysis techniques used were the same as those used in the study of the original pieces of music. In this case, the Cluster Analysis was particularly important. It was interesting to see how, here, musicians used to working with or, at least, listening to electronic music arranged the synthesized pieces differently from musicians used to working with traditional acoustic instruments, placing them differently, in a hypothetical semantic-musical space. The group of musicians used to electronic music, in fact, moved the stimuli towards the bright extremity of the bright-heavy axis. These tests, too, have shown that the subjects' replies agreed reasonably well. The MultiDimensional Scaling, in particular, showed how the stimuli had been differentiated. The factor analysis seemed, moreover, to ratify the listeners' effective recognition of the expressive intentions. The most encouraging result was, however, the fact that the same expressive types are emerged from these tests as demonstrated by the similar experiment on the original pieces. This would support the hypothesis that the main parameters which characterize the various interpretations have been correctly identified.

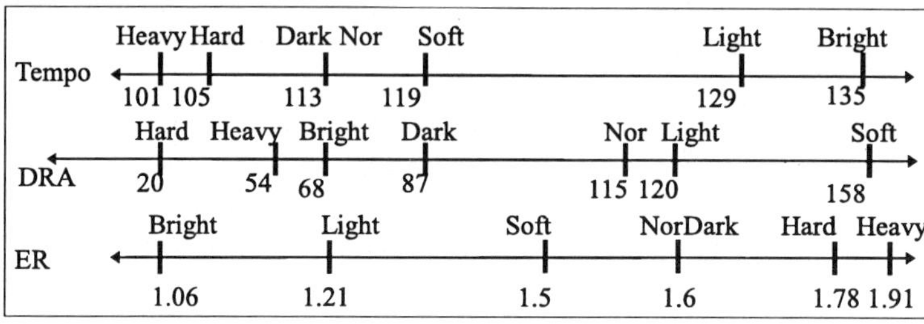

Fig. 4. Values of the main parameters used in the synthesis of Händel's *Sonata for Alto and Basso Continuo*. The factor loadings was used to determine the values of the parameters

6 Conclusions

Various studies on performance have led to suggesting models that could render synthesized music less monotonous and mechanical. An attempt has been made, here, to go beyond this, in that the authors have tried to identify an expressive model that could duplicate the various possible intentions of a human performer. Therefore, musical expressivity was analyzed in order to implement this when using electronic instruments. In order to do so, measurements of a perceptive nature were used. By analyzing these results, two quite distinct expressive directions were observed, one tending towards brightness and the other, towards softness of the pieces. The tests, in fact, made use of two factors linked, essentially, to sonological variables. From a musical point of view, in fact, the first axis sets rapid Tempo (bright and light pieces) against slow (heavy) or moderate (soft) Tempo. The second axis is connected with the average attack time of the notes of the musical phrase. Modifying these parameters would, therefore, change the relative position of the pieces along the respective directions. However, it should be remembered that it is absolutely impossible to think that there is any correspondence between the deduced factors and the sonological variables of the one on one type observed. Studies on expressivity have, in fact, shown (Repp, 1995) that some physical parameters depend on others, such as, for example, the expressive variations of the temporal microstructure with respect to the overall Tempo of the piece. Interpretation of the factors can not, therefore, exclude this. All the tests carried out up to the present lead, in any case, to the pieces being arranged along these two axes which have, consequently, been interpreted as two expressive types. It is interesting to note how it was also possible to insert the syntheses along these two directions, judging from the subjects' replies. This would demonstrate that the performer's expressive intentions had been captured and that they could be moved to other scores. Some sonological variables, correlated to factors deduced from the statistical analysis - opportunely amplified or diminished - seemed, in fact, to be able to move the stimuli towards one extreme or the other of some semantic guides that have been identified here with the chosen adjectives. However, it should be said, finally, that even though this study was limited to only one piece, other studies currently underway would seem to confirm the results discussed here.

Appendix

Sound Examples (CD-tracks 63-69)

– Sound Examples #1-7: Fixed waveform synthesis of seven interpretations of a fragment of G.F. Händel's *Sonata for Alto and Basso Continuo* according to the parameters (i.e. factor loadings) deduced from perceptual analysis of the following adjectives:
  - Bright
  - Dark
  - Hard
  - Light
  - Soft
  - Normal
  - Heavy

References

Canazza, S., De Poli, G., Rinaldin, S., & Vidolin, A. (1997). Sonological analysis of clarinet expressivity. In M. Leman (Ed.), *Music, Gestalt, and computing: Studies in cogntive and systematic musicology.* Berlin, Heidelberg: Springer-Verlag.

Friberg, A., Fryden, L., Bodin, L., & Sundberg, J. (1991). Performance rules for computer-controlled contemporary keybord music. *Computer Music Journal, 15*, 49–55.

Gabrielsson, A. (1973). Similarity ratings and dimensional analysis of auditory rhythm patterns. *Scandinavian Journal of Psychology, 14*, 138–160.

Gabrielsson, A. (1993). Intentional and emotional expression in music performance. In *Proceedings of the Stockholm Music Acoustics Conference* (pp. 108–111). Stockholm.

Gabrielsson, A. (1995). Expressive intention and performance. In R. Steinberg (Ed.), *Music and mind machine* (pp. 37–47). Berlin, Heidelberg: Springer-Verlag.

Nakamura, T. (1987). The communication of dynamics between musicians and listeners through musical performance. *Perception and Psychophysics, 41*, 525-533.

Namba, S., Kuwano, S., Hatoh, T., & Kato, M. (1991). Assessment of musical performance by using the method of continous judgment by selected description. *Music Perception, 8*, 251–276.

Repp, B. (1990). Patterns of expressive timing in performances of a Beethoven minuet by nineteen famous pianists. *The Journal of the Acoustical Society of America, 93*, 622–641.

Repp, B. (1995). Quantitative effects of global tempo on expressive timing in music performance: Some perceptual evidence. *Music Perception, 13*, 39–57.

Seashore, C. (1938). *Psychology of music.* New York, NY: McGraw-Hill.

Senju, M., & Ohgushi, K. (1987). How are the player's ideas conveyed to the audience? *Music Perception, 4*, 311–324.

Vercoe, B. (1993). *Csound: A manual for the audio processing system and supporting programs.* Cambridge, MA: The MIT Press.

Singing, Mind and Brain – Unit Pulse, Rhythm, Emotion and Expression

Eliezer Rapoport

Musicology Department, Bar - Ilan University, Ramat-Gan, Israel

Abstract. Singing, and especially emotional expression in opera and lied singing, can be described in terms of a basic unit, called here *unit pulse*, deduced from Fast Fourier Transform (FFT) spectrograms, and interpreted as a basic brain command to the vocal folds consisting of tightening followed by immediate tension release. This unit pulse, of 100-160 ms duration, is a measure of the degree of excitement in singing. It also defines a unit of time, and serves as a brain time counting mechanism (biological clock) in rhythm. It explains the origin of vibrato and of legato coloratura singing. Each vocal tone is composed of one or more unit pulses. Excitement and calmness can be defined on the microscopic level of a 160 ms time scale, and described in terms of three varieties: large pulse, small pulse, and zero pulse. The various ways of their arrangement within the structure of the tone determine the singer's ways of emotional expression in singing. This is a simplified, more basic code of emotional expression in singing than the high-level emotional expression language presented by the author in a previous work. Biologically, the three varieties of the unit pulse correspond to particular sequences of neural firings in specific patterns. Analysis of performance of vocal music in terms of the unit pulse allows quantitative measurement of the (momentary) degree of excitement for each individual tone along the sung melodic line, and gives an insight into the singers' intentions and personal ways of expression and interpretation.

1 Introduction: Unit Pulse in Singing, Rhythm Perception, and Human Motor Behavior

The present work is concerned with processes in the performing artist's mind and brain, particularly in opera and lied singing. These processes operate in translating the emotional messages of the text and the musical score into musical tones as uniquely shaped acoustic signals by characteristic commands sent from the brain as neuron firing sequences activating the musculature that controls the vocal folds tightening and vibrating frequency.

In recent papers (Rapoport, 1995, 1996) the author demonstrated that emotional messages in singing are encoded at the level of the single individual tone as characteristic temporal structures constructed from various characteristic expressive elements. These are actually timbre elements, as they use specific parts of, or add extra frequencies to the ensemble of frequencies contained in the vocal tone. The Fast Fourier Transform (FFT) analysis was shown to be very potent in the identification and deciphering of the emotional expression code in vocal tones. These expressive timbre elements are:

(1) singing in the frequency range corresponding to the lower formants here, labelled *phonation*, (2) excitation of the higher partials of the singing formant, (3) vibrato, (4) transition – a gradual pitch increase from the onset to the sustained state, (5) pitch change within the tone, (6) sforzando – an abrupt pitch increase at the onset of the tone, and (7) unit pulse.

Excitement or calmness were shown to be built into the vocal tones in such a way that the smaller the number, and the more gradually these aforementioned timbre or expressive elements enter into operation – the more calm the tone will be, and, inversely: the more abrupt, and the larger the number of the timbre elements operating simultaneously – the more excitement is built into the vocal tone.

On this basis, the author was able to classify the large number and variety of (temporal) structures or (acoustical) shapes of vocal tones encountered in singing of western art music (opera arias and lieder) into eight categories or tone families in a very well-defined hierarchical scheme.

The fundamental frequency (or pitch) of the vocal tone is determined by the length and mass, but predominantly by the momentary tension in the singer's vocal folds, in such a way that increasing tension leads to increasing pitch. In singing, the singer constantly changes the vocal folds tension in order to produce the various tones. The Fast Fourier Transform (FFT) spectrogram, displaying the frequencies-time diagram is thus equivalent to the vocal folds tension-time diagram. Rapoport (1996) identified a basic unit, appearing as the inverted letter U in FFT spectrograms, with a time duration of 100-160 ms . This unit, to be called *unit pulse*, is actually a frequency increase followed by frequency decrease, corresponding to vocal folds tension increase followed by tension release. This can be traced back to some basic unit command in singing sent from the brain (Davis, Zhang, Winkworth, & Bandler, 1996; Larson, 1988) to the vocal folds, via the central nervous system and the musculature activating the vocal folds. Excitement and calmness in singing can be defined on a neurophysiological level of tightening of the vocal folds: a strong and rapid tightening command corresponds to excitement, whereas weak and slow tightening corresponds to calmness or relaxation. Rapoport (1996) proposed a simple neuro-physiological unit pulse model for singing. The unit pulse was demonstrated in a number of opera arias and lieder, and was also interpreted as a brain pacemaker time-counting mechanism (biological clock) operative in rhythm.

The well-documented observation by many authors (d'Alessandro & Castellengo, 1994; Sundberg, 1987) that in legato coloratura singing the behavior of each individual tone consists of the unique and characteristic feature of initial pitch increase, with an overshoot above and beyond the nominal intended tonal pitch, followed by immediate pitch decrease bellow the intended nominal pitch, with the pitch perceived consisting of a time integration over this peculiar pitch-time trajectory (d'Alessandro & Castellengo, 1994), is consistent with the unit pulse model. Thus, legato coloratura singing consists of a sequence of unit pulses.

Puccini: Turandot, In questa Reggia - Maria Callas 11.8.96

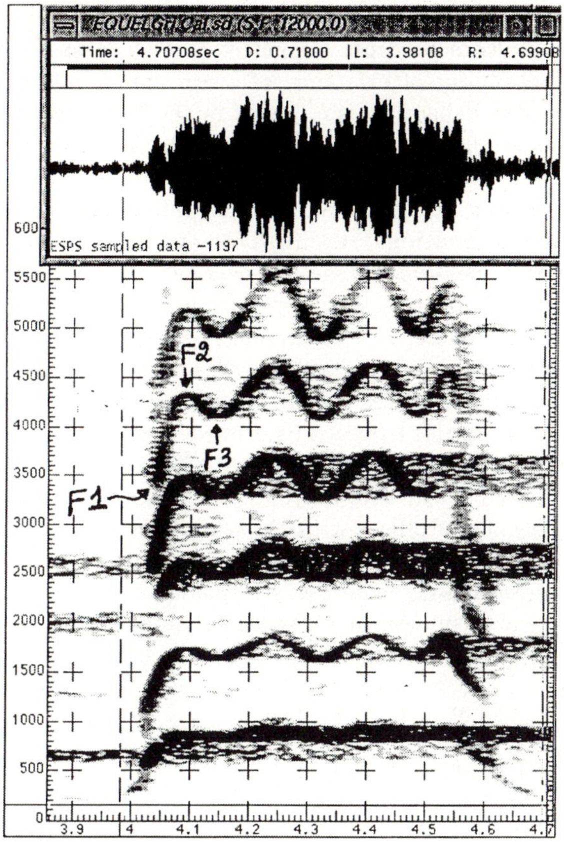

Fig. 1. An FFT spectrogram of a vocal tone (Maria Callas singing *In Questa Reggia* in *Turandot*) composed of four unit pulses. X-axis denotes time, and Y-axis frequency in Hz. F1, F2, and F3 in the fifth harmonic partial indicate the beginning, apex, and termination, respectively of the first unit pulse

Vibrato could also be interpreted as a pulse train of unit pulses. Some measurements on the behavior of individual vibrato pulses along a single tone and along a melodic phrase are to be presented and discussed here.

Recent physiological electromyographic (EMG) measurements (Hsiao, Solomon, Luschei, & Titze, 1994) demonstrated the correlation of vocal vibrato with characteristic pulse sequences of motor unit firings of well defined frequency modulation patterns in the cricothyroid muscles. This muscle activity is controlled by corresponding brain commands of neuron firings of well defined frequency modulation patterns. More recent neurological measurements on bird singing elucidate in more detail the brain mechanisms involved in bird singing (Yu & Margoliash, 1996).

The concept of *fastest pulse* or *elementary pulse*, as deduced from ethnomusicological studies of African polyrhythmic music, was discussed in a recent review article on rhythm perception (Seifert, Olk, & Schneider, 1995). Ideas concerning the possibility of existence of some unit of motor action in the field of human motor behavior also appear in the literature (Viviani, 1986). Thus, the unit pulse in singing might belong to a family of more general processes.

In the present study, unit pulse parameters in the frequency-time diagram, such as pulse height, duration, risetime, etc.. (to be defined in Sect.2), were measured along the melodic phrase for over 50 excerpts from opera arias and lieder, with the goal of seeking correlation with the level of excitement in singing. Some combination of these parameters might eventually yield a single parameter to serve as a *pulse strength* and be a quantitative measure of the degree of excitement. In the present report, three examples will be presented in which the pulse height is examined. A very simplified description of singing was found, in terms of three entities: strong pulse, weak pulse, and zero pulse, and a vocal tone beyond a certain excitement threshold is described as a sequence of pulses with the arrangement of either small pulses followed by increasingly larger pulses – this gradual structure is typical of calm and expressive singing, whereas a sequence of a very large first pulse followed by other large pulses, or by pulses of steadily diminishing strength is characteristic of excited singing.

2 Brief Summary of the Physiology of Singing and Definition of the Unit Pulse

Singing is the result of a combined operation of four physiological systems: (1) laryngeal, (2) respiratory, (3) resonatory, and (4) articulatory. Their coordination and synchronization is commanded by various centers in the brain (Davis et al., 1996; Larson, 1988). Altogether 196 different muscles are activated in opera singing (Scotto Di Carlo, 1991).

The physiology of singing is extensively treated in Sundberg (1987, Chaps. 2-4). Briefly summarized: Singing (and speech) is produced by two processes: (1) closure (adduction) and tightening of the vocal folds to the desired tension, and (2) buildup of the pulmonary (or subglottic) air pressure required to open the vocal folds and set them vibrating (Scotto Di Carlo, 1991; Sundberg, 1987).

Both the coordination and the synchronization of these two processes are programmed and controlled by the brain via specific commands as neural firing sequences transmitted to the operating musculature via the central nervous system. The pitch of the vocal tone is determined mainly by the tension in the vocal folds (increasing tension leading to increased pitch), whereas subglottic pressure determines airflow and hence the sound intensity, and the resonance cavities of the vocal tract (oral and nasal) determine the formants and the voice timbre.

In the present study we are concerned with the very onset of the vocal tone, on a time scale of about 160 ms. It was previously deduced from FFT spectrograms (Rapoport, 1996) that the elementary command to the vocal folds is a unit pulse consisting of tightening followed by immediate tension release. Typically, such pulse duration is around 100-160 ms , considered as an elementary beat – a unit of time counting in singing.

An FFT spectrogram of a vocal tone composed of four unit pulses is shown in Fig.1, where the X-axis denotes time, and the Y-axis denotes frequency. Referring to the fifth harmonic partial in Fig.1, the frequency of the beginning of the first unit pulse is marked F1, the apex is marked F2. This is the vocal folds tension increase stage. Beyond it is the tension release stage until the point F3 – the commencement of the second pulse.

The pulse height H (in semitones) is defined:

$$H = log(F2/F1)/log(1.05946) \tag{1}$$

The pulse rise time is the time interval elapsed on passing from F1 to F2.

Another important parameter is the pulse average rate of increase, HR, in semitones per second, corresponding to the velocity of tightening of the vocal folds. It is defined as:

$$HR = H/risetime \tag{2}$$

On the timescale of 160 ms, calmness and excitement (or relaxation and tension, respectively) in singing can be deduced from the manner of operation and degree of coordination of the two processes of vocal folds tension and lung pressure buildup. The emotional character of the vocal tone is determined at the very first beat by the character of the first unit pulse.

3 The First Unit Pulse, Strong Pulse/Weak Pulse, Excitement and Calmness

In singing with great excitement, vocal folds closure and tension buildup will take place very abruptly with simultaneous, abrupt subglottic pressure buildup. This first pulse will be recorded in the FFT spectrogram (which is essentially a vocal fold tension-time diagram) as a pulse starting well below the intended target pitch, and rising steeply in time. At its apex the target pitch along with the corresponding vocal folds tension and subglottic pressure are attained (usually with an overshoot), and then follows the tension release stage, completing the unit pulse. A time lag between vocal folds tension buildup and the start of subglottic pressure buildup accounts for the fact that the initial stage of the vocal folds tension will be unphonated, inaudible, and therefore not recorded in the FFT spectrogram. The resulting pulse will therefore be smaller. The longer the time lag the smaller will be the unit pulse. Longer time lag means a gradual buildup of phonation – a gradual development or buildup of a vocal tone corresponds to a tone of a more calm or relaxed character. Hence a big first pulse corresponds to excitement, and small pulse to calmness. The small first pulse can also be a result of a proper brain command to produce such a pulse in nonexcited singing. A third case is one in which attaining the required vocal folds tension was achieved first, and then followed by subglottic pressure buildup (a two-stage process). The resulting tone will thus start at the exact target frequency, and this is the most gradually built tone onset. This is therefore a tone of a most calm or relaxed nature, as are the N and C tones typically sung in Schubert's *Ave Maria* (Rapoport, 1995, 1996).

4 The Unit Pulse In Vibrato, Biological Clock, and Rhythm

The behavior of the unit pulse along a melodic line is demonstrated in the singing of Maria Callas of the first twenty one bars from the opening of the aria *In Questa Reggia* from *Turandot* by Puccini (EMI CDC 7 47966 2). Referring again to Fig.1, where a four-pulse (vibrato) tone from this aria is shown, we note that contrary to the common view of vibrato as an ornamental sinusoidal undulation around a mean pitch, the pulse nature of vibrato is well demonstrated: (1) The four pulses are different from each other. (2) The first has a large pulse height, short duration, and a steep rise, starting more than an octave below the target pitch, with a very fast risetime – indicating a very abrupt and rapid tension buildup in the vocal folds simultaneously with subglottic pressure buildup, as is indeed recorded in the FFT spectrogram. (3) The steep, abrupt rise at the tone onset, (the first pulse) is very significant, it is an accentuation (sforzando) at the tone beginning, and is quite audible. This indeed corresponds to emotional tension or excitement. (4) The two middle pulses are more typical *vibrato* pulses. (5) The fourth pulse is, again, an abrupt and short pulse with fast risetime and fast, abrupt tension release. (6) These four pulses might be overlapping: the one commences before the previous one terminated, indicating, as in many human motor processes, that the brain commands are faster than the operating muscles can follow.

The marked performance score (Rapoport, 1996) of these twenty one bars in the opening of the aria is presented in Fig.2, with the tone of Fig.1 marked therein. Each tone in the score is marked according the particular tone category (Rapoport, 1996), or family, to which it belongs: Tones marked Z or X belong to the *Excited* category, Z indicates that the tone starts with sforzando – a very large first pulse, such as the tone of Fig.1, marked Z4. Tones marked R, oR, or r belong to the *Expressive* category. C denotes the *Calm* category (the tone starts with *zero pulse*), N denotes the Neutral-soft category (tones consisting only of *zero pulses*). G, g, and K belong to the *Transitional-Multistage* (T) category, and m denotes one type of tone belonging to the *Short* (S) category. With regard to the degree of excitement in the classification scheme of the tone categories, the hierarchical order from the most calm to the most excited is given in equation 3:

$$N < C < R < T < X = S < Z \tag{3}$$

The reader is referred to Rapoport (1995, 1996) for a description of the tone categories.

Fig. 2. Marked performance score (according to the code of Rapoport, 1996, see also explanation in the text) of Maria Callas singing the opening of the aria *In Questa Reggia* from *Turandot* by Puccini

The numbers following the letters in Fig.2 denote the number of vibrato pulses the tone contains. The author observed in many FFT spectrograms that vocal tones always contain an integral number of pulses, and thus tone duration is counted in an integral number of pulses. In this way the unit pulse serves as a basic unit of time counting, as a brain pacemaker mechanism (biological clock). Time is thus divided on the neurophysiological level into units of about 100-160 ms duration. Like other biological clocks, such as a heart beat, its exact period is subject to changes according to the stress or excitement state of the particular person. This confirms a remark made by Vennard (1967) [1] that a good singer sets the tempo in multiples of his vibrato. If tempo is imposed, a singer can speed up or slow down the rate of his vibrato, in order to always keep an integer number of vibrato cycles in a note.

As a basic unit of time counting, the unit pulse also plays a role in rhythm and expressiveness in singing. Each vocal tone is composed of one or more unit pulses. In arias and lieder of rapid tempi and short duration tones the shortest note will be a single pulse, and the longer notes integral multiples of it, as was previously observed (Rapoport, 1996). When, as is quite often the case, there is a repetitive rhythmic unit: [eighth, eighth, quarter], [sixteenth, sixteenth, eighth], or [eighth, quarter], the shorter note is a single pulse and the longer a double pulse tone. More flexibility in tone duration for expressiveness is practiced by the singer, when singing long duration tones, as can be seen in Fig.2, where a sixteenth corresponds to a three pulse tone, but sometimes to a double or even four pulse tone, and a quarter to seven, eight, ten, or even an eleven-pulse tone.

The first two bars of the vocal part in the excerpt from *Turandot* (Fig.2) open in six notes of the very same pitch (D). Except for the fourth tone (Reg), all are Z type tones, with a very large first pulse – very much the same as the tone of Fig.1. They start about an octave below, with very abrupt vocal folds tension buildup (with simultaneous lung pressure buildup), and terminate in very abrupt, dramatic tension and pressure release, seemingly returning to a common baseline of relaxed vocal folds state at the end of each tone – to start afresh on the next tone. This is characteristic of very excited, dramatic singing.

This 21-bar excerpt from the aria consists of five phrases. Each of the first three is characterized by a dramatic opening (with Z tones), symbolizing the desperate cry of the tortured princess, then relaxing to R tones towards each phrase's end. The fourth phrase changes to a mood of tenderness with singing in R tones, much in accordance with the text. Even more delicate tenderness is expressed in the fifth phrase, upon pronouncing the name of the princess Lou-Ling, very dear to *Turandot*. This is indeed the spirit of the interpretation of Maria Callas in this performance, and also of Joan Sutherland, in another performance.

Figure 3a shows the pulse height for all the tones of the first phrase. The Z tones are indeed conspicuous, with very high values of the (first) pulse height. Fig.3b shows the pulse height for all the tones in the entire 21-tone excerpt. It can be seen that the figure is an exact parallel of the performance score, showing the dramatic parts (Z tones) separated by more relaxed (R tones)

[1] Quoted in Castellengo et al. (1989), to whom credit is due for recognizing the pulse nature of vibrato (they used the term *rate of pulsation* for vibrato frequency).

parts, with complete absence of Z tones in the fourth and fifth phrases. In addition, the gradual decrease of the average vibrato pulse height from around 2 semitones at the beginning towards final 1 semitone at the 5th phrase is clearly demonstrated.

5 Legato Coloratura Singing as a Sequence of Unit Pulses

Legato coloratura singing consists of a sequence of unit pulses, where each tone is one unit pulse. In the present study, a number of coloratura excerpts from opera arias and cantatas were studied, with the pulse height H, expressed in semitones, and other parameters, measured for each individual note along the melodic phrase. Figure 4 shows score excerpts of four representative examples. The first, Fig.4a, is from the second aria of the Queen of the Night, from *The Magic Flute* by Mozart. Fig.4b is an excerpt from the *Bell Song* from *Lakmé* by Delibes. In these two coloratura excerpts the virtuoso singing by Edita Gruberova, and by Sumi Jo were compared.

Fig.4c shows an excerpt from *Exsultate Jubilate* by Mozart. An excerpt from Händel's cantata *Lucrezia* is shown in Fig.4d – these two sung by Janet Baker. These excerpts, though of primarily ornamental character, are, nevertheless, in different emotional contexts. The first is at a moment of tenderness, before a phrase of virtuoso staccato coloratura is to commence. The second is in an enchanting atmosphere of the *Legend of the Pariah Maiden*, immersed in exotic fragrance of flowering mimosa and the Indian forest at moonlight, sung in timbral beauty and vocal artistic perfection. The third is Halleluja (alleluja), and the fourth is of utmost fury (Furioso) as due to the tragic, traumatic event in the tale.

The pulse height along these musical phrases shows rather complex and intricate patterns, as is also the case in many other excerpts studied. Globally, one can discern regions (or sub-phrases) with high H values (3-5 semitones), corresponding to excitement, and sub-phrases with low H values (1-2 semitones, sometimes 0-1 semitone), corresponding to calmness or tension release. At the microlevel the matter is more complex and delicate, involving *assignments* and *constraints*.

In these legato coloratura excerpts, the tones are legato-connected to each other, without breathing stops. The assignment is the interval in semitones between two successive tones in an ascending sequence (a value of zero is arbitrarily assigned to descending tones). The constraints are that the time average of the tightening-release trajectory should correspond to the tonal pitch. The pulse height should be compared to the assignment. Obviously a pulse height corresponding to an assignment of 2 semitones (two semitone interval) should be twice as large as the one corresponding to 1 semitone assignment. A pulse height close or only somewhat larger than the assignment means very little added excitement.

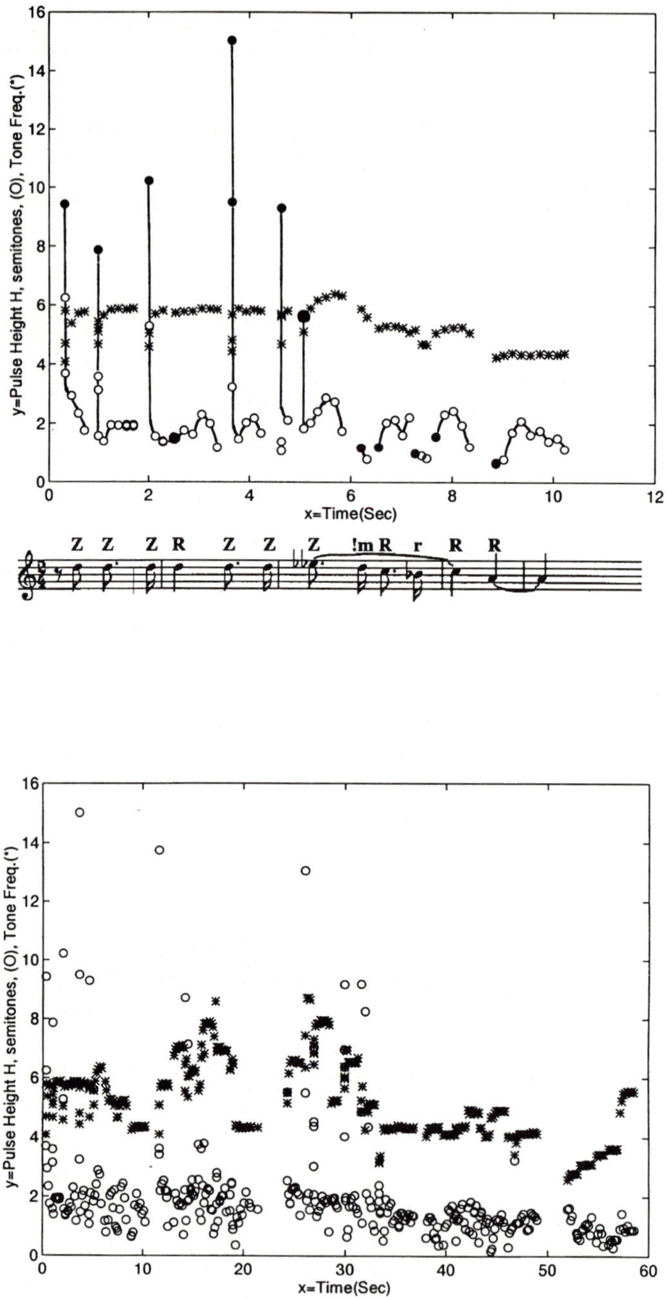

Fig. 3. (a) Unit pulse height (H parameter – open and full circles, in semitones) along the first phrase of Fig.2. Full circles denote the first pulse of each tone. The asterisks denote pitch in units of 100 Hz. (b) Unit pulse height for the entire 21 bars in Fig.2. H values larger than 3 semitones indicate Z pulses (great excitement)

(a)

(b)

(c)

(d)

Fig. 4. Four coloratura phrases studied in the present work: (a) from the second aria of the Queen of the Night from *The Magic Flute* by Mozart, (b) from the *Bell Song* from *Lakmé* by Delibes, (c) from *Exsultate Jubilate* by Mozart, (d) from the cantata *Lucrezia* by Händel

Many different cases were encountered, to quote only a few:

- In the excerpt of the Queen of the Night, Fig.4a, both Gruberova and Sumi Jo followed just about the assignment, with the pulse height very close to the assignment. Gruberova sings more softly, while Sumi Jo's singing is somewhat more vivid.
- The behavior of the unit pulse along the coloratura line of the excerpt from *Lakmé* in Fig.4b, is shown in Fig.5a, as sung by Sumi Jo (ERATO 4509-97239-2). This excerpt has a meter of 2/4. If we consider a grouping of four sixteenths as a unit or a *word* (or a *composite tone*), with the accent on the first sixteenth, then there are two such units in each bar. The first sixteenth of each *word* is marked as a large and full circle in Fig. 5a. We classify these *words* according to the H values (equation 1) of the first pulse. H values between zero and 1 correspond to the *Calm* category (C), between 1 and 2 correspond to the *Expressive* category (R), between 2-3 correspond to the *Excited* category, (to be marked X), and above 3 to be marked Z (the more excited tones of the *Excited* category). This convention enables us to mark the one-line performance score in Fig.5a in a similar manner to that of Fig.2.

  The phrasing, and degree of excitement in the singing of Sumi Jo can be seen, with a *word* sequence: R Z, Z X, X Z, Z R, Z R, Z R, Z. Sumi Jo sings mostly in the lower formant with only the fundamental and first harmonic partial present in the FFT spectrogram. The FFT spectrogram of Gruberova with six harmonic partials excited up to 6 kHz is more complex, particularly with regard to the morphology of her individual pulses.

  The HR parameter (equation 2), related to the speed of tightening of the vocal folds, is another good parameter useful in the description of the degree of excitement, as it also takes into account the time element and rate. Indeed, a linear relationship was found between the HR parameter (tightening speed) and the pulse height, computed from the data of Fig.5a, and shown in Fig.5b, and similarly in Fig.7, computed from the data of Fig.6a of Janet Baker. In these figures, a time unit of 166ms is used to express the values of the risetime (corresponding to pulse frequency of 6 Hz).
- In other cases of excited singing, the brain sends pulses of constant height (2 semitone pulse for 1 semitone assignment, and 4 semitone pulse for 2 semitone assignment) in the proper sub-phrase, with a transition from one type pulse to the other triggered by an event, such as a leap. Such case of excited singing is shown in Fig.6, and was also encountered in three coloratura excerpts from Händel's *Lucrezia* (Janet Baker singing).

On the micro scale of individual pulses one encounters a large variety of different patterns, an attempt at a systematic understanding of the various patterns is still under way. A large pulse obviously corresponds to a leap (assignment greater than two), and sometimes to a precursor, anticipating

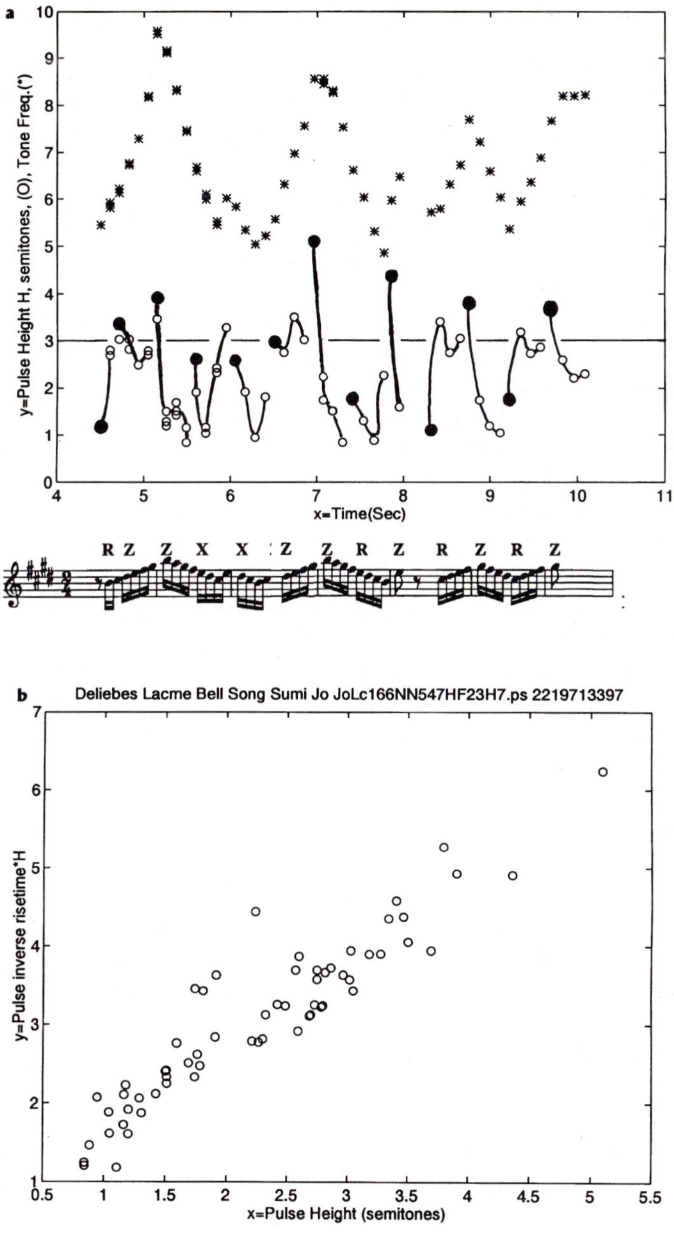

Fig. 5. (a) Unit pulse height H, as in Fig.3, for the coloratura phrase from *Lakmé*, in Fig.4b sung by Sumi Jo, (b) pulse rate of increase (parameter HR of equation 2) versus pulse height H from the data of (a), showing the proportionality between these two pulse parameters

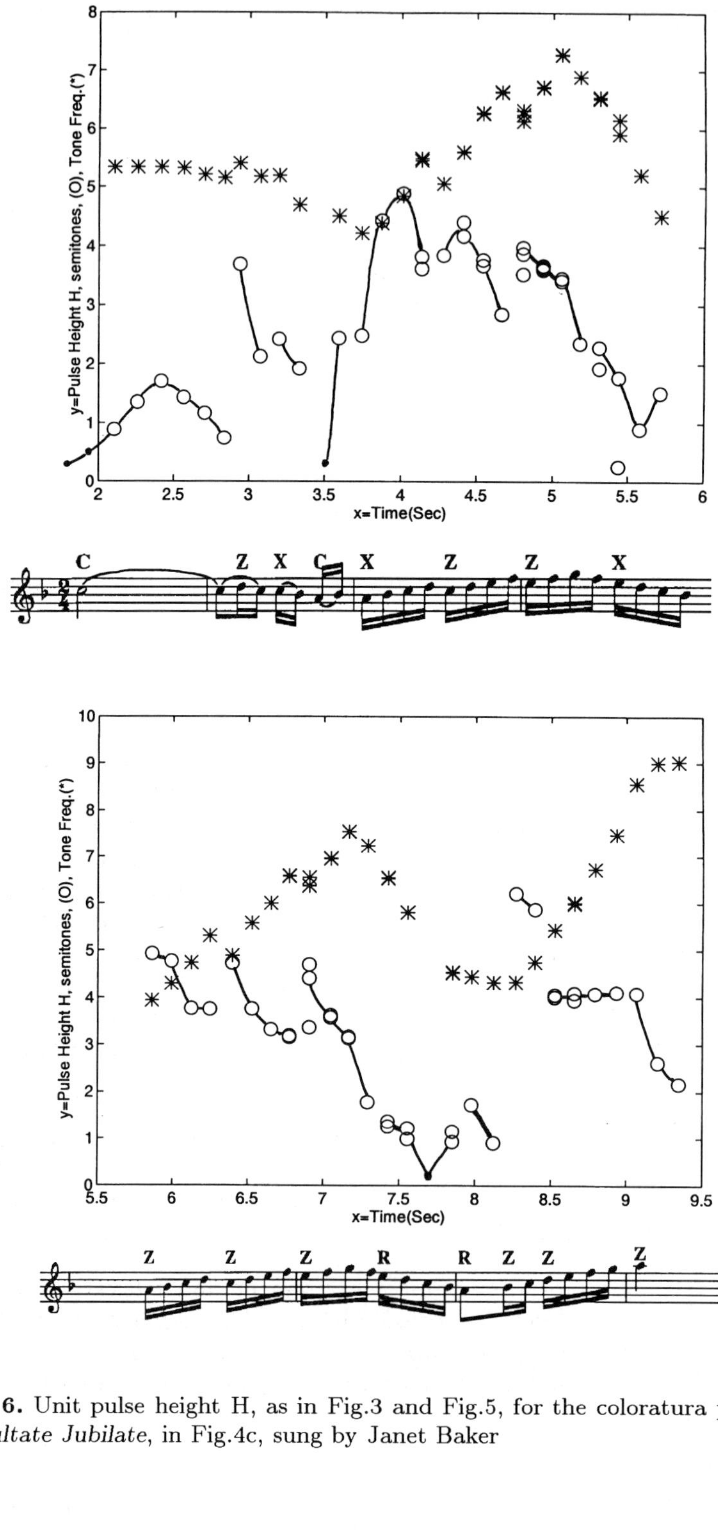

Fig. 6. Unit pulse height H, as in Fig.3 and Fig.5, for the coloratura phrase from *Exsultate Jubilate*, in Fig.4c, sung by Janet Baker

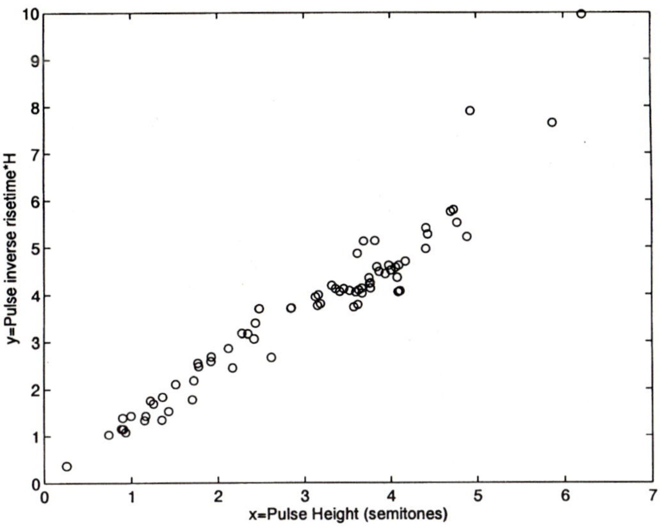

Fig. 7. Pulse parameter HR versus pulse height H from the singing of Janet Baker in Fig.6a

the coming leap, to an accented beat, to a beginning of an ascending phrase – reflecting the artist's desire to emphasize a certain tone or group of tones, or sometimes might even reflect the artist's momentary anxiety at the beginning of a difficult phrase, while at the end of such phrase, even on reaching a climax – apparently on (subconsciously) regaining confidence, the pulse height diminishes (relaxing). A pulse is also influenced by the previous one. One should realize that in these coloratura singing at fast tempi, the individual pulses are non-voluntary, and are controlled by brain mechanisms that reflect the emotional state of the singer, either personal, confronting difficult task, or acting the theatrical or literary role (Queen of the Night, The Pariah Maiden, or Lucrezia).

Figure 6 shows another pattern that can be marked as the sequence: C, Z X C; X Z, Z X; Z Z, Z R, R Z Z, Z. taken from the excerpt in Fig.4c from *Exsultate Jubilate* by Mozart, sung by Janet Baker (Erato ECD 88090), and Fig.7 shows the linear relationship between HR and H.

In conclusion, in coloratura phrases excitement and relaxation can be followed tone by tone along the melodic line, and by segmentation of the phrase into proper units (tone groupings) or *words*, legato coloratura singing can be described and analysed in terms of the emotional expression code (Rapoport, 1996).

6 Conclusions

The present work demonstrates that the individual vocal tone is composed of units, the basic unit being the unit pulse in its three varieties: zero pulse, small pulse, and large pulse. Each vocal tone contains one or more unit pulses, and always an integral number of pulses. The neurophysiological origin of the unit pulse was described. Excitement and relaxation is defined on neurophysiological grounds on the microlevel of 100-160 ms time scale by the mode of tightening of the vocal folds and the degree of synchronization with pulmonary pressure buildup. The unit pulse parameters, height H – expressed in semitones, risetime, and rate of increase HR, define the degree of excitement in singing – which is measurable quantitatively. The unit pulse originates from some brain pacemaker mechanism (biological clock), and its duration defines a unit of time counting in singing.

Vibrato in singing is described as a pulsetrain of unit pulses – its role in expressiveness and excitement is demonstrated in the singing of Maria Callas (from *Turandot* by Puccini).

Excitement and relaxation on the tone to tone level in legato coloratura singing was demonstrated in two examples, the one from the *Bell Song* from *Lakmé* by Delibes, where the virtuoso singing of the soprano Sumi Jo was analysed, and the other from *Exsultate Jubilate* by Mozart sung by Janet Baker.

The unit pulse is an ingenious brain device that serves several purposes simultaneously: (1) It provides momentary tension release after each act of vocal folds stretching (or tightening), and avoids having the vocal folds under maximum tension for long durations, thus guarding their health and assuring prolonged functioning. (2) It helps in defining the boundary between tones, improving the *distinctness* of the individual tone. (3) It is a clock mechanism, providing a means of time counting. (4) It is a measure of the degree of excitement. (5) It adds some timbral qualities to the vocal tone.

Acknowledgement

The FFT spectrograms on which this work is based were taken at the Computer laboratory of Prof. N. Tishby, at the Computer Science Institute, The Hebrew University, Jerusalem. The author expresses his deep gratitude to Prof. Tishby, whose kindness and hospitality made this work possible. The author also expresses his gratitude to Prof. E. Avitsur and Prof. E. Seroussi of the Musicology Department, Bar-Ilan University. The author expresses his thanks to Mrs S. Shatz (Physics Department, Bar-Ilan University) for her help in computer programming.

Appendix

Sound Examples (CD-tracks 70-76)

- Sound Example #1: The first phrase from the opening of the aria *In Questa Reggia* from *Turandot* by Puccini, Fig.2, sung by Maria Callas (EMI CDC 7 47966 2).
- Sound Examples #2-5: the second, third, fourth, and fifth phrases, respectively, from the same performance, and same excerpt of Fig.2.
- Sound Example #6: Coloratura phrase from the *Bell Song* from *Lakmé* by Delibes, score Fig.4b, sung by Sumi Jo (ERATO 4509-97239-2).
- Sound Example #7: Coloratura phrase from the *Exsultate Jubilate* by Mozart, Fig.4c, sung by Janet Baker. (ERATO ECD 88090).

References

Castellengo, M., Richard, G., & d'Alessandro, C. (1989). Study of vocal pitch vibrato perception using synthesis. In *13th International Congress on Acoustics, 1989.* Belgrade.

d'Alessandro, C., & Castellengo, M. (1994). The pitch of short-duration vibrato tones. *The Journal of the Acoustical Society of America, 95*, 1617–1630.

Davis, P., Zhang, S., Winkworth, A., & Bandler, R. (1996). Neural control of vocalization: respiratory and emotional influences. *Journal of Voice, 10*, 23–38.

Hsiao, T., Solomon, N., Luschei, E., & Titze, I. (1994). Modulation of fundamental frequency by laryngeal muscles during vibrato. *Journal of Voice, 8*, 224–229.

Larson, C. (1988). Brain mechanisms involved in the control of vocalization. *Journal of Voice, 2*, 301–311.

Rapoport, E. (1995). FFT-analysis and emotional expression in opera and lied singing. In L. Finarelli & F.Regazzi (Eds.), *Proceedings of the XIth Colloquio di Informatica Musicale, 1995.* Bologna.

Rapoport, E. (1996). Emotional expression code in opera and lied singing. *Journal of New Music Research, 25*, 109–149.

Scotto Di Carlo, N. (1991). La voi chantée. *La Recherche, 22*, 1016–1025.

Seifert, U., Olk, F., & Schneider, A. (1995). On rhythm perception: Theoretical issues, empirical findings. *Journal of New Music Research, 24*, 164–195.

Sundberg, J. (1987). *The science of the singing voice.* DeKalb, IL: Northern Illinois University Press.

Vennard, W. (1967). *Singing, the mechanism and the technique.* New York, NY: Fischer Ed.

Viviani, P. (1986). Do units of motor action really exist? *Experimental Brain Research Series, 15*, 205–216.

Yu, A., & Margoliash, D. (1996). Temporal hierarchical control of singing in birds. *Science, 273*, 1871–1875.

Emulating Gestalt Mechanisms by Combining Symbolic and Subsymbolic Information Processing Procedures

Udo Mattusch

Am Duffesbach 21, D-50677 Köln (Cologne), Germany

Abstract. Pattern matching based on Gestalt principles can be understood as a process of discovering an abstract *logical function*. Knowledge based processes are needed in order to handle musical information on higher levels and to describe the relations between patterns in special musical contexts, e.g. in harmonic perception. The article sketches a computational approach to a cognitive model of Gestalt perception integrating subsymbolic and symbolic methods of information processing. The model is based on the combination of a neural network (backpropagation net) with methods of inductive learning in artificial intelligence. For the special case of musical transposition, the transfer of information between subsymbolic and symbolic based methods is demonstrated. From this point of view, the recognition of musical transposition, being a special case of Gestalt perception, depends on the inference of *logical functions* which can not be achieved by standard methods of pattern matching.

1 Introduction

Natural hearing possesses characteristic attributes which can be described in terms of a series of typical Gestalt mechanisms. One of the most prominent features is the capacity to reconstruct patterns that are either incomplete or that have been altered in a regular fashion. This is achieved by means of a tendency to categorize, and a tendency to insert abstract *construction principles* based on previously acquired knowledge of the patterns. These achievements are to be understood as a special kind of pattern adjustment, characterized by the combined use of static passive representations and dynamic active levelling procedures. The invariance of pattern features resulting from this process of assimilation forms a basis for the assertiveness and stability of musical figures. The study of elementary phenomena of contour recognition and transposibility in mechanical information processing may contribute to a better understanding of natural perception. Patterns, or *Gestalts*, possess configurative features that remain constant when certain transformations are made. In the perception of melodies, certain forms of data abstraction are applied in the construction of contour and intervals. This is a prerequisite for being able to perceive musical structures as whole figures or Gestalts (Carterette, Kohl, & Pitt, 1986; Deutsch, 1982a, 1982b). The abstraction of features reduces a pattern's structural grid of inner relations to a prototypical outline that refers to a symbolic category which can then be applied and adapted to each individual case of pattern recognition. Especially when patterns are associated to each other as being *similar*, this leads to a certain latitude of appraisal which demands an interpretation that is based on knowledge of a subjective perceptual context.

Gestalt perception, as a special form of pattern recognition, is based on aspects of *unsharpness* in the formation of symbolic categories, and on the abstract representation of the formative laws of element configurations. In Gestalt formation, the activities of hierarchisation, abstraction and simplification need therefore be associated with the structure that constitutes it. In machine perception, processes are characterized by a dual character that resides (i) in a *digital* representation of knowledge describable through formal logic and, (ii) in *analog* features, especially in the area of pattern recognition which tolerates a certain amount of *mistakes*. The advantages and disadvantages of both approaches complement each other, so that computerized simulation of Gestalt processing offers the promise to produce a new quality type of cognitive oriented perception models. This becomes possible thanks to (i) the combination of fine tuned neural networks for transitions and (ii) a symbolic-abstract description of certain characteristics of subsymbolic and symbolic processes (Leman, 1989, 1993).

The following approach, in which selected features of Gestalt perception are simulated, pursues the goal of extending the available array of models for linear information processing. Aspects of Gestalt perception of musical structures must be taken into account by combining symbolic and subsymbolic procedures. It is assumed that Gestalt oriented assimilation processes result from the joint effect of knowledge based systems that operate on a logical basis (symbolically), as well as of systems that operate together in loose interconnection. When both procedures are combined, a certain amount of *unsharpness* can be handled in the formal description of certain categories that can be symbolically grasped and that belong to those procedures of pattern recognition that are based on former knowledge.

2 Musical Information Processing and Gestalt Mechanisms

The formation of *contour* and *intervals* can be understood as one of the steps in the abstraction processes of musical perception. It forms the foundation for the construction of prototypes and invariants that permit to extract regularities from unordered quantities of information. The following descriptions represent an attempt at modelling these features by combining symbolic and subsymbolic procedural levels that are meant to complement each other in their effect by exchanging information back and forth between themselves. The basic Gestalt features of contour and transposibility should serve as a starting point for elaborating a computer supported model of perception in which both symbolic as well as subsymbolic procedures are combined in a closely defined set of tasks. A symbolically operating system of models based on PROLOG (Lavrac & Dzeroski, 1993; Michalski, Carbonell, & Mitchell, 1983; Quinlan, 1979, 1986, 1990), and a processing system based on an artificial neural network (in particular a 3 layer backpropagation net, Pao, 1989) should be able to construct representations that can be linked and thus related to each other (Brause, 1991; Fu, 1993). Our approach aims to show the potential and limitations in exchanging information between these two systems.

2.1 The Learning of Contour Interval Patterns and their Transposition

To train the artificial neural network, contour interval patterns of only four *notes* are used. This restriction is applied in order to limit the amount of computation involved. The *notes* are encoded in MIDI codes that are to be associated as *bit patterns* to the corresponding encoding of contour and intervals. For reasons of efficiency, the nuanced gradations of pitch and the continuous pitch curves that we encounter in natural music are replaced in the experiment by the absolute pitches represented by the MIDI values. No interval grid is given in advance, so that the neural network may derive its own categories without any outside interference. Interval categories are understood to constitute implicit *knowledge* that evolves only during the course of learning processes. In order to avoid detrimental effects caused by linearly dependent vectors in the production of the weighting matrix, sparse coding is used inside the network (Brause, 1991). Thus, the weighting matrix of the network is expanded so that the vectors of the patterns that are to be learned possess as few common components as possible. Thanks to this step, it is possible to represent one *bit* by one *neuron*.

In order to obtain a set of training probes, random successions of notes are produced that possess a prototypical structure without repetition. Only half of all examples are actually used for training. The network can be subsequently tested with the other half of the examples. This overall procedure is necessary when modelling abstract achievements such as interval formation. Furthermore, it permits to observe the network's learning behavior. An artificial neural network extracts contour and intervals from the tone sequences by comparing, in each round of the experiment, the vector in the output layer with a goal vector in the pattern that is to be recognized. If, following a satisfactory number of training rounds, the *error* of this process has fallen below a certain minimum, then the network has become able not only to assimilate the tone sequences belonging to the training corpus, but also unknown sequences or their transpositions.

Figure 1 shows the input and output matrix of the network. The matrix INPUT contains two times four bit, for two sequences of four tones. The first sequence is located in lines 1 5 and the second sequence in lines 6 8. The matrix OUTPUT contains, in lines 1 6, seven bit for contour and inter-

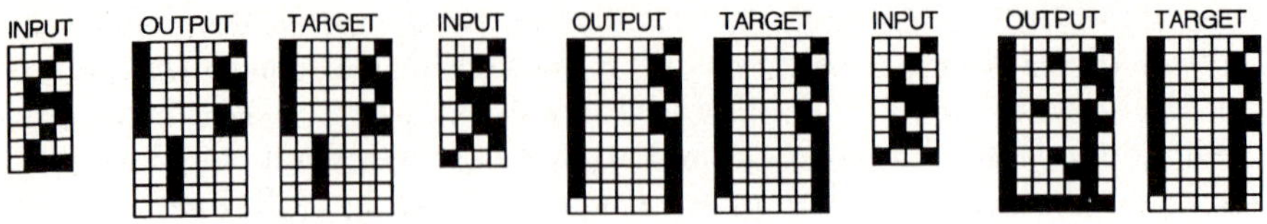

Fig. 1. (a) Input-Output-Target example of correctly recognized pattern from the training corpus. (b) Input-Output-Target example of correct assimilation of an unknown pattern. (c) Input-Output-Target example of incorrectly recognized unknown pattern

vals (three and four bit respectively) of both sequences as well as four times seven bit for the transposition intervals (in lines 7 10, for testing purposes). The eleventh line indicates the generalized transposition interval (one times seven bit) between the two sequences. The matrix TARGET contains the reference vector to which the vector of the artificial neural network's departure data is constantly compared during the learning process. A black-filled square indicates that the corresponding weighting's intensity has surpassed a certain threshold limit. In Fig.1a, for two identical tone sequences the value "0" (no bit placed), is given. In Fig.1b the value "1" (first bit placed), is given. The incorrect pattern in Fig.1c shows some characteristics of *tolerant* pattern recognition. When the training is left incomplete, intervals that are similar in size can be reconstructed up to a certain extent when the weights of the hidden layer encompass a certain interval class. A considerable amount of training is necessary in order to learn *abstract* operations such as transposition, since a representation of the functional relationships between the patterns must be synthesized in the network's hidden layers.

Transposition favorably fulfilled requires a connection between symbolic and subsymbolic procedures in a model. The pattern transformation that is carried out in transposition results in a special kind of identity of two patterns that seems self evident in natural perception. If we want to try to emulate it by technical means, we need to grasp the abstraction processes involved in it. The function of a transposition can be emulated with a neural network, and it can be symbolically described in a formally operative way using operators that include, for instance, a factor for the transposition interval. This entails the question as to what kind of communicative exchange processes can exist between these two areas. The network's error behavior can help us to describe the characteristics of the joint performance of subsymbolic and symbolic processes. The exchange of information between the artificial neural network and the PROLOG system is made possible by adapting the symbolic representation structures and the values in the network's weighting matrix. These values are incorporated into a directed labeled graph. The graphs of patterns that are similar to each other are plotted against each other and stand for the probability of the firing of neurons at the local level (Brause, 1991; von der Malsburg, 1988). By this method, the *activity* of neurons is represented at a level that can be processed further in PROLOG. The graph costs of the relationships between positive and negative antecedents serve to derive rules of the *if-premise-action* type (Gallant, 1988). Similarly to the procedures of exact and inexact reasoning in expert systems, the values indicate the confidence factor for the rules' attributes and premises. By conjuncting premises, rules of the type $A1 + (Value_1+), ...Am + (Value_m+), ...A1 - (Value_1-), ...An - (Value_n) - C(Value_C)$ can be built, in which the negative values indicate the degree of absence of an attribute. This simple procedure is meant to be part of a tryout phase and is subject to restrictions. For instance, disjunctions or negations such as *A1, A2 not(C)* are not possible. The transformation into PROLOG predicates is carried out by inductive learning and by conceptual clustering (Cheng & Fu, 1985; Fisher, 1987; Gennari, Langley, & Fisher, 1989), so that an induction tree and a set of predicates in PROLOG can be produced.

The results of the *logical* analysis are then transferred to the weighting matrix of the neural network by writing the costs found at the graph intersections. In doing so, all values that come into the neighborhood of localized minimum values are lowered according to a threshold of 75% of the minimum value. This imprints a grid in the weighting matrix, and thus certain patterns can readily be processed. However, in order for the network to be able to tolerate errors, overlearning is also needed so that it can refine the framework with which it has been previously imprinted. Subsequently, new transposition patterns are written into the weighting matrix, according to the symbolical values that have been derived by induction from the training results. Figure 2b shows the learning curve of the pattern PATTERN1 that has been imprinted: it was not among the training probes and it has a plateau in the error curve.

The network's learning mechanism has no weights that would enable it to separate the patterns from each other. As mentioned above, the step of imprinting resets the weights of the connections and interrupts what would otherwise become an endless matching process. The error specifically associated with this pattern can be immediately reduced, a learning effect cannot be observed until a context is achieved, consisting in the neighboring connections made to the other patterns. To assure that new patterns can be learned reliably, it is not enough to simply imprint the values into the matrix. When there are different amounts of pattern probes, in each case dependencies on the weights of the other training probes must be established so that new patterns can be trained while taking into account the context formed by the preceding ones. Figure 3b shows how a new set of patterns is learned. At first, there is a high degree of error which then constantly decreases: plateau-like adjustment stages can be observed. In order to fully adjust all weights, the network must go through all patterns once again (Fig.3c). After a constant

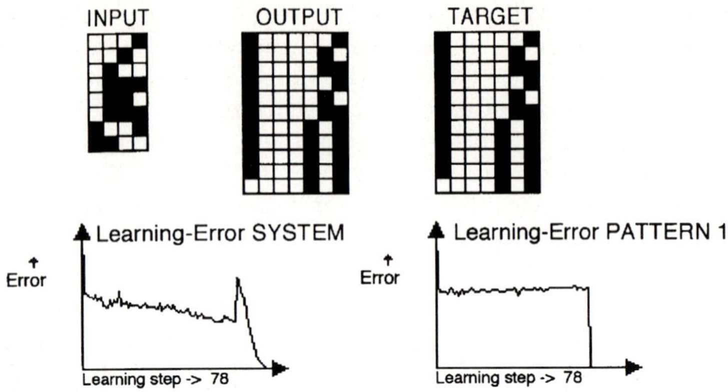

Fig. 2. (a) Error curve of an artificial neural network with an altered weighting matrix. (b) Error reduction with the inclusion of pattern PATTERN1.

decrease in error similar to that in normal training, a certain minimum of mistakes is reached that cannot be resolved. In Fig.3d, this is shown by a *pulsation*: the error quota cannot achieve an absolute minimum and swings back up to a level that had been attained at previously.

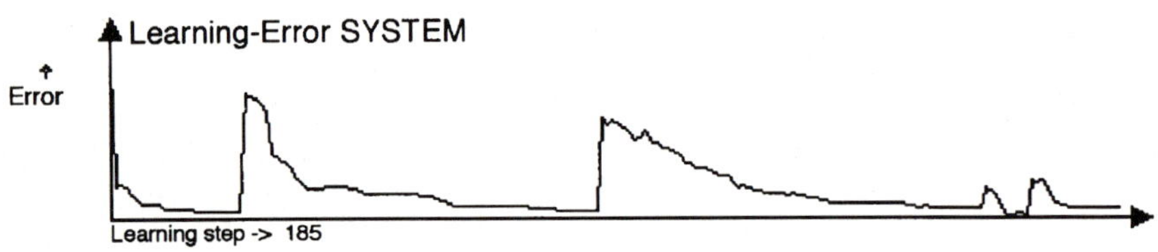

Fig. 3. The error evolution can be divided into several areas: (a) error curve in normal training, (b) manipulation of the weighting matrix, (c) repeated training with all probes, and (d) final adjustment processes

The exchange of information between the network and the PROLOG system is based on the selection of *traces* of already trained patterns in the weighting matrix, which is transformed into a graph and propositional representations in order to be processed symbolically. For the production of rules, a learning technique (conceptual clustering Fisher, 1987) is used that can derive simple PROLOG predicates for the description of the relationships that have been learned. At this stage of the research, it is not yet possible to derive combined or recursively interconnected organized functions, since this would require either the setting of certain rules in advance or constantly having to rearrange the network structure.

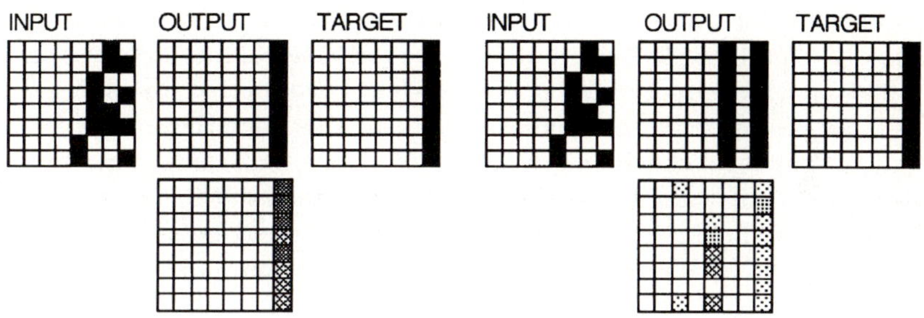

Fig. 4. (a) Input-Output-Target example of the *trace* of a learned transposition interval "1" (first bit placed from a total of eight bits; contour bits suppressed), as related to a melodic probe "1 8". (b) Input-Output-Target example of an erroneous pattern with weighting that is not yet adequate

Figure 4 represents a sector of the weighting matrix reaching from a neuron of the suppressed layer to the neurons of the output layer and containing the weights for the interval value "1". In order to simplify the illustration, the usual ordering of bits in data words was used. In the practical experiment, however, bit compression must be carried out for efficiency reasons and the *traces* are not represented in a graphical way. The intensity of the weights is symbolized by the degree of blackness in the filling pattern, so that, once training is over, one can produce a more or less clear image of the *bit pattern* that was learned. Such a *trace* of activated neurons can be transferred to the symbolical level and, after having been transformed into propositional representations, it can be translated into PROLOG predicates through inductive learning. A left-recursive version of the transposition predicate, slightly altered for readability *(transpose(InputList, OutputList, TranspositionFactor))* runs as follows:

```
transpose([],[],_):- true.
transpose([_4093|_4094],[_4095|_4096],_4097):-
      transpose(_4094,_4096,_4097),
```

```
      plus(_4093,_4097,_4095).

plus(0,Var,Var) :- true .
plus(Var,_4093,_4092) :-
      inc(_4094,_4092),
      plus(_4095,_4093,_4094),
      inc(_4095,Var).
```

In a future development stage of the model in the future, the possibility of producing left-recursive predicates should permit the use of more powerful grammar formalisms in order to describe the derived structures more adequately (Williams & Zipser, 1989). The predicate *plus/3* corresponds to a context specific representation of knowledge, connected by shifts to antecedent successor relations (clause *inc*/2). In this case they are applied to intervals at the level of whole numbers, but simple measures could be taken in order to permit them to describe other symbolic relations between signs. The learned predicate *transpose/3* has the characteristics of conventionally programmed PROLOG predicates, as the following command examples show:

```
?-transpose([1,2,3],Transposed_list,1).
Transposed_list = [2,3,4]
yes
?-

?-transpose([1,2,3],Transposed_list,4).
Transposed_list = [5,6,7]
yes
?-

?-transpose(Input_list,[5,6,7],4).
Input_list = [1,2,3]
yes
?-

?-transpose([1,2,3],[5,6,7],Transposition_value).
Transposition_value = 4
yes
?-
```

An input list can not only be transposed by a certain factor. Also the input pattern can be deduced from either the output pattern or the transposition factor. Such *pattern constructions* are only possible after considerable training and can only remain correspondingly reliable by remaining within the problem area that was trained.

3 Discussion

In order to point towards areas that could not yet be realized in the approach described here, we discuss some theoretical issues concerning further possible developments and some hypotheses about Gestalt oriented perception.

In a broad interpretation, the case of a transposition can be seen as a variant of the Gestalt law of *common fate* (Deutsch, 1982a). The law postulates that elements with the same characteristics are grouped together, for example the parallel movement of all elements in the same direction in the pitch domain. For modelling reasons, the matrix in Fig.4a contains in each case the interval difference related to an element; this easily permits grouping all pattern elements together since they have been transformed in the same neural network.

Applied to a hierarchical system of representations, this would permit a convenient disposal of structures and substructures along various representational levels. Especially the grouping together of elements according to the criterion of the smallest difference to a neighboring element in a newly focussed representational level leads to a convenient ordering of complex structures and to more rapid recognition. The alternative various shifts of parts of a musical pattern would entail a more complex hierarchical representation and a more complex pattern in the weighting matrix; this would demand a greater amount of processing and would make the pattern harder to recognize (Stoffer, 1981). In Fig.2, the rapid decrease of system error effectuated by directly imprinted the weights for a new pattern, showed that within certain limits, the new pattern could be recognized *immediately* even without the supporting context of the other patterns. Since additional *knowledge* from the symbolical level can be brought into the network's representation, this might offer a hypothetical explanation for the abrupt full presence effect of well known or especially assertive musical motifs. The example of transposition shows that different forms of *knowledge* that complement each other can be added. In the model, this is done by means of the predicate *plus/3* for shifting. Such *background knowledge* is characteristic of Gestalt-like perception and it is one of the preconditions, along with context specific representational structures, for the successful recognition of patterns in rapidly changing aural or visual environments.

4 Limitations

The combination of symbolic and subsymbolical operating processes as studied here, does have its limitations. The relearning process can require a considerable amount of computation. The rearrangement of the network structure for e.g. other encodings of the input and output level involves interferences and loss of information. In addition the network cannot apply what it has learned to element sequences of any length; it can only adjust learned values within certain limits. The greater the differences with respect to the values it has learned, the larger the error in its *estimates*. In order to recognize the probe function at the symbolical level, the structure of the network must be known in advance. The extension of the network's capacity by means of newly derived functions is thus limited to the supplementary training of those values that can not be reliably learned for the transposition factor. Another issue concerns the fact that the logical component can induce *factors* for all practical cases, but in order to achieve this it has to produce permutations of the element sequences offered, since otherwise the necessary formation laws could only be derived incompletely. The advantages of symbolical representations such as the logical inferability of missing information, the transferability of *functions* and other representational areas, as well as the possibility of connecting them syntactically/semantically to other types of knowledge, can all be applied to the network, within certain limits, by making use of correspondingly organized network layers. This entails that in future models the network would need to operate through a certain *time window*, the contents of which would be evaluated on the symbolic level in certain time intervals. This would also permit an evaluation of the relations between representations emerging at different points in time. The network arranges information temporarily as in short term memory and then triggers knowledge based *top-down* processing in the symbolic system, which in turn imprints its own results back into the net. The interplay of adjustment and exchange between both processing systems results in a kind of pendulum motion between upward and downward processes in both the symbolical and the subsymbolical working areas that have complementary features. The approach described here gives special emphasis to this dynamic component of Gestalt oriented perception processes and is thus quite different from models of pattern recognition that use pure mechanical matching only.

5 Conclusion

By designating transposition as a special case of Gestalt oriented perception, the approach presented here attempted to put into practice the basic possibilities of combining subsymbolic and symbolic procedures. An artificial neural network can learn operations such as transpositions. In doing so, it converges relatively quickly, but it needs to be trained considerably until reliable functionality is achieved. The correct *factor* cannot be learned for all transpositions as in conclusion by induction; rather, the values for each case must be painstakingly learned. The area that operates symbolically can logically derive *useful* representations by inductive learning and describe those relations that it has perceived using predicate logic.

The combination of a neural network with a symbolic processing system is supposed to reflect the dual character of musical figures. As opposed to technical data patterns, musical figures are functional wholes with certain dependencies that are represented and processed in qualitatively different forms. The adjustment procedures between the subsystems of the model simulate the interdependence of cognitive processing areas that are networked together and that have common access to an intermediary representation. The method of direct information exchange, as used here, represents an attempt at emulating a preliminary stage of *communicating* model stages at the level of functional isomorphisms. These adjustment processes are to be understood as a *to and fro motion* causing dynamic changes of state in all areas of a processing system and, as a limited form of self organization, they can optimize a system's performance. Such *to and fro motions* have an added quality that complements the features of the subsymbolic and symbolic processing systems and thus emulates an essential mechanism of the perception of musical structures.

References

Brause, R. (1991). *Neuronale Netze.* Stuttgart: B.G. Teubner.

Carterette, E., Kohl, D., & Pitt, M. (1986). Similarities among transformed melodies: The abstraction of invariants. *Music Perception, 3,* 393–410.

Cheng, Y., & Fu, K. (1985). Conceptual clustering in knowledge organisation. *IEEE Transactions an Pattern Analysis and Machine Intelligence, 7,* 592–598.

Deutsch, D. (1982a). Grouping mechanisms in music. In D. Deutsch (Ed.), *The psychology of music.* New York, NY: Academic Press.

Deutsch, D. (1982b). Organizational processes in music. In M. Clynes (Ed.), *Music, mind and brain: The neuropsychology of music.* London: Plenum Press.

Fisher, D. (1987). Knowledge acquisition via incremental conceptual clustering. *Machine Learning, 2,* 139–172.

Fu, L. (1993). Knowledge-based connectionism for revising domain theories. *IEEE Transactions on Systems, Man and Cybernetics, 23,* 173–182.

Gallant, S. (1988). Connectionist expert systems. *Communications of the ACM, 31,* 152–169.

Gennari, J., Langley, P., & Fisher, D. (1989). Models of incremental concept formation. *Artificial Intelligence, 40,* 11–61.

Lavrac, N., & Dzeroski, S. (1993). *Inductive logic programming.* New York, NY: Elis Horwood.

Leman, M. (1989). Symbolic and subsymbolic information processing in models of musical communication and cognition. *Interface – Journal of New Music Research, 18,* 141–160.

Leman, M. (1993). Symbolic and subsymbolic description of music. In G. Haus (Ed.), *Music processing.* Madison, WI: Oxford University Press, A-R Editions.

Michalski, R., Carbonell, J., & Mitchell, T. (1983). *Machine learning: An artificial intelligence approach.* Palo Alto, CA: Palo Alto.

Pao, Y. (1989). *Adaptive pattern recognition and neural networks.* Reading: Addison-Wesley.

Quinlan, J. (1979). Discovering rules by induction from large collections of examples. In D. Michie (Ed.), *Expert systems in the microelectronic age.* Edinburgh: Edinburgh University Press.

Quinlan, J. (1986). Induction of decision trees. *Machine Learning, 1,* 81–106.

Quinlan, J. (1990). Learning logical definitions from relations. *Machine Learning, 5,* 239–266.

Stoffer, T. (1981). *Wahrnehmung und Repräsentation musikalischer Strukturen. Funktionale und strukturelle Aspekte eines kognitiven Modells des Musikhärens.* Unpublished doctoral dissertation, Bochum.

von der Malsburg, C. (1988). Pattern recognition by labeled graph matching. *Neural Networks*, *1*, 141–148.

Williams, R., & Zipser, D. (1989). A learning algorithm for continually running fully recurrent neural networks. *Neural Computation*, *1*, 270.

Interactive Computer Music Systems and Concepts of Gestalt

Paul Modler

Liegnitzer Str. 22, D-10999 Berlin, Germany

Abstract. In the past few years of the development in the field of electro-acoustic applications, it turned out that new input and output options are needed to make a convenient control of technical and aesthetical processes possible. This particularly holds for live performances. As a consequence thereof, several different hard- and software developments emerged recently. Many of them are closely related to general improvements of the computational industry (such as mouse or joystick), others have been developed for a particular occasion, such as for a certain artist, a performance, or a concert. This paper investigates the potential effects of these new developments for musical applications. The paper is organized as follows: Starting from a description of interactive computer music systems, I will introduce Gestalt-theoretical concepts in order to allow for a characterization of the basic properties of these systems. Thereafter I will apply them to the interaction of man and music systems. Gestalt concepts will be introduced to describe this interaction. Finally, I will illustrate the criteria I am going to set up in this paper, taking a dataglove sound synthesis configuration as an example of such a system.

1 Introduction

The term *interactive* is used inconsistently in the different branches of the so-called new media and computer business. A common definition of interaction by Rowe (1993b, p.1) proposes that an acting object, here, a human being changes the behavior or the state of another object, here, an instrument, machine, or human being. Conducting a computer program with an input device, such as a mouse or keyboard, is often referred to as the prototype of an interactive system. But a musical instrument, which is played by a musician, can as well be viewed as such a system. In this section I introduce the theoretical background of interaction using Gestalt thinking.

2 Input Devices: Hardware

The term *interactivity* has received a great deal of attention in the computational field. This is mostly due to the impossibility of attaining interactivity in the past. There were many reasons that made such an interaction hard, such factors as a troublesome input and output (magnetic tape, punch cards) or a long reaction time of application programs.

As a consequence of more recent achievements, such as the enormous increase of the computational capacity, better input/output possibilities, and the improvements of modern software tools, it turned out that computers could behave like other machines and instruments, which are characterized by a high degree of interaction. Their degree of interaction can be compared to a typewriter or a bicycle, instruments whose interactive potential increased with the technical improvements they underwent or which are interactive by their structure.

Additionally, the need of an improvement of the input/output possibilities increased. Integration of speech, force and motion, high resolution screens are examples of achievements that were made in order to allow for a more comfortable and faster interaction with the computer.

This development corresponds with progresses in the area of computer music and electroacoustic music (Hunt, 1991; Behrman, 1991, 1991). Already since the early seventies, analogous musical devices that could be operated intuitively, have been common. Well-known examples for this are the Moog-synthesizer of 1970, and much earlier Theremin's instrument that generated sounds through ether waves (1924). From the point of view of interactivity, these devices were superior to the generation of digital devices that were introduced in the early eighties. A handicap of those digital or digitally controlled instruments, like the Yamaha DX7, were the reduced immediate control-possibilities. Editing a sound on the DX7 was an exhausting step by step procedure, whereas the sound editing of the older MOOG-Synthesizer was done directly using a potentiometer for each sound parameter. When the digital computer and other digital electronic devices were introduced, the potentiometer was replaced by a field of buttons. However, while the buttons were a cheap control device and appropriate for digital environments, the intuitive control over the sounds was to a large extend lost.

Further progress with digitalization also brought along a process of standardization. Sound and control data now appear in a digital shape and can be viewed as universal units of information (Balzano, 1989; Schrader, 1991). Therefore one should expect it to be easy to connect input devices to produce a digital stream of data to a digital sound generator. However, even with a well defined interface such as MIDI, the obstacles sometimes seem to be impossible to overcome.

Several developments in that area have to be distinguished:
One tendency is to construct input devices that try to imitate conventional instruments as good as possible. For example a keyboard imitating a piano, wind-controller imitating a saxophone or a flute, digital drum set imitating an acoustic drum set. This correlates with attempts to imitate the sound of acoustic instruments with synthesizers as closely as possible, like commercial sampling devices or sound modules which use physical modeling, like the Yamaha VL1.

Good imitation results have been achieved concerning instruments that already have some inherent discretization, such as a piano or a drum set. Instruments that either have a more continuous sound generation, like a trumpet or a violin or whose sound generation is connected to the human body, are more problematic, both with respect to the control-input as well with respect to the generation of sounds.

A different approach is the attempt to construct new input devices such as the hands of Waiszvisz (1993), violin controller of J.Rose (1993) the infrared motion sensor by Karkowski (Lee, 1996), the Biomuse system of Tanaka (1997) and so on, who all try to consequently exploit the idea of the possibility of coupling digital devices mentioned above. All these approaches share the experience of playing electro-acoustic and digital-acoustic instruments, as well as the knowledge of the possibilities of digital sound generation.

These factors explain the need to abandon the traditional button field and knob- or slider- potentiometers in order to achieve a faster, more intuitive access to the sound parameters. Additionally, the aesthetic component property of each of these instruments has to be considered too. The way a performer is perceived when playing his instrument as well as the way he is playing the instrument, have additional influence on sound generation and perception. An electronic sound generator will be played by a performer in different ways, as well as it will be perceived by the recipient in different ways, depending on the input device which is used. For example, there will be a different creative and perceptive sensation if a wind-controller like the WX7 or a standard keyboard controller is used to play a synthetic flute sound.

In between these two developments a third kind of approach aims at improving existing instruments with additional input potentials. This idea has been applied to electro-acoustic instruments of the consumer area, such as a ribbon-controller, interactive displays, and potentiometer fields, but also to conventional acoustic instruments, for example the MIDI accordion or the MIDI piano.

3 Data Transmitted Processing Software

Most of the data that is transmitted by electro-acoustic instruments is composed of either switching information or quasi-continuous information[1]. Some features of digital transmission of information, however are uncomfortable, especially in musical applications, such as the abrupt changing of behavior when leaving the transmittable range of values (overflow), and the quantization of the range of values. These shortcomings should be improved on, although some composers and musicians consciously use these puzzling sounds as stylistic effects in their music.

Closely related to the input device and the data stream is the software which produces the output of the system. First of all, models of sound generation, like frequency modulation, digital sampling, physical modeling, analog synthesis, and others, have to be mentioned in this context. Procedures of sound modulation, like a harmonizer effect or others are equally important.

[1] Like Note-On, Note-Off, Controllerdata.

Besides this, there are tone generation procedures (tone in this context represents both, a paradigm for a classical note as well as a MIDI note event) which are normally characterized as procedures of algorithmic composition, such as *M & Jam Factory* (Ameres, Zicarelli, Chadabe, Offenhartz, & Widorf, 1988), *"MIDI"-MAX* (Puckette & Zicarelli, 1988/1995), *Cypher* (Rowe, 1993a), *Autobusk* (Barlow, 1990), or others.

Although both areas, tone generation and sound generation, appear to be distinct, only little structural differences can be stated. Rowe (1993b) explicitly mentions different procedures of tone generation, but does not comment on interactive sound generation or sound modulation. Generally one can state that procedures of tone generation operate in a more abstract tone-space. But the usual hierarchization or rather, abstraction approach is only applicable with some restrictions. Compositions which are characterized by a precise sound space will require a sound generation with a corresponding complex hierarchical level.

Both procedures operate with user input in order to generate the output stream and use internal setups to shape this output stream. In sound generation environments these procedures are the sound parameters and the synthesis algorithm, in tone generation environments these are the algorithmic parameters.

Procedures of tone generation, however, use a sound generation device in order to make the tones audible that are produced by the tone generation algorithm. Some procedures of tone generation analyze the incoming datastream, like in programs as Cypher (Rowe, 1993a) or Digital Piano (1984). This analysis attempts to imitate the human musical perception, or rather to imitate a hearing behavior the composer or programmer want. Examples are the detection of phrases, tonal centers, meter, rhythm and so on. Therefore, the system should be granted some amount of intelligence. This analysis section of course depends to a high degree on the procedure of analysis and the data to be analyzed.

But such an analysis of incoming data can be done for procedures of sound generation as well. The motivation behind this is the desire to use an acoustic signal as an input. Spectra, formants and envelopes are potential targets for feature-extraction.

For future developments, special attention should be devoted to the intelligence of such systems. Camurri and Leman (1997, this book) propose the concept of *multimodal environments*. Their multimodal environment HARP aims at the integration of dance and gestures, audio signals and control equipment. These aims fit very well into the model of interaction that will be proposed below. The observation of the human output, for example dance, by sophisticated agents can be seen as intelligent decoding of the data transmitted by the human. There is a strange interaction between this observation and the aesthetic concepts of the performer: Although this observation is of high technical and aesthetic interest for the variable degrees of intelligence, at the same time it reflects the aesthetic concepts of the performer himself as well as the technical possibilities and individual features of the environment used.

4 Model of Interaction

In this section a simple schematized model of an interaction is discussed. The model represents a 2-channel (2-times simplex) communication system which can easily be transferred to any interactive system, for example, a musician playing an instrument, a user working on a computer system, or a driver conducting a car.

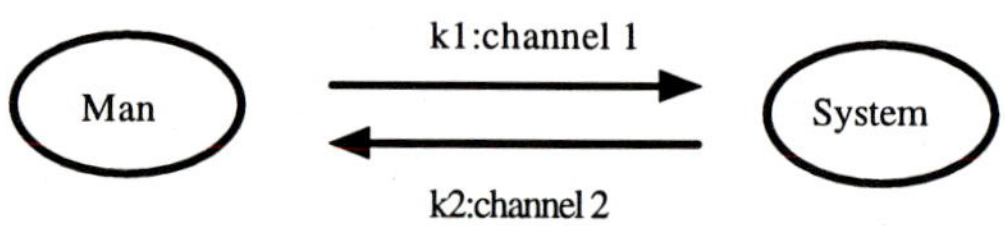

Fig. 1. General Model of an Interactive System

For a music system the model would be designed as shown in Fig.2. The information transmitted via the channels is used to describe and evaluate the communication. As a first step, let us assume that k1 is a data stream, which can be either MIDI or audio, and k2 is a system response in pure audio. A concrete example for this is the coupling of a slider to the pitch of a synthesis module, or rather to a stochastic process to control the random-depth of the process and thereby the feedback to the performer. Such a control is inevitable for an initial evaluation of the interactive communication, especially for evaluating questions such as the value of the bandwidth of the channels or the rate of transmission.

However, these criteria alone do not suffice as a basis for analyzing an interactive system. Therefore, the model will be modified in order to account for transmitting objects of an higher level of abstraction (and more animated from a musical point of view). With that, we have two revised definitions of the term *channel* at hand:

1. A channel is related to a mental idea (imagination) and the corresponding concept of its realization. k1 then stands for a human transmitting information to the system and intending Gestalt 1.
2. A channel is related to a perceivable change of the systems behavior. k2 then stands for a system transmitting information to a human and evoking Gestalt 2.

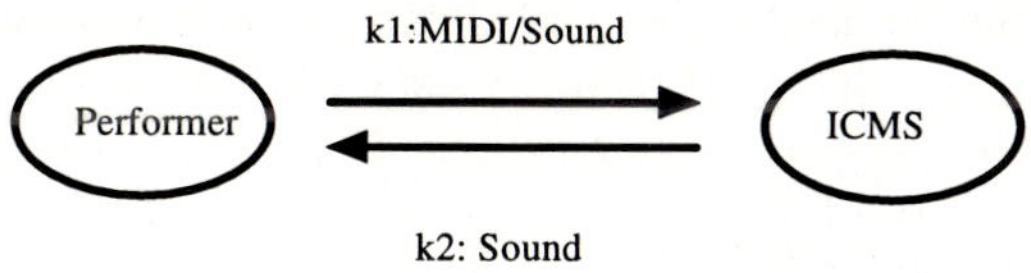

Fig. 2. Model of an Interactive Computer Music System

4.1 Concepts of Gestalt

This model does not look at the concrete stream of data transferred, as we did in the first step, but instead at a more abstract information content. For k1 (human being to system) several features for the channel-information should be combined. By this, an anticipated imagination of a sound can be connected to the imagination of a certain algorithm of sound generation, or to the changing of one or more certain parameters, but also to a certain handling of the instrument, or experiences which were made with that instrument. This imagination I call Gestalt 1.

For k2 (system to human being) a similar model holds, but one has to consider that the performer has to decode more abstract information out of the sound waves. In other words, both the intended Gestalt 1 and the Gestalt 2, which is evoked by the system, are produced within the performer.

Although Gestalt theory is normally used in the area of perception, it can be used here adequately in order to interpret the more abstract information mentioned above as a Gestalt-quality itself. Several inconsistent features (for example, the concept of auditory perception, the algorithm for the generation, the setting of parameters, body and device, position and motion, aesthetic features) are interpreted as one single Gestalt-quality (following von Ehrenfels, 1890/1974, p.38).

The transmitting channel k1 has to allow for an encoded transmission of such a Gestalt. The receiver (computer system) has to be able to interpret the information and to produce a corresponding output. I will not go into details about the Gestalt-theoretical properties that may follow from this assumption.

4.2 Oversummativity, Inverse Oversummativity, Assimilation

According to one of the basic principles of Gestalt theory the whole is greater than the sum of its parts. Like the other principles, this was set up on the basis of investigation of the perception. Other definitions of this principle say that the whole is different from the sum of its single parts (C.Pratt in Köhler, 1971, p.7). These differences don't matter here.

More importantly, in the area of aesthetic production one has to deal with a contrasting tendency by which Gestalt-building allows for an addition and abstraction of the divergent features. It prepares the performer to deal with control processes of a complex and temporal nature, in an intuitive and creative way. This works for all aspects of the performance, senso-motoric as well as intellectual, aesthetic, and emotional ones. An example for this is the performance of a virtuous piano-piece, which the performer can hardly produce as a series of single notes and chords, but rather as a sequence of more abstract groupings and parts which can have different features.

The term auditory images was introduced by Jackendoff (1987) in order to describe the mental events behind that. A similar abstraction is assumed by Rowe (1993b, p.121) for his concept of feature space. However, for more complex demands, the concepts introduced by Rowe (high and low volume, high and low pitch, and so on) will not suffice.

4.3 Group Formation - Inverse Group Formation

Proximity - Continuity The term continuity characterizes a feature which should be achieved in interactive systems, namely that one Gestalt associated to k1 can be transformed into a another Gestalt associated to k2 by continuous changes (cf. the term of controllability, Föllinger, 1984, p.302). In other words, the system should behave deterministic, that is, small input changes should trigger small changes of the result. For example it should be possible to change the mid-frequency of a filter continuously.

Discontinuity I employ the term discontinuity in order to characterize the opposite feature. Discontinuity allows for changing of Gestalt 1 to Gestalt 2 in an abrupt fashion.

Relatability (Similarity) Gestalts which can be produced in an interactive music system have to be located in a Gestalt-space in such a way that similar user inputs should result in similar system outputs. For example, the control-input at two different times should result in the same effect (i.e. the change of parameters) in a MAX-patch. The correspondence between the Gestalts of both k1 and k2 is defined by k1 and k2. A satisfying evaluation of the interaction demands a sufficient correspondence of Gestalt 1 with Gestalt 2, as well as a sufficient small interval of time in which the comparison of Gestalt 1 to Gestalt 2 can be carried out by the composer.

Responsiveness Responsiveness is an important feature of interactive systems. Responsiveness demands that the amount of time which an interactive system needs in order to respond to an information transmitted by k1 (mental idea and realization) be included in the evaluation. For example, a composer writing for an ensemble represents a system with low responsiveness, whereas a musician and the instrument he plays normally constitute a system with high responsiveness. Computational systems vary with respect to their responsiveness depending on their computing power and the complexity of the task. Besides this, it has to be considered that the transmission channel (i.e. Internet) requires some extra amount of time in computational systems. The reaction time has to be within the limits of aesthetic, senso-motoric and perceptive constraints. It can vary depending on the Gestalts to be transmitted. Processes that are immediately coupled in time (i.e. the transposition of a performed motive) require a reaction time below 10 ms. Still, processes with a correspondingly smaller reaction time can be perceived (beats). Concerning processes that are not temporally coupled, even longer reaction times can be tolerated (i.e. the echo of a performed motive). It is obvious that a successful interactive music system should grant real-time capacity.

Variability An interactive music system is expected to make different Gestalts and Gestalt properties possible. The spectrum of these potential Gestalts is part of the utility of the system in different performance modes. For example there should be various algorithmic potentials for analysis and synthesis available, like Cypher (Rowe, 1993a). Gestalt 1 can be interpreted as an abstract, multi-dimensional sign. The amount of signs which can be transmitted, as well as the possibility of a continuous change between these signs are indicators of the power of k1. In k2 the Gestalt 2 (system part, responding part) is evaluated analogously.

Autonomy A question that has to be accounted for is the possibility of a local intelligence within the musical system. The question, whether an interactive system is provided with intelligence or not, results from the properties of k1 (or Gestalt 1, respectively) in the sense of a musical, algorithmic autonomy. If the expected answer is not previously contained in the mental conception of the user, the system has to provide the rules and intelligence that is required in order to derive the resulting answer in k2. This is done in Cypher (Rowe, 1993a), M & Jam Factory (Ameres et al., 1988), as well as in Autobusk (Barlow, 1990)

4.4 Summary

As a summary of the discussion it has to be noted that the criteria described above only partially fulfill the requirements of Gestalt theory and they can only to a certain degree be derived from them. This is not surprising, since these criteria do not result from the point of view of perception analysis as Gestalt theory itself. Rather, my criteria result from the point of view of synthesis, in other words, the conduction of a music instrument or the aesthetic creation, respectively. Anyway, Gestalt phenomena are one important aspect of complex control and adjustment events in general and they have to be applied here as well.

5 Sensor Glove: An Example of a Modeling of Musical Gestalt

5.1 Configuration of the Equipment

The Sensor Glove is a data glove developed by F. Hofmann at the Institute for Real-Time Systems and Robotics of the Computer Science Department at the Technical University of Berlin and tested for musical applications at the Federal Institute for Music Research in Berlin in the TEMA project.

Inside the glove, inductive distance sensors are employed for the recording of movement and *Piezo* sensors are employed for the recording of acceleration. With respect to the definition of channels given above, the sensor glove is an efficient medium which can control synthesis procedures, like granular synthesis, physical modeling, as well as algorithmic processes.

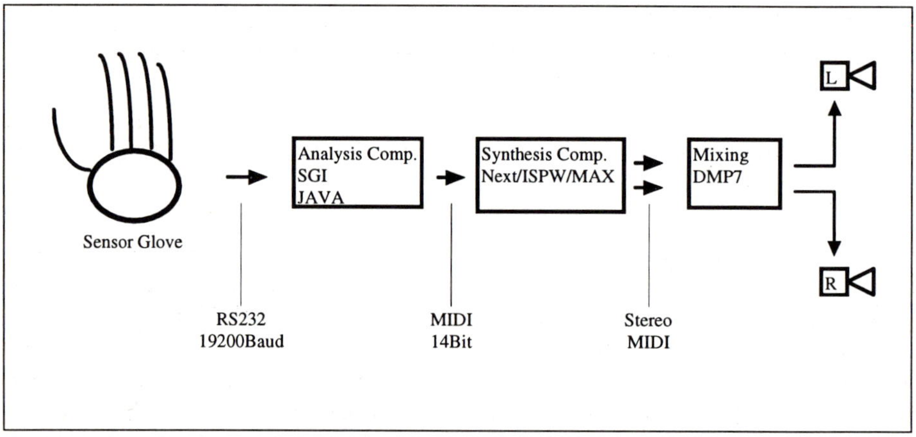

Fig. 3. Configuration of the DataGlove Sound Synthesis System

The data of the glove are transmitted via a serial interface RS232 with 19.200 baud. This theoretically allows for a resolution of 20 ms. These data are transmitted to an analysis computer which evaluates the protocol and is intended to prepare and analyze the data. Since it is easy to handle and it has plenty of object oriented and network possibilities. JAVA was chosen to serve as the environment for the design of the graphical surface. C-routines were integrated in JAVA or through communicating sockets in order to handle speed related or hardware related tasks. As a first application, the single data streams of the glove were visualized (the slider data display). Thereafter, the data were coupled to a MIDI controller via input panels. Each of the sensors can now handle an arbitrary MIDI-controller. In order to increase the spectrum range of 7 bit, which is the MIDI-standard but too restrictive for processes of sound synthesis, the format was extended to 14 bit for a single value.

These sensor-controller adjustments can be saved in scenes and are available for further demands.

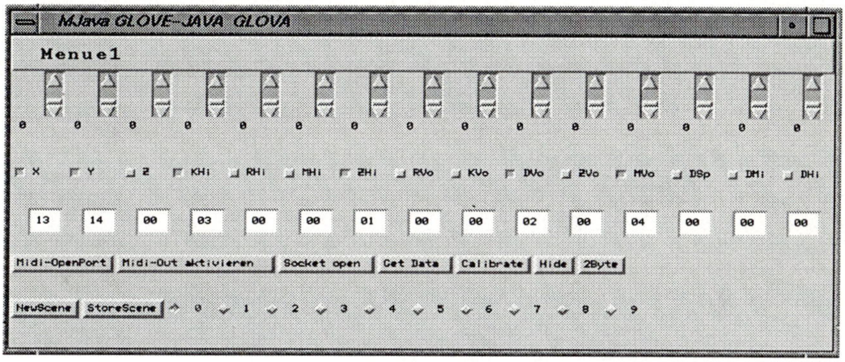

Fig. 4. User Interface for Data Acquisition of the SensorGlove

As a first test concerning synthesis, an FM algorithm was used that was extended by an effect section with distortion and Pan. This was realized as a MAX-patch on a NEXT computer with an ISPW card. Through the mixing console (DMP7), further effects, like reverb and delay that are subject to control, were integrated. This configuration can easily be transferred to other synthesis methods. Although the direct coupling of sensor data to the process of sound synthesis yielded satisfying results, a higher integration and abstraction of the sensor glove movements, or the control parameter of the synthesis respectively, is desirable. This will be granted by a gesture recognition for the range of the sensor glove.

The definition of alphabets accounts for an integration of the huge amount of control parameter of the synthesis (i.e. physical modeling). The signs are tested and defined in training units with the interactive computer music system. The mental representation of the alphabet is considered to be the Gestalt of the sound and the corresponding movement or position of the hand. Continuous switches between the learned Gestalts increase the power of the transferring channels.

5.2 Sensor Glove and Gestalt concepts

Both the sensor glove and the equipment described above allow for a remarkable intuitive control of the sound synthesis. The following remarks and

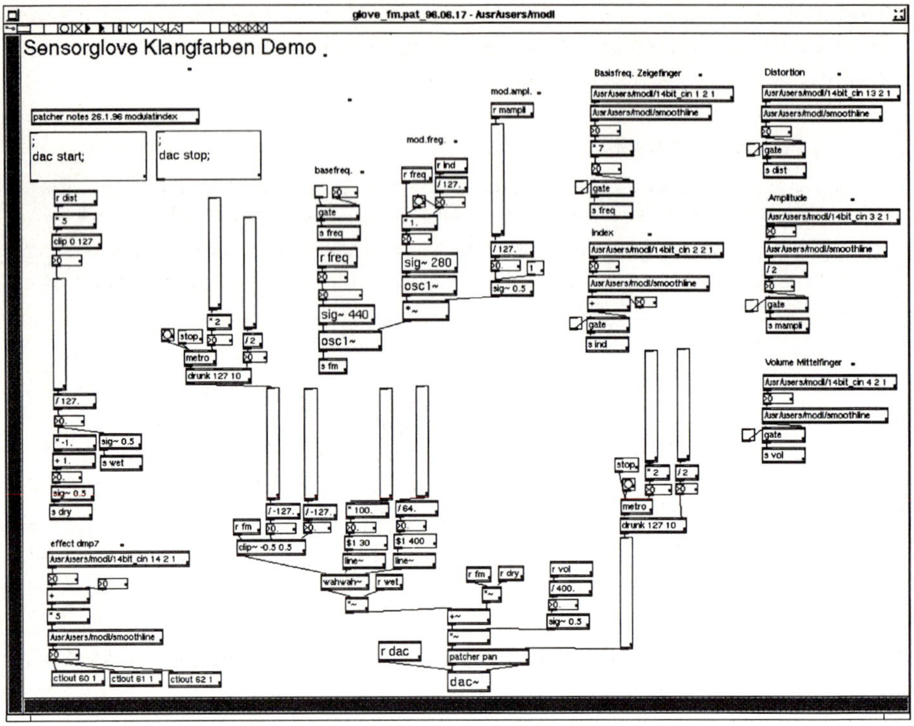

Fig. 5. Sound Generation as a Max Patch

observations concern the concepts for an evaluation of interactivity, which were introduced in the first part of the paper.

Assimilation The sensor glove and its equipment can be characterized as a Gestalt building medium. However, a graphic representation of the hand data as well as of the sound data would improve on that.

Group formation Single parameters can be controlled separately by a movement of the fingers, without affecting the basic position of the hand.

Continuity / Discontinuity Since the continuous value range of the sensors is directly connected to single sound parameters, soft or hard transitions between the sounds can be achieved, depending on the movement of the hand. However, there are some technical obstacles that partly constrain the value ranges or make them discontinuous (drift, overflow).

Relatability A remarkable relatability is achieved because the positioning of the hands is intuitive and can easily be learned and memorized. This is again constrained by technical restrictions.

Responsiveness The temporal resolution is around 60 ms, which could be further refined. At this stage, especially with fast movements of the hands, slight delays in the reaction can be observed. However, these effects can easily be removed in later versions of the sensor glove. As far as the synthesis is concerned, an immediate reaction is possible, granted by the high computational efficiency of the NEXT/ISPW-configuration.

Variability Since a standard interface is employed in the sensor glove, an arbitrary coupling to other synthesis computers or synthesizers is possible. Algorithmic methods of tone generation such as MIDI-MAX, M & Jam Factory, and others, can equally be controlled via this interface. A change of the sound environment could result from changes in the sensor-controller relation and the synthesis methods. A second, optimized input device such as a second sensor glove or a keyboard is possible.

Autonomy In the configuration described above, the method of synthesis is characterized by minor autonomy. A coupling to more complex methods of generation is yet to be examined.

6 Conclusion

As a conclusion it can be stated that the control of sound synthesizers equipped frequency modulations, granular synthesis, physical modeling as a part of an interactive computer music system, together with new input devices like the sensor glove will lead to new and different concepts within the user. As a consequence, new and revised musical frameworks based on Gestalts will have to be introduced.

The relationship between existing Gestalts and the methods applied for sound synthesis, tone synthesis and transformation will be the subject of further research. This research focuses particularly on gesture recognition and further methods for synthesis. Besides this, it will be necessary into account the psychological effects of an interactive computer music system in relation to Gestalt creation and Gestalt criteria.

Acknowledgment

I would like to thank M. Leman and an anonymous reviewer for helpful comments on an earlier version of this paper and F. Hoffmann and I. Zannos whose cooperation made this paper possible and U. Kleinhenz for proof-reading.

References

Ameres, E., Zicarelli, D., Chadabe, J., Offenhartz, J., & Widorf, A. (1988). *M & JAM-Factory.* Albany, NY: Intelligent Systems. (software program)

Balzano, G. (1989). Command performances, performance commands. *Contemporary Music Review, 4*, 437–446.

Barlow, C. (1990). Autobusk: An algoritmic real-time pitch & rhytm improvistation programme. In S. Arnold & G. Hair (Eds.), *Proceedings of the International Computer Music Conference (ICMC-90).* Glasgow.

Behrman, B. (1991). Observations on live electronics. *Contemporary Music Review, 6*, 151–157.

Behrman, D. (1991). Designing interactive computer-based music installations. *Contemporary Music Review, 6*, 139–142.

Föllinger, O. (1984). *Regelungstechnik.* Heidelberg: Hüthig.

Hunt, E. (1991). Interactive performance systems. *Contemporary Music Review, 6*, 131–138.

Jackendoff, R. (1987). *Consciousness and the computational mind.* Cambridge, MA: The MIT Press.

Köhler, W. (1971). *Die Aufgabe der Gestaltpsychologie.* Berlin: Walter de Gruyter.

Lee, C. (1996). DIGITAL AESTHETICS ONE in MESH 1996. *Journal of Experimenta Media Arts, 8/9.*

Puckette, M., & Zicarelli, D. (1988/1995). *MAX 3.0.* Palo Alto, CA: Opcode Systems / IRCAM. (software program)

Rose, J. (1993). Violine and electronics. In *STEIM de zoetgevooisde bliksem.* Amsterdam: STEIM.

Rowe, R. (1993a). *Cypher.* Cambridge, MA: The MIT Press. (software program)

Rowe, R. (1993b). *Interactive music systems.* Cambridge, MA: The MIT Press.

Schrader, B. (1991). Live/electro-acoustic music – a perspective from history and California. *Contemporary Music Review, 6*, 91–106.

Tanaka, A. (1997). *Biomuse.* (http://www.imaginet.fr/~atau/sensorband/index.html)

Teitelbaum, R. (1984). The digital piano and the patch control language system. In W. Buxton (Ed.), *Proceedings of the International Computer Music Conference (ICMC-84).* San Francisco: Computer Music Association.

von Ehrenfels, C. (1974). Über "Gestaltqualitäten". In F. Weinhandl (Ed.), *Gestalthaftes sehen.* Darmstadt: Wissenschaftliche Buchgesellschaft. (Original work published 1890)

Waisvisz, M. (1993). Twenty-five years of STEIM. In *STEIM de zoetgevooisde bliksem.* Amsterdam: STEIM.

Gestalt-Based Composition and Performance in Multimodal Environments

Antonio Camurri[1] and Marc Leman[2]

[1] DIST, University of Genova, Lab. of Musical Informatics, Viale Causa 13, I-16145, Genova, Italy
[2] IPEM, University of Ghent, Blandijnberg 2, B-9000 Ghent, Belgium

Abstract. The paper introduces the concept of multimodal environments in relation to Gestalt theory. Multimedia environments provide digital extensions for different human activities, such as movement, thinking, composing, listening, planning, etc. The environments discussed in this paper make use of the state-of-the art in sensoring technology and computing. They typically combine different methods of knowledge representation such as symbolic, iconic, and subsymbolic representations into a hybrid architecture. Movement detection, as well as beat induction and musical responses to action that happen on the scene all involve Gestalt notions. In this paper, we focus on a number of requirements for the application of multimodal environments in music and art. Applications are discussed and the basic architecture of an existing experimental platform is outlined.

1 Introduction

Multimodal environments consist of *agents* capable of establishing a creative multimodal user interaction. Agents are software modules, specialized in carrying out a certain task at a certain moment. Agents are capable of working independently from each other but the global reaction of the system to environmental input will consist of the collaborative actions of different agents. The reaction of the system, therefore, can be conceived of as emergent behavior. Agents exhibit dynamic, intelligent, real-time, and adaptive behavior. In a typical scenario, one or more users are immersed in an environment allowing them to communicate by means of body movement (dance and gestures), audio signals (singing, talking), and controlling equipment (playing on an instrument or keyboard). Users (and/or audience) get feedback from the environment in real time, in terms of sound, music, visual media, and actuators in general (e.g., movements of semi-autonomous mobile systems). Multimodal environments can be seen from two different perspectives: on one hand, they can be considered as digital environments that are used as a prolongation of the human mind, senses and actions; on the other hand, they can be seen as a population of artificial agents, each characterized by its own behavior, reasoning, and emotional state (Riecken, 1992; Camurri & Ferrentino, 1997). What is important, however, is that multimodal environments should be able (i) to interpret sensors data to extract high-level information with the aim of recognizing gesture and audio gestalts, (ii) to use this information for adapting their behavior and response to users. The paper is structured in two parts. In the first part we introduce multimodal environments with particular focus on the implementation of Gestalt-ideas. In the second part, a software model for the development of multimodal environments is introduced. Some examples of systems implemented at the DIST Laboratory of Musical Informatics has been used in several public events. Some examples are described as to illustrate the major theoretical statements.

2 Multimodal Environments, Virtual Environments and Hyper-Instruments

A multimodal environment is conceived of as an audio-visual environment which can be used to communicate with other humans and machines, other actors participating in the same event (including autonomous robots) or external spectators of the action. The system is driven, tuned, and moulded by the movements and sounds produced by the user(s), (actors, dancers, players, singers ...), using specific metaphors for reaching, grasping, turning, pushing, navigating, playing, communicating states or emotions, etc.

In some way, multimodal environments are related to existing state-of-the-art systems such as virtual environments and hyper-instruments (Machover & Chung, 1989). However, virtual environments and hyper-instruments are usually conceived of as static environments in the sense that they can be explored or navigated by users, or simply played as virtual musical instruments. They do not change their structure and behavior over time. They do not adapt themselves to users, nor do they try to guess what the user is doing or wanting. An exception is the ALIVE system (Maes, Blumberg, Darrel, Pentland, & Wexelblat, 1995) and, in some aspects also, *Cypher* (Rowe, 1993).

Most of the state-of-the-art systems fall into two main categories. In the first category, we have real-time systems which basically involve simple, low-level cause-effects mechanisms between modalities. The paradigm which is often used is that of a musical (hyper) instrument, where a fixed simple mapping between human movement and sound output is defined. For example, the hand goes up and the musical pitch raises, or it goes down and it lowers. In the second category, the paradigm is that of computer animation for films and computer games. The goal is here to track as precisely as possible the human movement and to use this information as-is in an off-line design process to animate believable artificial characters (Bates, 1994).

In contrast to these, multimedia environments aim to observe a user and to get sufficient information in real time in order to perform an overall evaluation of the actions (such as the style of the movement) and to perform an appropriate reaction, such as a communication with the user or a change in the behavior or the *emotional state* of the system. As a consequence, a main focus of research is concerned with the observation of context-dependent significant body movement features. Rather than detecting local movements, this approach focuses on the extraction global features in a sort of Gestalt approach. This is done, for example, through the integration of movement samples over time or through the observation of overall information by a combination of different sensors data. Our approach can therefore be put into relation with previous work by choreographers like R. Laban, as was already pointed out in (Camurri, Morasso, Tagliasco, & Zaccaria, 1986). In particular, Laban's theories of *space* and of *effort* are of particular relevance.

In short, the paradigm adopted here is that of a human observer of dance. The focus of attention is on different Gestalt properties and this changes dynamically according to the evolution of the dance itself and of the music produced. Multimodal environments should therefore be able to change their reactions, their social interaction and rules over time. In order to achieve this aim, multimodal environments embed multilevel representations of different media and modalities, as well as representations of communication metaphors and of analogies to integrate modalities. Moreover, multimodal environments should be able to build, apply and adapt plans in response to users actions and behavior.

3 Basic Requirements for Multimodal Environments

Multimodal environments can be classified as particular extensions of *augmented reality* environments integrating intelligent features. The concept of augmented reality pertains to digital extensions of humans acting in a natural reality. It includes multimedia interactive systems. Both multimodal environments and multimedia interactive systems are considered to be of high interest by a growing number of composers and artists (Machover & Chung, 1989; Rowe, 1993; Povall, 1995). In what follows, we give an overview of the basic requirements for such systems:

- First, there is the necessity of an effective on-body and spatial sensor systems technology which is wireless, low-cost, robust, and characterized by sufficient range and precision for observation both in time and space. In addition, different users of the system – from the director, the choreographer, and the composer in the off-line design phase, to the dancers in the performance – should not be forced to think of or be influenced by the technology they wear or interact with: they must be free to concentrate to their tasks and goals and technology should be transparent.
- Another point deals with the use and integration (in real time) of gestures, music, and possibly other modalities (e.g., visual) into a unified language. What is needed is a flexible, adaptive, effective language and real-time supervision system able to manage the situations happening in the environment. Such a system should also support a higher-level

interaction far more sophisticated than mere synchronization of modalities. Multimodal environments should therefore best be conceived of as a population of agents, each having its own goals and *character*, capable of establishing and conducting creative, adaptive, intelligent interaction with users in different application scenarios. As a direct consequence of this, it is necessary to have systems which are capable to model and utilize emotion potentials (Riecken, 1992). In the recent literature in brain science, emotions are believed to play an essential role in decision making and planning of actions. According to Damasio (1994), who speaks about *somatic markers*, learned emotions have been connected to predicted future outcomes of certain scenarios. When a negative emotion potential is related to a particular future outcome, the combination functions as an alarm bell. Emotional potentials allow decision making in terms of a biasing devise. So what is needed are systems that are able to manage the multifarious structure and levels of abstraction of perceptual spaces, of music and multimedia, from symbolic abstract representations (e.g., the symbol evoked by a body gesture, the style of the dance in a music section) to subsymbolic representations (e.g. a 3D trajectory of a part of the body; a signal perceived by the human ear, an auditory image). Different views of the same object are often necessary, according to the reasoning perspective, the goal to be reached, and the internal state (including emoting potentials) of the system. Multimedia object representations can vary from an atom in a symbolic high-level representation to a stream of low-level signals in the broadest view of the same material. More specifically in music, Leman (1989) distinguishes three main types of level: (i) the acoustic (signal) level, (ii) the auditory (subsymbolic) level, and (iii) the conceptual (symbolic) level. This can be extended to other media, and multimodal environments should be able to represent and manipulate in real time objects at all these levels.

- Analogies and metaphors are crucial issues in the representation and reasoning capabilities of multimodal environments. Metaphors are widely used in natural language. They provide a means for integrating concepts from different knowledge domains. In music, the metaphorical richness derived from real world dynamics is significant (see for example, Camurri et al., 1986). Terms and descriptions in one modality can be used to express intuitively similar concepts in other modalities. We therefore consider analogies and metaphors to be the basic glue for integrating different modalities, with particular regard to sound/music and movement/dance representations.

- Formalisms that can support users should provide mechanisms for reasoning on actions and plans, and for analyzing alternatives and strategies, starting from user requirements and goals. They should provide both formal and informal analysis capabilities for inspecting the objects represented.

- Other crucial issues are concerned with learning and adaptation, i.e. how to automatically update or adapt system knowledge to new information and to the nonlinear human behavior. Several solutions have been proposed in the AI literature, such as purely symbolic approaches and learning systems based on neural networks. As for human movement recognition, some learning problems can be translated into a dimensionality reduction problem of extracting principal components from a high-dimensional space of redundant degrees of freedom. In any case, we need learning because in multimodal systems there is not, in general, a simple one-to-one translation of signals and events into significant features.

4 A Multimodal Environment Architecture: HARP

Taking the previous requirements into consideration, HARP and its applications (Camurri, Innocenti, & Massari, 1995; Camurri, 1995, 1995; Camurri & Leman, 1997) are a contribution to the development of multimodal environments. The overall system architecture is a distributed network of agents, in certain aspects similar to *Cypher* (Rowe, 1993) and *M* (Riecken, 1994). The gross anatomy of the architecture consists of the following basic building blocks, sketched in Fig. 1:

- Input Mapping: The input mapping consists of a set of modules which are able to receive signals from sensors and to map them onto schemata (perceptual spaces). A schema can be a self-organizing neural network, e.g. one that has been trained to organize some specific statistical properties of context-dependent patterns into some global structure, or a manifold reduced representation of an input multi-dimensional space (Leman, 1995). The recorded signals may be mapped onto the schema in terms of movement trajectories. A crucial problem here concerns the reduction of the dimensionality of the multimodal input space when mapped into a schema.
- Output Mapping: The output mapping consists of a set of modules that manage the orders given by the cognitive processing modules. The orders may pertain to high level parameters. As stated before, the output may be of any kind and need not be restricted to a direct low-level causal transformation of the input. An example of output mapping is the space of the emotional parameters for controlling the expression of an artificial face, or the timbral space of a section of a composition.
- Cognitive Processing: Cognitive processing is about reasoning and planning. This layer deals with two kinds of activities, *subsymbolic* and *symbolic.* An example of subsymbolic reasoning is when the movement of a dancer is projected onto a topological map (e.g., a neural network classifying gestures) and its trajectory is influenced by attraction forces thus allowing to recognize input gestures. Symbolic reasoning can only perform symbolic inference and planning on the entities that emerge in

the symbolic database from subsymbolic reasoning and recognition processes. Each agent in a multimodal environment is characterized by a state whose dynamics is modeled in terms of the analogy with *emotional states*. Such state influences the reasoning activity. Details on the mathematical model of our emotion space model can be found in (Camurri & Ferrentino, 1997). In a few words, the emotion space for an agent living in a multimodal environment is modeled as a two-dimensional space. The horizontal axis represents a subjective component roughly in terms of a sadness-happyness axis. The vertical axis represents a component of the state regarding the relation of the subject with the external world. This is an asocial-sociable axis. A point in the emotion space then defines to the current emotional character of the agent.
Particular actions or events may cause changes to the agent's mood or emotional disposition. The latter is a relative vector with its origin in the point of the emotional character. This has the effect that, if the place is desert, the system may feel *happy* if it is asocial, because it does not meet anybody and does not want to, but it may feel *unhappy* if it wants to meet someone. The model presents other characteristics regarding its dynamics. Movements of the short-term component (mood or disposition) may influence the medium term component (character). The metaphor which models the dynamics of the emotional space is the viscous movement of a sphere (the character) in a fluid (the emotional space), and subject to abstract potential fields. Further, the position of the character influences the viscousness in which the sphere moves.
- Symbolic Database: The symbolic database is a high-level, symbolic representation of the domain space(s), e.g. music composition and performance spaces, movement and gesture spaces. It consists of a knowledge base management system, written in a standard representation language, a subset of first order predicates, which is structured in a long term and a short term memory. The long term memory stores ontologies related to general definitions of events, situations, facts, objects, and relations in the specific application domain. Instances of relevant facts about events, situations, objects, and related features are stored during a work session in the short term working memory. The agent updates its working memory according to its inputs and reasoning processes.

The dynamics of the system is quite complex. Let us consider an example about the role of the bidirectional links between (i) Symbolic Database and (ii) Subsymbolic Cognitive Processing and Input Mapping. The activity of the Subsymbolic Cognitive Processing modules causes modifications in the Symbolic Database (e.g., it adds assertions of particular situations conducing to a change of context). This starts a reasoning processes which is capable to change or tune the Input Mapping processes. For example, the focus of attention of the observer modules in the Input Mapping can be moved from one set of gesture-detectors to another set of gesture-detectors, or the input mapping sensitivity on certain gestures recognition can be tuned. In other words, the cognitive processing (i) feeds data to the Output Mapping modules, and (ii) can cause feedback to the perceptual processes in order to better capture the aspects of the external world of greater current interest or expectation for the agent. A similar aspect is also present, for example, in the system architecture described by Thorisson (1995).

5 The Software Architecture

The software system is structured in a development environment and in a run time environment. It is useful to make a distinction between the programmer mode, the user mode and the analysis mode. The development environment, at a first glance, may be conceived of as a sort of hybrid expert-system shell which – like any shell – allows the application programmer to do several things. One of the main goal of this module is to support the programmer in the design and implementation of HARP agents. To this aim, for each agent, the long term domain knowledge (the ontology) has to be stored in the symbolic database. The user does this by entering the programming mode and defining a set or structured labels (sometimes called *symbols* or *concepts*), their relations (or *roles*), some rules that apply on them, and this within the constraints of some temporal framework. Then, the programmer has to link the Interface and Task modules implementing the behavior of the agent. In case of special requirements, the programmer may intervene on the Reasoner module by augmenting its reasoning capabilities. The implementation of the body of the agent is made easier by a Wizard module which automatically builds the standard agent (C++) code parts.

Agents are program modules which, at run-time, perform certain tasks exhibiting skills such as music analysis, gesture analysis, interactive composition, perception of audio input and so on. Agents can collaborate to achieve overall goals or to better accomplish their single tasks. Therefore, an important aspect regards the communication among agents. Again, the programmer introduces the desired mechanisms of interaction, negotiation, cooperation.

6 Using HARP

In the run-time environment the user uses HARP as a prolongation of human activity within a digital environment. In this mode, the system typically provides low-level scheduling and coordination primitives to agents, i.e., it supports the creation, deletion of agent modules, and influences them according to the needs of that moment. In the example of an agent simulating a listening process, the system, at a certain moment, comes up with the recognition of a particular chord. It then can create a corresponding new situation in its domain world, describing that chord in symbolic terms, and it may possibly tell an analysis agent about this new information such as to allow new inferences about the harmony of the piece. The communication

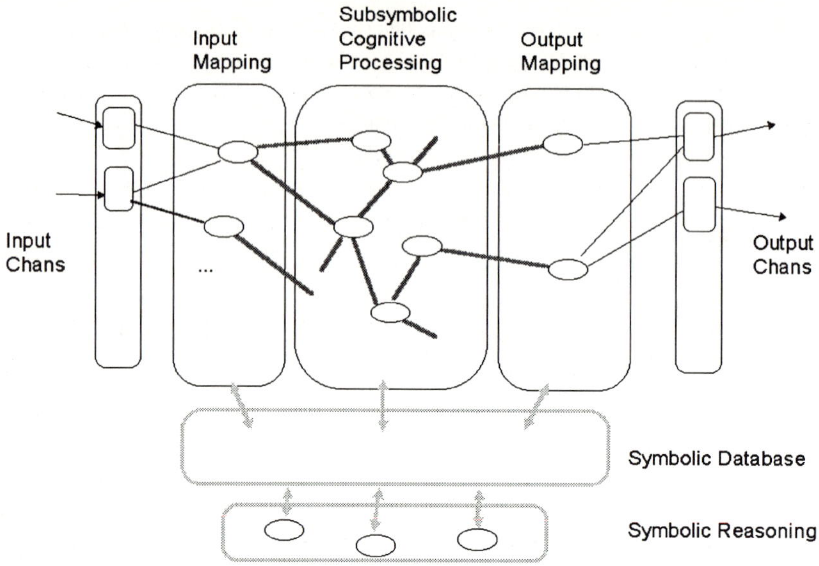

Fig. 1. A sketch of the conceptual model for a multimodal environment

of declarative information may wake up the analysis agent, which can start reasoning about the chord in relation to the context in which it appeared and the history maintained by the analysis agent.

In addition to the programmer and user modes, HARP allows the user to switch into an analysis mode in which the user can exploit the system to infer properties of its own actions. A query module, including a visual user interface, is available to the user by any agent. This allows the user to investigate what the agent has done during the performance of the system. This module can be invoked at any time. For example, it is possible to stop a performance and start an investigation on the current and past behavior of the involved agents. All this makes of HARP a quite complex but highly flexible system. Several system applications have been developed, in which the system has been configured and adapted, typically to reduce the complexity of interaction by compiling certain behaviors and customizing the user interface for the specific application to fulfill the needs of non-expert users.

7 Examples of Multimodal Environment Applications

In this section we give a short description of applications that have been based on the ideas and concepts described above:

7.1 HARP/Vscope

This HARP application uses an agent for the tracking of continuous human gestures and its integration with sound/music. The agent is designed to acquire information on the position of a number of wireless markers placed on the body of a user. A dancer can generate, filter, or control different levels of the output of acoustical instruments by means of movements. The system is composed of four main subsystems:

- human movement acquisition, based on the V-scope sensor system, including data pre-processing and filtering
- (input mapping) human movement and gesture recognition, including modules based on the force field metaphor, and capable of movement pattern classification by means of self-organizing neural networks
- (cognitive) symbolic reasoning, for the representation and supervision of contexts in music and gesture spaces
- (output mapping) mapping of internal data to sound and music output (including MIDI)

Figures 2 and 3 show the system at work during the performance of *En Voyage* (described below). The Vscope markers are in the palm of the hands of the dancer. By opening and closing of the hands, the dancer can start and stop the acquisition process of the corresponding markers. Other markers are positioned in the ankles and in the torso of the dancer.

Fig. 2. An example of performance involving HARP/Vscope

7.2 The SoundCage Interactive Music Machine

The SoundCage Interactive Music Machine (IMM) is a application developed by *SoundCage Srl.* The SoundCage hardware is composed by a set of proprietary active infrared and pressure sensor systems displaced in a sort of cage structure (the stage), and by special I/O boards. Pressure sensors are on the floor of the machine. In total, there are more than 60 spatial sensors. We designed the system such that the dancer is completely free of moving without any on-body sensors. The spatial sensors cover the three-dimensional space of a cube of less than three meters (the cage). The criteria adopted for the sensor distribution in the cage are based on the task to recognize human Gestalt movements, i.e., within given kinematic and joint constraints. Furthermore, we assume a single dancer inside each cage. Due to the previous assumptions, certain crucial areas in the cage have a more dense array of sensors. Roughly, there are four different horizontal layers which are more carefully observed into the cage, a floor layer, an ankles layer, an arms layer, and an over-the-head layer. The same approach to Gestalt gesture recognition is adopted here as well as in all the other movement recognition systems we developed.

7.3 The HARP/DanceWeb

DanceWeb is a proprietary low-cost sensor system, based on ultrasound technology. Up to 64 (in the current version) ultrasound sensors and up to 32 analog and digital I/O can be connected in a multimodal environment installation. The sensor system includes an external rack unit with a micro controller and electronics for the acquisition and preprocessing of sensors data, connected via a fast serial link to the supervision workstation. Low-level measurements include distance measures of parts of the body from sensors. Specific modules for the integration of multiple sensors data have been developed, and integrated within the HARP environment.

The HARP/DanceWeb has been utilized with other sensor systems in several public events. For example, in *Upper Space*, an interactive installation at Teatro Fondamenta Nuove in Venezia, september 1996, patronized by Biennale Architettura, a HARP/DanceWeb agent cooperates with a SoundCage agent to transform the whole theatre into a living multimodal environment. Such an environment, which has its own composition and choreography goals, can be influenced and moulded by the movement and voice of performing artists and the visitors.

8 Discussion

The exploration of multimodal environments in the musical domain is of great interest for the design of multimodal environments in different other environments, not necessarily restricted to music and art applications. One

Fig. 3. An example of performance involving HARP/Vscope

of the reasons is that music is a temporal art which imposes constraints on the real-time capabilities of the performances. Research in multimodal user interfaces can benefit from such experiments (Schomaker, 1995). Music is a crucial test domain for the development of systems which should exhibit a sort of intelligence based on cognitive processing, perception and emotional modeling. The latter is a relatively newly *discovered* feature which, when properly applied, has a high potential. In the coming years, this feature will be refined and combined with expressive features as well. In music research, we are now at a cross-point where actions on stage extends the dimensions and degrees of freedom of artists. Actions and choreographies can be immersed in a computer music artwork. This implies the opening of boundaries of music languages, their integration with dance languages and the extension of the role of the director of the performance in live electronics (Vidolin, 1987).

A more difficult problem is concerned with the communication of the causality of movement and sound during a performance. We call it the problem of *causality perception*. To some extend, audiences want to know what is going on at the scene. Is there a tape playing, or is the actor really exerting control over sound in real-time? To what extend can he exert control? What responses are caused by what kind of actions, etc... The (Western) human mind seems to be constantly in search after causes and effects. In a traditional theater setting, where the audience is not allowed to participate in the action, this may require particular solutions at the choreographical level in order to clarify the causes and effects more explicitly. A simple example of this problem is how to express that a particular section of a piece starts. We experimented that this can be reached by preceding such section by an unnatural movement/music causal association (e.g., a fast, sudden movement creating a slow, smooth sound object). This visual/listening disagreement correspond to a tension situation in the listener/observer, which in determinate conditions can cause an increase in the focus and level of attention, and it is reasonable that subsequent music/movement events can more clearly perceived. All these and other issues need to be further explored and developed, both from a theoretical and experimental point of view.

Acknowledgements

The research described in this paper has been partially supported by the Esprit Project 8579 - MIAMI (Multimodal Interaction for Advanced Multimedia Interfaces), by a special project of the Italian National Research Council (CNR) on interactive multimedia systems for art and entertainment, and by MURST projects. The SoundCage IMM has been developed by SoundCage Srl, the HARP system and the other MEs have been developed by the Laboratorio di Informatica Musicale at DIST.

References

Bates, J. (1994). The role of emotions in believable agents. *Communications of the ACM, 37*, 122–125.

Camurri, A. (1995). Interactive dance/music systems. In *Proceedings International Computer Music Conference (ICMC-95).* Banff: ICMA Press.

Camurri, A., & Ferrentino, P. (1997). *A computational model of artificial emotion and its application in the theatrical museal machine project* (Tech. Rep.). DIST Laboratory on Musical Informatics.

Camurri, A., Innocenti, C., & Massari, A. (1995). Toward an integrated agent architecture for real-time, multi-modal interaction in multimedia systems. In *First International Congress on Music and Artificial Intelligence (ICMAI-95).* Edinburgh.

Camurri, A., & Leman, M. (1997). Gestalt-based composition and performance in multimodal environments. In M. Leman (Ed.), *Music, Gestalt, and computing: Studies in cognitive and systematic musicology.* Berlin, Heidelberg: Springer-Verlag.

Camurri, A., Morasso, P., Tagliasco, V., & Zaccaria, R. (1986). Dance and movement notation. In P. Morasso & V.Tagliasco (Eds.), *Human movement understanding.* Amsterdam: North Holland.

Damasio, A. (1994). *Descatres error: Emotion, reason, and the human brain.* Chatham, Kent: Mackays of Chatham.

Leman, M. (1989). Symbolic and subsymbolic information processing in models of musical communication and cognition. *Interface – Journal of New Music Research, 18*, 141–160.

Leman, M. (1995). *Music and schema theory: Cognitive foundations of systematic musicology.* Berlin, Heidelberg: Springer-Verlag.

Machover, T., & Chung, J. (1989). Hyperinstruments: Musically intelligent and interactive performance and creativity systems. In *Proceedings International Computer Music Conference (ICMC-1989).* Columbus, Ohio, USA.

Maes, P., Blumberg, B., Darrel, T., Pentland, A., & Wexelblat, A. (1995). Modeling interactive agents in ALIVE. In *Proceedings International Joint Conference on Artificial Intelligence (IJCAI-95).* Montreal.

Povall, R. (1995). Compositional methods in interactive performance environments. *Journal of New Music Research*, 109–120.

Riecken, D. (1992). Wolfgang: A system using emoting potentials to manage musical design. In M. Balaban, K. Ebcioglu, & O. Laske (Eds.), *Understanding music with AI: Perspectives on music cognition.* Cambridge, MA: The MIT Press.

Riecken, D. (Ed.). (1994). *Special issue on intelligent agents.* (Communications of the ACM, July 1994)

Rowe, R. (1993). *Interactive music systems.* Cambridge, MA: The MIT Press.

Schomaker, L. (1995). *Taxonomy on multimodal interaction in multimedia systems.* MIAMI Esprit Project.

Thorisson, K. (1995). *Multimodal interface agents.* Unpublished doctoral dissertation, MIT Media Lab.

Vidolin, A. (1987). Ambienti esecutivi. In *I profili del suono.* Musica Verticale.

VI

List of Sound Examples

List of Sound Examples on the CD

The sound examples on the CD that accompanies this book have a scientific purpose and aim at illustrating what is explained in the articles.

Track 1-4:
Knowledge in Music Theory by Shapes of Musical Objects and Sound-Producing Actions (p.89).
Rolf Inge Godøy

1. Sound Example #1: A scale of 7 variants of a synthesized bass like sound, gradually increasing its *mass* (overall harmonic content) and *grain* (textural fluctuation) density. This sound was produced by modal synthesis of a bowed string, i.e. by physical modelling, by progressively adding vibrating modes in the model using the Modalys software from IRCAM.
2. Sound Example #2: The fourth variant sound of the previous example is here first presented unmodified, then with the application of a stationary and lastly a fluctuating formantic filtering, using the formantic envelope of a trombone wah-wah mute as the filter. This illustrates the projection of formantic shape on to a sound as well as the *allure* (slower fluctuation) of a sound in the last sound of this example. The signal processing was done with the Audiosculpt software from IRCAM.
3. Sound Example #3: A scale of 7 variant attacks on the same sound as in the previous example, proceeding from gradual to steep and illustrating the effect of the attack shape on the quality of the sound.
4. Sound Example #4: A series of variants of the C-major prelude by Chopin, starting out with the original, then continuing with variants in minor, phrygian, lydian and finally again in C-major but octave compressed by a modulo 12 operation. This allows for a comparison of the effect of changing the modality of a musical object while keeping the other features constant, as well as the effect of retaining the original modality but deforming the spatial layout in the last variant. The performance is by a MIDI sequencer playing a sampled piano.

Track 5-10:
"Verschmelzung", Tonal Fusion, and Consonance: Carl Stumpf Revisited (p.117).
Albrecht Schneider

5. Sound Example #1: Harmonic major chord, each of the components comprising of seven harmonics.
6. Sound Example #2: Additive synthesis of nineteen harmonics. The amplitudes of these partials increase with their respective number. The sounds resembles that of a tenor sax.

7. Sound Example #3: Pulse type high energy sound. Labeled for convenience cosine pulse.
8. Sound Example #4a: Some brief musical examples, played with a sampled synthesized sax sound
9. Sound Example #4b: Some brief musical examples, played with a sampled synthesized cosine pulse sound.
10. Sound Example #5: Stumpf's chords number 1, 2, 3, and 4 (each repeated three times).

Track 11-15:
Schema and Gestalt: Testing the Hypothesis of Psychoneural Isomorphism by Computer Simulation (p.144).
Marc Leman and Francesco Carreras

11. Sound Example #1: The chords of the C-major context (I IV V). This sound example, as well as the following sound examples are built using Shepard-spectra.
12. Sound Example #2: Context in C-major, followed by a C-major triad probe. Fig.5a shows the network output after processing this chord sequence.
13. Sound Example #3: Context in C-major, followed by a C♯-major triad probe. Fig.5b shows the network output after processing this chord sequence.
14. Sound Example #4: Context in C-major, followed by a C-minor triad probe.
15. Sound Example #5: Context in C-major, followed by a C♯-minor triad probe.

Track 16-29:
Problems of Shape and Background in Sounds with Inharmonic Spectra (p.214).
Jenö Keuler

16. Sound Example #1: A sound which is consonant with the same sound at interval distance of a major seventh, but *not* with its octave transposition.
17. Sound Example #2: The effects of consonance and dissonance. The sound of the example #1 is confronted with its consonant major seventh and dissonant octave.
18. Sound Example #3: The sound of the example #1 is first confronted with its consonant major seventh and dissonant minor seventh.
19. Sound Example #4: A short composition entitled *Major Seventh Type Octave.*
20. Sound Example #5: Enriched sound spectrum.
21. Sound Example #6: A succession of paradoxical sounds.
22. Sound Example #7: A melody built up of paradoxical sounds.

23. Sound Example #8: The paradoxical sounds are superimposed to form chords.
24. Sound Example #9: Sounds that do not have identical pitch quality with their major seventh, but with their minor ninth instead.
25. Sound Example #10: The consonance of the minor ninth, and dissonance of the octave and the major ninth.
26. Sound Example #11: Inharmonic complex sound based on the tempered minor ninth.
27. Sound Example #12: A succession of octaves is played upwards, while a chromatic movement is heard downwards.
28. Sound Example #13: A succession of octaves is played downwards while the chromatic movements are perceived upwards.
29. Sound Example #14: A short three part composition, having as title *Pseudo Twelve Tone Tunings.*

Track 30:
Optimizing Self-Organizing Timbre Maps: Two Approaches (p.337).
Petri Toiviainen

30. The 27 tone stimuli used in the experiments.

Track 31-38:
Karl Erich Schumann's Principles of Timbre as a Helpful Tool in Stream Segregation Research (p.362).
Christoph Reuter

Each sound is played twice.

Instruments with corresponding formant positions (bassoon and valve horn):

31. Sound Example #1: bn/cor: bassoon (first) and valve horn (second).
32. Sound Example #2: cor/bn: valve horn (first) and bassoon (second).

Instruments with non-corresponding formant positions (bassoon and trumpet):

33. Sound Example #3: bn/tp: bassoon (first) and trumpet (second).
34. Sound Example #4: tp/bn: trumpet (first) and bassoon (second).

Instruments with formants and with fluctuations (bassoon and celli):

35. Sound Example #5: bn/vc: bassoon (first) and celli (second).
36. Sound Example #6: vc/bn: celli (first) and bassoon (second).

Instruments with fluctuations (violins and celli):

37. Sound Example #7: vl/vc: violins (first) and celli (second).
38. Sound Example #8: vc/vl: celli (first) and violins (second).

Track 39-44:
Cross-Synthesis Using Interverted Principal Harmonic Sub-Spaces (p.375).
Thierry Rochebois and Gérard Charbonneau

39. Sound exapmple #1: Grey's Flute Tone. This sound is presented four times. The first time with its original length. The second, third and fourth time, it is time-stretched by factors 2, 4 and 8.

    The next examples are cross synthesis examples, based on four sounds exerpted from the McGill University sound database CDROM. These four sounds include: a flute with tremolo, an oboe, a cello with vibrato, and a piano.
40. Sound exapmple #2: First the original sounds are presented:
    flute
    oboe
    cello
    piano
41. Sound exapmple #3: The first hybridization:
    flute
    oboe
    first hybrid tone: flute sub-space, oboe trajectory
    second hybrid tone: oboe sub-space, flute trajectory
42. Sound exapmple #4: The second hybridization is presented:
    cello
    piano
    first hybrid tone: cello sub-space, piano trajectory
    second hybrid tone: piano sub-space, cello trajectory
43. Sound exapmple #5: The third hybridization:
    flute
    piano
    first hybrid tone: flute sub-space, piano trajectory
    second hybrid tone: piano sub-space, flute trajectory
44. Sound exapmple #6: The fourth hybridization:
    oboe
    cello
    first hybrid tone: oboe sub-space, cello trajectory
    second hybrid tone: cello sub-space, oboe trajectory

Track 45-48:
Technology of Interpretation and Expressive Pulses (p.409).
Andranick Tanguiane

45. Sound Example #1: The theme of Vivaldi's five Roland M-OC1 synthesized violin voices Voice # 214 (Solo Vn 1)

Voice # 218 (Stereo Vn)
Voice # 216 (Chorus Vn)
Voice # 215 (Solo Vn 2)
Voice # 217 (Agitato Vn)

46. Sound Example #2: Computer performance with no nuances by Roland M-OC1 voice # 214 (Solo Vn 1) and Kurzweil RG 200 voice Grand Piano 1.
47. Sound Example #3: Computer performance with dynamics changes, tempo deviations, and articulation variety by using the five violin voices listed.
48. Sound Example #4: Violin part: M. Tangian, 8 years old. Piano part: A. Tangian (Tanguiane). The recording has been realized at the Fern Universität, ZFS-studios on July 31, 1996,

Track 49-62:
Sonological Analysis of Clarinet Expressivity (p.431).
Sergio Canazza, Giovanni De Poli, Stefano Rinaldin and Alvise Vidolin

Sound Examples #1-7: Mozart's *Concert for Clarinet* K622: original recordings. Each performance was inspired giving the player a sensorial type adjective in order to suggest various kinds of *expressive intentions*. Performances according to the following adjectives:

49. Bright
50. Dark
51. Hard
52. Light
53. Soft
54. Normal
55. Heavy

Sound Examples #8-14: Mozart's *Concert for Clarinet* K622: Physical Model Synthesis. Physical modeling synthesis, based on the values of the parameters measured in the performances. Synthesized sounds according to the performance parameters of the following adjectives:

56. Bright
57. Dark
58. Hard
59. Light
60. Soft
61. Normal
62. Heavy

Track 63-69:
Perceptual Analysis of the Musical Expressive Intention in a Clarinet Performance (p.441).
Sergio Canazza, Giovanni De Poli and Alvise Vidolin

Sound Examples #1-7: Fixed waveform synthesis of seven interpretations of a fragment of G.F.Händel's *Sonata for Alto and Basso Continuo* according to the parameters (i.e. factor loadings) deduced from perceptual analysis of the following adjectives:

63. Bright
64. Dark
65. Hard
66. Light
67. Soft
68. Normal
69. Heavy

Track 70-76:
Singing, Mind and Brain – Unit Pulse, Rhythm, Emotion and Expression (p.451).
Eliezer Rapoport

70. Sound Example #1: The first phrase from the opening of the aria *In Questa Reggia* from *Turandot* by Puccini, Fig.2, sung by Maria Callas (EMI CDC 7 47966 2).
71. Examples #2-5: the second, third, fourth, and fifth phrases, respectively, from the same performance, and same excerpt of Fig.2.
72. Sound Example #6: Coloratura phrase from the *Bell Song* from *Lakmé* by Delibes, score Fig.4b, sung by Sumi Jo (ERATO 4509-97239-2).
73. Sound Example #7: Coloratura phrase from the *Exsultate Jubilate* by Mozart, Fig.4c, sung by Janet Baker. (ERATO ECD 88090).

Name Index

Subject Index

Lecture Notes in Artificial Intelligence (LNAI)

Vol. 1159: D.L. Borges, C.A.A. Kaestner (Eds.), Advances in Artificial Intelligence. Proceedings, 1996. XI, 243 pages. 1996.

Vol. 1160: S. Arikawa, A.K. Sharma (Eds.), Algorithmic Learning Theory. Proceedings, 1996. XVII, 337 pages. 1996.

Vol. 1168: I. Smith, B. Faltings (Eds.), Advances in Case-Based Reasoning. Proceedings, 1996. IX, 531 pages. 1996.

Vol. 1171: A. Franz, Automatic Ambiguity Resolution in Natural Language Processing. XIX, 155 pages. 1996.

Vol. 1177: J.P. Müller, The Design of Intelligent Agents. XV, 227 pages. 1996.

Vol. 1187: K. Schlechta, Nonmonotonic Logics. IX, 243 pages. 1997.

Vol. 1188: T.P. Martin, A.L. Ralescu (Eds.), Fuzzy Logic in Artificial Intelligence. Proceedings, 1995. VIII, 272 pages. 1997.

Vol. 1193: J.P. Müller, M.J. Wooldridge, N.R. Jennings (Eds.), Intelligent Agents III. XV, 401 pages. 1997.

Vol. 1195: R. Trappl, P. Petta (Eds.), Creating Personalities for Synthetic Actors. VII, 251 pages. 1997.

Vol. 1198: H. S. Nwana, N. Azarmi (Eds.), Software Agents and Soft Computing: Towards Enhancing Machine Intelligents. XIV, 298 pages. 1997.

Vol. 1202: P. Kandzia, M. Klusch (Eds.), Cooperative Information Agents. Proceedings, 1997. IX, 287 pages. 1997.

Vol. 1208: S. Ben-David (Ed.), Computational Learning Theory. Proceedings, 1997. VIII, 331 pages. 1997.

Vol. 1209: L. Cavedon, A. Rao, W. Wobcke (Eds.), Intelligent Agent Systems. Proceedings, 1996. IX, 188 pages. 1997.

Vol. 1211: E. Keravnou, C. Garbay, R. Baud, J. Wyatt (Eds.), Artificial Intelligence in Medicine. Proceedings, 1997. XIII, 526 pages. 1997.

Vol. 1216: J. Dix, L. Moniz Pereira, T.C. Przymusinski (Eds.), Non-Monotonic Extensions of Logic Programming. Proceedings, 1996. XI, 224 pages. 1997.

Vol. 1221: G. Weiß (Ed.), Distributed Artificial Intelligence Meets Machine Learning. Proceedings, 1996. X, 294 pages. 1997.

Vol. 1224: M. van Someren, G. Widmer (Eds.), Machine Learning: ECML-97. Proceedings, 1997. XI, 361 pages. 1997.

Vol. 1227: D. Galmiche (Ed.), Automated Reasoning with Analytic Tableaux and Related Methods. Proceedings, 1997. XI, 373 pages. 1997.

Vol. 1228: S.-H. Nienhuys-Cheng, R. de Wolf, Foundations of Inductive Logic Programming. XVII, 404 pages. 1997.

Vol. 1236: E. Maier, M. Mast, S. LuperFoy (Eds.), Dialogue Processing in Spoken Language Systems. Proceedings, 1996. VIII, 220 pages. 1997.

Vol. 1237: M. Boman, W. Van de Velde (Eds.), Multi-Agent Rationality. Proceedings, 1997. XII, 254 pages. 1997.

Vol. 1244: D. M. Gabbay, R. Kruse, A. Nonnengart, H.J. Ohlbach (Eds.), Qualitative and Quantitative Practical Reasoning. Proceedings, 1997. X, 621 pages. 1997.

Vol. 1249: W. McCune (Ed.), Automated Deduction – CADE-14. Proceedings, 1997. XIV, 462 pages. 1997.

Vol. 1257: D. Lukose, H. Delugach, M. Keeler, L. Searle, J. Sowa (Eds.), Conceptual Structures: Fulfilling Peirce's Dream. Proceedings, 1997. XII, 621 pages. 1997.

Vol. 1263: J. Komorowski, J. Zytkow (Eds.), Principles of Data Mining and Knowledge Discovery. Proceedings, 1997. IX, 397 pages. 1997.

Vol. 1266: D.B. Leake, E. Plaza (Eds.), Case-Based Reasoning Research and Development. Proceedings, 1997. XIII, 648 pages. 1997.

Vol. 1265: J. Dix, U. Furbach, A. Nerode (Eds.), Logic Programming and Nonmonotonic Reasoning. Proceedings, 1997. X, 453 pages. 1997.

Vol. 1285: X. Jao, J.-H. Kim, T. Furuhashi (Eds.), Simulated Evolution and Learning. Proceedings, 1996. VIII, 231 pages. 1997.

Vol. 1286: C. Zhang, D. Lukose (Eds.), Multi-Agent Systems. Proceedings, 1996. VII, 195 pages. 1997.

Vol. 1297: N. Lavrač, S. Džeroski (Eds.), Inductive Logic Programming. Proceedings, 1997. VIII, 309 pages. 1997.

Vol. 1299: M.T. Pazienza (Ed.), Information Extraction. Proceedings, 1997. IX, 213 pages. 1997.

Vol. 1303: G. Brewka, C. Habel, B. Nebel (Eds.), KI-97: Advances in Artificial Intelligence. Proceedings, 1997. XI, 413 pages. 1997.

Vol. 1307: R. Kompe, Prosody in Speech Understanding Systems. XIX, 357 pages. 1997.

Vol. 1314: S. Muggleton (Ed.), Inductive Logic Programming. Proceedings, 1996. VIII, 397 pages. 1997.

Vol. 1317: M. Leman (Ed.), Music, Gestalt, and Computing. IX, 524 pages. 1997.

Vol. 1321: M. Lenzerini (Ed.), AI*IA 97: Advances in Artificial Intelligence. Proceedings, 1997. XII, 459 pages. 1997.

Lecture Notes in Computer Science

Lancaster University Library
64009846